Universitext

For other titles in this series, go to
http://www.springer.com/series/223

James J. Dudziak

Vitushkin's Conjecture for Removable Sets

James J. Dudziak
Lyman Briggs College
Michigan State University
East Lansing, MI 48825
USA
dudziak@msu.edu

ISBN 978-1-4419-6708-4 e-ISBN 978-1-4419-6709-1
DOI 10.1007/978-1-4419-6709-1
Springer New York Dordrecht Heidelberg London

Library of Congress Control Number: 2010930973

Mathematics Subject Classification (2010): 30H05

Printed on acid-free paper

Springer is part of Springer Science+Business Media (www.springer.com)

Contents

Preface: Painlevé's Problem

Let K be a compact subset of the complex plane. Call K *removable for bounded analytic functions*, or more concisely *removable*, if for each open superset U of K in the complex plane, each function that is bounded and analytic on $U \setminus K$ extends across K to be analytic on the whole of U. In 1888 Paul Painlevé became the first to seriously investigate the nature of removable sets in his thesis [PAIN]. Because of this the removable subsets of the complex plane are often referred to as *Painlevé null sets* and the task of giving them a "geometric" characterization has come to be known as *Painlevé's Problem.* In addition to being an academic, Painlevé was also a politician and statesman who served as War Minister and Prime Minister of France at various times in his life. For more on this interesting and multifaceted individual see Section 6 of Chapter 5 of [PAJ2].

The notion of "geometric" here is unavoidably vague and intuitive. On the one hand, a necessary but not sufficient condition for such a characterization is that it should make no reference to analytic functions. On the other hand, a sufficient but not necessary condition for such a characterization is that it be couched in terms of the cardinality of K or the topological, metric, or rectifiability properties of K. At the very end of this book the following question will command our attention: Should a characterization involving totally arbitrary measures be counted as "geometric"?

The goal of this book is to present a complete proof of the recent affirmative resolution of a special case of Painlevé's Problem known as *Vitushkin's Conjecture*. This conjecture states that a compact set with finite linear Hausdorff measure is removable if and only if it intersects every rectifiable curve in a set of zero arclength measure. We note in passing that arclength measure here can be replaced by linear Hausdorff measure since the two have the same zero sets among subsets of rectifiable curves. More importantly, we note that the forward implication of Vitushkin's Conjecture is equivalent to an earlier conjecture about a still more special case of Painlevé's Problem known as *Denjoy's Conjecture*. This conjecture states that a compact subset of a rectifiable curve with positive arclength measure is non-removable. So to prove Vitushkin's Conjecture, we must also prove Denjoy's Conjecture.

To understand this book a prospective reader should have a firm grasp of the first 14 chapters of Walter Rudin's *Real and Complex Analysis*, 3rd Edition (hereafter referred to as [RUD]). Indeed, the author has somewhat eccentrically sought to make this book, when used in conjunction with [RUD], entirely self-contained. Thus any standard result of analysis which is needed but is not contained in [RUD] is proved in this book (e.g., Besicovitch's Covering Lemma), and conversely, any standard result of analysis which is needed and is contained in [RUD] is always given a citation from [RUD] (e.g., Lebesgue's Dominated Convergence Theorem). Another eccentricity of the book is a deliberate exclusion of figures but an equally deliberate inclusion of verbal descriptions precise enough to enable an attentive reader to reconstruct the excluded figures. To a great extent the author wrote this book to convince himself of the truth of Vitushkin's Conjecture "beyond a reasonable doubt" and so has elected to err on the side of too much detail rather than too little. Finally, the author believes his notation is fairly standard or obvious but has nevertheless spelled out the meaning of a number of symbols upon first use and appended a symbol glossary and list to the back of the book for the reader's convenience.

We now turn to detailing the contents of the book, chapter by chapter.

Chapter 1 introduces and then proves various standard elementary results about the notions of removability and analytic capacity. The analytic capacity of a compact subset K of the complex plane is a nonnegative number $\gamma(K)$ which can be thought of as a quantitative measure of removability/nonremovability since K is removable if and only if $\gamma(K) = 0$. This result does *not* solve Painlevé's Problem since $\gamma(K)$ is *not* a geometric quantity – its definition (see Section 1.2) involves suping over a space of bounded analytic functions!

Chapter 2 introduces the notions of s-dimensional Hausdorff measure $\mathcal{H}^s$ and Hausdorff dimension $\dim_{\mathcal{H}}$ – these are not dealt with in [RUD] – and then relates them to removability. It turns out that a result of Painlevé implies that a compact K is removable whenever $\dim_{\mathcal{H}}(K) < 1$ and a result of Frostman implies that a compact K is nonremovable whenever $\dim_{\mathcal{H}}(K) > 1$. So Painlevé's Problem is reduced to determining the removability of those compact K for which $\dim_{\mathcal{H}}(K) = 1$. At the end of this chapter a natural conjecture presents itself which would finish off Painlevé's Problem if true. It is couched in terms of $\mathcal{H}^1$ but is summarily slain by a counterexample!

Chapter 3 proves a special case of Garabedian duality needed for our proof of Denjoy's Conjecture. Analytic capacity, whose definition involves suping over a space of bounded analytic functions, is an L^∞ object. It has an L^2 analog and Garabedian duality asserts that these two capacities, one L^∞ and the other L^2, are related in a manner that makes it clear that they vanish for the same sets. The importance of Garabedian duality is that it thus allows us to use Hilbert space methods to study an L^∞ problem – it is frequently easier to estimate an L^2 norm than it is to estimate an L^∞ norm.

Chapter 4 introduces the notion of the Melnikov curvature of a measure and the notion of a measure with linear growth. Garabedian duality is then used to prove a result called Melnikov's Lower Capacity Estimate. Given a compact set supporting a nontrivial positive Borel measure with finite Melnikov curvature and linear growth, this estimate gives a positive lower bound on the analytic capacity of the set in terms of the Melnikov curvature, the linear growth bound, and the mass of the measure. Of course this quantitative result trivially implies a qualitative one: a compact set which supports a nontrivial positive Borel measure with finite Melnikov curvature and linear growth is nonremovable. A Fourier transform argument due to Mark Melnikov and Joan Verdera is then given that shows that Lipschitz graphs support many such measures. After some preliminaries dealing with arclength and arclength measure, these two results combine to give a nice proof of Denjoy's Conjecture. At the end of this chapter a natural conjecture presents itself which would finish off Painlevé's Problem if true. It is couched in terms of rectifiable curves but meets the same fate as the earlier conjecture, i.e., it is summarily slain by a counterexample!

Chapter 5 is a grab bag of the measure theory needed to carry us forward. Amazingly, up to this point in the book it has sufficed to just know that s-dimensional Hausdorff measure is an outer measure defined on all subsets of the complex plane! Not so for what follows where we must know that it is an honest-to-god measure on a σ-algebra of subsets containing the Borel sets. The chapter has more in it than one would expect. The reason is that measures in [RUD] are typically obtained via the Riesz Representation Theorem and, in consequence, always put finite mass on any compact set. This is a property that s-dimensional Hausdorff measure on the complex plane has only when $s = 2$. So we cannot simply rely on [RUD] here for our measure theory.

Chapter 6 has a proof of Vitushkin's Conjecture *modulo* two difficult results. The next two chapters, comprising roughly half the book, are taken up with proving these results.

Chapter 7 has a proof of the first difficult result, a $T(b)$ theorem due to Fedor Nazarov, Sergei Treil, and Alexander Volberg for measures that need not satisfy a doubling condition. The complexity of this proof precludes us from saying anything enlightening about it just now.

Chapter 8 has a proof of the second difficult result, a curvature theorem for arbitrary measures due to Guy David and Jean-Christophe Léger. The complexity of this proof precludes us from saying anything enlightening about it just now.

With the end of Chapter 8, the goal of this book, the presentation of a complete proof of Vitushkin's Conjecture, has been achieved. But Vitushkin's Conjecture, although a big part of Painlevé's Problem, is not all of it. With the affirmative resolution of Vitushkin's Conjecture, Painlevé's Problem has been reduced to determining the removability of those compact sets K for which $\dim_{\mathcal{H}}(K) = 1$ but $\mathcal{H}^1(K) = \infty$. A Postscript following Chapter 8 seeks to shed some light on these sets. This Postscript deals with two items: first, the extension of Vitushkin's Conjecture to compact sets that are σ-finite for $\mathcal{H}^1$, and second, a conjecture due to Melnikov which essentially says that the qualitative consequence of Melnikov's Lower Capacity Estimate mentioned a few paragraphs ago is reversible. Both of these matters are resolved affirmatively with the aid of a quite recent and deep theorem, which we state but do not prove, due to Xavier Tolsa.

In writing this book the author has found three useful sources on Hausdorff measure and dimension: [ROG], [FALC], and [MAT3]. These items have been listed in order of increasing depth. For the purposes of this book [FALC] proved to be ideal. The author was also helped by several excellent survey articles dealing with the status of Painlevé's Problem and the various subproblems it has spawned. These are, in chronological order: [MARSH], [VER], [MAT5], [MAT6], [DAV2], [DAV3], [TOL4], and [PAJ3]. The author is also indebted to two books that are of a much more comprehensive scope than this one but deal with Painlevé's Problem: [GAR2], from the pre-Melnikov-curvature era, and [PAJ2], from the post-Melnikov-curvature era. Finally, it should be noted that [MAT3], a very comprehensive and deep book on Hausdorff measure and rectifiability that appeared at the cusp between the two eras, has an excellent chapter devoted to the status of Painlevé's Problem at that time. These sources also have superb and complete bibliographies. The bibliography of this book, being restricted solely to those articles and books that the author found necessary to cite, is spare by comparison.

The author would like to express his gratitude to the University of Tennessee at Knoxville where, as a visitor during the 2002–2003 academic year, he was able to present some of the material that made its way into this book in a faculty seminar. At various times during the composition of Chapter 8, Jean-Christophe Léger kindly responded to email inquiries about fine points of the proof of the curvature theorem bearing his and David's name. His responses were prompt, gracious, and, most importantly, very helpful, thus earning the author's heartfelt thanks. Last but certainly not least, the author would like to thank his wife for many things, just one of which is her making possible a leave of absence from teaching duties in order to engage in the last push to the finish line with this book!

East Lansing, MI James J. Dudziak

Chapter 1
Removable Sets and Analytic Capacity

1.1 Removable Sets

For now and forevermore, let K be a compact subset of the complex plane $\mathbb{C}$. This will be restated for emphasis many times in what follows but just as often will be tacitly assumed and not mentioned. For the sake of those readers who skip prefaces, we repeat a definition: K is *removable for bounded analytic functions*, or more concisely *removable*, if for each open superset U of K in $\mathbb{C}$, each function that is bounded and analytic on $U \setminus K$ extends across K to be analytic on the whole of U. These analytic extensions must be bounded on the whole of U since they are continuous and so also bounded on K. Thus the definition may be equivalently restated as follows: K is *removable* if for each open superset U of K in $\mathbb{C}$, each element of $H^\infty(U \setminus K)$ extends to an element of $H^\infty(U)$. Of course, for any open set V of $\mathbb{C}$, $H^\infty(V)$ denotes the Banach algebra of all functions bounded and analytic on V. What can we say about a removable K?

First, a removable K must have no interior. For if there were a point z_0 in the interior of K, then the function $z \mapsto 1/(z - z_0)$ would be a function nonconstant, bounded, and analytic on $\mathbb{C} \setminus K$ which, by Liouville's Theorem [RUD, 10.23], would not extend analytically to all of $\mathbb{C}$.

An immediate consequence of this first observation is that the analytic extension of any element of $H^\infty(U \setminus K)$ to an element of $H^\infty(U)$ is unique and of the same supremum norm, i.e., $H^\infty(U \setminus K)$ and $H^\infty(U)$ are isometrically isometric. This explains the terminology: "removing" K from U has made no difference to $H^\infty(U)$.

Second, a removable K must have connected complement. For if $\mathbb{C} \setminus K$ had more than one component, then the function which is one on the unbounded component and zero on all the bounded components would be a function nonconstant, bounded, and analytic on $\mathbb{C} \setminus K$ which, by Liouville's Theorem [RUD, 10.23], would not extend analytically to all of $\mathbb{C}$.

Third, a removable K must be *totally disconnected*, i.e., a removable K can contain no nontrivial connected subset. To see this we suppose otherwise and deduce a contradiction. So let C be a nontrivial connected subset of a removable K. Replacing C with its closure, we may assume that C is closed. Let U_∞ be the component of $\mathbb{C}^* \setminus C$ containing ∞. Of course, $\mathbb{C}^* = \mathbb{C} \cup \{\infty\}$ denotes the extended complex plane (also known as the Riemann sphere). By our second observation, U_∞ contains all

J.J. Dudziak, *Vitushkin's Conjecture for Removable Sets*, Universitext,
DOI 10.1007/978-1-4419-6709-1_1, © Springer Science+Business Media, LLC 2010

of $\mathbb{C}^* \setminus K$. It is an exercise in point-set topology, which we leave to the reader, to show that $\mathbb{C}^* \setminus U_\infty$ is a nontrivial connected subset of $\mathbb{C}^*$. Fix $z_0 \in \mathbb{C}^* \setminus U_\infty$ and set $g(z) = 1/(z - z_0)$. Then $g(U_\infty)$ is a proper subregion of $\mathbb{C}$ for which $\mathbb{C}^* \setminus g(U_\infty) = g(\mathbb{C}^* \setminus U_\infty)$ is connected. Hence, by the many equivalences to simple connectivity and the Riemann Mapping Theorem [RUD, 13.11 and 14.8], there exists a one-to-one analytic mapping f of $g(U_\infty)$ onto the open unit disc at the origin. It follows that $f \circ g$ is a bounded analytic function on $\mathbb{C} \setminus K$. By the removability of K and Liouville's Theorem [RUD, 10.23], $f \circ g$ is constant on $\mathbb{C} \setminus K$, and so too on $U_\infty \setminus \{\infty\}$ by the uniqueness property of analytic functions [RUD, 10.18]. Clearly then, f does *not* map $g(U_\infty)$ onto the open unit disc at the origin. With this contradiction we are done.

As an aside, we note that the third observation subsumes the first two since any totally disconnected K must have no interior and a connected complement. While the first part of this assertion is trivial, the reader may find verifying the second part one of those exercises in "mere" point-set topology that is a wee bit frustrating!

Turning to concrete examples, any single point is removable since bounded analytic functions can be analytically continued across isolated singularities. This assertion is simple [RUD, 10.20], but it is instructive to reproduce its proof. So suppose f is bounded and analytic on $\operatorname{int} B(z_0; r) \setminus \{z_0\}$, a punctured open disc about z_0. Using the boundedness of f, one easily sees that $(z - z_0)^2 f(z)$ extended to be zero at z_0 is differentiable there with derivative zero. Thus $(z - z_0)^2 f(z)$ is analytic on $\operatorname{int} B(z_0; r)$ and so has a power series expansion there [RUD, 10.16]:

$$(z - z_0)^2 f(z) = a_0 + a_1(z - z_0) + a_2(z - z_0)^2 + a_3(z - z_0)^3 + \cdots .$$

Since this extension and its derivative vanish at z_0, $a_0 = 0$ trivially and $a_1 = 0$ by [RUD, 10.6]. Thus we may divide out a factor of $(z - z_0)^2$ in the above to conclude that f, extended to be a_2 at z_0, has the following power series expansion on $\operatorname{int} B(z_0; r)$:

$$f(z) = a_2 + a_3(z - z_0) + a_4(z - z_0)^2 + a_5(z - z_0)^3 + \cdots .$$

But then, by [RUD, 10.6] again, our extended f is differentiable at z_0 with derivative a_3 there, and so f has been extended analytically across z_0.

Of course it follows that any finite set of points is removable. A little more nontrivially, any countable compact set K is removable. One proof involves a transfinite process which starting at $K_0 = K$ and any $f_0 \in H^\infty(U \setminus K)$, generates transfinite sequences $\{K_\alpha\}$ and $\{f_\alpha\}$ by stripping an isolated point off K_β after extending f_β across the isolated point when one is at a successor ordinal $\alpha = \beta + 1$, while intersecting all previous sets K_β, $\beta < \alpha$, and patching together all previous extensions f_β, $\beta < \alpha$, when one is at a limit ordinal α. Note that we always have $f_\alpha \in H^\infty(U \setminus K_\alpha)$. This process must break down at some ordinal α since K is not a proper class. A little thought shows that this can only happen at a successor ordinal $\alpha = \beta + 1$ and only then when K_β has no isolated points. Since any nonempty countable compact subset of the plane must have an isolated point by the Baire

Category Theorem [RUD, 5.7], it follows that we must have some $K_\beta = \emptyset$. But then f has been extended to $f_\beta \in H^\infty(U)$ and so K is removable. We note in passing that this ordinal β must be less than ω_1, the first uncountable ordinal, since K is countable. Those readers uncomfortable with transfinite induction/recursion will be relieved to know that the removability of countable compact sets is a simple corollary of Proposition 1.7 below whose proof does not use these tools from set theory.

This is as far as one can go with topology and cardinality alone since the next proposition shows that some uncountable, totally disconnected, compact subsets of the complex plane are removable while others are not. The simplest such sets are the linear Cantor sets. The standard middle-third's Cantor set is removable, but not all linear Cantor sets are. The next proposition shows this and indeed settles the question of removability for all linear compact sets. It also shows that a nonanalytic characterization of removability, if one is to be had, must involve metric notions which measure the "size" of the set. Later we shall see that metric "size" alone does not always suffice and that in certain situations the "rectifiability structure" of the set is decisive.

Proposition 1.1 *Let K be a linear compact subset of $\mathbb{C}$. Then K is removable if and only if the linear Lebesgue measure of K is zero.*

Proof Without loss of generality, let the line in which K lies be the real line $\mathbb{R}$.

Suppose that the linear Lebesgue measure of K is zero. Let U be any open superset of K in $\mathbb{C}$ and consider any $f \in H^\infty(U \setminus K)$. A very useful general fact, which we shall employ many times in this book, is that given any open superset U in $\mathbb{C}$ of any compact subset K of $\mathbb{C}$, there always exists a cycle Γ in $U \setminus K$ with winding number 1 about every point of K and 0 about every point of $\mathbb{C} \setminus U$ [RUD, proof of 13.5]. Letting Γ be such a cycle for the U and K now under consideration, set V equal to the union of the collection of components of $\mathbb{C} \setminus \Gamma$ which intersect K. Then V is an open superset of K. Since the winding number of Γ about every point of V is 1 [RUD, 10.10], V is a subset of U. Define a function g on V by

$$g(z) = \frac{1}{2\pi i} \int_\Gamma \frac{f(\zeta)}{\zeta - z}\, d\zeta.$$

Clearly g is analytic on V, so to show that f extends analytically across K it suffices to show that $f = g$ on $V \setminus K$.

Fixing $z \in V \setminus K$, let $\varepsilon > 0$ be smaller than the distance of K to $(\mathbb{C} \setminus V) \cup \{z\}$. Then, since K is compact and has linear Lebesgue measure zero, K can be covered by a finite number of open intervals of the real axis whose lengths sum to less than ε. By amalgamating intervals which intersect one another and then discarding any intervals which miss K, we may assume that these intervals are pairwise disjoint and intersect K. Upon each such interval describes the counterclockwise circle having that interval as a diameter. Let Γ_ε be the cycle consisting of the circles so produced. Clearly the length of Γ_ε is less than $\pi\varepsilon$. Applying Cauchy's Integral Theorem [RUD, 10.35] to the cycle $\Gamma - \Gamma_\varepsilon$ in $U \setminus K$, one has that

$$f(z) = \frac{1}{2\pi i} \int_{\Gamma - \Gamma_\varepsilon} \frac{f(\zeta)}{\zeta - z}\, d\zeta = g(z) - \frac{1}{2\pi i} \int_{\Gamma_\varepsilon} \frac{f(\zeta)}{\zeta - z}\, d\zeta.$$

The absolute value of the second integral above is at most

$$\frac{1}{2\pi} \cdot \frac{\|f\|_\infty}{\operatorname{dist}(z, K) - \varepsilon} \cdot \pi\varepsilon$$

which converges to zero as ε does. Thus $f(z) = g(z)$, and so K is removable.

Now suppose that the linear Lebesgue measure of K, denoted l, is positive. Define a function h on $\mathbb{C} \setminus K$ by

$$h(z) = \frac{1}{2} \int_K \frac{1}{z - t}\, dt.$$

(As an aside, the factor of a half in the definition of h is not necessary for this proof; however, we will be reusing h in the proof of Proposition 1.19 below and there it will be necessary!) Clearly h is analytic on $\mathbb{C} \setminus K$. Since as $z \to \infty$, $h(z) \to 0$ yet $zh(z) \to l/2 \neq 0$, h is nonconstant. For $z = x + iy$ with $y \neq 0$,

$$|\operatorname{Im} h(z)| \leq \frac{1}{2} \int_K \frac{|y|}{(x - t)^2 + y^2}\, dt < \frac{1}{2} \int_{-\infty}^{+\infty} \frac{|y|}{u^2 + y^2}\, du = \frac{\pi}{2}.$$

In consequence, $\exp(ih)$ is a nonconstant element of $H^\infty(\mathbb{C} \setminus K)$. By Liouville's Theorem [RUD, 10.23] such an element cannot be extended analytically to all of $\mathbb{C}$. Hence K is nonremovable. □

Our next goal is to state and prove a number of equivalences for removability. To do this however, we first need a few words about *analyticity at* ∞ and a lemma.

Consider a function f analytic and bounded on $\{z \in \mathbb{C} : |z| > R\}$, a punctured neighborhood of ∞. Note that $g(w) = f(1/w)$ is bounded and analytic on $\{w \in \mathbb{C} : 0 < |w| < 1/R\}$, a punctured neighborhood of 0. Since we have shown single points to be removable, g extends analytically to $\{w \in \mathbb{C} : |w| < 1/R\}$. By [RUD, 10.16],

$$g(w) = a_0 + a_1 w + a_2 w^2 + a_3 w^3 + \cdots$$

with the convergence being absolute and uniform for all w with $|w| \leq 1/R'$ for any $R' > R$. Clearly then,

$$f(z) = a_0 + \frac{a_1}{z} + \frac{a_2}{z^2} + \frac{a_3}{z^3} + \cdots$$

with the convergence being absolute and uniform for all z with $|z| \geq R'$ for any $R' > R$. Set

$$f(\infty) = g(0) = a_0 = \lim_{z \to \infty} f(z)$$

and

$$f'(\infty) = g'(0) = a_1 = \lim_{z \to \infty} z\{f(z) - f(\infty)\}.$$

We describe this situation by saying that f extends to be "analytic" at ∞ with value $f(\infty)$ and "derivative" $f'(\infty)$ there. Another typical example of our use of language would be to say that f is "analytic at ∞" since it can be written as a "power series about ∞".

The last paragraph applies to any element of $H^\infty(\mathbb{C} \setminus K)$ and extends it to a function bounded and "analytic" on $\mathbb{C}^* \setminus K$. These extentions form a Banach algebra which we denote $H^\infty(\mathbb{C}^* \setminus K)$. Of course, it is isometrically isomorphic to $H^\infty(\mathbb{C} \setminus K)$. We will frequently be cavalier about the distinction between these two algebras.

Let us now back up a bit and consider any function f analytic on $\mathbb{C} \setminus K$ and bounded on a deleted neighborhood of ∞. Then for Γ a cycle in $\mathbb{C} \setminus K$ with winding number 1 about every point of K and C a counterclockwise circular path centered at the origin encircling K and Γ, one has

$$f'(\infty) = \frac{1}{2\pi i} \int_C f(\zeta)\, d\zeta = \frac{1}{2\pi i} \int_\Gamma f(\zeta)\, d\zeta.$$

The first equality follows by plugging the power series about ∞ above into the integral and then integrating term-by-term while the second inequality follows from Cauchy's Integral Theorem [RUD, 10.35]. If, in addition, z_0 is any point of K, then

$$f(\infty) = \frac{1}{2\pi i} \int_C \frac{f(\zeta)}{\zeta - z_0}\, d\zeta = \frac{1}{2\pi i} \int_\Gamma \frac{f(\zeta)}{\zeta - z_0}\, d\zeta.$$

To see this, apply the integral representations for $f'(\infty)$ above to the function $g(z) = f(z)/(z - z_0)$, noting that $g(\infty) = \lim_{z\to\infty} f(z)/(z - z_0) = f(\infty) \cdot 0 = 0$ and so $g'(\infty) = \lim_{z\to\infty} zf(z)/(z - z_0) = f(\infty) \cdot 1 = f(\infty)$.

Given U a punctured neighborhood of ∞, let $\{f_n\}$ be a sequence of functions analytic and uniformly bounded on U that converges uniformly on compact subsets of U to a function f. Of course f is then analytic and bounded on U [RUD, 10.28] and so all functions here are analytic at ∞. Our integral representations now make it clear that $f_n(\infty) \to f(\infty)$ and $f_n'(\infty) \to f'(\infty)$.

Lemma 1.2 *Let U be an open superset of a compact subset K of $\mathbb{C}$. Then any analytic function f on $U \setminus K$ can be written uniquely as $g + h$ where g is analytic on U and h is analytic on $\mathbb{C}^* \setminus K$ with $h(\infty) = 0$. Moreover, if f is bounded on $U \setminus K$, then g and h are bounded on U and $\mathbb{C}^* \setminus K$ respectively.*

Proof Given any $z \in U$, choose a cycle $\Gamma_g(z)$ in $U \setminus (K \cup \{z\})$ with winding number 1 about every point of $K \cup \{z\}$ and 0 about every point of $\mathbb{C} \setminus U$. Define

$$g(z) = \frac{1}{2\pi i} \int_{\Gamma_g(z)} \frac{f(\zeta)}{\zeta - z}\, d\zeta.$$

Cauchy's Integral Theorem [RUD, 10.35] implies that this integral is independent of the particular $\Gamma_g(z)$ chosen. Thus g is well defined and, given any $\Gamma_g(z)$, we have

$$g(w) = \frac{1}{2\pi i} \int_{\Gamma_g(z)} \frac{f(\zeta)}{\zeta - w} \, d\zeta$$

for all w in the component of $\mathbb{C} \setminus \Gamma_g(z)$ containing z since $\Gamma_g(z)$ will serve as $\Gamma_g(w)$ for such w [RUD, 10.10]. We may thus differentiate under the integral in the last displayed equation to conclude that g is analytic on U.

Given any $z \in \mathbb{C} \setminus K$, choose a cycle $\Gamma_h(z)$ in $(U \setminus \{z\}) \setminus K$ with winding number one about every point of K and zero about every point of $\mathbb{C} \setminus (U \setminus \{z\})$. Define

$$h(z) = -\frac{1}{2\pi i} \int_{\Gamma_h(z)} \frac{f(\zeta)}{\zeta - z} \, d\zeta.$$

Cauchy's Integral Theorem [RUD, 10.35] implies that this integral is independent of the particular $\Gamma_h(z)$ chosen. Thus h is well defined and an argument similar to that just given for g shows that h is analytic on $\mathbb{C} \setminus K$. Let Γ be a cycle in $U \setminus K$ with winding number one about every point of K and zero about every point of $\mathbb{C} \setminus U$. Then

$$h(z) = -\frac{1}{2\pi i} \int_{\Gamma} \frac{f(\zeta)}{\zeta - z} \, d\zeta$$

for all z in the unbounded component of $\mathbb{C} \setminus \Gamma$ since Γ serves as $\Gamma_h(z)$ for such z [RUD, 10.10]. The last displayed equation makes it clear that h is bounded in a deleted neighborhood of ∞ and so analytic at ∞. Moreover, letting $z \to \infty$ in this equation, we see that $h(\infty) = 0$.

Cauchy's Integral Theorem [RUD, 10.35] applied to the cycle $\Gamma_g(z) - \Gamma_h(z)$ implies that $f = g + h$ on $U \setminus K$, so existence has been shown. Consider another representation $f = \tilde{g} + \tilde{h}$ as desired. Liouville's Theorem [RUD, 10.23] applied to $g - \tilde{g} = \tilde{h} - h$ gives us uniqueness.

To finish, suppose f is bounded on $U \setminus K$. Choose an open set V with compact closure such that $K \subseteq V \subseteq \operatorname{cl} V \subseteq U$. Since h is bounded on $\mathbb{C} \setminus V$, $g = f - h$ is bounded on $U \setminus V$. But then, since g is bounded on $\operatorname{cl} V$, g is bounded on U. Finally, $h = f - g$ is bounded on $U \setminus K$ and so also on $\mathbb{C} \setminus K$. □

Now to all but one of the promised equivalences for removability.

Proposition 1.3 *For a compact subset K of $\mathbb{C}$, the following are equivalent:*

(a) *K is removable.*
(b) *There exists an open superset U of K in $\mathbb{C}$ such that each function that is bounded and analytic on $U \setminus K$ extends across K to be analytic on the whole of U.*

(c) *The only elements of* $H^\infty(\mathbb{C}^* \setminus K)$ *are the constant functions.*
(d) *For every* $f \in H^\infty(\mathbb{C}^* \setminus K)$, *one has* $f'(\infty) = 0$.

Proof (a) $\Rightarrow$ (b): Trivial.

(b) $\Rightarrow$ (c): Given $f \in H^\infty(\mathbb{C}^* \setminus K)$, clearly one also has $f \in H^\infty(U \setminus K)$ and so f extends analytically across K. Since the extension is still bounded, Liouville's Theorem [RUD, 10.23] now implies that f is constant.

(c) $\Rightarrow$ (d): Trivial.

(d) $\Rightarrow$ (c): We suppose that there exists a nonconstant $g \in H^\infty(\mathbb{C}^* \setminus K)$ and construct a function $f \in H^\infty(\mathbb{C}^* \setminus K)$ with $f'(\infty) \neq 0$. Since g is nonconstant, there exists a point $z_0 \in \mathbb{C} \setminus K$ such that $g(z_0) \neq g(\infty)$. Set $f(z) = \{g(z) - g(z_0)\}/(z - z_0)$. Then $f \in H^\infty(\mathbb{C}^* \setminus K)$ and $f(\infty) = \lim_{z\to\infty} f(z) = 0$. In consequence, $f'(\infty) = \lim_{z\to\infty} zf(z) = g(\infty) - g(z_0) \neq 0$.

(c) $\Rightarrow$ (a): Given any open superset U of K in $\mathbb{C}$ and any $f \in H^\infty(U \setminus K)$, get g and h as in the previous lemma. Then h is constant, and since $h(\infty) = 0$, this constant must be 0. Thus g is an analytic extension of f to U. □

Our last equivalence for removability is a surprising apparent strengthening of the way we have defined the concept. It has been stated separately because it requires a finicky bit of topology which is the content of ...

Lemma 1.4 *Let X be a totally disconnected, compact, Hausdorff space. Suppose C_1 and C_2 are disjoint closed subsets of X. Then X can be written as the disjoint union of two closed subsets X_1 and X_2 such that $C_1 \subseteq X_1$ and $C_2 \subseteq X_2$.*

Proof Recall that a *clopen* subset of a topological set is one that is both closed and open.

First Claim. Given any point $x \in X$, let E_x denote the intersection of all clopen subsets of X containing x. Then E_x is connected. (Note: The total disconnectedness of X is not used here!)

Indeed, supposing E_x is the disjoint union of two closed subsets E_1 and E_2 with x contained in E_1 say, it suffices to show that E_2 is empty. Since X is compact and Hausdorff, there exist two disjoint open sets U_1 and U_2 such that $E_1 \subseteq U_1$ and $E_2 \subseteq U_2$. Since $E_x \subseteq U_1 \cup U_2$ and E_x is an intersection of clopen sets, finitely many of these sets must have intersection $\widetilde{E}$ contained in $U_1 \cup U_2$ by compactness. But then $\widetilde{E}$, being a finite intersection of clopen sets, is itself clopen. Hence $\widetilde{E} \cap U_1 = \widetilde{E} \setminus U_2$ is also clopen and it clearly contains x. Thus by the definition of E_x, $E_x \subseteq \widetilde{E} \setminus U_2$ and so $E_2 \cap U_2 = \emptyset$. Since $E_2 \subseteq U_2$ also, E_2 must be empty.

Second Claim. For any distinct points $x, y \in X$, there exists a clopen subset E of X such that $x \in E$ and $y \notin E$. (Note: The total disconnectedness of X is used here!)

This claim can be rephrased as $E_x = \{x\}$ where E_x is as in the first claim. It thus follows immediately from the first claim since the total disconnectedness of X just means that the only nonempty connected subsets of X are the singletons.

Third Claim. For any closed subset C_1 of X and any point $y \in X \setminus C_1$, there exists a clopen subset E of X such that $C_1 \subseteq E$ and $y \notin E$.

The proof of this claim is a simple compactness argument using the second claim.

Fourth Claim. For C_1 and C_2 disjoint closed subsets of X, there exists a clopen subset E of X such that $C_1 \subseteq E$ and $C_2 \cap E = \emptyset$.

The proof of this claim is a simple compactness argument using the third claim.

The proposition now follows by setting $X_1 = E$ and $X_2 = X \setminus E$ where E is as in the fourth claim. □

Proposition 1.5 *A compact subset K of $\mathbb{C}$ is removable if and only if for each open subset U of $\mathbb{C}$, each function that is bounded and analytic on $U \setminus K$ extends across K to be analytic on the whole of U.*

Note that the backward implication is trivial. The forward implication is also trivial if K is totally contained in or totally disjoint from U. Thus the meat of the proposition is when K is removable and "half" in/out of U.

Proof Suppose K is removable, U is open, and $f \in H^\infty(U \setminus K)$.

Given $\varepsilon > 0$, set $C_1(\varepsilon) = \{z \in K : \operatorname{dist}(z, \mathbb{C} \setminus U) \geq \varepsilon\}$ and $C_2 = K \setminus U$. By Lemma 1.4, K can be written as the disjoint union of two closed subsets $K_1(\varepsilon)$ and $K_2(\varepsilon)$ such that $C_1(\varepsilon) \subseteq K_1(\varepsilon)$ and $C_2 \subseteq K_2(\varepsilon)$. The equivalence of (a) with (d) in Proposition 1.3 makes it clear that any compact subset of a removable set is removable. Thus $K_1(\varepsilon)$ is removable. Then, since $f \in H^\infty(\{U \setminus K_2(\varepsilon)\} \setminus K_1(\varepsilon))$ and $U \setminus K_2(\varepsilon)$ is an open superset of $K_1(\varepsilon)$, f extends to a function f_ε that is analytic on $U \setminus K_2(\varepsilon)$.

Given any $z \in U$, note that $f_\varepsilon(z)$ is defined whenever $0 < \varepsilon < \operatorname{dist}(z, \mathbb{C} \setminus U)$. Since K is removable, it has no interior. Because of this, all the values of $f_\varepsilon(z)$ for these various values of ε are equal since they are uniquely determined as the limit of $f(w)$ as $w \in U \setminus K$ approaches z. Thus we may properly define an extension f_0 of f to U by setting $f_0(z) = f_\varepsilon(z)$ for any ε such that $0 < \varepsilon < \operatorname{dist}(z, \mathbb{C} \setminus U)$. Clearly f_0 is analytic on all of U and so we are done. □

The next result establishes the existence of a *nonremovable kernel* so-to-speak for compact sets with no interior.

Lemma 1.6 *Suppose that K is a compact subset of $\mathbb{C}$ with no interior. Then there exists a compact subset K^* of K with the following property: for each compact subset J of K, every element of $H^\infty(\mathbb{C}^* \setminus K)$ extends to an element of $H^\infty(\mathbb{C}^* \setminus J)$ if and only if $K^* \subseteq J$.*

Proof Call a subset J of K *good* if it is compact and every element of $H^\infty(\mathbb{C}^* \setminus K)$ extends to an element of $H^\infty(\mathbb{C}^* \setminus J)$. Let K^* be the intersection of all good subsets of K. Note that we have just made the forward implication of the equivalence we wish to prove true by definition. Also note that a compact subset of K which contains a good subset is clearly good. Thus to prove the backward implication it suffices to show that K^* itself is good.

So let $f \in H^\infty(\mathbb{C}^* \setminus K)$ and $z \notin \mathbb{C}^* \setminus K^*$. Then $z \notin J$ for at least one and possibly many good subsets J of K. There exist extensions $f_J \in H^\infty(\mathbb{C}^* \setminus J)$ of f for these J. Now take note of our assumption that K has no interior. Because of this, all the values of $f_J(z)$ for these various subsets J are equal since they are uniquely determined as the limit of $f(w)$ as $w \in \mathbb{C}^* \setminus K$ approaches z. Thus we may properly define an extension f_* of f to $\mathbb{C}^* \setminus K^*$ by setting $f_*(z) = f_J(z)$ for any good subset J of K that does not contain z. Clearly $f_* \in H^\infty(\mathbb{C}^* \setminus K^*)$ and so we are done. □

The final result of this section follows. Since single points are removable it has as a corollary that every countable compact subset of $\mathbb{C}$ is removable. It will also come in useful in the Postscript when we consider whether Vitushkin's Conjecture extends to compact subsets of $\mathbb{C}$ with infinite, but σ-finite, linear Hausdorff measure.

Proposition 1.7 *Let $\{K_n\}$ be a sequence of removable compact subsets of $\mathbb{C}$ whose union K is also compact. Then K is removable.*

Proof Each K_n, being removable, has no interior in $\mathbb{C}$. An argument by contradiction using the Baire Category Theorem [RUD, 5.7] now shows that K also has no interior in $\mathbb{C}$. Thus the last lemma applies and it suffices to show that K^*, the nonremovable kernel of K, is empty.

So we suppose that K^* is nonempty and get a contradiction. Since K^* is the countable union of $K_n \cap K^*$, some $K_n \cap K^*$ must have nonempty interior in K^* by the Baire Category Theorem [RUD, 5.7]. Thus there exists an open subset U of $\mathbb{C}$ such that $U \cap K^* \neq \emptyset$ and $U \cap K^* \subseteq K_n \cap K^*$.

Consider now any $f \in H^\infty(\mathbb{C}^* \setminus K)$. Via the last lemma extend it to an element of $H^\infty(\mathbb{C}^* \setminus K^*)$ which we will also denote by f. Note that $K_n \cap K^*$, being a compact subset of the removable set K_n, is itself removable. Thus by Proposition 1.5, f restricted to $U \setminus K^* = U \setminus \{K_n \cap K^*\}$ has an analytic extension g to U ...which is also bounded since f is and K^* has no interior. Let h denote the function on $\mathbb{C}^* \setminus \{K^* \setminus U\} = \{\mathbb{C}^* \setminus K^*\} \cup U$ which is equal to f on $\mathbb{C}^* \setminus K^*$ and g on U. This function is well defined since $f = g$ on $\{\mathbb{C}^* \setminus K^*\} \cap U = U \setminus K^*$. Clearly h is a bounded analytic extension of f to $\mathbb{C}^* \setminus \{K^* \setminus U\}$.

The last paragraph has shown that every element of $H^\infty(\mathbb{C}^* \setminus K)$ extends to an element of $H^\infty(\mathbb{C}^* \setminus \{K^* \setminus U\})$. Thus by the last lemma we must have $K^* \subseteq K^* \setminus U$, i.e., $U \cap K^* = \emptyset$. This contradiction finishes the proof. □

1.2 Analytic Capacity

For K a compact subset of $\mathbb{C}$, the number

$$\gamma(K) = \sup\{|f'(\infty)| : f \in H^\infty(\mathbb{C}^* \setminus K) \text{ with } \|f\|_\infty \leq 1\}$$

is called the *analytic capacity* of K. Thus (d) of Proposition 1.3 could just as well have been phrased as "$\gamma(K) = 0$" and Painlevé's Problem formulated as the task of giving a geometric characterization of the compact sets K for which $\gamma(K) = 0$. Indeed, for the rest of the book we shall treat the two phrases "$\gamma(K) = 0$" and "K is removable" as synonymous. The notion of analytic capacity was introduced in 1947 by Lars Ahlfors in [AHL] where he proved the equivalence of removability with analytic capacity zero [(a) $\Leftrightarrow$ (d) of Proposition 1.3 above] and also introduced the Ahlfors function (see Proposition 1.14 below). Beyond this, analytic capacity has turned out to be of great importance for rational approximation theory (see Chapter VIII of [GAM1], [VIT2], and/or [ZALC]). Unfortunately it is difficult to work with, a fact to which this book is an indirect testimonial. The last decade has seen progress in understanding it. This section is devoted to an exposition of the classical elementary properties and estimates of analytic capacity.

The proofs of the first three propositions below are simple and left to the reader.

Proposition 1.8 *Analytic capacity is monotone, i.e.,* $\gamma(K_1) \leq \gamma(K_2)$ *whenever* $K_1 \subseteq K_2$.

Proposition 1.9 *For* α *and* β *complex numbers,* $\gamma(\alpha K + \beta) = |\alpha|\gamma(K)$.

Proposition 1.10 *The analytic capacity of* K *depends only on the unbounded component of* $\mathbb{C}^* \setminus K$. *Thus, letting* $\widehat{K}$ *denote the union of* K *with all the bounded components of* $\mathbb{C}^* \setminus K$,

$$\gamma(K) = \gamma(\partial K) = \gamma(\widehat{K}) = \gamma(\partial \widehat{K}).$$

The slightly different characterization of capacity provided by the next proposition will be used without mention in what follows.

Proposition 1.11 *Suppose* $g \in H^\infty(\mathbb{C}^* \setminus K)$ *satisfies* $\|g\|_\infty \leq 1$. *Then there exists an* $f \in H^\infty(\mathbb{C}^* \setminus K)$ *with* $\|f\|_\infty \leq 1$, $f(\infty) = 0$, *and* $|f'(\infty)| \geq |g'(\infty)|$. *Consequently,*

$$\gamma(K) = \sup\{|f'(\infty)| : f \in H^\infty(\mathbb{C}^* \setminus K) \text{ with } \|f\|_\infty \leq 1 \text{ and } f(\infty) = 0\}.$$

Proof On the one hand, if $|g(\infty)| = 1$, then g is constant on the unbounded component of $\mathbb{C}^* \setminus K$ by the Maximum Modulus Principle [RUD, 12.1 modified to take into account regions containing ∞]. Clearly $f \equiv 0$ works in this case.

On the other hand, if $|g(\infty)| < 1$, then the function

$$f(z) = \frac{g(z) - g(\infty)}{1 - \overline{g(\infty)}g(z)}$$

is in $H^\infty(\mathbb{C}^* \setminus K)$ with $\|f\|_\infty \leq 1$ and $f(\infty) = 0$ [RUD, 12.4]. Moreover,

$$|f'(\infty)| = \lim_{z \to \infty} |zf(z)| = \frac{|g'(\infty)|}{1 - |g(\infty)|^2} \geq |g'(\infty)|,$$

so we are done. □

Proposition 1.12 *Suppose* K *is a nontrivial continuum in* $\mathbb{C}$. *Let* f *denote the one-to-one analytic mapping of the unbounded component of* $\mathbb{C}^* \setminus K$ *onto the open unit disc at the origin for which* $f(\infty) = 0$ *and* $f'(\infty) > 0$. *Then* $\gamma(K) = f'(\infty)$.

Existence here is a consequence of the Riemann Mapping Theorem [RUD, 14.8] which we are assuming to have been modified in the obvious way to encompass simply connected regions in $\mathbb{C}^*$. Recall that a *continuum* is a connected compact set.

Proof Extending f to be identically 0 on the bounded components of $\mathbb{C}^* \setminus K$, we clearly have $f \in H^\infty(\mathbb{C}^* \setminus K)$ with $\|f\|_\infty = 1$. Thus $f'(\infty) = |f'(\infty)| \le \gamma(K)$.

For the reverse inequality, consider any $g \in H^\infty(\mathbb{C}^* \setminus K)$ with $\|g\|_\infty \le 1$ and $g(\infty) = 0$. It suffices to show that $|g'(\infty)| \le |f'(\infty)|$. Note that the function $g \circ f^{-1}$ is an analytic mapping of the open unit disc at the origin into itself which fixes the origin. Thus by Schwarz's Lemma [RUD, 12.2], $|g(f^{-1}(w))| \le |w|$ whenever $|w| < 1$, so $|g(z)| \le |f(z)|$ whenever $|z|$ is big enough. To finish, multiply this inequality by $|z|$, let $z \to \infty$, and then take the supremum over all the gs. □

Corollary 1.13 *The analytic capacity of a closed ball is its radius and the analytic capacity of a closed line segment is a quarter of its length.*

Proof Considering closed balls first, by Proposition 1.9 we need only consider the special case of K equal to the closed unit ball at the origin. Clearly the function $f(z) = z^{-1}$ is a one-to-one analytic mapping of $\mathbb{C}^* \setminus K$ onto the open unit disc at the origin for which $f(\infty) = \lim_{z\to\infty} f(z) = 0$ and $f'(\infty) = \lim_{z\to\infty} zf(z) = 1$. Thus f is the function of Proposition 1.12 for this K and so $\gamma(K) = f'(\infty) = 1$.

Turning to closed line segments, by Proposition 1.9 we need to only consider the special case of $K = [-2, 2]$. Set $g(w) = w + w^{-1}$ for $w \in \mathbb{C} \setminus \{0\}$. An application of the quadratic formula shows that for any $z \in \mathbb{C}$, $z = g(w)$ for, and only for, the values $w_\pm = \{z \pm \sqrt{z^2 - 4}\}/2$. Note that these values are reciprocals of one another. Thus either both $|w_+|$ and $|w_-|$ are equal to 1 or exactly one of $|w_+|$ and $|w_-|$ is strictly smaller than 1. In the first case, $w_\pm = e^{\pm i\theta}$ for some real θ and so $z = g(e^{\pm i\theta}) = 2\cos\theta \in [-2, 2] = K$. Hence for $z \in \mathbb{C} \setminus K$ we must be in the second case. But then clearly $z = g(w)$ for exactly one w satisfying $|w| < 1$. We conclude that g is a one-to-one analytic mapping of the punctured unit disc at the origin onto $\mathbb{C} \setminus K$. By [RUD, 10.33], g restricted to the punctured unit disc at the origin has a one-to-one analytic inverse f that maps $\mathbb{C} \setminus K$ onto the punctured unit disc at the origin. By our discussion of behavior at ∞ just after Proposition 1.1, f extends to be analytic at ∞ with $f(\infty) = \lim_{z\to\infty} f(z) = \lim_{w\to 0} w = 0$ and $f'(\infty) = \lim_{z\to\infty} zf(z) = \lim_{w\to 0} g(w)\, w = 1$. Thus f is the function of Proposition 1.12 for this K and so $\gamma(K) = f'(\infty) = 1$. □

Proposition 1.14 *Let K be a compact subset of $\mathbb{C}$. Then there is a unique function f analytic and bounded by one on the unbounded component of $\mathbb{C}^* \setminus K$ such that $f(\infty) = 0$ and $f'(\infty) = \gamma(K)$.*

The unique function whose existence is guaranteed by this proposition is called the *Ahlfors function* of K. Note that if K is a nontrivial continuum, then the Ahlfors function of K is just the Riemann map of Proposition 1.12. We will frequently consider the Ahlfors function of K to be an element of $H^\infty(\mathbb{C}^* \setminus K)$ by canonically extending it to be zero on any bounded components $\mathbb{C}^* \setminus K$ may have. The very clever and elegant proof of uniqueness given below is taken from [FISH].

Proof We may choose a sequence of functions $\{f_n\}$ analytic and bounded by one on the unbounded component of $\mathbb{C}^* \setminus K$ such that each $f_n(\infty) = 0$ and $f_n'(\infty) \to \gamma(K)$. Since these functions form a normal family [RUD, 14.6], by extracting a subsequence we may assume that $\{f_n\}$ converges uniformly on compact subsets of the unbounded component of $\mathbb{C} \setminus K$ to a function f analytic and bounded by one on this same unbounded component. Then $f_n(\infty) \to f(\infty)$ and $f_n'(\infty) \to f'(\infty)$. Hence $f(\infty) = 0$ and $f'(\infty) = \gamma(K)$. This settles existence.

To establish uniqueness, suppose both f and g are analytic and bounded by one on the unbounded component of $\mathbb{C}^* \setminus K$ with values at ∞ equal to 0 and derivatives at ∞ equal to $\gamma(K)$. Setting $h = (f + g)/2$ and $k = (f - g)/2$, we have both functions analytic and bounded by one on the unbounded component of $\mathbb{C}^* \setminus K$ with $h(\infty) = 0$, $h'(\infty) = \gamma(K)$, $k(\infty) = 0$, and $k'(\infty) = 0$. Since $f = h + k$ and $g = h - k$, it suffices to show that $k = 0$.

Noting that $|h|^2 + |k|^2 \pm 2 \operatorname{Re} h\overline{k} = |h \pm k|^2 \leq 1$, we must have $|h|^2 + |k|^2 \leq 1$. Thus

$$|h| + \frac{1}{2}|k|^2 \leq |h| + \frac{1}{2}(1 - |h|^2) \leq |h| + \frac{1}{2}(1 + |h|)(1 - |h|) \leq |h| + (1 - |h|) = 1.$$

If k is nonzero, then

$$\frac{1}{2}k^2 = \frac{a_n}{z^n} + \frac{a_{n+1}}{z^{n+1}} + \cdots$$

where $a_n \neq 0$. Since $k(\infty) = 0$, $n \geq 2$. Choose $\varepsilon > 0$ sufficiently small so that $\varepsilon|a_n||z|^{n-1} \leq 1$ on a neighborhood U of K. Set $\tilde{f} = h + \varepsilon\overline{a_n}z^{n-1}k^2/2$ and note that $\tilde{f}$ is analytic on the unbounded component of $\mathbb{C}^* \setminus K$. Then $|\tilde{f}| \leq |h| + |k|^2/2 \leq 1$ on $U \setminus K$ and so on this same unbounded component by the Maximum Modulus Principle [RUD, 12.1 modified to take into account regions containing ∞]. Hence $|\tilde{f}'(\infty)| \leq \gamma(K)$. However,

$$\tilde{f}'(\infty) = h'(\infty) + \varepsilon|a_n|^2 > \gamma(K).$$

Because of this contradiction, we must have $k = 0$. □

Scholium 1.15 *An inspection of the proof of Proposition* 1.11 *shows that one actually has the stronger conclusion* $|f'(\infty)| > |g'(\infty)|$ *when* $g(\infty) \neq 0$ *and* $g'(\infty) \neq 0$. *In consequence, when* $\gamma(K) > 0$, *one may drop the requirement that* $f(\infty) = 0$ *from the definition of the Ahlfors function and still retain its uniqueness. When* $\gamma(K) = 0$, *this however fails (consider the constant functions with modulus less than 1).*

Proposition 1.16 *Let* $\{K_n\}$ *be a decreasing sequence of compact subsets of* $\mathbb{C}$ *with intersection* K. *Then* $\gamma(K_n) \to \gamma(K)$.

Proof By Proposition 1.8, the sequence $\{\gamma(K_n)\}$ is decreasing and bounded below by $\gamma(K)$. Thus

$$\gamma(K) \leq \lim_{n\to\infty} \gamma(K_n) = \liminf_{n\to\infty} \gamma(K_n).$$

Let f_n denote the Ahlfors function of K_n. Since these functions form a normal family on each $\mathbb{C} \setminus K_n$, we may apply [RUD, 14.6] and a diagonalization argument to conclude that a subsequence of $\{f_n\}$ converges uniformly on compact subsets of $\mathbb{C} \setminus K$ to a function f analytic and bounded by one on $\mathbb{C} \setminus K$. Clearly then

$$\liminf_{n\to\infty} \gamma(K_n) = \liminf_{n\to\infty} f_n'(\infty) \leq f'(\infty) = |f'(\infty)| \leq \gamma(K).$$

Hence $\lim_{n\to\infty} \gamma(K_n) = \gamma(K)$. □

Scholium 1.17 *With a little more work, one can exploit the uniqueness of the Ahlfors function to show that in the situation above the Ahlfors functions of the sets K_n converge uniformly on compact subsets of $\mathbb{C}^* \setminus K$ to the Ahlfors function of K.*

We finish this section with a number of classical estimates of the analytic capacity of a set in terms of its diameter, length (when the set is linear), and area. Recall that the *diameter* of a subset E of $\mathbb{C}$ is

$$|E| = \sup\{|z - w| : z, w \in E\}.$$

Proposition 1.18 *For K a compact subset of $\mathbb{C}$,*

$$\gamma(K) \leq |K|.$$

If K is also connected, and thus a continuum, then

$$\gamma(K) \geq \frac{|K|}{4}.$$

Proof Clearly any subset of the plane can be contained in a closed ball whose radius is the diameter of the set. Thus the first estimate follows from Proposition 1.8 and Corollary 1.13.

With regard to the second estimate, we may assume that K is nontrivial. Let f be the Riemann map of Proposition 1.12. Given $z_1 \in K$, set $g(w) = \gamma(K)/(f^{-1}(w) - z_1)$. Then g is a one-to-one analytic map on the open unit disc at the origin such that

$$g(0) = \lim_{w\to 0} g(w) = \lim_{w\to 0} \frac{\gamma(K)}{f^{-1}(w) - z_1} = \lim_{z\to\infty} \frac{\gamma(K)}{z - z_1} = 0$$

and

$$g'(0) = \lim_{w\to 0} \frac{g(w) - g(0)}{w} = \lim_{w\to 0} \frac{\gamma(K)}{w(f^{-1}(w) - z_1)} = \lim_{z\to\infty} \frac{\gamma(K)}{f(z)(z - z_1)}$$
$$= \frac{\gamma(K)}{f'(\infty) - f(\infty) \cdot z_1} = 1.$$

Hence, by the Koebe One-Quarter Theorem [RUD, 14.14], the range of g contains the open disc of radius one-quarter centered at the origin. For $z_2 \in K \setminus \{z_1\}$, $\gamma(K)/(z_2 - z_1)$ is not in the range of g. Thus $\gamma(K)/|z_2 - z_1| \geq 1/4$, i.e., $\gamma(K) \geq |z_2 - z_1|/4$. Since $z_1, z_2 \in K$ with $z_1 \neq z_2$ are otherwise arbitrary, $\gamma(K) \geq |K|/4$. □

Jung's Theorem states that any subset of the plane can be contained in a closed ball whose radius is $1/\sqrt{3}$ times the diameter of the set. Many proofs of this can be found in the delightful book [YB]. Thus the first estimate of the above proposition is not sharp and can be improved to $\gamma(K) \leq |K|/\sqrt{3}$. While the factor of $1/\sqrt{3}$ is known to be sharp in Jung's Theorem (consider an equilateral triangle), it is not sharp in the capacity estimate here. See the two paragraphs following the proof of Theorem 2.6 for a demonstration of this and an identification of the sharpest constant possible. With regard to the second estimate of the above proposition, consideration of a line segment and use of Corollary 1.13 show the factor of $1/4$ to be sharp.

Given a linear subset E of $\mathbb{C}$, we denote the "length", i.e., *linear Lebesgue measure*, of E by $\mathcal{L}^1(E)$ (see Section 2.1 for our official definition of this).

Proposition 1.19 *Let K be a linear compact subset of $\mathbb{C}$. Then*

$$\frac{\mathcal{L}^1(K)}{4} \leq \gamma(K) \leq \frac{\mathcal{L}^1(K)}{\pi}.$$

Proof Without loss of generality, let the line in which K lies be the real line $\mathbb{R}$.

The function h in the proof of Proposition 1.1 is analytic off K with $h(\infty) = 0$ and $h'(\infty) = \mathcal{L}^1(K)/2$. While h is not bounded off K, its imaginary part is bounded by $\pi/2$ off K. So consider the one-to-one analytic map g of the horizontal strip $\{|\text{Im}\, w| < \pi/2\}$ onto the open unit disc at the origin given by

$$g(w) = \frac{e^w - 1}{e^w + 1}.$$

The function $f = g \circ h$ is then an element of $H^\infty(\mathbb{C} \setminus K)$ with norm at most one such that

$$f(\infty) = \lim_{z \to \infty} g(h(z)) = g(0) = 0$$

and

$$f'(\infty) = \lim_{z \to \infty} zg(h(z)) = \lim_{z \to \infty} zh(z) \cdot \frac{e^{h(z)} - 1}{h(z)} \cdot \frac{1}{e^{h(z)} + 1} = h'(\infty) \cdot 1 \cdot \frac{1}{2}$$
$$= \frac{\mathcal{L}^1(K)}{4}.$$

Hence $\gamma(K) \geq f'(\infty) = \mathcal{L}^1(K)/4$.

Now consider any $f \in H^\infty(\mathbb{C}^* \setminus K)$ with $\|f\|_\infty \leq 1$. Given $\varepsilon > 0$, by the definition of linear Lebesgue measure and the compactness of K there exists a finite collection of pairwise disjoint, open intervals covering K whose lengths sum to less than $\mathcal{L}^1(K) + \varepsilon/2$. Upon each such interval describe a rectangle having that interval as a bisector. Let Γ_ε be the cycle consisting of the counterclockwise boundary paths of all the rectangles so produced. Clearly, if we make the thickness of our rectangles small enough, then the length of Γ_ε can be made less than $2\mathcal{L}^1(K) + 2\varepsilon$. Then

$$|f'(\infty)| = \left|\frac{1}{2\pi i}\int_{\Gamma_\varepsilon} f(\zeta)\,d\zeta\right| \leq \frac{2\mathcal{L}^1(K) + 2\varepsilon}{2\pi}.$$

Suping over all our fs and then letting $\varepsilon \downarrow 0$, we get $\gamma(K) \leq \mathcal{L}^1(K)/\pi$. □

A result of Christian Pommerenke (see [POM], or Section 6 of Chapter I of [GAR2]) states that the analytic capacity of any linear compact set is exactly equal to a quarter of its length. We are content with the weaker estimates above since they suffice for removability and most other considerations while Pommerenke's proof would cost us too much effort.

Given a subset E of $\mathbb{C}$, we denote the "area," i.e., *planar Lebesgue measure*, of E by $\mathcal{L}^2(E)$ (see Section 5.3 for our official definition of this).

Lemma 1.20 *Let K be a compact subset of $\mathbb{C}$. Then for each $z \in \mathbb{C} \setminus K$,*

$$\left|\iint_K \frac{1}{\zeta - z}\,d\mathcal{L}^2(\zeta)\right| \leq \sqrt{\pi\mathcal{L}^2(K)}.$$

Proof Fix $z \in \mathbb{C} \setminus K$. By translating K and then rotating the resulting set by a unimodular constant, we may assume that $z = 0$ and that the integral of the lemma is nonnegative. Thus

$$\left|\iint_K \frac{1}{\zeta - z}\,d\mathcal{L}^2(\zeta)\right| = \left|\iint_K \frac{1}{\zeta}\,d\mathcal{L}^2(\zeta)\right| = \mathrm{Re}\iint_K \frac{1}{\zeta}d\mathcal{L}^2(\zeta) = \iint_K \mathrm{Re}\frac{1}{\zeta}d\mathcal{L}^2(\zeta)$$

and so it suffices to show that

$$\iint_K \mathrm{Re}\,\frac{1}{\zeta}\,d\mathcal{L}^2(\zeta) \leq \sqrt{\pi\mathcal{L}^2(K)}.$$

Without loss of generality, $\mathcal{L}^2(K) > 0$. Choose $a > 0$ such that $\pi a^2 = \mathcal{L}^2(K)$. Set $B = B(a; a)$. Then $\mathcal{L}^2(K \setminus B) = \mathcal{L}^2(B \setminus K)$. Writing $\zeta = re^{i\theta}$, we see that $\zeta \in B \Leftrightarrow r \leq 2a\cos\theta \Leftrightarrow \mathrm{Re}(1/\zeta) = (\cos\theta)/r \geq 1/2a$. Thus

$$\iint_{K\setminus B} \mathrm{Re}\,\frac{1}{\zeta}\,d\mathcal{L}^2(\zeta) \leq \frac{\mathcal{L}^2(K \setminus B)}{2a} = \frac{\mathcal{L}^2(B \setminus K)}{2a} \leq \iint_{B\setminus K} \mathrm{Re}\,\frac{1}{\zeta}\,d\mathcal{L}^2(\zeta)$$

and so, adding the integral of $\mathrm{Re}(1/\zeta)$ over $K \cap B$ to both sides of this inequality, we get

$$\iint_K \mathrm{Re}\,\frac{1}{\zeta}\,d\mathcal{L}^2(\zeta) \le \iint_B \mathrm{Re}\,\frac{1}{\zeta}\,d\mathcal{L}^2(\zeta).$$

Finally, using polar coordinates we have

$$\iint_B \mathrm{Re}\,\frac{1}{\zeta}\,d\mathcal{L}^2(\zeta) = \int_{-\pi/2}^{+\pi/2}\int_0^{2a\cos\theta} \frac{\cos\theta}{r}\,r\,dr\,d\theta = \pi a = \sqrt{\pi\mathcal{L}^2(K)}.$$

□

Proposition 1.21 *For K a compact subset of $\mathbb{C}$,*

$$\gamma(K) \ge \sqrt{\frac{\mathcal{L}^2(K)}{\pi}}.$$

Proof Define a function f on $\mathbb{C}\setminus K$ by

$$f(z) = \iint_K \frac{1}{\zeta - z}\,d\mathcal{L}^2(\zeta).$$

Clearly f is analytic on $\mathbb{C}\setminus K$. The lemma implies that f is bounded on $\mathbb{C}\setminus K$ with modulus less than or equal to $\sqrt{\pi\mathcal{L}^2(K)}$ there. Lastly, $f(\infty) = 0$ and so $f'(\infty) = \lim_{z\to\infty} zf(z) = -\mathcal{L}^2(K)$. Hence

$$\mathcal{L}^2(K) = |f'(\infty)| \le \gamma(K)\|f\|_\infty \le \gamma(K)\sqrt{\pi\mathcal{L}^2(K)}$$

which leads to the desired inequality. □

Consideration of a closed ball and use of Corollary 1.13 show this estimate to be sharp.

If one wishes to show $\gamma(K) > 0$, i.e., K nonremovable, one must construct a nonconstant bounded analytic function on the complement of K. Propositions 1.1, 1.19, and 1.21 exhibit the most common technique for doing this: one considers the *Cauchy transform*

$$z \in \mathbb{C} \mapsto \hat{\mu}(z) = \int_{\mathbb{C}} \frac{1}{\zeta - z}\,d\mu(\zeta)$$

of an appropriately chosen nontrivial finite Borel measure μ supported on K. While not well defined at every point of $\mathbb{C}$, the Cauchy transform is always defined and analytic off the support of μ and so on the complement of K. It is nonconstant since its derivative at infinity is $-\mu(K) \ne 0$. The catch is that the Cauchy transform need not be bounded on the complement of K! The need to ensure boundedness, or to somehow get around unboundedness, accounts for the phrase "appropriately chosen" four sentences ago.

Another technique for showing nonremovability is by means of the Riemann maps introduced in Proposition 1.12. The lower estimate on capacity in Proposition 1.18 is an example of this. The author hazards to state that Cauchy transforms and Riemann maps are ultimately the only means known to mortals for showing nonremovability!

Chapter 2
Removable Sets and Hausdorff Measure

2.1 Hausdorff Measure and Dimension

At a fuzzy intuitive level, removable sets have small "size" and nonremovable sets big "size." A precise notion of "size" applicable to arbitrary subsets of $\mathbb{C}$ and appropriate to our problem is given by Hausdorff measure (and Hausdorff dimension). So in this section we will simply introduce Hausdorff measure as a gauge of the smallness of a set and as a necessary preliminary for another such gauge, Hausdorff dimension. Surprisingly, the assertions 2.1 through 2.4 below are enough to get us through to the end of Chapter 4. It is only after, in Section 5.1, that we shall need to take up the fact that Hausdorff measure is indeed a positive measure defined on a σ-algebra containing the Borel subsets of $\mathbb{C}$!

Given an arbitrary subset E of $\mathbb{C}$ and $\delta > 0$, a *δ-cover* of E is simply a countable collection of subsets $\{U_n\}$ of $\mathbb{C}$ such that $E \subseteq \bigcup_n U_n$ and $0 < |U_n| < \delta$ for each n. For any $s \geq 0$, define

$$\mathcal{H}^s_\delta(E) = \inf \left\{ \sum_n |U_n|^s : \{U_n\} \text{ is a } \delta\text{-cover of } E \right\}.$$

Clearly $\mathcal{H}^s_\delta(E)$ increases as δ decreases and so converges to a limit in $[0, \infty]$ as $\delta \downarrow 0$. This limit is called the *s-dimensional Hausdorff measure* (or *s-dimensional Hausdorff–Besicovitch measure*) of E and denoted $\mathcal{H}^s(E)$. Thus

$$\mathcal{H}^s(E) = \lim_{\delta \downarrow 0} \mathcal{H}^s_\delta(E) = \sup_{\delta > 0} \mathcal{H}^s_\delta(E).$$

Since the diameters of a set, its convex hull, its closure, and its closed convex hull are the same, $\mathcal{H}^s(E)$ may be computed by restricting ones attention to δ-covers of E by convex, closed, or closed convex sets. Similarly, since any nonempty set U is contained in the open set $\{z : \text{dist}(z, U) < \varepsilon\}$ whose diameter is $|U| + 2\varepsilon$ and since $\{z : \text{dist}(z, U) < \varepsilon\}$ is convex whenever U is convex, $\mathcal{H}^s(E)$ may be computed by restricting ones attention to δ-covers of E by open or open convex sets. Lastly, when

J.J. Dudziak, *Vitushkin's Conjecture for Removable Sets*, Universitext,
DOI 10.1007/978-1-4419-6709-1_2,

E is compact, $\mathcal{H}^s(E)$ may be computed by restricting ones attention to δ-covers of E by a finite number of open or open convex sets.

The following results are fairly immediate from the definition of Hausdorff measure.

Proposition 2.1 *For any $s \geq 0$, $\mathcal{H}^s$ is a set function defined on all subsets of $\mathbb{C}$ and taking values in $[0, \infty]$ such that*

$$\mathcal{H}^s(\emptyset) = 0,$$

$$\mathcal{H}^s(E) \leq \mathcal{H}^s(F) \textit{ whenever } E \subseteq F \subseteq \mathbb{C},$$

and

$$\mathcal{H}^s\left(\bigcup_n E_n\right) \leq \sum_n \mathcal{H}^s(E_n) \textit{ whenever } \{E_n\} \textit{ is a countable collection}$$

of subsets of $\mathbb{C}$.

Proposition 2.2 *Let E be a subset of $\mathbb{C}$ and let f be a mapping of E into $\mathbb{C}$ such that there exists a constant $c \geq 0$ for which $|f(z) - f(w)| \leq c|z - w|$ whenever $z, w \in E$. Then $\mathcal{H}^s(f(E)) \leq c^s\mathcal{H}^s(E)$.*

Corollary 2.3 *For α and β complex numbers and E a subset of $\mathbb{C}$, $\mathcal{H}^s(\alpha E + \beta) = |\alpha|^s\mathcal{H}^s(E)$.*

If $t > s \geq 0$, $0 < \delta < 1$, and $\{U_n\}$ is a δ-cover of E, then $\sum_n |U_n|^t \leq \sum_n |U_n|^s$. By infing over all δ-covers and then letting $\delta \downarrow 0$, we see that $\mathcal{H}^t(E) \leq \mathcal{H}^s(E)$. Thus $\mathcal{H}^s(E)$ decreases as s increases. But more can be said in this situation: $\sum_n |U_n|^t \leq \delta^{t-s}\sum_n |U_n|^s$, and so by infing over all δ-covers,

$$\mathcal{H}^t_\delta(E) \leq \delta^{t-s}\mathcal{H}^s_\delta(E).$$

Letting $\delta \downarrow 0$, we see that if $\mathcal{H}^s(E) < \infty$, then $\mathcal{H}^t(E) = 0$ for all $t > s$, and that if $\mathcal{H}^t(E) > 0$, then $\mathcal{H}^s(E) = \infty$ for all $s < t$. In consequence, there exists a unique nonnegative number called the *Hausdorff dimension* (or *Hausdorff–Besicovitch dimension*) of E and denoted $\dim_{\mathcal{H}}(E)$ such that

$$\mathcal{H}^s(E) = \begin{cases} \infty \text{ when } s < \dim_{\mathcal{H}}(E) \\ 0 \text{ when } s > \dim_{\mathcal{H}}(E). \end{cases}$$

What happens at $s = \dim_{\mathcal{H}}(E)$? In this case, $\mathcal{H}^s(E)$ can be 0, ∞, or anything in between. When $\mathcal{H}^s(E) = \infty$, it can even be the case that E is non-σ-finite for $\mathcal{H}^s$, i.e., E cannot be expressed as a countable union of sets of finite $\mathcal{H}^s$-measure. An example of this is the Joyce–Mörters set at the end of Chapter 4 (see Proposition 4.34).

Almost all of the following is fairly immediate from the definition of Hausdorff dimension and Proposition 2.1.

Proposition 2.4 *The Hausdorff dimension* $\dim_{\mathcal{H}}$ *is a function defined on all subsets of* $\mathbb{C}$ *and taking values in* $[0, 2]$ *such that*

$$\dim_{\mathcal{H}}(\emptyset) = 0,$$

$$\dim_{\mathcal{H}}(E) \leq \dim_{\mathcal{H}}(F) \text{ whenever } E \subseteq F \subseteq \mathbb{C},$$

and

$$\dim_{\mathcal{H}}\left(\bigcup_n E_n\right) = \sup_n \dim_{\mathcal{H}}(E_n) \text{ whenever } \{E_n\} \text{ is a countable collection of subsets of } \mathbb{C}.$$

The nonimmediate item is that $\dim_{\mathcal{H}}(E) \leq 2$ always. Because of this $\mathcal{H}^s(E) = 0$ for any subset of $\mathbb{C}$ whenever $s > 2$. Thus we will have no interest in $\mathcal{H}^s$ for $s > 2$.

Proof To show that $\dim_{\mathcal{H}}(E) \leq 2$ always, it suffices to show that $\dim_{\mathcal{H}}(Q) \leq 2$ for any closed square Q in $\mathbb{C}$ whose sides have length 1. Given $\delta > 0$, let n be a positive integer such that $\sqrt{2}/n < \delta$. Cut up Q into n^2 squares whose sides have length $1/n$ in the obvious way. Then $\mathcal{H}^2_\delta(Q) \leq n^2(\sqrt{2}/n)^2 = 2$. Upon letting $\delta \downarrow 0$, we get $\mathcal{H}^2(Q) \leq 2$. Hence, $\dim_{\mathcal{H}}(Q) \leq 2$. □

We close this section with a few examples to give the reader a feel for what Hausdorff measure and dimension are.

When $s = 0$, $\mathcal{H}^s$ is easily seen to be counting measure, i.e., $\mathcal{H}^0(E)$ is just the cardinality of E.

When $s = 1$, $\mathcal{H}^s$ of a linear subset of $\mathbb{C}$ is just the linear Lebesgue measure of the subset. To see this recall that for E a subset of a line L, we have by definition that

$$\mathcal{L}^1(E) = \inf\left\{\sum_n |I_n| : \{I_n\} \text{ is a countable cover of } E \text{ by intervals in } L\right\}.$$

Also recall that in computing $\mathcal{H}^1_\delta(E)$ we need only consider δ-covers of E by convex sets in the plane. Now an interval in L is a convex set in the plane and the intersection of a convex set in the plane with L gives us an interval in L with smaller diameter than the convex set. Hence

$$\mathcal{H}^1_\delta(E) = \inf\left\{\sum_n |I_n| : \{I_n\} \text{ is a } \delta\text{-cover of } E \text{ by intervals in } L\right\}.$$

Since any interval can be written as a finite union of intervals whose diameters are all less than δ and which sum to the diameter of the original interval, one clearly then has $\mathcal{H}^1_\delta(E) = \mathcal{L}^1(E)$. Letting $\delta \downarrow 0$ yields $\mathcal{H}^1(E) = \mathcal{L}^1(E)$.

More generally, $\mathcal{H}^1$ of a subset of a rectifiable arc in $\mathbb{C}$ is just the subset's "arclength measure." This and more will be shown later (Sections 4.5 and 5.2) when we consider rectifiable curves and define arclength measure on them. Thus one-dimensional Hausdorff measure, $\mathcal{H}^1$, may be thought of as a "generalized length" defined for any subset of $\mathbb{C}$ and will be referred to more concisely as *linear Hausdorff measure* in what follows.

When $s = 2$, $\mathcal{H}^s$ of any subset of $\mathbb{C}$ is just a multiple of the area, i.e., the planar Lebesgue measure, of the subset. More precisely, $\mathcal{H}^2(E) = (4/\pi)\mathcal{L}^2(E)$ for any subset E of $\mathbb{C}$. This will be shown later (Section 5.3) after we prove Vitali's Covering Lemma and the Isodiametric Inequality. Clearly then $\dim_{\mathcal{H}}(E) = 2$, whenever $\mathcal{L}^2(E) > 0$; in particular, whenever E has nonempty interior.

To get an example of a set of "fractional" Hausdorff dimension, consider the standard middle-thirds Cantor set K. Recall that K is the countable intersection of closed sets C_n where $C_0 = [0, 1]$, $C_1 = [0, 1/3] \cup [2/3, 1]$, $C_3 = [0, 1/9] \cup [2/9, 1/3] \cup [2/3, 7/9] \cup [8/9, 1]$, etc. Note that each C_n is the union of 2^n closed intervals of length 3^{-n} and that each C_{n+1} is obtained from C_n by removing the open middle-third from each of these constituent intervals of C_n.

Considering the obvious covering of K by the constituent intervals of C_n, we see that for any $\delta > 3^{-n}$, $\mathcal{H}^s_\delta(K) \le 2^n(3^{-n})^s = (2 \times 3^{-s})^n$. We would like to let $\delta \downarrow 0$, i.e., $n \to \infty$, and have this upper estimate for $\mathcal{H}^s_\delta(K)$ remain bounded. The smallest value of s for which this happens occurs when $2 \times 3^{-s} = 1$, i.e., when $s = \ln 2/\ln 3$. So fixing s at this value and letting $\delta \downarrow 0$, we have that $\mathcal{H}^s(K) \le 1$.

To show that $\mathcal{H}^s(K) \ge 1/2$ and so that $\dim_{\mathcal{H}}(K) = s = \ln 2/\ln 3$, it suffices to prove that

$$\sum_{I \in \mathcal{C}} |I|^s \ge \frac{1}{2}$$

for any finite covering $\mathcal{C}$ of K by open intervals. If this inequality fails, then among all coverings for which it fails, there is one with a minimal number of intervals. Denoting this minimal failing covering by $\mathcal{C}$, we will obtain a contradiction by constructing a failing covering $\mathcal{C}'$ with fewer intervals.

For any interval I of length $1/3$ or more, $|I|^s \ge 3^{-s} = 1/2$. Thus each interval of $\mathcal{C}$ must have length less than $1/3$ and so cannot intersect both $K_l = K \cap [0, 1/3]$ and $K_r = K \cap [2/3, 1]$. Let $\mathcal{C}_l$ and $\mathcal{C}_r$ consist of those intervals of $\mathcal{C}$ intersecting K_l and K_r respectively. Then $\mathcal{C}_l$ and $\mathcal{C}_r$ are nonempty, have no common intervals, and have cardinality strictly less than that of $\mathcal{C}$. Moreover, $\mathcal{C}_l$ and $\mathcal{C}_r$ are coverings of K_l and K_r respectively. If

$$\sum_{I \in \mathcal{C}_l} |I|^s \le \sum_{I \in \mathcal{C}_r} |I|^s,$$

then, since the dilation $x \mapsto 3x$ maps K_l onto K, $\mathcal{C}' = \{3I : I \in \mathcal{C}_l\}$ is the desired failing covering of K with fewer intervals. It fails because $\mathcal{C}$ fails and

$$\sum_{I \in \mathcal{C}} |I|^s \geq \sum_{I \in \mathcal{C}_l} |I|^s + \sum_{I \in \mathcal{C}_r} |I|^s \geq 2 \sum_{I \in \mathcal{C}_l} |I|^s = \sum_{I \in \mathcal{C}_l} 3^s |I|^s = \sum_{I' \in \mathcal{C}'} |I'|^s.$$

If $\mathcal{C}_r$ has the smaller associated sum, then proceed similarly using the dilation $x \mapsto 3x - 2$ of K_r onto K instead. In either case, we have produced the desired $\mathcal{C}'$ and so are done.

The elegant analysis above is taken from 8.2.22 of [KK]. A more refined analysis, as in the proof of Theorem 1.14 of [FALC], which gives more attention to the lengths of the constituent intervals and the distances between them, leads to the conclusion that for this K and s, $\mathcal{H}^s(K) = 1$ exactly.

Note that the upper estimate on the Hausdorff measure of K was easy to come by, whereas the lower estimate was difficult to establish. This is no accident but typical of most sets and in the nature of things. According to the definition of Hausdorff measure, to establish an upper estimate one need only come up with a single economical δ-covering, whereas to establish a lower estimate one needs to compute an infimum over all possible δ-coverings.

The analysis just given for K can be used to estimate the Hausdorff measure and obtain the Hausdorff dimension of other Cantor sets. The energetic reader may take the following two paragraphs as exercises to be worked out.

Given α strictly between 0 and 1/2, let K_α be the linear Cantor set gotten by the same procedure used to generate the standard middle-thirds set only now split each constituent interval from a generation, each "parent," into two constituent intervals of the next generation, two "children," each with length α times that of the parent interval (so the standard middle-thirds Cantor set just considered is, appropriately enough, $K_{1/3}$). Then $(1 - 2\alpha)^s \leq \mathcal{H}^s(K_\alpha) \leq 1$ for $s = \ln 2/\ln(1/\alpha)$ and so $\dim_{\mathcal{H}}(K_\alpha) = \ln 2/\ln(1/\alpha)$. This shows that for any s strictly between 0 and 1, there exist linear Cantor sets of Hausdorff dimension s. (The more careful analysis in the proof of Theorem 1.14 of [FALC] can easily be adapted to show that for these K_α and s, $\mathcal{H}^s(K_\alpha) = 1$ exactly.)

Given α strictly between 0 and 1/2, let C_0 be the closed unit square $[0, 1] \times [0, 1]$. Let C_1 be the union of the four closed squares contained in C_0 each of which contains a corner of C_0 and has edges of length α. Let C_2 be the union of the sixteen closed squares contained in C_1 each of which contains a corner of a constituent square of C_1 and has edges of length α^2. Continue in this fashion to generate a sequence of nested closed sets $\{C_n\}$, each C_n consisting of the disjoint union of 4^n closed squares with edges of length α^n. Set $K_\alpha = \bigcap_n C_n$. Then $(1 - 2\alpha)^s \leq \mathcal{H}^s(K_\alpha) \leq \sqrt{2^s}$ for $s = \ln 4/\ln(1/\alpha)$ and so $\dim_{\mathcal{H}}(K_\alpha) = \ln 4/\ln(1/\alpha)$. This shows that for any s strictly between 0 and 2, there exist planar Cantor sets of Hausdorff dimension s. (A more careful analysis which can be found in [McM] shows that $\mathcal{H}^1(K_{1/4}) = \sqrt{2}$ exactly. We shall return to this set $K_{1/4}$ and examine it more carefully in the last section of this chapter.)

2.2 Painlevé's Theorem

The main concern of this section, Painlevé's Theorem from [PAIN], will enable us to dispose of sets of Hausdorff dimension less than one – they will all be shown removable! Painlevé's Theorem will follow easily from a more general quantitative result which bounds $\gamma(K)$ by $\mathcal{H}^1_\infty(K)$. (Note that $\mathcal{H}^1_\infty(K)$ is just $\mathcal{H}^1_\delta(K)$ with $\delta = \infty$! It is usually referred to as the *linear Hausdorff content* of K.) For this we need a lemma which constructs cycles surrounding K with length just a little more than $2\pi\mathcal{H}^1_\infty(K)$. We give two proofs of this lemma: one, short but loose, sufficient for those who are willing to take certain topological assertions as obvious, and another, more rigorous but longer, addressed to those who are finicky about such things.

Lemma 2.5 *Let K be a compact subset of $\mathbb{C}$ and $\varepsilon > 0$. Then there exists a cycle Γ in $\mathbb{C} \setminus K$ of length at most $2\pi\mathcal{H}^1_\infty(K) + \varepsilon$ with winding number one about each point of K.*

Proof (Loose) The definition of $\mathcal{H}^1_\infty(K)$ provides us with a finite cover $\{U_n\}$ of K by bounded open sets such that $\sum_n |U_n| < \mathcal{H}^1_\infty(K) + \varepsilon/2\pi$. Without loss of generality, each U_n intersects K. Replace each U_n by an open disk of radius $|U_n|$ centered anywhere on U_n. Let the cycle Γ be the counterclockwise boundary of the union of the disks so produced. The reader may "easily" verify that Γ works. □

The next proof is a slight modification of the proof of [RUD, 13.5].

Proof (More Rigorous) The definition of $\mathcal{H}^1_\infty(K)$ provides us with a finite cover $\{U_n\}$ of K by bounded open sets such that $\sum_n |U_n| < \mathcal{H}^1_\infty(K) + \varepsilon/4$. Without loss of generality, each U_n intersects K. Choose $\eta > 0$ so that $4\sum_n |U_n| + 8N\eta < 4\mathcal{H}^1_\infty(K) + \varepsilon$ where N is the number of sets U_n in our finite cover. Cover the plane with a grid of closed squares whose interiors are nonintersecting and whose edges have length η. Let R_n be the smallest closed rectangle whose interior contains U_n and whose sides lie in the lines formed by our grid. Clearly the perimeter of R_n is at most $4(|U_n| + 2\eta)$. Let $Q_1, \ldots, Q_M$ be those squares of our grid contained in $R_1 \cup \cdots \cup R_N$. Each square Q_m has four boundary intervals. Direct these four intervals so that Q_m's boundary is given a counterclockwise orientation. Let Σ be the collection of the $4M$ directed intervals that result.

Then Σ is a *balanced* collection of directed intervals, which means that for every point p of the plane, the number of intervals of Σ with p as initial point is equal to the number of intervals of Σ with p as terminal point. Remove those intervals from Σ whose oppositely oriented intervals are also in Σ and denote the collection of remaining intervals of Σ by Φ. The collection Φ is still balanced and each interval of Φ is contained in the boundary of some R_n.

Construct a cycle Γ from Φ as follows. Pick any γ_1 from Φ. Since Φ is balanced, there is an interval γ_2 in $\Phi\setminus\{\gamma_1\}$ whose initial point is the terminal point of γ_1. Since Φ is balanced, there is an interval γ_3 in $\Phi\setminus\{\gamma_1, \gamma_2\}$ whose initial point is the terminal point of γ_2. Continue in this manner until you reach an interval γ_k whose endpoint is the initial point of γ_1. This must occur since Φ is finite and balanced. Then the intervals $\gamma_1, \gamma_2, \ldots, \gamma_k$ form a closed path. The collection $\Phi\setminus\{\gamma_1, \gamma_2, \ldots, \gamma_k\}$ is still balanced so, if it is nonempty, we may repeat our procedure again. Continuing in this manner until Φ is empty, we see that the intervals of Φ form a chain consisting of closed paths, i.e., the intervals of Φ form a cycle, Γ.

If an edge of some Q_m intersects K, then it must be common to two adjacent Q_m's since $K \subseteq \operatorname{int} \bigcup_{m=1}^{M} Q_m$. Hence Σ contains the two oppositely oriented intervals determined by the edge and so these intervals are not in Φ. In consequence, Γ is a cycle in $\mathbb{C} \setminus K$.

Clearly the length of Γ is at most the sum of the perimeters of the rectangles R_n and so less than $4\mathcal{H}^1_\infty(K) + \varepsilon$.

For any z in K but not on the boundary of any Q_m, the winding number of Γ about z, being the sum of the winding numbers of the boundaries of the Q_m's about z, is 1. This and the constancy of the winding number of a cycle on each component of the complement of its range [RUD, 10.10] make the last assertion of the lemma clear. □

Note that the more rigorous proof has produced a cycle with a better estimate on its length: the 2π has been reduced to 4. We will say more about this after proving the main result of this section which now follows quite easily ...

Theorem 2.6 $\gamma(K) \le \mathcal{H}^1_\infty(K)$.

Proof Let $f \in H^\infty(\mathbb{C}^* \setminus K)$ be such that $\|f\|_\infty \le 1$. Given $\varepsilon > 0$, obtain a cycle Γ as in the previous lemma. Then, by the integral representation for $f'(\infty)$ in the discussion just after Proposition 1.1,

$$|f'(\infty)| = \left| \frac{1}{2\pi i} \int_\Gamma f(\zeta)\, d\zeta \right| \le \frac{2\pi \mathcal{H}^1_\infty(K) + \varepsilon}{2\pi}.$$

Letting $\varepsilon \downarrow 0$ and then suping over the fs in question, we see that $\gamma(K) \le \mathcal{H}^1_\infty(K)$. □

Of course what has been shown is that if the cycle Γ of the lemma has length at most $c\mathcal{H}^1_\infty(K) + \varepsilon$, then $\gamma(K) \le \kappa \mathcal{H}^1_\infty(K)$ where $\kappa = c/2\pi$. The loose proof led to $c = 2\pi$ and so $\kappa = 1$; while the more rigorous proof led to $c = 4$ and so $\kappa = 2/\pi = 0.6366\cdots$. If in the loose proof one uses Jung's Theorem (recall the paragraph after the proof of Proposition 1.18) instead of simply enclosing each U_n in a ball of radius $|U_n|$, then one has $c = 2\pi/\sqrt{3}$ and so $\kappa = 1/\sqrt{3} = 0.5773\cdots$. It turns out that it is possible to take $c = \pi$ and so $\kappa = 1/2 = 0.5$. To obtain this one must do three things however: first, extend ones complex integration theory from chains of segments and circular arcs to chains of rectifiable arcs; second, show that the boundary of a bounded open convex set is rectifiable; and third, show a (sharp) "peri-diametric inequality" to the effect that the perimeter of a bounded open convex set is at most π times its diameter. We dispense with these complications and content ourselves with the cruder estimate enunciated in the theorem above since from the point of view of removability it suffices simply to have $\kappa < \infty$!

However, before passing to Painlevé's Theorem, let us take the estimate $(\star)$ $\gamma(K) \le \mathcal{H}^1_\infty(K)/2$ as established and play with it to see a few things. By using the trivial covering of K by itself we have $(\star\star)$ $\mathcal{H}^1_\infty(K) \le |K|$ always. Thus $(\star\star\star)$ $\gamma(K) \le |K|/2$ always, which is a sharper form of the first estimate of Proposition 1.18 and beats out the improvement of it via Jung's Theorem noted immediately after its proof. Next, invoking Corollary 1.13 we see that for K a closed ball of radius r,

$$r = \gamma(K) \le \frac{\mathcal{H}^1_\infty(K)}{2} \le \frac{|K|}{2} = r,$$

and so $(\star)$, $(\star\star)$, and $(\star\star\star)$ are all as sharp as possible.

The theorem which gives this section its title is more in the nature of a corollary to the main result above since one trivially has $\mathcal{H}^1_\infty(K) \le \mathcal{H}^1(K)$, but we label it a theorem due to its importance.

Theorem 2.7 (Painlevé) *If* $\mathcal{H}^1(K) = 0$, *then* K *is removable.*

Since any set of Hausdorff dimension less than 1 has linear Hausdorff measure 0, we immediately obtain the following.

Corollary 2.8 *If* $\dim_{\mathcal{H}}(K) < 1$, *then* K *is removable.*

2.3 Frostman's Lemma

Painlevé's Theorem (2.7) has disposed of sets with dimension less than 1. The main concern of this section, a potential-theoretic lemma from the thesis of Otto Frostman, will enable us to dispose of sets with dimension greater than one – they will all be shown nonremovable! We prove only a simple version of this lemma sufficient for our purposes. Throughout this book we shall denote the radius of a closed ball B by $\mathrm{rad}(B)$. (Note that $\mathcal{H}^s_\infty(K)$ in the next lemma is just $\mathcal{H}^s_\delta(K)$ with $\delta = \infty$! It is usually referred to as the *s-dimensional Hausdorff content* of K.)

Lemma 2.9 (Frostman) *Given* $1 < s \le 2$, *set* $M = 9 \times 4^s$. *Then for any compact subset* K *of* $\mathbb{C}$, *there exists a regular positive Borel measure* μ *on* K *with mass at least* $\mathcal{H}^s_\infty(K)$ *such that* $\mu(B) \le M\,\mathrm{rad}(B)^s$ *for all closed balls* B *in* $\mathbb{C}$.

Proof Call a square *half-closed* when it is obtained from its closure by removing the top and right closed edges. Note that any half-closed square can be decomposed, in the obvious way, into four pairwise disjoint, half-closed squares whose diameters are half that of the original. We call these four squares the *children* of the original square. In what follows we shall also use the terms *parent*, *grandparent*, and *ancestor* in a self-evident manner given the way the term *children* has just been defined.

Choose any half-closed square Q_0 containing K in its interior and set $\mathcal{G}_0 = \{Q_0\}$. Let $\mathcal{G}_1$ consist of the four children of the single square of $\mathcal{G}_0$. Let $\mathcal{G}_2$ consist of the 16 children of the 4 squares of $\mathcal{G}_1$. Continuing in this manner, we obtain the generations $\mathcal{G}_0, \mathcal{G}_1, \mathcal{G}_2, \ldots$ of the dyadic grid $\mathcal{G} = \bigcup_{n\ge 0} \mathcal{G}_n$.

Fixing n for the moment, let ν_n be the regular positive Borel measure on Q_0 whose restriction to each $Q \in \mathcal{G}_n$ is equal to κ times the restriction of $\mathcal{L}^2$ to Q

where $\kappa = |Q|^s/\mathcal{L}^2(Q)$ if Q intersects K and 0 otherwise. Clearly $\nu_n(Q) \le |Q|^s$ for every $Q \in \mathcal{G}_n$ and $\nu_n(Q) = |Q|^s$ for every $Q \in \mathcal{G}_n$ which intersects K.

Let ν_{n-1} be the regular positive Borel measure on Q_0 whose restriction to each $Q \in \mathcal{G}_{n-1}$ is equal to κ times the restriction of ν_n to Q where $\kappa = |Q|^s/\nu_n(Q)$ if $|Q|^s < \nu_n(Q)$ and 1 otherwise. Note that the measure ν_{n-1} just defined has been gotten by uniformly reducing ν_n or by leaving it undisturbed on each square of $\mathcal{G}_{n-1}$. In consequence, $\nu_{n-1}(Q) \le |Q|^s$ for every $Q \in \mathcal{G}_{n-1} \cup \mathcal{G}_n$. Furthermore, given any square from $\mathcal{G}_n$ which intersects K, we have $\nu_{n-1}(Q) = |Q|^s$ for Q the square or its parent.

Let ν_{n-2} be the regular positive Borel measure on Q_0 whose restriction to each $Q \in \mathcal{G}_{n-2}$ is equal to κ times the restriction of ν_{n-1} to Q where $\kappa = |Q|^s/\nu_{n-1}(Q)$ if $|Q|^s < \nu_{n-1}(Q)$ and 1 otherwise. Note that the measure ν_{n-2} just defined has been gotten by uniformly reducing ν_{n-1} or by leaving it undisturbed on each square of $\mathcal{G}_{n-2}$. In consequence, $\nu_{n-2}(Q) \le |Q|^s$ for every $Q \in \mathcal{G}_{n-2} \cup \mathcal{G}_{n-1} \cup \mathcal{G}_n$. Furthermore, given any square from $\mathcal{G}_n$ which intersects K, we have $\nu_{n-2}(Q) = |Q|^s$ for Q the square, its parent, or its grandparent.

Continuing in this manner, we obtain measures ν_n, ν_{n-1}, ν_{n-2}, ..., ν_1, and ν_0. Denote the measure ν_0 by μ_n. Then $(\star)$ $\mu_n(Q) \le |Q|^s$ for every Q in $\mathcal{G}_0 \cup \mathcal{G}_1 \cup \cdots \cup \mathcal{G}_n$. Furthermore, given any square from $\mathcal{G}_n$ which intersects K, we have $\mu_n(Q) = |Q|^s$ for Q the square or one of its ancestors. In consequence, each point of K is contained in a unique largest square Q_z in $\mathcal{G}_0 \cup \mathcal{G}_1 \cup \cdots \cup \mathcal{G}_n$ for which $\mu_n(Q) = |Q|^s$. The collection $\{Q_z : z \in K\}$ is actually finite and so can be written $\{Q_1, Q_2, \ldots, Q_M\}$. Since these squares are pairwise disjoint and K is contained in their union, we have on the one hand

$$(\star\star)\ \mu_n(\mathrm{cl}\, Q_0) \ge \sum_{m=1}^{M} \mu_n(Q_m) = \sum_{m=1}^{M} |Q_m|^s \ge \mathcal{H}^s_\infty(K),$$

while on the other hand we obviously have $(\star\star\star)$ $\mu_n(\mathrm{cl}\, Q_0) \le |Q_0|^s$.

Unfix n now and realize that, by $(\star\star\star)$, we have produced a sequence $\{\mu_n\}$ of uniformly bounded linear functionals on the Banach space of continuous functions on $\mathrm{cl}\, Q_0$. Moreover, this space is separable (consider the polynomials in z and $\bar{z}$ with rational coefficients and invoke Stone-Weierstrass). We may thus apply [RUD, 11.29] and [RUD, 6.19] to obtain a subsequence $\{\mu_{n_k}\}$ of $\{\mu_n\}$ which converges weakly to some regular complex Borel measure μ on $\mathrm{cl}\, Q_0$.

Given a compact subset C of $\mathrm{cl}\, Q_0$, consider the functions $f_N(z) = \max\{1 - N \operatorname{dist}(z, C), 0\}$. By the weak convergence and positiveness of the measures μ_{n_k}, $\int f_N \, d\mu \ge 0$ for all $N \ge 1$. Letting $N \to \infty$ and invoking Lebesgue's Dominated Convergence Theorem [RUD, 1.34], we conclude that $\mu(C) \ge 0$. If C is also disjoint from K, then the same argument shows that $\mu(C) = 0$ since each μ_n is supported on the closed $|Q_0|/2^n$-neighborhood of K. The inner regularity of μ now implies that μ is a positive measure supported on K. Of course $(\star\star)$ and the weak convergence imply that μ has mass at least $\mathcal{H}^s_\infty(K)$. So it only remains to verify the growth condition on μ.

We first show that for any n, $\mu_n(B) \le M\,\mathrm{rad}(B)^s$ for all closed balls B such that $|B \cap Q_0| > |Q_0|/2^{n+1}$. Let m be the integer between 0 and n such that

$$\frac{|Q_0|}{2^{m+1}} < |B \cap Q_0| \le \frac{|Q_0|}{2^m}.$$

For this m, $B \cap Q_0$ is contained in the union of at most nine squares from $\mathcal{G}_m$. Thus by ($\star$) we have

$$\mu_n(B) = \mu_n(B \cap Q_0) \le 9\left\{\frac{|Q_0|}{2^m}\right\}^s \le 9\{2|B \cap Q_0|\}^s \le M\,\mathrm{rad}(B)^s.$$

Now consider any closed ball B. Without loss of generality, B intersects cl Q_0. Given $N \ge 1$, set $g_N(z) = \max\{1 - N\,\mathrm{dist}(z, B), 0\}$ and let B_N be the closed ball concentric with B and of radius $\mathrm{rad}(B) + 1/N$. Then

$$\begin{aligned}\mu(B) \le \int g_N\,d\mu = \lim_{k\to\infty}\int g_N\,d\mu_{n_k} &\le \limsup_{k\to\infty}\mu_{n_k}(B_N) \le M\,\mathrm{rad}(B_N)^s \\ &= M\{\mathrm{rad}(B) + 1/N\}^s.\end{aligned}$$

Letting $N \to \infty$, we are done. □

The second ingredient needed for the main result of this section is an estimate on the Newtonian potential of a regular positive Borel measure satisfying a growth condition as in Frostman's Lemma.

Lemma 2.10 *Let $0 < M < \infty$ and $1 < s \le 2$. Suppose μ is a nontrivial regular positive Borel measure on a compact subset K of $\mathbb{C}$ such that $\mu(B) \le M\,\mathrm{rad}(B)^s$ for all closed balls B in $\mathbb{C}$. Then for any $z \in \mathbb{C} \setminus K$,*

$$\int \frac{1}{|\zeta - z|}\,d\mu(\zeta) \le \frac{s}{s-1}M^{1/s}\mu(K)^{1-1/s}.$$

Proof By Fubini's Theorem [RUD, 8.8],

$$\begin{aligned}\int \frac{1}{|\zeta - z|}\,d\mu(\zeta) &= \int\left\{\int_{|\zeta-z|}^{\infty}\frac{1}{t^2}\,dt\right\}d\mu(\zeta) \\ &= \int\left\{\int_0^{\infty}\frac{1}{t^2}\mathcal{X}_{\{|\zeta-z|\le t\}}\,dt\right\}d\mu(\zeta) \\ &= \int_0^{\infty}\left\{\int\frac{1}{t^2}\mathcal{X}_{\{|\zeta-z|\le t\}}\,d\mu(\zeta)\right\}dt \\ &= \int_0^{\infty}\frac{\mu(B(z;t))}{t^2}\,dt.\end{aligned}$$

We now wish to plug in our growth estimate on $\mu(B(z;t))$, but only for t not too large. Obviously past some point it is better to estimate $\mu(B(z;t))$ simply by $\mu(K)$ rather than by Mt^s. This change-over point is the T such that $MT^s = \mu(K)$, i.e., $T = \{\mu(K)/M\}^{1/s}$. This leads to

$$\int \frac{1}{|\zeta - z|}\, d\mu(\zeta) \leq \int_0^T \frac{Mt^s}{t^2}\, dt + \int_T^\infty \frac{\mu(K)}{t^2}\, dt = \frac{MT^{s-1}}{s-1} + \frac{\mu(K)}{T}.$$

Substituting in our choice for T and doing a bit of algebra, we obtain what we want. □

The main result of this section now follows quite easily.

Theorem 2.11 *For* $1 < s \leq 2$, $\gamma(K) \geq \dfrac{s-1}{4s}\left\{\dfrac{\mathcal{H}^s_\infty(K)}{9}\right\}^{1/s}$.

Proof Without loss of generality, $\mathcal{H}^s_\infty(K) > 0$. (As an aside, we note that $\mathcal{H}^s_\infty(K) \leq |K|^s < \infty$ always.) Get a measure μ as in Frostman's Lemma (2.9). Clearly μ is nontrivial since $\mu(K) \geq \mathcal{H}^s_\infty(K)$. Consider the function f equal to $\hat{\mu}$, the Cauchy transform of μ. Thus

$$f(z) = \int \frac{1}{\zeta - z}\, d\mu(\zeta).$$

Clearly f is defined and analytic on $\mathbb{C} \setminus K$ since μ is supported on K. By the last lemma,

$$\|f\|_\infty \leq \frac{s}{s-1}\{9 \times 4^s\}^{1/s}\mu(K)^{1-1/s} \leq \frac{4s}{s-1}\left\{\frac{9}{\mathcal{H}^s_\infty(K)}\right\}^{1/s}\mu(K).$$

But $f(\infty) = \lim_{z\to\infty} f(z) = 0$, so $|f'(\infty)| = \lim_{z\to\infty} |zf(z)| = \mu(K)$. Thus

$$\gamma(K) \geq \frac{|f'(\infty)|}{\|f\|_\infty} \geq \frac{s-1}{4s}\left\{\frac{\mathcal{H}^s_\infty(K)}{9}\right\}^{1/s}.$$

□

Corollary 2.12 *If* $\dim_{\mathcal{H}}(K) > 1$, *then* K *is nonremovable.*

Proof Let s be any number strictly between 1 and $\dim_{\mathcal{H}}(K)$. Then $\mathcal{H}^s(K) = \infty$, so there exists a $\delta > 0$ such that $\mathcal{H}^s_\delta(K) > 0$. Then for any countable cover of K by bounded sets $\{U_n\}$ we must have $\sum_n |U_n|^s \geq \mathcal{H}^s_\delta(K)$ if $\{U_n\}$ is a δ-cover and $\sum_n |U_n|^s \geq \delta^s$ otherwise. Thus $\mathcal{H}^s_\infty(K) \geq \min\{\mathcal{H}^s_\delta(K), \delta^s\} > 0$. By the theorem we are done. □

Note how the theorem is quantitative: given $\mathcal{H}^s_\infty(K) > 0$, it allows us to conclude $\gamma(K) > 0$ and even to estimate how much bigger than zero $\gamma(K)$ is in terms of $\mathcal{H}^s_\infty(K)$. The corollary however is merely qualitative: although we see that $\gamma(K) > 0$ when $\dim_{\mathcal{H}}(K) > 1$, we do not obtain an effective estimate of how much bigger than zero $\gamma(K)$ is in terms of $\dim_{\mathcal{H}}(K)$. This is to be expected since under a similitude $\gamma(K)$ scales linearly (Proposition 1.9) while $\dim_{\mathcal{H}}(K)$ clearly remains invariant. In the proof this shows up as a lack of effective control over the quantity $\min\{\mathcal{H}^s_\delta(K), \delta^s\}$; although positive it could be arbitrarily small since δ could be arbitrarily small!

2.4 Conjecture and Refutation: The Planar Cantor Quarter Set

The results of the last two sections, which settle the question of removability for all compact sets except those of dimension one, are all consistent with the conjecture that Painlevé's Theorem (2.7) is reversible. If true, one would have that a compact subset K of $\mathbb{C}$ is removable if and only if $\mathcal{H}^1(K) = 0$. This beautiful conjecture, which would end our quest in a quite tidy manner, is false however! The first example of a removable compact set with positive, and even finite, linear Hausdorff measure was due to Anatoli Vitushkin (see [VIT1], or Section 3 of Chapter IV of [GAR2]). His example is quite complicated, slaying other conjectures than just the one of interest to us here. Later John Garnett (see [GAR1], or Section 2 of Chapter IV of [GAR2]) realized that a planar Cantor quarter set is a much simpler example of a set with positive finite linear Hausdorff measure but zero analytic capacity. Since then other proofs that this set works have been published (see [JON1], [MAT1], and/or [MUR]). We give Garnett's original proof below, it being the most elementary.

So consider the *planar Cantor quarter set*, i.e., the set $K_{1/4}$ of the last paragraph in Section 2.1. Recall that $K_{1/4} = \bigcap_n C_n$ where C_0 is the closed unit square $[0, 1] \times [0, 1]$, C_1 is the union of the four closed squares contained in C_0 each of which contains a corner of C_0 and has edges of length $1/4$, C_2 is the union of the 16 closed squares contained in C_1 each of which contains a corner of a constituent square of C_1 and has edges of length $1/16$, etc. As noted in Section 2.1, the coverings of $K_{1/4}$ by the 4^n closed constituent squares of C_n with edges of length $1/4^n$ lead easily to the upper estimate $\mathcal{H}^1(K_{1/4}) \le \sqrt{2} = 1.4142\cdots$, while the more difficult self-similarity argument given there leads to the lower estimate $\mathcal{H}^1(K_{1/4}) \ge 1/2 = 0.5$. As an easy and nice application of Proposition 2.2, we now substantially improve this lower estimate. Note that the orthogonal projection of the plane onto the line $y = x/2$ takes each of the sets C_n, and so too $K_{1/4}$, onto the closed segment with endpoints $(0, 0)$ and $(6/5, 3/5)$. Since orthogonal projection onto any fixed line decreases distances, it can only decrease linear Hausdorff measure by Proposition 2.2. Recall that in Section 2.1 we also showed the coincidence of $\mathcal{H}^1$ and $\mathcal{L}^1$ for linear sets. Hence we obtain our substantial improvement: $\mathcal{H}^1(K_{1/4}) \ge 3\sqrt{5}/5 = 1.3416\cdots$. Of course, as noted already in Section 2.1, there is a more careful and difficult analysis in [McM] which shows that $\mathcal{H}^1(K_{1/4}) = \sqrt{2}$ exactly.

The rest of this section is devoted to the task of showing that $\gamma(K_{1/4}) = 0$. A number of lemmas are required. The first is a version of Cauchy's Integral Theorem for functions analytic at ∞. We use $n(\Gamma; z)$ to denote the winding number of a cycle Γ in $\mathbb{C}$ about a point z in $\mathbb{C} \setminus \Gamma$.

Lemma 2.13 *Let K be a compact subset of $\mathbb{C}$ and let f be an analytic function on $\mathbb{C}^* \setminus K$. Suppose that Γ is a cycle in $\mathbb{C} \setminus K$ with winding number one about every point of K and that z is a point off $K \cup \Gamma$.*

Then

$$f(\infty) - \frac{1}{2\pi i} \int_\Gamma \frac{f(\zeta)}{\zeta - z}\, d\zeta = \begin{cases} f(z) \text{ if } & n(\Gamma; z) = 0 \\ 0 \text{ if } & n(\Gamma; z) = 1 \end{cases}.$$

Proof Consider any R large enough so that $K \cup \Gamma \cup \{z\}$ is encircled by C_R, the counterclockwise circle about the origin of radius R. Cauchy's Integral Theorem [RUD, 10.35] applied to the cycle $C_R - \Gamma$ implies that

$$\frac{1}{2\pi i} \int_{C_R} \frac{f(\zeta)}{\zeta - z}\, d\zeta - \frac{1}{2\pi i} \int_\Gamma \frac{f(\zeta)}{\zeta - z}\, d\zeta = \begin{cases} f(z) \text{ if } & n(\Gamma; z) = 0 \\ 0 \text{ if } & n(\Gamma; z) = 1 \end{cases}.$$

Thus, to finish we need only show that the first integral in the above equation converges to $f(\infty)$ as $R \to \infty$. Since $n(C_R; z) = 1$, we have

$$\begin{aligned} \left| \frac{1}{2\pi i} \int_{C_R} \frac{f(\zeta)}{\zeta - z}\, d\zeta - f(\infty) \right| &= \left| \frac{1}{2\pi i} \int_{C_R} \frac{f(\zeta) - f(\infty)}{\zeta - z}\, d\zeta \right| \\ &\le \frac{1}{2\pi} \cdot \frac{\sup\{|f(\zeta) - f(\infty)| : \zeta \in C_R\}}{R - |z|} \cdot 2\pi R \\ &= \frac{R}{R - |z|} \cdot \sup\{|f(\zeta) - f(\infty)| : \zeta \in C_R\}. \end{aligned}$$

Since the last quantity converges to 0 as $R \to \infty$, we are done. (An aside: Being the only term involving R in the penultimate displayed equation, the integral over C_R does not really depend on R. Thus we have shown not merely that this integral converges to $f(\infty)$ as $R \to \infty$ but that it is actually equal to it for all large R!) □

When we proved Painlevé's Theorem (2.7) we essentially verified (d) of Proposition 1.3 for sets of linear Hausdorff measure 0. As an exercise the energetic reader may reprove Painlevé with the help of this last lemma by verifying (c) of Proposition 1.3 for sets of linear Hausdorff measure 0!

For convenience's sake, denote $K_{1/4}$ simply by K for the remainder of this section. Let $\mathcal{G}_0$ denote the collection consisting of the single square which is C_0, let $\mathcal{G}_1$ denote the collection consisting of the four squares whose union comprise C_1, let $\mathcal{G}_2$ denote the collection consisting of the 16 squares whose union comprise C_2, etc. Next, let $\mathcal{G}$ be the collection of all squares appearing in one of these generations, so $\mathcal{G} = \bigcup_n \mathcal{G}_n$. Continuing, for every square Q from $\mathcal{G}$, set $K_Q = K \cap Q$, $\mathcal{G}(Q) = \{R \in \mathcal{G} : R \subseteq Q\}$, and $\mathcal{G}_n(Q) = \{R \in \mathcal{G}_n : R \subseteq Q\}$. Note that each K_Q

is geometrically similar to K with similarity ratio $l(Q)$, the common length of the edges of Q. Finally, given a square Q from $\mathcal{G}$, a square R from $\mathcal{G}(Q)$, and a function f from $H^\infty(\mathbb{C}^* \setminus K_Q)$ which vanishes at ∞, define a function $f_{Q,R}$ on $\mathbb{C}^* \setminus K_R$ by

$$f_{Q,R}(z) = \frac{-1}{2\pi i} \int_{\Gamma_{Q,R}} \frac{f(\zeta)}{\zeta - z} \, d\zeta$$

where $\Gamma_{Q,R}$ is any cycle in $\mathbb{C} \setminus (K_Q \cup \{z\})$ with winding number 1 about every point of K_R and 0 about every point of $(K_Q \setminus K_R) \cup \{z\}$.

Lemma 2.14 *With the conventions and definitions as above, we have*

(a) *Each $f_{Q,R}$ is a well-defined element of $H^\infty(\mathbb{C}^* \setminus K_R)$ which vanishes at ∞.*
(b) *$\|f_{Q,R}\|_\infty \leq M \|f\|_\infty$ where $M = 1 + 6/\pi$.*
(c) *Fix $n \geq m$, where m is the integer such that $\mathcal{G}_m$ contains Q. Then on the set $\mathbb{C} \setminus K_Q$ one has*

$$f = \sum_{R \in \mathcal{G}_n(Q)} f_{Q,R}.$$

(d) *Let φ be a similitude of the complex plane onto itself involving no rotation, so $\varphi(z) = \alpha z + \beta$ where $\alpha > 0$ and β is a complex number. Suppose there are squares Q^* and R^* from $\mathcal{G}$ such that $\varphi(Q^*) = Q$ and $\varphi(R^*) = R$. Then*

$$f_{Q,R} \circ \varphi = (f \circ \varphi)_{Q^*,R^*}.$$

(e) *For any square S from $\mathcal{G}(R)$, $(f_{Q,R})_{R,S} = f_{Q,S}$.*

Proof (a) The functions $f_{Q,R}$ are well defined because of Cauchy's Integral Theorem [RUD, 10.35]. Analyticity and vanishing at infinity are clear. Boundedness follows from the next item.

(b) Let Γ denote the counterclockwise boundary of the square concentric with R and with 3 times the diameter. Then by Cauchy's Integral Theorem [RUD, 10.35] applied to the cycle $\Gamma - \Gamma_{Q,R}$, for z close to but not on K_R we have

$$f(z) = \frac{1}{2\pi i} \int_\Gamma \frac{f(\zeta)}{\zeta - z} \, d\zeta + f_{Q,R}(z).$$

Suppose z, not on K_R, is within a distance $\varepsilon > 0$ of K_R. Then

$$|f_{Q,R}(z)| \leq |f(z)| + \frac{1}{2\pi} \cdot \frac{\|f\|_\infty}{l(R) - \varepsilon} \cdot 12 l(R) \leq \left\{ 1 + \frac{6}{\pi} \cdot \frac{l(R)}{l(R) - \varepsilon} \right\} \|f\|_\infty.$$

By the Maximum Modulus Principle [RUD, 12.1 modified to take into account regions containing ∞],

$$\|f_{Q,R}\|_\infty \le \left\{1 + \frac{6}{\pi}\cdot\frac{l(R)}{l(R)-\varepsilon}\right\}\|f\|_\infty.$$

Letting $\varepsilon \downarrow 0$, we have our estimate.

(c) Apply the last lemma to the cycle which is the sum of the cycles $\Gamma_{Q,R}$ for $R \in \mathcal{G}_n(Q)$.

(d) Let z^* be such that $\varphi(z^*) = z$. Simply use the change of variables $\zeta = \varphi(\xi)$ in the equation defining $f_{Q,R}(z) = (f_{Q,R} \circ \varphi)(z^*)$ and then notice that we are done since the cycle $\varphi^{-1}(\Gamma_{Q,R})$ will serve as a cycle Γ_{Q^*,R^*} as in the definition of $(f \circ \varphi)_{Q^*,R^*}(z^*)$.

(e) What follows is a pain but (e) is absolutely crucial to our success! Fixing $z \in \mathbb{C} \setminus K_S$, we must show that $(f_{Q,R})_{R,S}(z) = f_{Q,S}(z)$.

Choose a cycle $\Gamma_{R,S}$ in $\mathbb{C} \setminus (K_Q \cup \{z\})$ with winding number 1 about every point of K_S and 0 about every point of $(K_Q \setminus K_S) \cup \{z\}$. Then $\Gamma_{R,S}$ is a cycle in $\mathbb{C} \setminus (K_R \cup \{z\})$ with winding number 1 about every point of K_S and 0 about every point of $(K_R \setminus K_S) \cup \{z\}$ and so

$$(\star)\ (f_{Q,R})_{R,S}(z) = \frac{-1}{2\pi i}\int_{\Gamma_{R,S}} \frac{f_{Q,R}(\zeta)}{\zeta - z}\,d\zeta.$$

Let U be the open set of points in the plane where the winding number of $\Gamma_{R,S}$ is 1. Choose a cycle $\Gamma_{Q,S}$ in $U \setminus K_S$ with winding number 1 about every point of K_S and 0 about every point of $\mathbb{C}\setminus U$. Then $\Gamma_{Q,S}$ is a cycle in $\mathbb{C}\setminus(K_Q \cup \{z\})$ with winding number 1 about every point of K_S and 0 about every point of $(K_Q \setminus K_S) \cup \{z\}$ and so

$$(\star\star)\ f_{Q,S}(z) = \frac{-1}{2\pi i}\int_{\Gamma_{Q,S}} \frac{f(\xi)}{\xi - z}\,d\xi.$$

Let V be the open set of points in the plane where the winding number of $\Gamma_{R,S}$ is 0. Choose a cycle Γ in $\{V \setminus (K_Q \setminus K_R)\} \setminus (K_R \setminus K_S)$ with winding number 1 about every point of $K_R \setminus K_S$ and zero about every point of $\mathbb{C} \setminus \{V \setminus (K_Q \setminus K_R)\}$. Then $\Gamma_{Q,R} = \Gamma_{Q,S} + \Gamma$ is a cycle in $\mathbb{C} \setminus (K_Q \cup \Gamma_{R,S})$ with winding number 1 about every point of K_R and 0 about every point of $(K_Q \setminus K_R) \cup \Gamma_{R,S}$ and so for every ζ in $\Gamma_{R,S}$,

$$(\star\star\star)\ f_{Q,R}(\zeta) = \frac{-1}{2\pi i}\int_{\Gamma_{Q,R}} \frac{f(\xi)}{\xi - \zeta}\,d\xi.$$

Lastly, note that by simple algebra,

$$\frac{1}{2\pi i}\int_{\Gamma_{R,S}} \frac{1}{(\zeta - \xi)(\zeta - z)}\,d\zeta = \frac{1}{\xi - z}\left\{\frac{1}{2\pi i}\int_{\Gamma_{R,S}} \frac{1}{\zeta - \xi}\,d\zeta - \frac{1}{2\pi i}\int_{\Gamma_{R,S}} \frac{1}{\zeta - z}\,d\zeta\right\}.$$

The two integrals on the right-hand side of this equation are winding numbers. Taking into account the way we have constructed our cycles, we thus obtain

$$(\star\star\star\star)\quad \frac{1}{2\pi i}\int_{\Gamma_{R,S}} \frac{1}{(\zeta-\xi)(\zeta-z)}\,d\zeta = \begin{cases} \dfrac{1}{\xi - z} & \text{when } \xi \in \Gamma_{Q,S} \\ 0 & \text{when } \xi \in \Gamma. \end{cases}$$

The desired equality is now just a computation: substitute $(\star\star\star)$ into $(\star)$, use Fubini's Theorem [RUD, 8.8] on the double integral obtained, simplify by means of $(\star\star\star\star)$, and then recognize the result as what we want via $(\star\star)$. □

Denote the southwest corner and the center of any square S from $\mathcal{G}$ by z_S and c_S, respectively.

Lemma 2.15 *For $Q \in \mathcal{G}_m$, the following assertions hold:*

(a) *The sequence* $A_n(Q) = \sum_{R\in\mathcal{G}_n(Q)} \frac{l(R)^2}{|c_R - z_Q|^2}$, *defined for $n \geq m$, is bounded by 3.*

(b) *The sequence* $B_n(Q) = \sum_{R\in\mathcal{G}_n(Q)} \frac{l(R)\cdot \mathrm{Re}(c_R - z_Q)}{|c_R - z_Q|^2}$, *defined for $n \geq m$, is unbounded.*

Proof (a) Let Q_{sw} denote the southwest square of $\mathcal{G}_{m+1}(Q)$, so Q_{sw} is the unique square in $\mathcal{G}_{m+1}(Q)$ which contains z_Q. Then for $n > m$,

$$A_n(Q) = \sum_{R\in\mathcal{G}_n(Q_{\mathrm{sw}})} \frac{l(R)^2}{|c_R - z_Q|^2} + \sum_{R\in\mathcal{G}_n(Q)\setminus\mathcal{G}_n(Q_{\mathrm{sw}})} \frac{l(R)^2}{|c_R - z_Q|^2}.$$

Let R^* denote the result of applying the similitude with center z_Q and similarity ratio 4 to a square R. Then $l(R)^2/|c_R - z_Q|^2 = l(R^*)^2/|c_{R^*} - z_Q|^2$ and as the Rs range over all of $\mathcal{G}_n(Q_{\mathrm{sw}})$, the R^*s range over all of $\mathcal{G}_{n-1}(Q)$. Thus the first sum in the above is just $A_{n-1}(Q)$.

The number of Rs involved in the second sum in the above is $(3/4)4^{n-m}$ and for each of these Rs, $|c_R - z_Q| \geq (1/2)l(Q) = 4^{-m}/2$, so the second sum in the above is at most $(3/4)4^{n-m}\cdot(4^{-n})^2/(4^{-m}/2)^2 = (3\times 4^m)4^{-n}$.

Hence $A_n(Q) \leq A_{n-1}(Q) + (3\times 4^m)4^{-n}$ for all $n > m$, which implies that the sequence $A_n(Q)$, $n \geq m$, is bounded and bounded by 3.

(b) For $n > m$, split $B_n(Q)$ up into two sums just as we splitted up $A_n(Q)$. Handle the resulting first sum in the same manner to conclude that it is $B_{n-1}(Q)$. With the resulting second sum throw away the terms, all positive, arising from Rs lying above Q_{sw}, i.e., Rs lying in Q_{nw} so to speak. For each of the Rs remaining, i.e., Rs in Q_{ne} or Q_{se} so to speak, we have $\mathrm{Re}(c_R - z_Q) \geq (1/2)l(Q) = 4^{-m}/2$ and $|c_R - z_Q| \leq \sqrt{2}\,l(Q) = \sqrt{2}\,4^{-m}$, so the resulting second sum is at least $(1/2)4^{n-m}\cdot(4^{-n})(4^{-m}/2)/(\sqrt{2}\,4^{-m})^2 = 1/8$.

Hence $B_n(Q) \geq B_{n-1}(Q) + 1/8$ for all $n > m$, which implies that the sequence $B_n(Q)$, $n \geq m$, is unbounded. □

Given Q, R, and f as in Lemma 2.14, define $g_{Q,R}(z) = f'_{Q,R}(\infty)/(c_R - z)$. (Warning: An ambiguous expression like "$f'_{Q,R}(\infty)$" is to be interpreted as "$(f_{Q,R})'(\infty)$" and *not* as "$(f')_{Q,R}(\infty)$"!) For $n \geq m$ where m is the integer such that $\mathcal{G}_m$ contains Q, define

$$g_{Q,n}(z) = \sum_{R \in \mathcal{G}_n(Q)} g_{Q,R}(z) = \sum_{R \in \mathcal{G}_n(Q)} \frac{f'_{Q,R}(\infty)}{c_R - z}.$$

Lemma 2.16 *For $Q \in \mathcal{G}_m$, the sequence $g_{Q,n}(z_Q)$, defined for $n \geq m$, is bounded.*

Proof Set $h_{Q,R} = f_{Q,R} + g_{Q,R}$ and let $B = B(c_R; l(R))$. Using (b) of Lemma 2.14 and Proposition 1.9, one sees that for any $z \in B \setminus R$,

$$|(c_R - z)^2 f_{Q,R}(z)| \leq l(R)^2 \|f_{Q,R}\|_\infty \leq l(R)^2 M \|f\|_\infty$$

and

$$\begin{aligned}|(c_R - z)^2 g_{Q,R}(z)| &= |(c_R - z) f'_{Q,R}(\infty)| \leq l(R)\gamma(K_R)\|f_{Q,R}\|_\infty \\ &\leq l(R)^2 \gamma(K) M \|f\|_\infty.\end{aligned}$$

So, setting $M_1 = \{1 + \gamma(K)\}M$ on $B \setminus R$, one has $|(c_R - z)^2 h_{Q,R}(z)| \leq l(R)^2 M_1 \|f\|_\infty$.

By the way $g_{Q,R}$ has been defined, $h_{Q,R} \in H^\infty(\mathbb{C}^* \setminus R)$, $h_{Q,R}(\infty) = 0$, and $h'_{Q,R}(\infty) = 0$. Hence the function $(c_R - z)^2 h_{Q,R}(z)$, besides being analytic on $\mathbb{C} \setminus R$, is also still analytic at ∞. Thus, by the Maximum Modulus Principle [RUD, 12.1 modified to take into account regions containing ∞], the supremum norm of this function on $\mathbb{C}^* \setminus R$ is the same as its supremum norm on $B \setminus R$. We conclude that for any $z \in \mathbb{C}^* \setminus R$,

$$(\star)\ |h_{Q,R}(z)| \leq \frac{l(R)^2 M_1 \|f\|_\infty}{|c_R - z|^2}.$$

For z any point strictly southwest of z_Q, note that $(\star\star)$ $|c_R - z| \geq |c_R - z_Q|$ for all $R \in \mathcal{G}(Q)$. Hence by (c) of Lemma 2.14, $(\star)$, $(\star\star)$, and (a) of Lemma 2.15, we have

$$\begin{aligned}|g_{Q,n}(z)| &\leq \sum_{R \in \mathcal{G}_n(Q)} |h_{Q,R}(z)| + |f(z)| \leq \sum_{R \in \mathcal{G}_n(Q)} \frac{l(R)^2 M_1 \|f\|_\infty}{|c_R - z_Q|^2} + \|f\|_\infty \\ &\leq M_2 \|f\|_\infty\end{aligned}$$

where $M_2 = 3M_1 + 1$. Finally, letting $z \to z_Q$, we are done. □

Lemma 2.17 *Suppose $Q \in \mathcal{G}_m$ and $f \in H^\infty(\mathbb{C}^* \setminus K_Q)$ with $f(\infty) = 0$. If $f'(\infty) \neq 0$, then for some $R \in \mathcal{G}(Q)$, we must have*

$$f'_{Q,R}(\infty) \neq \frac{l(R)}{l(Q)} f'(\infty).$$

Proof We proceed by contradiction. Assuming the nonexistence of Rs as in the lemma's conclusion, we have

$$g_{Q,n}(z_Q) = \sum_{R \in \mathcal{G}_n(Q)} \frac{l(R)}{l(Q)} \frac{f'(\infty)}{c_R - z_Q}$$

and so

$$\begin{aligned}
\operatorname{Re}\left\{\frac{l(Q)}{f'(\infty)} g_{Q,n}(z_Q)\right\} &= \operatorname{Re}\left\{\sum_{R \in \mathcal{G}_n(Q)} \frac{l(R)}{c_R - z_Q}\right\} \\
&= \sum_{R \in \mathcal{G}_n(Q)} \frac{l(R) \cdot \operatorname{Re}(c_R - z_Q)}{|c_R - z_Q|^2} = B_n(Q).
\end{aligned}$$

We are now in a pickle: on the one hand, the leftmost item in the above must be bounded by the last lemma, while on the other hand, the rightmost item in the above must be unbounded by (b) of the next to last lemma. □

Lemma 2.18 *For $M < \infty$ and $a > 0$, there exists a $\delta > 0$ such that for any $Q \in \mathcal{G}$ and any $f \in H^\infty(\mathbb{C}^* \setminus K_Q)$, if $\|f\|_\infty \leq M$, $f(\infty) = 0$, and $|f'(\infty)|/l(Q) \geq a$, then there exists an $R \in \mathcal{G}(Q)$ such that*

$$\frac{|f'_{Q,R}(\infty)|}{l(R)} \geq (1+\delta)\frac{|f'(\infty)|}{l(Q)}.$$

Proof Given $M < \infty$ and $a > 0$, we first find a $\delta > 0$ that works only for the largest square in $\mathcal{G}$, i.e., the square $Q_0 = C_0$ for which $l(Q_0) = 1$, $K_{Q_0} = K$, $\mathcal{G}(Q_0) = \mathcal{G}$, and $\mathcal{G}_n(Q_0) = \mathcal{G}_n$. If the lemma fails for this square Q_0, then there exist sequences $\delta_k \downarrow 0$ and $\{f_k\}$ from $H^\infty(\mathbb{C}^* \setminus K)$ such that $\|f_k\|_\infty \leq M$, $f_k(\infty) = 0$, and $|f'_k(\infty)| \geq a$, and yet for all k and all $R \in \mathcal{G}$

$$\frac{|(f_k)'_{Q_0,R}(\infty)|}{l(R)} \leq (1+\delta_k)|f'_k(\infty)|.$$

Since the functions $\{f_k\}$ form a normal family [RUD, 14.6], a subsequence of them converges to a function $f \in H^\infty(\mathbb{C}^* \setminus K)$ satisfying $\|f\|_\infty \leq M$, $f(\infty) = 0$, $|f'(\infty)| \geq a$, and $|f'_{Q_0,R}(\infty)| \leq l(R)|f'(\infty)|$ for all $R \in \mathcal{G}$. Now there are 4^n squares $R \in \mathcal{G}_n = \mathcal{G}_n(Q_0)$ and for each of these squares we have $l(R) = 4^{-n}$. Hence by (c) of Lemma 2.14, we have for every n that

$$|f'(\infty)| = \left| \sum_{R \in \mathcal{G}_n(Q_0)} f'_{Q_0,R}(\infty) \right| \leq \sum_{R \in \mathcal{G}_n(Q_0)} |f'_{Q_0,R}(\infty)|$$
$$\leq \sum_{R \in \mathcal{G}_n(Q_0)} 4^{-n} |f'(\infty)| = |f'(\infty)|.$$

This forces $f'_{Q_0,R}(\infty) = 4^{-n} f'(\infty) = \{l(R)/l(Q_0)\} f'(\infty)$ for all $R \in \bigcup_n \mathcal{G}_n(Q_0) = \mathcal{G}(Q_0)$, which, being a contradiction to the last lemma, shows that there exists a $\delta > 0$ that works for Q_0.

Now we show that the $\delta > 0$ just gotten for Q_0 actually works for any $Q \in \mathcal{G}$. To this end, let $f \in H^\infty(\mathbb{C}^* \setminus K_Q)$ be such that $\|f\|_\infty \leq M$, $f(\infty) = 0$, and $|f'(\infty)|/l(Q) \geq a$. For the similitude $\varphi(z) = l(Q)z + z_Q$, the square Q^* for which $\varphi(Q^*) = Q$ is just Q_0. Thus $f \circ \varphi \in H^\infty(\mathbb{C}^* \setminus K)$ is such that $\|f \circ \varphi\|_\infty \leq M$, $(f \circ \varphi)(\infty) = 0$, and $|(f \circ \varphi)'(\infty)| = |f'(\infty)|/l(Q) \geq a$. So, by the last paragraph, for some square $R^* \in \mathcal{G}$ we must have

$$\frac{|(f \circ \varphi)'_{Q_0,R^*}(\infty)|}{l(R^*)} \geq (1+\delta)|(f \circ \varphi)'(\infty)|.$$

By (d) of Lemma 2.14, $(f \circ \varphi)_{Q_0,R^*} = f_{Q,R} \circ \varphi$ for $R = \varphi(R^*)$. We also have $(f_{Q,R} \circ \varphi)'(\infty) = f'_{Q,R}(\infty)/l(Q)$, $l(R^*) = l(R)/l(Q)$, and $(f \circ \varphi)'(\infty) = f'(\infty)/l(Q)$. Applying all these equalities to the inequality above and doing a bit of algebra, we obtain our conclusion. □

Theorem 2.19 *The planar Cantor quarter set has positive finite linear Hausdorff measure, yet is removable.*

Proof We only need to suppose that $\gamma(K) > 0$ and deduce a contradiction. So let $f \in H^\infty(\mathbb{C}^* \setminus K)$ be such that $\|f\|_\infty \leq 1$, $f(\infty) = 0$, and $|f'(\infty)| > 0$. Let $\delta > 0$ be gotten from the last lemma with $M = 1 + 6/\pi$ [as in (b) of Lemma 2.14] and $a = |f'(\infty)|$. Since the last lemma applies to Q_0 and f, we get $Q_1 \in \mathcal{G}(Q_0)$ such that

$$\frac{|f'_{Q_0,Q_1}(\infty)|}{l(Q_1)} \geq (1+\delta)|f'(\infty)|.$$

Since the last lemma applies to Q_1 and f_{Q_0,Q_1} [to see this use (a) and (b) of Lemma 2.14 here], we get $Q_2 \in \mathcal{G}(Q_1)$ such that

$$\frac{|(f_{Q_0,Q_1})'_{Q_1,Q_2}(\infty)|}{l(Q_2)} \geq (1+\delta)\frac{|f'_{Q_0,Q_1}(\infty)|}{l(Q_1)}.$$

Combining the last two inequalities and using (e) of Lemma 2.14, we have

$$\frac{|f'_{Q_0,Q_2}(\infty)|}{l(Q_2)} \geq (1+\delta)^2 |f'(\infty)|.$$

Continuing in this manner, we obtain a sequence of squares $\{Q_n\}$ from $\mathcal{G}$ such that

$$\frac{|f'_{Q_0,Q_n}(\infty)|}{l(Q_n)} \geq (1+\delta)^n |f'(\infty)|.$$

However, by Proposition 1.9 and (b) of Lemma 2.14,

$$|f'_{Q_0,Q_n}(\infty)| \leq \gamma(K_{Q_n}) \| f_{Q_0,Q_n} \|_\infty \leq l(Q_n)\gamma(K)M.$$

From the last two inequalities we immediately obtain $(1+\delta)^n |f'(\infty)| \leq M\gamma(K)$ for all integers n. This is patently absurd given that $\delta > 0$ and $|f'(\infty)| > 0$! $\square$

Chapter 3
Garabedian Duality for Hole-Punch Domains

3.1 Statement of the Result and an Initial Reduction

In the last chapter we learned from Painlevé's Theorem (2.7) that linear Hausdorff measure zero implies removability and from the planar Cantor quarter set that the converse fails in general (Theorem 2.19). However, by one of our early results, Proposition 1.1, the converse does hold if the compact set in question lies in a line. A line, or rather line segment, is the simplest example of a rectifiable curve. One can define arclength measure on a rectifiable curve and show that its zero sets coincide with those of linear Hausdorff measure on the curve (see Sections 4.5 and 5.2). This leads to what is known as *Denjoy's Conjecture*: A compact subset of a rectifiable curve is removable if and only if it has arclength measure, i.e., linear Hausdorff measure, zero. In 1909 Arnaud Denjoy claimed to have proved this in [DEN]. His proof however had a gap in it, thus giving rise to the eponymous conjecture which was resolved affirmatively only in 1977 by Alberto Calderón's famous and justly celebrated paper [CAL] on the L^2-boundedness of the Cauchy integral operator on Lipschitz curves. The interested reader may consult [MARSH] for a proof of Denjoy's Conjecture from Calderón's result and more history on this topic.

Much later Mark Melnikov introduced the notion of the curvature of a measure in [MEL] which he and Joan Verdera then immediately used to provide an alternative proof of Denjoy's Conjecture in [MV]. Their proof is more geometric in nature and does not rely on singular integral theory. In addition, the notion of curvature it employs has proven enormously useful in the further study of analytic capacity. Thus their proof is the one we will present in the next chapter.

Necessary to this proof is an alternative characterization of analytic capacity known as Garabedian duality that is of interest in itself and thus the subject of this chapter. We will only establish Garabedian duality for a restricted class of domains, the so-called *hole-punch domains*, since these domains suffice for our purposes and let us avoid technicalities concerning conformal maps and their boundary values. The proof mainly follows [GAR2] with changes inspired by [GAM2] and [GS]. These changes allow us to avoid the notion of an H^p space, $1 \leq p < \infty$, on a multiply connected (or even simply connected) domain.

J.J. Dudziak, *Vitushkin's Conjecture for Removable Sets*, Universitext, DOI 10.1007/978-1-4419-6709-1_3,

Notational Conventions. For the course of this chapter, the following setup will be in force. First, K denotes the union of a finite number of pairwise disjoint closed balls $B_j = B(c_j; r_j)$, $j = 1, 2, \ldots, N$. Second, the open subset $U = \mathbb{C}^* \setminus K$ of the extended complex plane will be called, and is, our *hole-punch domain*. Third, Γ denotes the cycle comprising the Γ_j for $1 \le j \le N$ where Γ_j is the counterclockwise circular boundary of B_j. This last bit of notation will be abused somewhat to also denote the ranges of the corresponding cycles. Fourth, Δ denotes the interior of K. Thus $\mathbb{C}^* = U \cup K = U \cup \Gamma \cup \Delta$ disjointly. Fifth, consider a point $z \in \Gamma$, so $z = c_j + r_j e^{i\theta}$ for some j and θ. Given $\alpha \ge 1$, define z_α to be $c_j + \alpha r_j e^{i\theta}$. Sixth, for a subset (or subpath) E of Γ, define $E_\alpha = \{z_\alpha : z \in E\}$. Lastly, for the rest of the chapter choose and fix a number α_0 bigger than but close enough to 1 so that the closed discs with boundary circles $(\Gamma_j)_{\alpha_0}$ are still pairwise disjoint.

Throughout this book, given a topological space X, we shall use $C(X)$ to denote the algebra of continuous functions on X.

Let $A(U)$ denote the Banach algebra of continuous functions on the closure of U whose restrictions to U are analytic. Thus $A(U) = C(\operatorname{cl} U) \cap H^\infty(U)$. The subalgebra of those functions in $A(U)$ which vanish at ∞ will be denoted $A_0(U)$. Equip both these spaces with the supremum norm. The *continuous analytic capacity* of our K is defined by

$$\alpha(K) = \sup\{|f'(\infty)| : f \in A_0(U) \text{ with } \|f\|_\infty \le 1\}.$$

Note the parallel with the analytic capacity of our K which can be written, by Proposition 1.11, as

$$\gamma(K) = \sup\{|f'(\infty)| : f \in H_0^\infty(U) \text{ with } \|f\|_\infty \le 1\}.$$

Here $H_0^\infty(U)$ is, of course, the subalgebra of those functions in $H^\infty(U)$ which vanish at ∞. It is trivially true that $\alpha(K) \le \gamma(K)$.

We remark that the definitions just given of $A(U)$, $A_0(U)$, and $H_0^\infty(U)$ make sense for U any open subset of $\mathbb{C}^*$ containing ∞ and not just a hole-punch domain. Consequently the definition just given of $\alpha(K)$ also makes sense for any compact subset K of $\mathbb{C}$ if we set $U = \mathbb{C}^* \setminus K$. However, there is a last quantity to be introduced which really requires K to be more than just a compact subset of $\mathbb{C}$. In order to make sense, this definition also requires K to have a rectifiable boundary. Clearly this is the case for $K = \mathbb{C}^* \setminus U$ with U a hole-punch domain. This quantity is the L^2 *continuous analytic capacity* of our K which is defined by

$$\alpha_2(K) = \sup\{|f'(\infty)| : f \in A_0(U) \text{ with } \|f\|_2 \le 1\}.$$

Here the L^2 norm in question is given by

$$\|f\|_2^2 = \frac{1}{2\pi} \int_\Gamma |f(\zeta)|^2 \, |d\zeta|.$$

There is also the notion of the L^2 *analytic capacity* of our K, denoted $\gamma_2(K)$. However, this requires the introduction of H^2 spaces over our multiply connected U and the assignment of nontangential boundary values to functions from these spaces so that the L^2 integral over Γ continues to makes sense. Since $\alpha_2(K)$ will suffice for our purposes, we will dispense with $\gamma_2(K)$ in order to keep things as simple as possible. This chapter is devoted to proving the following.

Theorem 3.1 (Garabedian Duality) *Let* $K = \mathbb{C}^* \setminus U$ *where* U *is a hole-punch domain. Then*

$$\gamma(K) = \alpha(K) = \alpha_2^2(K).$$

A more general version of Garabedian duality has the boundary circles Γ_j replaced by analytic loops and adds a fourth term, $\gamma_2^2(K)$, to the chain of equalities in the conclusion (see [GAR2]).

Useful for much to come is the following Cauchy Integral Formula for $A(U)$.

Proposition 3.2 *For* $f \in A(U)$, $(\star)$ $f(\infty) - \dfrac{1}{2\pi i}\displaystyle\int_\Gamma \frac{f(\zeta)}{\zeta - z}\,d\zeta = \begin{cases} f(z) \text{ when } z \in U \\ 0 \quad \text{when } z \in \Delta \end{cases}.$

Moreover, $(\star\star)$ $\dfrac{1}{2\pi i}\displaystyle\int_\Gamma f(\zeta)\,d\zeta = f'(\infty).$

Proof By Lemma 2.13, we know that $(\star)$ holds with Γ replaced by Γ_α for all α between 1 and α_0 (and less than $1 + \operatorname{dist}(z, K)/r_j$ for $1 \le j \le N$ in the case $z \in U$). The same can be said of $(\star\star)$; indeed, we showed this after Proposition 1.1 when discussing analyticity at ∞. Letting $\alpha \downarrow 1$, the proposition now follows by the uniform continuity on $\operatorname{cl} U$ of any f from $A(U)$. □

We will see a proof of the first equality of Theorem 3.1 soon; it is not hard to show and will end up being a by-product of our construction of something called the Garabedian function for K. Just now we concentrate on the second equality of Theorem 3.1 which is the real meat of the result. It is a consequence of the following.

Reduction 3.3 *There exists a function* $\varphi \in A(U)$ *such that* $\varphi(\infty) = 1$ *and* $\|\varphi\|_2^2 = \alpha(K)$.

Proof of the Second Equality of Garabedian Duality (3.1) *from the Reduction* (3.3) For any $f \in A_0(U)$, we have

$$f'(\infty) = f'(\infty)\varphi(\infty) = \lim_{z\to\infty} zf(z)\varphi(z) = (f\varphi)'(\infty) = \frac{1}{2\pi i}\int_\Gamma f(\zeta)\varphi(\zeta)\,d\zeta$$

with the last equality following from the last proposition. Thus, by the Cauchy–Schwarz Inequality [RUD, 3.5],

$$|f'(\infty)| \le \|f\|_2\,\|\varphi\|_2 = \|f\|_2\sqrt{\alpha(K)}.$$

Hence, suping over all $f \in A_0(U)$ with $\|f\|_2 \le 1$, we conclude that $\alpha_2(K) \le \sqrt{\alpha(K)}$. Moreover,

$$|f'(\infty)| = |(f\varphi)'(\infty)| \le \alpha_2(K)\,\|f\varphi\|_2 \le \alpha_2(K)\,\|f\|_\infty\,\|\varphi\|_2 = \alpha_2(K)\,\|f\|_\infty\,\sqrt{\alpha(K)}.$$

Hence, suping over all $f \in A_0(U)$ with $\|f\|_\infty \le 1$, we conclude that $\alpha(K) \le \alpha_2(K)\,\sqrt{\alpha(K)}$. Thus $\sqrt{\alpha(K)} \le \alpha_2(K)$ and so we are done. □

The function φ of Reduction 3.3 will be gotten, after much hard work, as an analytic square root of something called the Garabedian function for K. Construction of the Garabedian function necessitates two interludes, to which we now turn.

3.2 Interlude: Boundary Correspondence for $H^\infty(U)$

To every function of $H^\infty(U)$ a function of $L^\infty(\Gamma)$ will be associated in a natural and well-behaved way. This will be achieved by relying on the pointwise bounded density of $A(U)$ in $H^\infty(U)$ which is the content of the following.

Lemma 3.4 *There is a number $M > 0$ such that for any $f \in H^\infty(U)$, there exists a sequence of functions $\{f_n\}$ from $A(U)$ such that $\|f_n\|_\infty \le M\|f\|_\infty$ for each n and $f_n(z) \to f(z)$ for each $z \in U$.*

Proof We will use $\mathbb{D}$ to denote the open unit disc of $\mathbb{C}$ centered at the origin. Given $f \in H^\infty(\mathbb{D})$, consideration of $f_n(z) = f(\{1 - 1/n\}z)$ shows that $A(\mathbb{D})$ is strongly pointwise boundedly dense in $H^\infty(\mathbb{D})$, i.e., pointwise boundedly dense in $H^\infty(\mathbb{D})$ with $M = 1$.

Setting $U_j = \mathbb{C}^* \setminus B_j$, consideration of the mappings $\varphi_j : \mathrm{cl}\,\mathbb{D} \mapsto \mathrm{cl}\,U_j$ defined by $z \mapsto r_j/z + c_j$ now shows that each $A(U_j)$ is strongly pointwise boundedly dense in $H^\infty(U_j)$.

Let $\delta > 0$ denote the smallest distance between the pairwise disjoint closed balls B_k, $1 \le k \le N$. Given $f \in H^\infty(U)$ and $z \in U_j$, choose α greater than 1 but smaller than α_0 so that $(\alpha - 1)r_k < \delta/4$ for all $1 \le k \le N$ and $(\alpha - 1)r_j < \mathrm{dist}(z, B_j)$. The last condition ensures that z lies outside the closed ball with counterclockwise boundary circle $(\Gamma_j)_\alpha$. Set

$$f_j(z) = \frac{-1}{2\pi i}\int_{(\Gamma_j)_\alpha} \frac{f(\zeta)}{\zeta - z}\,d\zeta.$$

By Cauchy's Integral Theorem [RUD, 10.35], each f_j so defined is independent of the particular α chosen and thus well defined on U_j. Clearly each f_j is analytic on U_j and vanishes at ∞. By Lemma 2.13, all the f_js sum to $f - f(\infty)$ on U.

Consider any $z \in U_j$ within a distance $\delta/4$ of B_j. Then

$$|f_j(z)| \le |f(z) - f(\infty)| + \sum_{k \ne j} |f_k(z)| \le 2\|f\|_\infty + \sum_{k \ne j} \frac{1}{2\pi}\cdot\frac{\|f\|_\infty}{\delta - \{\delta/4 + \delta/4\}}\cdot 2\pi\alpha r_k.$$

Letting $\alpha \downarrow 1$, we see that for such z,

$$|f_j(z)| \leq 2\left\{1+\sum_{k\neq j}\frac{r_k}{\delta}\right\}\|f\|_\infty.$$

By Maximum Modulus [RUD, 12.1 modified to take into account regions containing ∞], this bound extends to all $z \in U_j$. We conclude that each f_j is an element of $H_0^\infty(U_j)$ with norm at most $2\{1+\sum_k r_k/\delta\}\|f\|_\infty$.

The conclusion of the lemma now follows with $M = 2N\{1+\sum_k r_k/\delta\}+1$. □

A bit of notation. Given a function f defined on a set E and $\varepsilon > 0$, let $\operatorname{osc}(f, E; \varepsilon)$, denote the *oscillation* of f on E with *gauge* ε. Thus

$$\operatorname{osc}(f, E; \varepsilon) = \sup\{|f(z) - f(w)| : z, w \in E \text{ with } |z-w| \leq \varepsilon\}.$$

Another bit of notation. We use $\operatorname{spt}(f)$ and $\operatorname{spt}(\mu)$ to denote the support of a function f or measure μ, respectively.

Lemma 3.5 *Let μ be a regular complex Borel measure on $\mathbb{C}$ and V be an open subset of $\mathbb{C}$. If $\hat{\mu} = 0$ $\mathcal{L}^2$-a.e. on V, then $|\mu|(V) = 0$.*

Proof Given a continuously differentiable, compactly supported function f on $\mathbb{C}$ and z any point of $\mathbb{C}$, we have the following representation (an easy consequence of Green's Theorem, or see [RUD, 20.3] for a simple direct proof sans Green):

$$f(z) = -\frac{1}{\pi}\iint_{\mathbb{C}} \frac{(\bar{\partial} f)(\zeta)}{\zeta - z}\, d\mathcal{L}^2(\zeta)$$

where

$$\bar{\partial} f = \frac{1}{2}\left(\frac{\partial f}{\partial x} + i\frac{\partial f}{\partial y}\right).$$

Integrating this representation with respect to μ and using Fubini's Theorem [RUD, 8.8], we see that

$$\int f\, d\mu = \frac{1}{\pi}\iint_{\mathbb{C}} (\bar{\partial} f)\hat{\mu}\, d\mathcal{L}^2.$$

Thus the hypothesis of the lemma implies that μ annihilates all the continuously differentiable functions compactly supported in V. However, to get the conclusion of the lemma it would suffice to show that μ annihilates every continuous function compactly supported in V [RUD, 6.18 and 6.19]. So one way to finish is to show that any continuous function f compactly supported in V is a uniform limit of continuously differentiable functions f_n compactly supported in V. This we do via a standard approximate identity argument which we now present.

To this end, define

$$g_n(\zeta) = \begin{cases} (3n^2/\pi)\left(1 - n^2|\zeta|^2\right)^2 & \text{when } |\zeta| \le 1/n \\ 0 & \text{when } |\zeta| > 1/n \end{cases}.$$

Note that each g_n is a continuously differentiable function compactly supported in $B(0; 1/n)$ which is nonnegative on $\mathbb{C}$ and whose integral over $\mathbb{C}$ with respect to $\mathcal{L}^2$ is equal to 1. Next define

$$f_n(z) = \int_{\mathbb{C}} f(z - \zeta)\, g_n(\zeta)\, d\mathcal{L}^2(\zeta) = \int_{\mathbb{C}} f(\xi)\, g_n(z - \xi)\, d\mathcal{L}^2(\xi).$$

Using the first integral defining f_n, we see that

$$\begin{aligned} |f(z) - f_n(z)| &= \left| \int_{\mathbb{C}} \{f(z) - f(z - \zeta)\} g_n(\zeta)\, d\mathcal{L}^2(\zeta) \right| \\ &\le \sup\{f(z) - f(z - \zeta)| : |\zeta| \le 1/n\} \end{aligned}$$

for any $z \in \mathbb{C}$. Thus, taking the supremum over all $z \in \mathbb{C}$, we see that

$$\|f - f_n\|_\infty \le \operatorname{osc}(f, \mathbb{C}; 1/n).$$

But $\operatorname{osc}(f, \mathbb{C}; 1/n) \to 0$ as $n \to \infty$ since any continuous function compactly supported on $\mathbb{C}$ is uniformly continuous there. Thus the uniform convergence of f_n to f has been shown.

Since g_n is continuously differentiable with compact support, difference quotients of g_n with respect of x converge boundedly to the x partial of g_n by the Mean Value Theorem (similar comments will hold for the y partial). Thus Lebesgue's Dominated Convergence Theorem [RUD, 1.34] allows us to differentiate under the second integral defining f_n to conclude that f_n has an x partial given by

$$\frac{\partial f_n}{\partial x}(z) = \int_{\mathbb{C}} f(\xi)\, \frac{\partial g_n}{\partial x}(z - \xi)\, d\mathcal{L}^2(\xi).$$

Moreover, this representation shows that

$$\left| \frac{\partial f_n}{\partial x}(z) - \frac{\partial f_n}{\partial x}(w) \right| \le \|f\|_\infty \operatorname{osc}\left(\frac{\partial g_n}{\partial x}, \mathbb{C}; |z - w| \right) \mathcal{L}^2(B(0; 1/n)).$$

But $\operatorname{osc}(\partial g_n/\partial x, \mathbb{C}; |z - w|) \to 0$ as $|z - w| \to 0$ since any continuous function compactly supported on $\mathbb{C}$ is uniformly continuous there. Thus the continuous differentiability of f_n has been shown.

Finally, note that $\operatorname{spt}(f_n) \subseteq \operatorname{spt}(f) + B(0; 1/n)$. Thus f_n is compactly supported in V as soon as n is large enough so that $1/n < \operatorname{dist}(\operatorname{spt}(f), \mathbb{C} \setminus V)$. □

When we speak of "$L^\infty(\Gamma)$", and later of "$L^1(\Gamma)$", the measure involved is understood to be arclength measure on Γ, which is simply the sum of arclength measure on the finitely many boundary circles Γ_j. To be precise and totally unambiguous, the arclength measure of a subset E of Γ is just

$$\sum_{j=1}^{N} r_j \mathcal{L}^1(\{\theta \in [0, 2\pi) : c_j + r_j e^{i\theta} \in E\}).$$

Proposition 3.6 *For $f \in H^\infty(U)$, there exists a unique $f^* \in L^\infty(\Gamma)$ such that*

$$(\star)\ f(\infty) - \frac{1}{2\pi i}\int_\Gamma \frac{f^*(\zeta)}{\zeta - z}\,d\zeta = \begin{cases} f(z) & \text{when } z \in U \\ 0 & \text{when } z \in \Delta \end{cases}.$$

This correspondence is an isometric algebra homomorphism on $H^\infty(U)$ that is the identity on $A(U)$.

Moreover,

$$(\star\star)\ \frac{1}{2\pi i}\int_\Gamma f^*(\zeta)\,d\zeta = f'(\infty).$$

Proof Existence and ($\star\star$). Given $f \in H^\infty(U)$, let $\{f_n\}$ be as in Lemma 3.4. Recall that the dual space of $L^1(\Gamma)$ is $L^\infty(\Gamma)$ [RUD, 6.16]. Thus we may view $\{f_n\}$ as a sequence of linear functionals on $L^1(\Gamma)$ whose norms are uniformly bounded by $M\|f\|_\infty$. By the Stone–Weierstrass Theorem, the polynomials in z and $\bar{z}$ with rational coefficients are uniformly dense in $C(\Gamma)$. Since $C(\Gamma)$ is dense in $L^1(\Gamma)$ [RUD, 3.14], it follows that $L^1(\Gamma)$ is a separable Banach space. Thus we may apply [RUD, 11.29] to $L^1(\Gamma)$. This authorizes us to assume, after dropping to a subsequence, that $\{f_n\}$ converges weakly to an element $f^* \in L^\infty(\Gamma)$ with norm at most $M\|f\|_\infty$. The relations ($\star$) and ($\star\star$) of the proposition now follow by letting $n \to \infty$ in the relations ($\star$) and ($\star\star$) of Proposition 3.2 applied to the f_ns.

Uniqueness. Suppose that $f^{**} \in L^\infty(\Gamma)$ satisfies the relation ($\star$) of the proposition (with f^{**} in place of f^* of course). Then the measure $d\mu(\zeta) = \{f^*(\zeta) - f^{**}(\zeta)\}d\zeta_\Gamma$ has $\hat{\mu} = 0$ everywhere off Γ. By the last lemma, $\mu = 0$, i.e., $f^* = f^{**}$.

Linearity and Identicality on $A(U)$. Exploit the Uniqueness that has just been shown.

Multiplicativity. Given $g \in H^\infty(U)$, get $\{g_m\}$ whose relation to g is as that of $\{f_n\}$ to f in Existence and ($\star\star$) above. Letting first $m \to \infty$ and then $n \to \infty$ in relation ($\star$) of Proposition 3.2 applied to the $f_n g_m$s, we obtain relation ($\star$) of the proposition for fg with f^*g^* in place of $(fg)^*$. By Uniqueness above, we must have $(fg)^* = f^*g^*$.

Isometricality. We know from our proof of Existence and ($\star\star$) above that $\|g^*\|_\infty \le M\|g\|_\infty$ for all g; in particular for $g = f^n$. Thus, by Multiplicativity above,

$$\|f^*\|_\infty^n = \|(f^*)^n\|_\infty = \|(f^n)^*\|_\infty \le M\|f^n\|_\infty = M\|f\|_\infty^n.$$

Taking n^{th} roots and then letting $n \to \infty$, we see that $\|f^*\|_\infty \le \|f\|_\infty$.

A similar argument will establish the other direction. Turning to the details, fix $z \in U$ and $z_0 \in \Delta$. Then, applying ($\star$) to z and z_0, we have

$$\begin{aligned}
|f(z)| &\le |f(z) - f(\infty)| + |f(\infty)| \\
&= \left| -\frac{1}{2\pi i}\int_\Gamma \frac{f^*(\zeta)}{\zeta - z}\,d\zeta \right| + \left| \frac{1}{2\pi i}\int_\Gamma \frac{f^*(\zeta)}{\zeta - z_0}\,d\zeta \right| \\
&\le \frac{1}{2\pi}\frac{\|f^*\|_\infty}{\operatorname{dist}(z,\Gamma)}\sum_j 2\pi r_j + \frac{1}{2\pi}\frac{\|f^*\|_\infty}{\operatorname{dist}(z_0,\Gamma)}\sum_j 2\pi r_j \\
&= \left\{ \frac{\sum_j r_j}{\operatorname{dist}(z,\Gamma)} + \frac{\sum_j r_j}{\operatorname{dist}(z_0,\Gamma)} \right\} \|f^*\|_\infty.
\end{aligned}$$

Replacing f by f^n, utilizing Multiplicativity, taking n^{th} roots, and then letting $n \to \infty$, we see that $|f(z)| \le \|f^*\|_\infty$. Suping over all $z \in U$, we obtain $\|f\|_\infty \le \|f^*\|_\infty$. □

Scholium 3.7 *For any $f \in H^\infty(U)$, there exists a sequence of functions $\{f_n\}$ from $A(U)$ such that $\|f_n\|_\infty \le M\|f\|_\infty$ for each n, $f_n(z) \to f(z)$ for each $z \in U$, and $f_n \to f^*$ weakly in $L^\infty(\Gamma)$. (The constant M here is as in Lemma* 3.4.)

Given $\alpha \ge 1$ and a function f defined on Γ_α, define a new function f_α on Γ by $f_\alpha(z) = f(z_\alpha)$. We finish this section with a proposition giving another slant on our boundary correspondence. Necessary for its proof is the following.

Lemma 3.8 *Let $\{\alpha_n\}$ be a sequence decreasing to* 1. *Suppose that $f \in H^\infty(U)$ and $h \in L^\infty(\Gamma)$ are such that $f_{\alpha_n} \to h$ weakly in $L^\infty(\Gamma)$. Then for any function g defined and continuous on a neighborhood of Γ,*

$$\int_{\Gamma_{\alpha_n}} f(\zeta)g(\zeta)\,d\zeta \to \int_\Gamma h(\zeta)g(\zeta)\,d\zeta.$$

Proof A simple computation shows that $\int_{\Gamma_\alpha} f(\zeta)g(\zeta)\,d\zeta = \alpha \int_\Gamma f_\alpha(\zeta)g_\alpha(\zeta)\,d\zeta$, so it suffices to show

$$(\star)\quad \int_\Gamma f_{\alpha_n}(\zeta)g_{\alpha_n}(\zeta)\,d\zeta \to \int_\Gamma h(\zeta)g(\zeta)\,d\zeta.$$

Let V be the neighborhood of Γ on which g is defined and continuous. Choose an open set W with compact closure such that $\Gamma \subseteq W \subseteq \operatorname{cl} W \subseteq V$. Given our hypotheses, ($\star$) is now a simple $\varepsilon/2$-argument since, for n large enough, the absolute value of the difference of the two integrals in ($\star$) is, after adding and subtracting $\int_\Gamma f_{\alpha_n}(\zeta)g(\zeta)\,d\zeta$ and performing some routine estimates, less than or equal to

$$\|f\|_\infty \cdot \mathrm{osc}(g;\, \mathrm{cl}\, W;\, (\alpha_n - 1)\max_j r_j) \cdot \sum_j 2\pi r_j + \left| \int_\Gamma f_{\alpha_n}(\zeta)g(\zeta)\,d\zeta - \int_\Gamma h(\zeta)g(\zeta)\,d\zeta \right|.$$

The first term can be made smaller than $\varepsilon/2$ since g, being continuous on the compact set cl W, is uniformly continuous there. The second term can be made smaller than $\varepsilon/2$ by the assumed weak convergence of f_{α_n} to h. □

Proposition 3.9 *For $f \in H^\infty(U)$, $f_\alpha \to f^*$ weakly in $L^\infty(\Gamma)$ as $\alpha \downarrow 1$.*

Proof If not, then there exists a sequence $\{\alpha_n\}$ decreasing to 1, a function $g \in L^1(\Gamma)$, and a number $\varepsilon > 0$ such that for all n,

$$(\star) \quad \left| \int_\Gamma f_{\alpha_n}(\zeta)g(\zeta)\,d\zeta - \int_\Gamma f^*(\zeta)g(\zeta)\,d\zeta \right| \geq \varepsilon.$$

Then, using [RUD, 11.29] as in the first paragraph of the proof of Proposition 3.6, we may assume, after dropping to a subsequence and relabeling, that for some $h \in L^\infty(\Gamma)$

$$(\star\star) \quad f_{\alpha_n} \to h \text{ weakly in } L^\infty(\Gamma).$$

Next, by the last lemma applied to the function $g(\zeta) = 1/(\zeta - z)$ where $z \notin \Gamma$, we have

$$\int_{\Gamma_{\alpha_n}} \frac{f(\zeta)}{\zeta - z}\,d\zeta \to \int_\Gamma \frac{h(\zeta)}{\zeta - z}\,d\zeta.$$

Applying Lemma 2.13 to Γ_{α_n} and letting $n \to \infty$, one thus concludes that

$$f(\infty) - \frac{1}{2\pi i}\int_\Gamma \frac{h(\zeta)}{\zeta - z}\,d\zeta = \begin{cases} f(z) & \text{when } z \in U \\ 0 & \text{when } z \in \Delta \end{cases}.$$

Uniqueness in Proposition 3.6 now implies that $h = f^*$. This makes ($\star$) and ($\star\star$) patently contradictory. □

3.3 Interlude: An F. & M. Riesz Theorem

This section for the most part follows the treatment of Section 7 of Chapter II of [GAM1]. We will use $\mathbb{T}$ to denote the boundary of $\mathbb{D}$. For convenience we will write dm in place of $d\theta/2\pi$ on $\mathbb{T}$. The following properties of $A(\mathbb{D})$, known as the disc algebra, will be needed:

(a) For any $f \in A(\mathbb{D})$, $\int f\, dm = f(0)$.
(b) For any regular complex Borel measure μ on $\mathrm{cl}\,\mathbb{D}$, $\mu \perp A(\mathbb{D})$, i.e., $\int f\, d\mu = 0$ for every $f \in A(\mathbb{D})$, if and only if $\int z^n\, d\mu(z) = 0$ for all integers $n \geq 0$.
(c) Re $A(\mathbb{D})$ is uniformly dense in the space of real-valued continuous functions on $\mathbb{T}$.

To prove both (a) and (b) one needs to know that the functions analytic in a neighborhood of $\mathrm{cl}\,\mathbb{D}$ are uniformly dense in $A(\mathbb{D})$. (To see this, given $f \in A(\mathbb{D})$, consider the functions $f_n(z) = f(\{1 - 1/n\}z)$.) With this fact in mind, (a) and (b) now follow from Cauchy's Integral Theorem [RUD, 10.35] and the representability of analytic functions as power series [RUD, 10.16], respectively. To establish (c) it suffices to show that any *real* annihilating measure μ of $A(\mathbb{D})$ which is supported on $\mathbb{T}$ must be the zero measure. But for such a μ one clearly has $\int z^n\, d\mu(z) = 0$ for all $n \geq 0$. Upon taking complex conjugates one also has $\int \bar{z}^n\, d\mu(z) = 0$ for all $n \geq 0$. By Stone–Weierstrass, μ now annihilates $C(\mathrm{cl}\,\mathbb{D})$ and so must be 0.

Lemma 3.10 (Forelli) *Suppose $\{K_j\}$ is a sequence of compact subsets of $\mathbb{T}$ such that $m(K_j) = 0$ for each j. Then there exists a sequence $\{f_n\}$ of functions from $A(\mathbb{D})$ such that $\|f_n\|_\infty \leq 1$ for each n, $f_n(z) \to 0$ for each $z \in \bigcup_{j=1}^{\infty} K_j$, but $f_n \to 1$ m-a.e. on $\mathbb{T}$.*

Proof Choose $\varphi_n \in C(\mathbb{T})$ such that $\varphi_n \geq 0$ on $\mathbb{T}$, $\varphi_n = n$ on $\bigcup_{j=1}^{n} K_j$, and $\int \varphi_n\, dm \leq 1/n^2$. The following will do nicely for N_n chosen sufficiently large:

$$\varphi_n(z) = n \left\{ 1 - \frac{1}{2}\mathrm{dist}(z, \bigcup_{j=1}^{n} K_j) \right\}^{N_n}.$$

Then, by (c), choose $h_n \in A(\mathbb{D})$ such that $|\mathrm{Re}\, h_n - \varphi_n| \leq 1/n^2$ on $\mathbb{T}$ and set $g_n = h_n + 1/n^2 - i\, \mathrm{Im} \int h_n\, dm$. Note that $g_n \in A(\mathbb{D})$, $\varphi_n \leq \mathrm{Re}\, g_n \leq \varphi_n + 2/n^2$ on $\mathbb{T}$, $\mathrm{Im} \int g_n\, dm = 0$, and $0 \leq \mathrm{Re} \int g_n\, dm \leq 3/n^2$.

Set $f_n = \exp(-g_n)$. Clearly $f_n \in A(\mathbb{D})$ and $|f_n| = \exp(-\mathrm{Re}\, g_n) \leq \exp(0) = 1$ on $\mathbb{T}$ and so too on $\mathrm{cl}\,\mathbb{D}$ by the Maximum Modulus Principle [RUD, 12.1]. Given $n \geq j$, $|f_n| = \exp(-\mathrm{Re}\, g_n) \leq \exp(-n)$ on K_j and so $f_n(z) \to 0$ for each $z \in \bigcup_{j=1}^{\infty} K_j$. Using (a) and the fact that $g_n(0) = \int g_n\, dm$ has imaginary part 0, we see that

$$\mathrm{Re} \int f_n dm = \mathrm{Re} f_n(0) = \mathrm{Re}\; e^{-g_n(0)} = e^{-\mathrm{Re} \int g_n\, dm} \geq 1 - \mathrm{Re} \int g_n dm \geq 1 - 3/n^2.$$

Hence

$$\int |1 - f_n|^2\, dm = 1 + \int |f_n|^2\, dm - 2\mathrm{Re} \int f_n\, dm \leq 2\{1 - \mathrm{Re} \int f_n\, dm\} \leq 6/n^2$$

and so

$$\int \sum_n |1 - f_n|^2 \, dm = \sum_n \int |1 - f_n|^2 \, dm \le \sum_n \frac{6}{n^2} < \infty.$$

We conclude that $\sum_n |1 - f_n|^2 < \infty$ m-a.e. Hence $f_n \to 1$ m-a.e. on $\mathbb{T}$. □

Lemma 3.11 (F. & M. Riesz Minus for $\mathbf{A}(\mathbb{D})$) *Let μ be a regular complex Borel measure on $\mathbb{T}$ that annihilates $A(\mathbb{D})$. Suppose $\mu = \mu_a + \mu_s$ where $\mu_a \ll m$ and $\mu_s \perp m$. Then both μ_a and μ_s annihilate $A(\mathbb{D})$.*

Proof Choose a sequence of compact subsets $\{K_j\}$ of $\mathbb{T}$ such that μ_s is concentrated on the union of the K_js and each $m(K_j) = 0$. Get $\{f_n\}$ as in Forelli (3.10). For any $g \in A(\mathbb{D})$, $gf_n \in A(\mathbb{D})$ and so $\int gf_n d\mu_a + \int gf_n d\mu_s = \int gf_n d\mu = 0$. Since $\mu_a \ll m$, $gf_n \to g$ μ_a-a.e. and boundedly, and so $\int gf_n \, d\mu_a \to \int g \, d\mu_a$ by Lebesgue's Dominated Convergence Theorem [RUD, 1.34]. Since μ_s is concentrated on the union of the K_js, $gf_n \to 0$ μ_s-a.e. and boundedly, and so $\int gf_n \, d\mu_s \to 0$. We conclude that $\int g \, d\mu_a = 0$ and thus $\int g \, d\mu_s = 0$ also. □

Proposition 3.12 (F. & M. Riesz for $\mathbf{A}(\mathbb{D})$) *Let μ be a regular complex Borel measure on $\mathbb{T}$ that annihilates $A(\mathbb{D})$. Then $\mu \ll m$.*

Proof Call a nonnegative integer N *nice* if and only if $z^{-N} \mu_s \perp A(\mathbb{D})$. By the last lemma, $N = 0$ is nice.

Assume N is nice and set $\tilde{\mu} = z^{-(N+1)} \mu_s - c\, m$ where $c = \int z^{-(N+1)} \, d\mu_s(z)$. By our choice of c, $\int z^0 \, d\tilde{\mu}(z) = 0$. But then for $n \ge 1$, using the niceness of N we see that $\int z^n \, d\tilde{\mu}(z) = \int z^{(n-1)-N} \, d\mu_s - c \int z^n \, dm = 0$. By (b), $\tilde{\mu} \perp A(\mathbb{D})$. By the last lemma, $\tilde{\mu}_s \perp A(\mathbb{D})$. However, $\tilde{\mu}_s = z^{-(N+1)} \mu_s$, so we have shown that $N + 1$ is nice.

Induction now shows every nonnegative integer N to be nice. In particular, since $\bar{z} = z^{-1}$ on $\mathbb{T}$, $\int \bar{z}^N \, d\mu_s(z) = \int z^{-N} \, d\mu_s(z) = 0$ for all nonnegative integers N. By the Stone–Weierstrass Theorem, the linear span of powers of z and $\bar{z}$ is dense in $C(\mathbb{T})$. Hence $\mu_s = 0$ and so $\mu = \mu_a \ll m$. □

Corollary 3.13 (F. & M. Riesz Plus for $\mathbf{A}(\mathbb{D})$) *Let μ be a regular complex Borel measure on* $\operatorname{cl} \mathbb{D}$ *that annihilates $A(\mathbb{D})$. Suppose $\mu = \nu + \sigma$ where ν is concentrated on $\mathbb{D}$ and σ is concentrated on $\mathbb{T}$. Then $\sigma \ll m$.*

Proof Let C be $\mathbb{T}$ traversed counterclockwise. Given $f \in A(\mathbb{D})$, Cauchy's Integral Theorem [RUD, 10.35] and a little limiting argument based on the uniform continuity of f on $\operatorname{cl} \mathbb{D}$ show that

$$f(z) = \frac{1}{2\pi i} \int_C \frac{f(\zeta)}{\zeta - z} \, d\zeta.$$

Integrating both sides of this with respect to ν and then using Fubini's Theorem [RUD, 8.8], we see that

$$\int f(z)\,d\nu(z) = \frac{-1}{2\pi i}\int_C f(\zeta)\hat{\nu}(\zeta)\,d\zeta.$$

Adding $\int f(z)\,d\sigma(z)$ to both sides, we conclude that the measure $\frac{-1}{2\pi i}\hat{\nu}(\zeta)\,d\zeta_C + d\sigma(\zeta)$ annihilates $A(\mathbb{D})$ since μ does. But this measure, unlike μ, lives on $\mathbb{T}$, and so by the last proposition is absolutely continuous with respect to m. It follows that $\sigma \ll m$. □

Theorem 3.14 (F. & M. Riesz for A(U)) *Let μ be a regular complex Borel measure on Γ that annihilates $A(U)$. Then μ is absolutely continuous with respect to arclength on Γ.*

Proof Consider the mapping $\varphi : \mathrm{cl}\,\mathbb{D} \mapsto \mathrm{cl}(\mathbb{C}^* \setminus B_1)$ defined by $z \mapsto c_1 + r_1/z$. Define a measure $\tilde{\mu}$ on $\mathrm{cl}\,\mathbb{D}$ by $\tilde{\mu}(\widetilde{E}) = \mu(\varphi(\widetilde{E}))$. Split μ up into the part of itself, ν, that lives on $\mathbb{C}^* \setminus B_1$ and the part of itself, σ, that lives on Γ_1. Obtain $\tilde{\nu}$ and $\tilde{\sigma}$ from ν and σ just as $\tilde{\mu}$ was obtained from μ. Note that $\tilde{\mu}$ annihilates $A(\mathbb{D})$ since, by the usual approximation argument that starts with characteristic functions, $\int \tilde{f}\,d\tilde{\mu} = \int \tilde{f}\circ\varphi^{-1}\,d\mu$ for any $\tilde{f}$ continuous on $\mathrm{cl}\,\mathbb{D}$. Also note that $\tilde{\mu} = \tilde{\nu}+\tilde{\sigma}$, $\tilde{\nu}$ is concentrated on $\mathbb{D}$, and $\tilde{\sigma}$ is concentrated on $\mathbb{T}$. Thus, by the last corollary, $\tilde{\sigma} \ll m$.

The arclength of a subset E of Γ_1 is just $r_1\, m(\varphi^{-1}(E))$. Hence, if E has arclength zero in Γ_1, then $m(\varphi^{-1}(E)) = 0$, and so $\tilde{\sigma}(\varphi^{-1}(E)) = 0$ too. But $\tilde{\sigma}(\varphi^{-1}(E)) = \mu(E)$, so $\mu(E) = 0$. Thus we have shown that μ restricted to Γ_1 is absolutely continuous with respect to arclength on Γ_1. A similar consideration of the other boundary circles of Γ finishes off the job. □

3.4 Construction of the Boundary Garabedian Function

The proof of Proposition 1.11 shows that in defining $\alpha(K)$ one can leave out the condition that the functions in $A(U)$ being suped over vanish at ∞. Thus $\alpha(K)$ is simply the norm of the continuous linear functional $f \in A(U) \mapsto f'(\infty) \in \mathbb{C}$. Because of the Maximum Modulus Theorem [RUD, 12.1 modified to take into account regions containing ∞], we may view $A(U)$ as a closed subspace of $C(\Gamma)$. Thus, by the Hahn–Banach and Riesz Representation Theorems [RUD, 5.16 and 6.19], there exists a regular complex Borel measure μ on Γ with $|\mu|(\Gamma) = \alpha(K)$ such that $\int f\,d\mu = f'(\infty)$ for all $f \in A(U)$. But then the measure $d\mu - \frac{1}{2\pi i}d\zeta_\Gamma$ annihilates $A(U)$ by Proposition 3.2 and so must be absolutely continuous with respect to arclength measure on Γ by Proposition 3.14. We conclude that $d\mu \ll d\zeta_\Gamma$. By the Radon–Nikodym Theorem [RUD, 6.10 and 6.12], $d\mu(\zeta) = \psi^*(\zeta)\,d\zeta_\Gamma$ for some $\psi^* \in L^1(\Gamma)$. This function ψ^* is called the *boundary Garabedian function* for K. Its crucial properties are the following:

$$\int_\Gamma f(\zeta)\psi^*(\zeta)\,d\zeta = f'(\infty) \text{ for all } f \in A(U) \text{ and } \int_\Gamma |\psi^*(\zeta)|\,|d\zeta| = \alpha(K)$$

with the last item being a consequence of [RUD, 6.13]. It easily follows from Scholium 3.7 that

$$\int_\Gamma f^*(\zeta)\psi^*(\zeta)\,d\zeta = f'(\infty) \text{ for all } f \in H^\infty(U).$$

Proof of the First Equality of Garabedian Duality (3.1) Let f be the Ahlfors function for K; so $f \in H_0^\infty(U)$, $\|f\|_\infty = 1$, and $f'(\infty) = \gamma(K)$. Then

$$\begin{aligned}\gamma(K) &= \int_\Gamma f^*(\zeta)\psi^*(\zeta)\,d\zeta \le \int_\Gamma |f^*(\zeta)|\,|\psi^*(\zeta)|\,|d\zeta| \\ &\le \|f^*\|_\infty \int_\Gamma |\psi^*(\zeta)|\,|d\zeta| = \alpha(K).\end{aligned}$$

Since the reverse inequality is trivial, we must have $\gamma(K) = \alpha(K)$. □

Since the chain above has its end members equal, the two inequalities of the chain must actually be equalities. This lets us extract a little more information which will prove crucial in what follows!

Scholium 3.15 *For f^* the boundary Ahlfors function for K and ψ^* the boundary Garabedian function for K, we have*

$$(\star)\ f^*(\zeta)\psi^*(\zeta)d\zeta_\Gamma \ge 0 \text{ and } (\star\star)\ |f^*(\zeta)| = 1 \quad |\psi^*(\zeta)||d\zeta_\Gamma|\text{-a.e.}$$

The first inequality of the chain being an equality gives us $(\star)$ by [RUD, proofs of 1.33 and 1.39(c)]. The second inequality of the chain being an equality gives us $(\star\star)$ by [RUD, 1.39(a)].

3.5 Construction of the Interior Garabedian Function

Eventually we wish to prove that ψ^*, defined on Γ, can be extended analytically to a neighborhood of $\text{cl}\, U = U \cup \Gamma$. For this paragraph *only* we assume such an analytic extension exists in order to see how its definition on U, which we will denote by ψ, is forced. Thus the function that is ψ on U and ψ^* on Γ is an element of $A(U)$ and so Proposition 3.2 applies to it. The part of the proposition dealing with $z \in U$ forces

$$\psi(z) = \psi(\infty) - \frac{1}{2\pi i}\int_\Gamma \frac{\psi^*(\zeta)}{\zeta - z}\,d\zeta.$$

So it only remains to assertain what $\psi(\infty)$ is. However, the part of the proposition dealing with $z \in \Delta$ forces

$$\psi(\infty) = \frac{1}{2\pi i}\int_\Gamma \frac{\psi^*(\zeta)}{\zeta - z}\,d\zeta.$$

Recall that ψ^* integrated against a function from $A(U)$ yields the function's derivative at ∞. But for $z \in \Delta$, the function $\zeta \mapsto 1/(\zeta - z)$ is in $A(U)$ with derivative at ∞ equal to 1. Thus $\psi(\infty)$ must be $1/2\pi i$!

Our motivational session is now ended and has led to a definition: Let ψ denote the analytic function on U defined by

$$\psi(z) = \frac{1}{2\pi i} - \frac{1}{2\pi i} \int_\Gamma \frac{\psi^*(\zeta)}{\zeta - z}\, d\zeta.$$

This function ψ is called the *interior Garabedian function* for K.

Proposition 3.16 *For* $f \in H^\infty(U)$,

$$(\star)\ f(\infty)\psi(\infty) - \frac{1}{2\pi i} \int_\Gamma \frac{f^*(\zeta)\psi^*(\zeta)}{\zeta - z}\, d\zeta = \begin{cases} f(z)\psi(z) & \text{when } z \in U \\ 0 & \text{when } z \in \Delta \end{cases}.$$

Moreover,

$$(\star\star)\ \int_\Gamma f^*(\zeta)\psi^*(\zeta)\, d\zeta = f'(\infty).$$

Proof By Scholium 3.7, it suffices to prove the proposition for $f \in A(U)$. But $(\star\star)$ for such f is just one of the crucial defining properties of ψ^*, so only $(\star)$ remains.

Fix $z \in U$. Clearly the function $\zeta \mapsto \{f(\zeta) - f(z)\}/(\zeta - z)$ is in $A(U)$ with derivative at ∞ equal to $f(\infty) - f(z)$. Thus, by the same crucial property of ψ^*, we must have

$$\int_\Gamma \frac{f(\zeta) - f(z)}{\zeta - z} \psi^*(\zeta)\, d\zeta = f(\infty) - f(z).$$

Breaking the integral up into two pieces and using a little algebra along with the definition of ψ, we obtain $(\star)$ for $z \in U$.

To obtain $(\star)$ for $z \in \Delta$, one proceeds in a similar manner using the function $\zeta \mapsto f(\zeta)/(\zeta - z)$ instead. □

3.6 A Further Reduction

We observe the usual convention when we speak of a function specified only almost everywhere as being continuous: we mean that it becomes continuous when changed appropriately on a set of measure zero.

Reduction 3.17 *Let* ψ *and* ψ^* *denote the interior and boundary Garabedian functions for* K. *Then the function defined on* $\operatorname{cl} U$ *which is* ψ *on* U *and* ψ^* *on* Γ *is in* $A(U)$ *and even extends analytically to a neighborhood of* $\operatorname{cl} U$. *Moreover, this extension has an analytic logarithm.*

Proof of Garabedian Duality (3.1) *from this Further Reduction* We only need to construct a function φ as in Reduction 3.3. Let Φ denote the analytic logarithm spoken of in Reduction 3.17. Set $\varphi = \alpha \exp(\Phi/2)$ where α is a complex number with modulus $\sqrt{2\pi}$ and with argument chosen so that $\varphi(\infty) > 0$. Since Φ is analytic in a neighborhood of cl U, clearly $\varphi \in A(U)$. Also, $|\varphi(\infty)|^2 = |\alpha|^2\, |\exp \Phi(\infty)| = 2\pi\, |\psi(\infty)| = 1$, which forces $\varphi(\infty) = 1$. Lastly,

$$\|\varphi\|_2^2 = \frac{1}{2\pi} \int_\Gamma |\alpha|^2\, |\exp \Phi(\zeta)|\, |d\zeta| = \int_\Gamma |\psi^*(\zeta)|\, |d\zeta| = \alpha(K).$$

□

The rest of the chapter is devoted to proving Reduction 3.17. This necessitates an interlude, to which we now turn.

3.7 Interlude: Some Extension and Join Propositions

Notational Conventions. For the duration of this section, γ will denote an open proper subarc of a circle. Thus we may write $\gamma = \{c_0 + r_0 e^{i\theta} : \theta_1 < \theta < \theta_2\}$. The endpoints of γ are clearly $\zeta_1 = c_0 + r_0 e^{i\theta_1}$ and $\zeta_2 = c_0 + r_0 e^{i\theta_2}$. The measure $d\zeta_\gamma$ will be oriented from ζ_1 to ζ_2. The sets R_+ and R_- are to be open "bent" rectangles abutting γ with each one being the Schwarz reflection of the other. Thus for some $\alpha_0 > 1$, we may write

$$R_+ = \{c_0 + re^{i\theta} : r_0 < r < \alpha_0 r_0 \text{ and } \theta_1 < \theta < \theta_2\}$$

and

$$R_- = \{c_0 + re^{i\theta} : \alpha_0^{-1} r_0 < r < r_0 \text{ and } \theta_1 < \theta < \theta_2\}.$$

We set $R = R_+ \cup \gamma \cup R_-$ and note that R is Schwarz symmetric about γ.

In what follows, we will see contour integrals beginning $\int_a^b$ where a and b are points of cl R. The question arises of which path from a to b is here meant. We will consistently use a canonical path defined as follows: starting from a, move steadily along the circle that passes through a and is centered at c_0 until one reaches the ray through b with vertex c_0 (one proceeds around the circle in the direction that lets one always stay in cl R); then move steadily along this ray in the direction that will bring one to b until one does indeed reach b.

The first main result we are shooting for is the following.

Proposition 3.18 (Continuous Extension) *Let g be an element of $L^1(\gamma)$ and σ be a regular complex Borel measure putting no mass on R. Define a continuous function h on γ and a regular complex Borel measure ν by*

$$h(\xi) = \int_{\zeta_1}^{\xi} g(\zeta)\,d\zeta \text{ and } d\nu(\zeta) = \frac{-1}{2\pi i} g(\zeta)\,d\zeta_\gamma + d\sigma(\zeta).$$

Suppose that $\hat{\nu} = 0$ on R_-. Choose H analytic on R_+ such that $H' = \hat{\nu}$ there (this can always by done by [RUD, 13.11]). Then there exists a constant c such that the function which is defined to be $H + c$ on R_+ and h on γ is continuous on $R_+ \cup \gamma$.

The proof of this result will require three lemmas, but first some notation is necessary. For a point $z \in \mathbb{C} \setminus \{c_0\}$, let z^* denote the Schwarz refection of z through the circle that γ lies in. Thus for $z = c_0 + re^{i\theta}$, we have $z^* = c_0 + (r_0^2/r)e^{i\theta}$. Given any $\zeta \in \mathbb{C}$, let

$$I(z;\zeta) = \begin{cases} \displaystyle\int_z^{z^*} \frac{dw}{w-\zeta} & \text{when } \zeta \notin [z, z^*] \\ 0 & \text{when } \zeta \in [z, z^*] \end{cases}.$$

Lemma 3.19 *For z and $\tilde{z} \in R_+$, let ξ and $\tilde{\xi}$ denote the intersection points of γ with $[z, z^*]$ and $[\tilde{z}, \tilde{z}^*]$, respectively. Then*

$$H(z) - H(\tilde{z}) = h(\xi) - h(\tilde{\xi}) + \int I(z;\zeta)\,d\nu(\zeta) - \int I(\tilde{z};\zeta)\,d\nu(\zeta).$$

Proof Let C be the cycle that is the sum of four of our canonical paths: the one from $\tilde{z}$ to z, the one from z to z^*, the one from z^* to $\tilde{z}^*$, and the one from $\tilde{z}^*$ to $\tilde{z}$. Then

$$\begin{aligned} H(z) - H(\tilde{z}) &= \int_{\tilde{z}}^{z} H'(w)\,dw = \int_{\tilde{z}}^{z} \hat{\nu}(w)\,dw = \int_{\tilde{z}}^{z} \hat{\nu}(w)\,dw + \int_{z^*}^{\tilde{z}^*} \hat{\nu}(w)\,dw \\ &= \int \left\{ \int_{\tilde{z}}^{z} \frac{dw}{\zeta - w} + \int_{z^*}^{\tilde{z}^*} \frac{dw}{\zeta - w} \right\} d\nu(\zeta) \\ &= \int \left\{ \int_C \frac{dw}{\zeta - w} - \int_z^{z^*} \frac{dw}{\zeta - w} - \int_{\tilde{z}^*}^{\tilde{z}} \frac{dw}{\zeta - w} \right\} d\nu(\zeta) \\ &= \int \{-2\pi i\, n(C;\zeta) + I(z;\zeta) - I(\tilde{z};\zeta)\}\,d\nu(\zeta) \\ &= \int_\gamma n(C;\zeta)\,g(\zeta)\,d\zeta - 2\pi i \int n(C;\zeta)\,d\sigma(\zeta) \\ &\quad + \int I(z;\zeta)\,d\nu(\zeta) - \int I(\tilde{z};\zeta)\,d\nu(\zeta) \end{aligned}$$

$$= \int_{\tilde{\xi}}^{\xi} g(\zeta)\,d\zeta - 0 + \int I(z;\zeta)\,d\nu(\zeta) - \int I(\tilde{z};\zeta)\,d\nu(\zeta)$$

$$= h(\xi) - h(\tilde{\xi}) + \int I(z;\zeta)\,d\nu(\zeta) - \int I(\tilde{z};\zeta)\,d\nu(\zeta).$$

Sweeping around from ζ_1 to ζ_2, one encounters either ξ first, $\tilde{\xi}$ first, or both simultaneously. The reader should check that the penultimate equality above works out correctly in all cases. □

Lemma 3.20 *Suppose that $\{z_n\}$ is a sequence in R_+ that converges to a point η on γ. Then $\int I(z_n;\zeta)\,d\sigma(\zeta) \to 0$.*

Proof Since σ puts no mass on R, it suffices to show that $I(z_n;\zeta) \to 0$ *uniformly* for $\zeta \in \mathbb{C} \setminus R$. The hypothesis on the sequence $\{z_n\}$ implies that there exists a number $\delta > 0$ such that $|w - \zeta| > \delta$ for all n, all $w \in [z_n, z_n^*]$, and all $\zeta \in \mathbb{C} \setminus R$. Thus $|I(z_n;\zeta)| \le |z_n - z_n^*|/\delta \to |\eta - \eta|/\delta = 0$ and the convergence is clearly uniform for $\zeta \in \mathbb{C} \setminus R$. □

Lemma 3.21 *Suppose that $\{z_n\}$ is a sequence in R_+ that converges to a point η on γ. Then $\int_\gamma I(z_n;\zeta)g(\zeta)\,d\zeta \to 0$.*

Proof By Lebesgue's Dominated Convergence Theorem [RUD, 1.34], it suffices to show that $I(z_n;\zeta) \to 0$ for all $\zeta \in \gamma \setminus \{\eta\}$ and that there exists a number M such that $|I(z_n;\zeta)| \le M$ for all n and all $\zeta \in \gamma$. The first assertion is clear: for any $\zeta \ne \eta$, $|I(z_n;\zeta)| \le |z_n - z_n^*|/\mathrm{dist}(\zeta, [z_n, z_n^*]) \to |\eta - \eta|/\mathrm{dist}(\zeta,\eta) = 0$.

To tackle the second assertion concerning boundedness, note that by translating and dilating (or contracting), we may assume that $\gamma \subseteq \mathbb{T}$. We may also assume that $1 < |z_n| < e$ always. Since $I(z;\zeta) = I(|z|; \zeta e^{-i \arg z})$, it thus suffices to find a number M such that $|I(r; e^{i\theta})| \le M$ whenever $1 < r < e$ and $0 \le \theta < 2\pi$. The case $\theta = 0$ poses no problems because $I(r;1) = 0$; so, without loss of generality, $0 < \theta < 2\pi$.

Since $r^* = 1/r$, the path over which we are integrating in the definition of $I(r; e^{i\theta})$ is just $[1/r, r]$, a segment in the real axis. Under the change of variables $w = \varphi(x) = (1+x)/(1-x)$, the reader may check that $r = \varphi(a)$ and $r^* = \varphi(-a)$ where $a = (r-1)/(r+1)$, that $e^{i\theta} = \varphi(iy)$ where $y = \tan(\theta/2)$, and that $dw = 2dx/(1-x)^2$. Thus

$$I(r; e^{i\theta}) = \int_{r^*}^{r} \frac{1}{w - \zeta}\,dw = \int_{-a}^{a} \frac{1}{\left\{\frac{1+x}{1-x} - \frac{1+iy}{1-iy}\right\}} \cdot \frac{2}{(1-x)^2}\,dx$$

$$= \int_{-a}^{a} \frac{1}{x - iy} \cdot \frac{1-iy}{1-x}\,dx.$$

From this it follows that

$$I(r; e^{i\theta}) - \int_{-a}^{a} \frac{1}{x - iy}\,dx = \int_{-a}^{a} \frac{1}{x - iy}\left\{\frac{1-iy}{1-x} - \frac{1-x}{1-x}\right\}dx$$

$$= \int_{-a}^{a} \frac{1}{1-x}\, dx = \ln\left(\frac{1+a}{1-a}\right) = \ln r.$$

However,

$$\int_{-a}^{a} \frac{1}{x-iy}\, dx = \int_{-a}^{a} \frac{x+iy}{x^2+y^2}\, dx = 2i \int_{0}^{a} \frac{y}{x^2+y^2}\, dx = 2i \arctan(a/y).$$

Putting this all together (and expressing a and y in terms of r and θ), we conclude that

$$I(r; e^{i\theta}) = \ln r + 2i \arctan\left\{\left(\frac{r-1}{r+1}\right)\cot(\theta/2)\right\}.$$

The boundedness of $I(r; e^{i\theta})$ for $1 < r < e$ and $0 < \theta < 2\pi$ is now clear ($M = \sqrt{1+\pi^2}$ will do). □

Proof of the Continuous Extension Proposition (3.18) Fix $\tilde{z} \in R_+$ and let $\tilde{\xi}$ denote the intersection point of γ with $[\tilde{z}, \tilde{z}^*]$. Set $c = -H(\tilde{z}) + h(\tilde{\xi}) + \int I(\tilde{z}; \zeta)\, d\nu(\zeta)$. Given any sequence $\{z_n\}$ in R_+ that converges to a point η on γ, it suffices to show that $H(z_n) + c \to h(\eta)$.

Let ξ_n denote the intersection point of γ with $[z_n, z_n^*]$. Applying Lemma 3.19 to z_n and $\tilde{z}$, we have $H(z_n) + c = h(\xi_n) + \int I(z_n; \zeta)\, d\nu(\zeta)$. Note that $\{\xi_n\}$, a sequence in γ, converges to η, a point in γ, and thus, since h is continuous on γ, we have $h(\xi_n) \to h(\eta)$. Finally, by Lemmas 3.20 and 3.21, we have $\int I(z_n; \zeta)\, d\nu(\zeta) \to 0$. □

The following lemma enunciates the property of Schwarz reflection that we need; it is elementary, but does not appear in [RUD] in just the form we need, so we prove it!

Lemma 3.22 *Suppose that $f(z)$ is analytic at $z_0 \in \mathbb{C} \setminus \{c_0\}$. Then $F(z) = \overline{f(z^*)}$ is analytic at z_0^*.*

Proof We begin with the following *Claim*: If $f(z)$ is analytic at z_0, then $F(z) = \overline{f(\overline{z})}$ is analytic at $\overline{z_0}$. To see this note that the analyticity of f at z_0 implies that

$$f(z) = \sum_{n \geq 0} a_n (z - z_0)^n$$

for all z in some neighborhood of z_0 [RUD, 10.16]. But then

$$F(z) = \sum_{n \geq 0} \overline{a_n} (z - \overline{z_0})^n$$

for all z in some neighborhood of $\overline{z_0}$, and so F is analytic at $\overline{z_0}$ [RUD, 10.6].

Consider the two biholomorphic maps

$$\varphi_1 : z \in \mathbb{C} \setminus \{i\} \mapsto \frac{1 - iz}{1 + iz} \in \mathbb{C} \setminus \{-1\} \text{ and } \varphi_2 : z \in \mathbb{C} \mapsto c_0 \pm r_0 z \in \mathbb{C}$$

where + or − is chosen in the definition of φ_2 to ensure that $\varphi_2^{-1}(z_0) \neq -1$. Clearly the function $f(\varphi_2(\varphi_1(z)))$ is analytic at the point $\varphi_1^{-1}(\varphi_2^{-1}(z_0))$. Then by the Claim the function $\overline{f(\varphi_2(\varphi_1(\bar{z})))}$ is analytic at the point $\overline{\varphi_1^{-1}(\varphi_2^{-1}(z_0))}$. Our hypothesis $z_0 \neq c_0$ implies that $\overline{\varphi_1^{-1}(\varphi_2^{-1}(z_0))} \neq i$. Hence the function $\overline{f\left(\varphi_2\left(\varphi_1\left(\overline{\varphi_1^{-1}(\varphi_2^{-1}(z))}\right)\right)\right)}$ is analytic at the point $\varphi_2\left(\varphi_1\left(\overline{\varphi_1^{-1}(\varphi_2^{-1}(z_0))}\right)\right)$. But this last assertion is actually what we want to prove since a computation shows that for $z = c_0 + re^{i\theta}$,

$$\varphi_2\left(\varphi_1\left(\overline{\varphi_1^{-1}(\varphi_2^{-1}(z))}\right)\right) = c_0 + \frac{r_0^2}{r} e^{i\theta} = z^*.$$

□

Proposition 3.23 (Analytic Extension) *Given the situation of the Continuous Extension Proposition* (3.18), *suppose in addition that* $g(\zeta)\, d\zeta_\gamma$ *is a real measure. Then* $\hat{\nu}$ *on* R_+ *extends to an analytic function* G *defined on all of* R. *Moreover,* $G(\zeta) = g(\zeta)\ \ d\zeta_\gamma$*-a.e.*

Proof Our extra assumption on g implies that the function h of the Continuous Extension Proposition is real, i.e., $\bar{h} = h$. Define a function $\widetilde{H}$ on all of R as follows:

$$\widetilde{H}(z) = \begin{cases} H(z) + c & \text{when } z \in R_+ \\ h(z) = \overline{h(z^*)} & \text{when } z \in \gamma \\ \overline{H(z^*) + c} & \text{when } z \in R_- \end{cases}.$$

By the Continuous Extension Proposition, $\widetilde{H}$ is continuous on $R_+ \cup \gamma$. But then $\widetilde{H}$ is continuous on $R_- \cup \gamma$, and thus on all of R. Our H being analytic on R_+, the lemma implies that $\widetilde{H}$ is analytic on $R_+ \cup R_-$. A routine use of Morera's Theorem [RUD, 10.17] now shows that $\widetilde{H}$ is analytic on all of R. It follows that $G = \widetilde{H}'$ is an analytic extension of $\hat{\nu}$ from R_+ to R.

To tackle the second assertion concerning the coincidence of G and g on γ, note that by translating and dilating (or contracting), we may assume that $\gamma \subseteq \mathbb{T}$ with endpoints $\zeta_1 = e^{i\theta_1}$ and $\zeta_2 = e^{i\theta_2}$. Then for $\theta \in (\theta_1, \theta_2)$, we have

$$G(e^{i\theta}) = \widetilde{H}'(e^{i\theta}) = \lim_{\zeta \to e^{i\theta}} \frac{\widetilde{H}(\zeta) - \widetilde{H}(e^{i\theta})}{\zeta - e^{i\theta}} = \lim_{\varphi \to \theta} \frac{\widetilde{H}(e^{i\varphi}) - \widetilde{H}(e^{i\theta})}{e^{i\varphi} - e^{i\theta}}$$

$$= \lim_{\varphi \to \theta} \frac{h(e^{i\varphi}) - h(e^{i\theta})}{\varphi - \theta} \Big/ \frac{e^{i\varphi} - e^{i\theta}}{\varphi - \theta} = \frac{d}{d\theta} h(e^{i\theta}) \Big/ \frac{d}{d\theta} e^{i\theta}$$

$$= \frac{d}{d\theta}\int_{\zeta_1}^{e^{i\theta}} g(\zeta)\,d\zeta \Big/ ie^{i\theta} \;=\; \frac{d}{d\theta}\int_{\theta_1}^{\theta} g(e^{i\varphi})\,ie^{i\varphi}d\varphi \Big/ ie^{i\theta}$$

$$= g(e^{i\theta})\,ie^{i\theta}/ie^{i\theta} \;=\; g(e^{i\theta}).$$

Note that the first equality of the last line holds for $\mathcal{L}^1$-a.e. $\theta \in (\theta_1, \theta_2)$ [RUD, 7.11]. It follows that $G(\zeta) = g(\zeta)\ \ d\zeta_\gamma$-a.e. □

Proposition 3.24 (Analytic Join) *Let g be an element of $L^1(\gamma)$ and σ_+ and σ_- be two regular complex Borel measures putting no mass on R. Define regular complex Borel measures ν_+ and ν_- by*

$$d\nu_+(\zeta) = \frac{-1}{2\pi i} g(\zeta)\,d\zeta_\gamma + d\sigma_+(\zeta) \text{ and } d\nu_-(\zeta) = \frac{+1}{2\pi i} g(\zeta)\,d\zeta_\gamma + d\sigma_-(\zeta).$$

Suppose that $\hat{\nu}_+ = 0$ on R_- and $\hat{\nu}_- = 0$ on R_+. Then the function which is $\hat{\nu}_+$ on R_+ and $\hat{\nu}_-$ on R_- extends to an analytic function G defined on all of R. Moreover, $G(\zeta) = g(\zeta)\ \ d\zeta_\gamma$-a.e.

Proof Choose H_+ analytic on R_+ such that $H_+' = \hat{\nu}_+$ there. Then, by the Continuous Extension Proposition (3.18), there exists a constant c_+ such that the function which is defined to be $H_+ + c_+$ on R_+ and h on γ is continuous on $R_+ \cup \gamma$. Choose H_- analytic on R_- such that $H_-' = \hat{\nu}_-$ there. There is an obvious variant of the Continuous Extension Proposition gotten by systematically interchanging the R_+s and the R_-s. It is valid provided one also changes the factor of $-1/2\pi i$ in the definition of ν to a factor of $+1/2\pi i$ (this change is necessary in order to insure that the conclusion of Lemma 3.19 remains true as stated). By this obvious variant, there exists a constant c_- such that the function which is defined to be $H_- + c_-$ on R_- and h on γ is continuous on $R_- \cup \gamma$.

Define a function $\widetilde{H}$ on all of R as follows:

$$\widetilde{H}(z) = \begin{cases} H_+(z) + c_+ & \text{when } z \in R_+ \\ h(z) & \text{when } z \in \gamma \\ H_-(z) + c_- & \text{when } z \in R_- \end{cases}.$$

Clearly $\widetilde{H}$ is continuous on all of R and analytic on $R_+ \cup R_-$. A routine use of Morera's Theorem [RUD, 10.17] now shows that $\widetilde{H}$ is analytic on all of R. It follows that $G = \widetilde{H}'$ is the analytic extension we seek.

The assertion concerning the coincidence of G and g on γ follows as in the proof of the Analytic Extension Proposition (3.23). □

3.8 Analytically Extending the Ahlfors and Garabedian Functions

Notational Conventions. For the course of this section, f and ψ will denote the interior Ahlfors and Garabedian functions of K, respectively. Recall that both are defined on U and have boundary values f^* and ψ^* defined arclength-a.e. on Γ. Let γ denote an arbitrary open proper subarc of one of the boundary circles making up Γ. We carry over all the notation associated with γ in the last section. Lastly, let the parameter α_0 involved in the definition of the regions R_+ and R_- associated with γ be the number α_0 associated with K chosen at the beginning of this chapter.

We wish to show that each of the functions f and ψ extends analytically across Γ with its extension being equal to its boundary values $d\zeta_\Gamma$-a.e. As a first step, we show the same for their product.

Proposition 3.25 *The function $f\psi$ extends analytically across Γ. Moreover, denoting the extension by G, we have $G(\zeta) = f^*(\zeta)\psi^*(\zeta)\ \ d\zeta_\Gamma$-a.e.*

Proof It suffices to prove our proposition locally, i.e., it suffices to prove our proposition for γ instead of Γ. Set $g = f^*\psi^*$ on γ and $d\sigma(\zeta) = \frac{-1}{2\pi i} f^*(\zeta)\psi^*(\zeta)\, d\zeta_{\Gamma\setminus\gamma}$. Then the measure ν of the Continuous and Analytic Extension Propositions (3.18 and 3.23) is just $d\nu(\zeta) = \frac{-1}{2\pi i} f^*(\zeta)\psi^*(\zeta)\, d\zeta_\Gamma$. By Proposition 3.16, $\hat{\nu} = 0$ on $\Delta \supseteq R_-$ and $f\psi$ on $U \supseteq R_+$ (recall that $f(\infty) = 0$). Also, $g(\zeta)\, d\zeta_\Gamma$ is a real measure, positive even, by ($\star$) of Scholium 3.15. Hence, the Analytic Extension Proposition (3.23) applies and gives us what we want. □

Examination of this last proof along with a consideration of how α_0 was chosen show that the extension G of $f\psi$ just obtained is defined on the set

$$U(\alpha_0) = \mathbb{C}^* \setminus \bigcup_{j=1}^{N} B(c_j; \alpha_0^{-1} r_j).$$

By decreasing α_0 slightly, we may, and do, assume that G is defined in a neighborhood of the closure of $U(\alpha_0)$ (which is needed in the proof of the next proposition).

In what follows, in expressions like "γ_α^*" or "ζ_α^*," the Schwarz reflection operation "*" is to be considered as applied after the expansion operation "$_\alpha$" (in the other order the Schwarz reflection operation would be superfluous since it is the identity on the circle containing γ). Recall that the "*" in "g^*" has nothing to do with Schwarz reflection: "g^*" just denotes the boundary values of a function g from $H^\infty(U)$! To extend the Garabedian function we need the following.

Lemma 3.26 *Suppose that $g \in H^\infty(U)$. Then for any function h defined and continuous on a neighborhood of the closure of the subarc γ,*

$$\int_{\gamma_\alpha^*} \overline{g(\zeta^*)} h(\zeta)\, d\zeta \to \int_\gamma \overline{g^*(\zeta)} h(\zeta)\, d\zeta$$

as $\alpha \downarrow 1$.

Proof A simple computation shows that $\int_{\gamma_\alpha^*} \overline{g(\zeta^*)}h(\zeta)\,d\zeta = \alpha^{-1}\int_\gamma \overline{g_\alpha(\zeta)}h(\zeta_\alpha^*)\,d\zeta$, so it suffices to show

$$(\star)\quad \int_\gamma \overline{g_\alpha(\zeta)}h(\zeta_\alpha^*)\,d\zeta \to \int_\gamma \overline{g^*(\zeta)}h(\zeta)\,d\zeta$$

as $\alpha \downarrow 1$. Let V be the neighborhood of the closure of γ on which h is defined and continuous. Choose an open set W with compact closure such that $\mathrm{cl}\,\gamma \subseteq W \subseteq \mathrm{cl}\,W \subseteq V$. Given our hypotheses, $(\star)$ is now a simple $\varepsilon/2$-argument since, for α close enough to 1, the absolute value of the difference of the two integrals in $(\star)$ is, after adding and subtracting $\int_\gamma \overline{g_\alpha(\zeta)}h(\zeta)\,d\zeta$ and performing some routine estimates, less than or equal to

$$\|g\|_\infty\cdot\mathrm{osc}(h;\mathrm{cl}\,W;(1-\alpha^{-1})r_0)\cdot r_0(\theta_2-\theta_1)+\left|\int_\gamma \overline{g_\alpha(\zeta)}h(\zeta)\,d\zeta - \int_\gamma \overline{g^*(\zeta)}h(\zeta)\,d\zeta\right|.$$

The first term can be made smaller than $\varepsilon/2$ since h, being continuous on the compact set $\mathrm{cl}\,W$, is uniformly continuous there. The second term can be made smaller than $\varepsilon/2$ since $\overline{g_\alpha} \to \overline{g^*}$ weakly in $L^\infty(\Gamma)$ as $\alpha \downarrow 1$ by Proposition 3.9. □

Proposition 3.27 *The function ψ extends analytically across Γ. Moreover, denoting the extension by Ψ, we have $\Psi(\zeta) = \psi^*(\zeta)$ $d\zeta_\Gamma$-a.e.*

Proof It suffices to prove our proposition locally, i.e., it suffices to prove our proposition for γ instead of Γ. Set $g = \psi^* - \frac{1}{2\pi i}$ on γ and $d\sigma_+(\zeta) = \frac{-1}{2\pi i}\{\psi^*(\zeta) - \frac{1}{2\pi i}\}\,d\zeta_{\Gamma\setminus\gamma}$. Then the measure ν_+ of the Analytic Join Proposition (3.24) is just $d\nu_+(\zeta) = \frac{-1}{2\pi i}\{\psi^*(\zeta) - \frac{1}{2\pi i}\}\,d\zeta_\Gamma$. By Proposition 3.16, $\hat{\nu}_+ = 0$ on $\Delta \supseteq R_-$ and $\psi - \frac{1}{2\pi i}$ on $U \supseteq R_+$. To finish, by the Analytic Join Proposition (3.24) it suffices to come up with a measure σ_- which puts no mass on R and for which the corresponding $\hat{\nu}_- = 0$ on R_+.

Given $\alpha \geq 1$, let $C(\alpha)$ be the sum of the following three paths: the straight line segment from $\zeta_{2,\alpha}^* = (\zeta_2)_\alpha^*$ to ζ_{2,α_0}^*, the path $-\gamma_{\alpha_0}^*$, and the straight line segment from ζ_{1,α_0}^* to $\zeta_{1,\alpha}^*$. Define $d\sigma_-(\zeta) = \frac{1}{2\pi i}\{\overline{f(\zeta^*)}G(\zeta) - \frac{1}{2\pi i}\}\,d\zeta_{C(1)}$. Fix z in R_+. Then the assertion $\hat{\nu}_-(z) = 0$, which we must verify, is just

$$(\star)\quad \frac{1}{2\pi i}\int_\gamma \frac{\psi^*(\zeta) - \frac{1}{2\pi i}}{\zeta - z}\,d\zeta + \frac{1}{2\pi i}\int_{C(1)} \frac{\overline{f(\zeta^*)}G(\zeta) - \frac{1}{2\pi i}}{\zeta - z}\,d\zeta = 0.$$

The function $\overline{f(\zeta^*)}G(\zeta)$ is analytic on the loop $\gamma_\alpha^* + C(\alpha)$ and the region it encloses by Lemma 3.22 and the previous proposition. Hence Cauchy's Integral Theorem [RUD, 10.35] implies that

$$(\star\star)\quad \frac{1}{2\pi i}\int_{\gamma_\alpha^*} \frac{\overline{f(\zeta^*)}G(\zeta) - \frac{1}{2\pi i}}{\zeta - z}\,d\zeta + \frac{1}{2\pi i}\int_{C(\alpha)} \frac{\overline{f(\zeta^*)}G(\zeta) - \frac{1}{2\pi i}}{\zeta - z}\,d\zeta = 0$$

for any α between 1 and α_0. We obtain ($\star$) by letting $\alpha \downarrow 1$ in ($\star\star$). The only limit that is nonobvious is

$$(\star\star\star) \quad \int_{\gamma_\alpha^*} \frac{\overline{f(\zeta^*)}G(\zeta)}{\zeta - z} d\zeta \to \int_\gamma \frac{\psi^*(\zeta)}{\zeta - z} d\zeta \text{ as } \alpha \downarrow 1.$$

Applying Lemma 3.26 with $g(\zeta) = f(\zeta)$ and $h(\zeta) = G(\zeta)/(\zeta - z)$, we obtain

$$\int_{\gamma_\alpha^*} \frac{\overline{f(\zeta^*)}G(\zeta)}{\zeta - z} d\zeta \to \int_\gamma \frac{\overline{f^*(\zeta)}G(\zeta)}{\zeta - z} d\zeta \text{ as } \alpha \downarrow 1.$$

But this is just ($\star\star\star$) since $\overline{f^*(\zeta)}G(\zeta) = |f^*(\zeta)|^2\psi^*(\zeta) = \psi^*(\zeta) \quad d\zeta_\Gamma$-a.e. by Proposition 3.25 and ($\star\star$) of Scholium 3.15. □

Note that our extension Ψ is defined on $U(\alpha_0)$.

Proposition 3.28 *The function f extends analytically across Γ. Moreover, denoting the extension by F, we have $|F| = 1$ everywhere on Γ.*

Proof Clearly G/Ψ is a *meromorphic* extension of f across Γ which is analytic at all points of Γ except possibly those where Ψ vanishes. Since Γ is compact and Ψ is analytic on Γ, these points are finite in number and result in isolated singularities of G/Ψ that are at worst poles [RUD, 10.18]. Let $z_0 \in \Gamma$ be such that $\Psi(z_0) = 0$. For any $z \in U$, $|G(z)/\Psi(z)| = |f(z)| \leq 1$ and so $|G(z)/\Psi(z)| \not\to \infty$ as $z \to z_0$. It follows that the singularity of G/Ψ at z_0 is not a pole and so must be removable [RUD, 10.21 and the paragraph after it]. Thus G/Ψ is an analytic extension of f across Γ which we henceforth denote by F.

Since Ψ has only finitely many zeros on Γ and $\Psi = \psi^* \quad d\zeta_\Gamma$-a.e., we have $|\psi^*| > 0 \quad d\zeta_\Gamma$-a.e. Thus $F(\zeta) = G(\zeta)/\Psi(\zeta) = f^*(\zeta)\psi^*(\zeta)/\psi^*(\zeta) = f^*(\zeta)$ $d\zeta_\Gamma$-a.e. and ($\star\star$) of Scholium 3.15 can be strengthened to $|f^*(\zeta)| = 1 \quad d\zeta_\Gamma$-a.e. Hence $|F(\zeta)| = 1 \quad d\zeta_\Gamma$-a.e. Since F is continuous on Γ, it follows that $|F| = 1$ everywhere on Γ. □

Notational Conventions. Our extension F is meromorphic on $U(\alpha_0)$ but only analytic on a neighborhood of the closure of U. By decreasing α_0 slightly, we may, and do, assume that both F and Ψ are analytic on the neighborhood $U(\alpha_0)$ of the closure of U. For the rest of the chapter, we simply use f to denote each of f, f^*, and F. Similarly, we simply use ψ to denote each of ψ, ψ^*, and Ψ. Summarizing the results of our work to this point, we have the following.

Scholium 3.29 *The Ahlfors function f for K and the Garabedian function ψ for K extend analytically across Γ to the neighborhood $U(\alpha_0)$ of the closure of U in such a way that*

$$(\star)\ f(\zeta)\psi(\zeta)d\zeta_\Gamma \geq 0 \text{ and } (\star\star)\ |f| = 1 \text{ everywhere on } \Gamma.$$

3.9 Interlude: Consequences of the Argument Principle

In this section any subarc of a circle occurring in a line integral is to be traversed in the counterclockwise direction. Denote the number of zeros (counted according to multiplicity) of an analytic function g on a set E by $Z(g; E)$. Our first goal is to establish a version of the Argument Principle valid for our hole-punch domain U which takes into account zeros at ∞ and on Γ.

Lemma 3.30 *Suppose that g is a function analytic at z_0 and having an isolated zero there of multiplicity n. For all sufficiently small $\varepsilon > 0$, let the arcs $\gamma(\varepsilon) = \{z_0 + \varepsilon e^{i\theta} : \theta_1(\varepsilon) \le \theta \le \theta_2(\varepsilon)\}$ be such that $\theta_2(\varepsilon) - \theta_1(\varepsilon) \to \pi$ as $\varepsilon \downarrow 0$. Then*

$$\lim_{\varepsilon\downarrow 0} \frac{-1}{2\pi i}\int_{\gamma(\varepsilon)} \frac{g'(\zeta)}{g(\zeta)}\,d\zeta = -\frac{n}{2}.$$

Proof For some $\varepsilon_0 > 0$, we may write $g(\zeta) = (\zeta - z_0)^n h(\zeta)$ where h is analytic and nonzero on a neighborhood of $B(z_0; \varepsilon_0)$. By the Product Rule, $g'(\zeta)/g(\zeta) = h'(\zeta)/h(\zeta) + n/(\zeta - z_0)$, and so

$$(\star)\quad \frac{-1}{2\pi i}\int_{\gamma(\varepsilon)} \frac{g'(\zeta)}{g(\zeta)}\,d\zeta = \frac{-1}{2\pi i}\int_{\gamma(\varepsilon)} \frac{h'(\zeta)}{h(\zeta)}\,d\zeta - n\cdot\frac{\theta_2(\varepsilon)-\theta_1(\varepsilon)}{2\pi}$$

for any ε greater than 0 but smaller than ε_0. There exists a positive finite number M such that $|h'(\zeta)/h(\zeta)| \le M$ for all $\zeta \in B(z_0; \varepsilon_0)$. Thus

$$(\star\star)\quad \left|\frac{1}{2\pi i}\int_{\gamma(\varepsilon)} \frac{h'(\zeta)}{h(\zeta)}\,d\zeta\right| \le \frac{1}{2\pi}\cdot M\cdot 2\pi\varepsilon = M\varepsilon$$

for any ε greater than 0 but smaller than ε_0. Letting $\varepsilon \downarrow 0$ in $(\star)$, we obtain what we want with the help of $(\star\star)$ and the hypothesis on $\theta_2(\varepsilon) - \theta_1(\varepsilon)$. □

Proposition 3.31 *Suppose g is a function analytic on a neighborhood of the closure of U whose distinct zeros on Γ are $z_1, z_2, \ldots, z_m$. For all sufficiently small $\varepsilon > 0$, let $\Gamma(\varepsilon)$ denote the chain of closed circular subarcs of Γ comprising the set $\Gamma \setminus \bigcup_{k=1}^{m} \operatorname{int} B(z_k; \varepsilon)$. Then*

$$Z(g; U) + \frac{1}{2}Z(g; \Gamma) = \lim_{\varepsilon\downarrow 0} \frac{-1}{2\pi i}\int_{\Gamma(\varepsilon)} \frac{g'(\zeta)}{g(\zeta)}\,d\zeta.$$

Of course, if g has no zeros on Γ, then

$$Z(g; U) = \frac{-1}{2\pi i}\int_{\Gamma} \frac{g'(\zeta)}{g(\zeta)}\,d\zeta.$$

Proof We prove only the first assertion. Let C_R denote the counterclockwise circle about the origin of radius R. For all sufficiently small $\varepsilon > 0$, let $\widetilde{\Gamma}(\varepsilon)$ denote the chain of circular subarcs of the $\partial B(z_k;\varepsilon)$s comprising the set $\mathrm{cl}\, U \cap \bigcup_{k=1}^{m} \partial B(z_k;\varepsilon)$. By the usual Argument Principle [RUD, 10.43 (a) – it is stated for closed paths but the proof works for cycles] applied to the cycle $C_R - \widetilde{\Gamma}(\varepsilon) - \Gamma(\varepsilon)$, we obtain

$$(\star)\ Z(g; U \cap \operatorname{int} B(0;R)) = \frac{1}{2\pi i}\int_{C_R} \frac{g'(\zeta)}{g(\zeta)}\,d\zeta - \frac{1}{2\pi i}\int_{\widetilde{\Gamma}(\varepsilon)} \frac{g'(\zeta)}{g(\zeta)}\,d\zeta - \frac{1}{2\pi i}\int_{\Gamma(\varepsilon)} \frac{g'(\zeta)}{g(\zeta)}\,d\zeta.$$

for all sufficiently large R and all sufficiently small $\varepsilon > 0$. Let n be the order of the zero of g at ∞. Thus $g(\zeta) = \zeta^{-n}h(\zeta)$ with h analytic and nonzero on a neighborhood of ∞. By the Product Rule, $g'(\zeta)/g(\zeta) = h'(\zeta)/h(\zeta) - n/\zeta$, and so

$$(\star\star)\ \frac{1}{2\pi i}\int_{C_R} \frac{g'(\zeta)}{g(\zeta)}\,d\zeta = \frac{1}{2\pi i}\int_{C_R} \frac{h'(\zeta)}{h(\zeta)}\,d\zeta - n.$$

Now $h(\zeta) = h(\infty) + h'(\infty)/\zeta + a_2/\zeta^2 + \cdots$, so $h'(\zeta) = 0 - h'(\infty)/\zeta^2 - 2a_2/\zeta^3 + \cdots$. Hence

$$\frac{\zeta^2 h'(\zeta)}{h(\zeta)} = \frac{-h'(\infty) - 2a_2/\zeta + \cdots}{h(\zeta)} \to \frac{-h'(\infty)}{h(\infty)}$$

as $\zeta \to \infty$. It follows that there exists a positive number M such that $|\zeta^2 h'(\zeta)/h(\zeta)| \le M$ for all ζ with modulus large enough. Thus for all R sufficiently large,

$$(\star\star\star)\ \left|\frac{1}{2\pi i}\int_{C_R} \frac{h'(\zeta)}{h(\zeta)}\,d\zeta\right| \le \frac{1}{2\pi}\cdot\frac{M}{R^2}\cdot 2\pi R = \frac{M}{R}.$$

By the last lemma,

$$(\star\star\star\star)\ \lim_{\varepsilon\downarrow 0} \frac{-1}{2\pi i}\int_{\widetilde{\Gamma}(\varepsilon)} \frac{g'(\zeta)}{g(\zeta)}\,d\zeta = -\frac{1}{2}Z(g;\Gamma).$$

Letting $R \to \infty$ and $\varepsilon \downarrow 0$ in $(\star)$, we obtain with the help of $(\star\star)$, $(\star\star\star)$, and $(\star\star\star\star)$, the result

$$Z(g; U \setminus \{\infty\}) = -n - \frac{1}{2}Z(g;\Gamma) - \lim_{\varepsilon\downarrow 0}\frac{1}{2\pi i}\int_{\Gamma(\varepsilon)} \frac{g'(\zeta)}{g(\zeta)}\,d\zeta.$$

Since $Z(g;U) = Z(g; U\setminus\{\infty\}) + n$, we are done. □

Lemma 3.32 *Suppose that g is a function analytic and never vanishing on a neighborhood of an arc* $\gamma = \{c_0 + r_0 e^{i\theta} : \theta_1 \le \theta \le \theta_2\}$. *If* $g(\zeta)\,d\zeta_\gamma \ge 0$, *then*

$$\frac{-1}{2\pi i}\int_\gamma \frac{g'(\zeta)}{g(\zeta)}\,d\zeta = \frac{1}{2\pi}\left\{(\theta_2-\theta_1)+i\ln\left|\frac{g(c_0+r_0e^{i\theta_2})}{g(c_0+r_0e^{i\theta_1})}\right|\right\}.$$

Proof The hypothesis $g(\zeta)\,d\zeta_\gamma \geq 0$ unwinds to $g(c_0+r_0e^{i\theta})(ir_0e^{i\theta}d\theta) \geq 0$ for $\theta \in [\theta_1, \theta_2]$. From this one sees that

$$(\star)\ |g(c_0+r_0e^{i\theta})| = g(c_0+r_0e^{i\theta})(ie^{i\theta})$$

for $\theta \in [\theta_1, \theta_2]$. Define $h(\theta) = \ln|g(c_0+r_0e^{i\theta})| - i\theta$ for $\theta \in [\theta_1, \theta_2]$. Then, after using $(\star)$ to obtain an absolute-value-free expression for h, one may differentiate and verify that

$$\frac{d}{d\theta}h(\theta) = \frac{g'(c_0+r_0e^{i\theta})}{g(c_0+r_0e^{i\theta})}(ir_0e^{i\theta}).$$

Hence

$$\begin{aligned}\frac{-1}{2\pi i}\int_\gamma \frac{g'(\zeta)}{g(\zeta)}\,d\zeta &= \frac{-1}{2\pi i}\int_{\theta_1}^{\theta_2} \frac{g'(c_0+r_0e^{i\theta})}{g(c_0+r_0e^{i\theta})}(ir_0e^{i\theta}d\theta)\\ &= \frac{-1}{2\pi i}\int_{\theta_1}^{\theta_2}\frac{d}{d\theta}h(\theta)\,d\theta\\ &= \frac{-1}{2\pi i}\{h(\theta_2)-h(\theta_1)\}\\ &= \frac{1}{2\pi}\left\{(\theta_2-\theta_1)+i\ln\left|\frac{g(c_0+r_0e^{i\theta_2})}{g(c_0+r_0e^{i\theta_1})}\right|\right\}.\end{aligned}$$

□

Proposition 3.33 *Suppose g is a function analytic on a neighborhood of the closure of U such that $g(\zeta)\,d\zeta_\Gamma \geq 0$. Then*

$$Z(g;U)+\frac{1}{2}Z(g;\Gamma) = N.$$

Recall that N is the number of holes we had to punch out of the extended complex plane $\mathbb{C}^*$ to get our hole-punch domain U.

Proof Since the left-hand side of the conclusion of Proposition 3.31 is clearly real, we may restate its conclusion as

$$Z(g;U)+\frac{1}{2}Z(g;\Gamma) = \lim_{\varepsilon\downarrow 0}\mathrm{Re}\,\frac{-1}{2\pi i}\int_{\Gamma(\varepsilon)}\frac{g'(\zeta)}{g(\zeta)}\,d\zeta.$$

The chain $\Gamma(\varepsilon)$ appearing in Proposition 3.31 is comprised of nonoverlapping circular subarcs. Applying the last lemma to each of these subarcs, we then get

$$Z(g;U) + \frac{1}{2}Z(g;\Gamma) = \lim_{\varepsilon \downarrow 0} \sum_{\gamma \text{ a Subarc of } \Gamma(\varepsilon)} \frac{\text{Angular Spread of } \gamma}{2\pi}.$$

As $\varepsilon \downarrow 0$, these subarcs from $\Gamma(\varepsilon)$ expand to fill out all of the N boundary circles making up Γ with the possible exception of only finitely many points (the zeros of g on Γ). Thus the sum of the angular spreads of these subarcs from $\Gamma(\varepsilon)$ converges to $2\pi N$ as $\varepsilon \downarrow 0$. □

Lemma 3.34 *Let γ be a counterclockwise circle. Suppose that g is a function analytic on a neighborhood of γ with $|g| = 1$ on γ and $|g| < 1$ just outside γ. Then*

$$\frac{-1}{2\pi i}\int_\gamma \frac{g'(\zeta)}{g(\zeta)}\,d\zeta \geq 1.$$

Proof Using the change of variables $\xi = g(\zeta)$, we see that

$$\frac{1}{2\pi i}\int_\gamma \frac{g'(\zeta)}{g(\zeta) - z}\,d\zeta = \frac{1}{2\pi i}\int_{g\circ\gamma} \frac{1}{\xi - z}\,d\xi.$$

The second line integral in the above is just the winding number $n(g\circ\gamma; z)$ of the path $g\circ\gamma$ about the point z. Thus the line integral of the conclusion of the lemma is just the negative of $n(g\circ\gamma; 0)$, an integer. Hence it suffices to prove that $n(g\circ\gamma; 0) \neq 0$ and that $n(g\circ\gamma; 0) \leq 0$.

Since range$(g\circ\gamma) \subseteq \mathbb{T}$, it follows by elementary properties of winding numbers [RUD, 10.10] that if $n(g\circ\gamma; 0) = 0$, then $n(g\circ\gamma; z) = 0$ for all $z \in \mathbb{C}\setminus\mathbb{T}$. But $n(g\circ\gamma; z) = \hat{\mu}(z)$ for the measure $d\mu(\xi) = d\xi_{g\circ\gamma}/2\pi i$. Thus by Lemma 3.5, $\mu = 0$, i.e., g is constant on γ. But then g is constant near γ by the Uniqueness Property of Analytic Functions [RUD, 10.18]. This contradicts our hypotheses on g, so $n(g\circ\gamma; 0) \neq 0$.

Being an integer, $n(g\circ\gamma; 0)$ is real, and so

$$(\star)\ n(g\circ\gamma; 0) = \text{Re}\,\frac{1}{2\pi i}\int_\gamma \frac{g'(\zeta)}{g(\zeta)}\,d\zeta = \text{Re}\,\frac{1}{2\pi}\int_\gamma \frac{-ig'(\zeta)\overline{g(\zeta)}}{|g(\zeta)|^2}\,d\zeta$$
$$= \frac{1}{2\pi}\int_\gamma \text{Re}\,\{-ig'(\zeta)\overline{g(\zeta)}\,d\zeta\}$$

with the last equality using the hypothesis that $|g| = 1$ on γ. Suppose that γ is parametrized as $\gamma(\theta) = c_0 + r_0e^{i\theta}$ for $0 \leq \theta \leq 2\pi$. Let u and v denote the real and imaginary parts of g. In what follows when we write u and v or their partials, they are all to be understood as being evaluated at $c_0 + r_0e^{i\theta}$. Plugging in our parametrization, we see that

$$(\star\star)\ \text{Re}\,\{-ig'(\zeta)\overline{g(\zeta)}\,d\zeta\} = \text{Re}\,\{-i(u_x + iv_x)(u - iv)ir_0(\cos\theta + i\sin\theta)d\theta\}$$
$$= \{(u_xu + v_xv)\cos\theta + (u_xv - v_xu)\sin\theta\}r_0d\theta.$$

For any fixed θ, consider the real function $r > 0 \mapsto |g(c_0 + re^{i\theta}|^2$. Since $|g| < 1$ just outside γ, this function has a nonpositive derivative at $r = r_0$. Hence

$$(2uu_x + 2vv_x)\cos\theta + (2uu_y + 2vv_y)\sin\theta \leq 0.$$

Dividing by 2 and using the Cauchy–Riemann equations [RUD, 11.2], this may be rewritten as

$$(\star\star\star)\ (u_x u + v_x v)\cos\theta + (u_x v - v_x u)\sin\theta \leq 0.$$

From $(\star)$, $(\star\star)$, and $(\star\star\star)$, we conclude that $n(g \circ \gamma; 0) \leq 0$. □

Proposition 3.35 *Suppose that g is a function analytic on a neighborhood of the closure of U with $|g| = 1$ on Γ and $|g| < 1$ on U. Then*

$$Z(g; U) \geq N.$$

Proof By Proposition 3.31 and Lemma 3.34 applied to the Γ_js, we have

$$Z(g; U) = \frac{-1}{2\pi i}\int_\Gamma \frac{g'(\zeta)}{g(\zeta)}\, d\zeta = \sum_{j=1}^{N} \frac{-1}{2\pi i}\int_{\Gamma_j} \frac{g'(\zeta)}{g(\zeta)}\, d\zeta \geq \sum_{j=1}^{N} 1 = N.$$

□

3.10 An Analytic Logarithm of the Garabedian Function

Lemma 3.36 *For f and ψ the Ahlfors and Garabedian functions for K, respectively, we have $Z(f; U) = Z(f; \mathrm{cl}\, U) = N$ and $Z(\psi; U) = Z(\psi; \mathrm{cl}\, U) = 0$. Moreover,*

$$\int_{\Gamma_j} \frac{\psi'(\zeta)}{\psi(\zeta)}\, d\zeta = 0$$

for $j = 1, 2, \ldots, N$.

Proof On the one hand, setting $g = f\psi$ in Proposition 3.33 and invoking $(\star)$ of Scholium 3.29, we have

$$(\star)\ Z(f\psi; U) + \frac{1}{2}Z(f\psi; \Gamma) = N.$$

On the other hand, by $(\star\star)$ of Scholium 3.29, we know that f, which is analytic in U, extends continuously to $\mathrm{cl}\, U$ with $|f| = 1$ everywhere on ∂U. Also, $f(\infty) = 0$ by Proposition 1.14, so f is clearly not constant. Thus $|f| < 1$ everywhere on U by Maximum Modulus [RUD, 12.1 modified to take into account regions containing ∞]. Hence, setting $g = f$ in Proposition 3.35, we have

$$(\star\star)\ Z(f; U) \geq N.$$

The relations $(\star)$ and $(\star\star)$ now force the conclusions we seek concerning the numbers of zeros of f and ψ on U and $\mathrm{cl}\, U$, so it only remains to verify the conclusion concerning the integral of ψ'/ψ over the Γ_js.

Setting $g = f$ in Proposition 3.31, we have

$$\sum_{j=1}^{N} \frac{-1}{2\pi i} \int_{\Gamma_j} \frac{f'(\zeta)}{f(\zeta)}\, d\zeta = \frac{-1}{2\pi i} \int_{\Gamma} \frac{f'(\zeta)}{f(\zeta)}\, d\zeta = Z(f; U) = N.$$

But by Lemma 3.34 applied to $g = f$ and $\gamma = \Gamma_j$, each of the terms in the above sum is greater than or equal to 1. This forces

$$(\star\star\star)\ \frac{-1}{2\pi i} \int_{\Gamma_j} \frac{f'(\zeta)}{f(\zeta)}\, d\zeta = 1$$

for each j. By Lemma 3.32 applied to $g = f\psi$ and $\gamma = \Gamma_j$, we have

$$\frac{-1}{2\pi i} \int_{\Gamma_j} \frac{(f\psi)'(\zeta)}{(f\psi)(\zeta)}\, d\zeta = 1$$

for each j. But by the Product Rule, $(f\psi)'/(f\psi) = f'/f + \psi'/\psi$, so

$$(\star\star\star\star)\ \frac{-1}{2\pi i} \int_{\Gamma_j} \frac{f'(\zeta)}{f(\zeta)}\, d\zeta + \frac{-1}{2\pi i} \int_{\Gamma_j} \frac{\psi'(\zeta)}{\psi(\zeta)}\, d\zeta = 1$$

for each j.

The relations $(\star\star\star)$ and $(\star\star\star\star)$ now force the integral of ψ'/ψ over each Γ_j to be 0 concluding the proof. □

Completion of the Proof of Reduction 3.17 It only remains to show that ψ has an analytic logarithm on a neighborhood of the closure of U. By the previous lemma, ψ has no zeros on a neighborhood of the closure of U. Thus, by decreasing α_0 slightly, we may, and do, assume that

$(\star)$ ψ is analytic and zero-free on a neighborhood of the closure of $U(\alpha_0)$.

In case the reader has forgotten it, the set $U(\alpha_0)$, another hole-punch domain "concentric" with U, was defined just after the proof of Proposition 3.25. The boundary circle of $U(\alpha_0)$ that corresponds to the boundary circle $\Gamma_j = \partial B(c_j; r_j)$ of U is given by $\widetilde{\Gamma}_j = \partial B(c_j; \alpha_0^{-1} r_j)$. Note that $\widetilde{\Gamma}_j$ and Γ_j have the same winding number about every point of the complement of the domain of analyticity of ψ. Thus by Cauchy's Integral Theorem [RUD, 10.35] and the previous lemma, we have

$$(\star\star)\quad \int_{\widetilde{\Gamma}_j} \frac{\psi'(\zeta)}{\psi(\zeta)}\,d\zeta = \int_{\Gamma_j} \frac{\psi'(\zeta)}{\psi(\zeta)}\,d\zeta = 0 \text{ for } j = 1, 2, \ldots, N.$$

Fix a point $z_0 \in U(\alpha_0)$ and choose c so that $e^c = \psi(z_0)$. Define a function Φ on $U(\alpha_0)$ by

$$\Phi(z) = \int_{\gamma_z} \frac{\psi'(\zeta)}{\psi(\zeta)}\,d\zeta + c$$

where γ_z is any path in $U(\alpha_0)$ beginning at z_0 and ending at z. Since $U(\alpha_0)$ is path connected, the requisite paths γ_z exist in abundance. By $(\star)$, the integral of ψ'/ψ over any such γ_z is absolutely convergent and so exists nicely. To know that $\Phi(z)$ is well defined, i.e., to know that the integral of ψ'/ψ over any such γ_z is actually independent of the particular γ_z chosen, it suffices to know that the integral of ψ'/ψ over any cycle γ in $U(\alpha_0)$ is 0. But this is now easy. Note that γ and $\sum_j n(\gamma; c_j)\,\widetilde{\Gamma}_j$ have the same winding number about every point of the complement of the domain of analyticity of ψ. Thus by Cauchy's Integral Theorem [RUD, 10.35] and $(\star\star)$, we have

$$\int_{\gamma} \frac{\psi'(\zeta)}{\psi(\zeta)}\,d\zeta = \sum_j n(\gamma; c_j) \int_{\widetilde{\Gamma}_j} \frac{\psi'(\zeta)}{\psi(\zeta)}\,d\zeta = 0.$$

Fixing a point $z \in U(\alpha_0)$, we have

$$\Phi'(z) = \lim_{\xi \to z} \frac{\Phi(\xi) - \Phi(z)}{\xi - z} = \lim_{\xi \to z} \frac{1}{\xi - z} \int_{[\xi, z]} \frac{\psi'(\zeta)}{\psi(\zeta)}\,d\zeta = \frac{\psi'(z)}{\psi(z)}$$

with the last equality holding by the continuity of ψ'/ψ at z. Thus, by the Product Rule,

$$\{\exp(-\Phi)\cdot\psi\}' = \exp(-\Phi)\cdot\psi' + \psi\cdot\exp(-\Phi)\left\{-\frac{\psi'}{\psi}\right\} = 0.$$

Since $U(\alpha_0)$ is path connected, $\exp(-\Phi)\cdot\psi$ must be constant there. Setting $z = z_0$, we see that this constant is 1. Hence, $\exp\Phi = \psi$ on $U(\alpha_0)$, a neighborhood of the closure of U, i.e., Φ is the sought-after analytic logarithm of ψ. □

Chapter 4
Melnikov and Verdera's Solution to the Denjoy Conjecture

4.1 Menger Curvature of Point Triples

The *Menger curvature* of a *noncollinear* triple of points in the complex plane is simply the curvature of the circle that passes through the three points; put in other words, it is simply the reciprocal of the radius of the circumcircle of the triangle determined by the points. A result of elementary geometry states that this circumcircle is unique and nontrivial. The Menger curvature of any noncollinear triple is thus well defined and positive. We consider any line to be a circle of infinite radius and so zero curvature. The *Menger curvature* of a *collinear* triple of points is thus appropriately defined to be zero. With this last stipulation the Menger curvature of any triple (ζ, η, ξ) of points has been defined and will be denoted $c(\zeta, \eta, \xi)$. The following proposition summarizes various alternative geometric descriptions of Menger curvature for the noncollinear case.

Proposition 4.1 *Let T denote the triangle determined by noncollinear points ζ, η, and ξ of the complex plane. We adopt the following labeling conventions: the measures of the angles of T at the vertices ζ, η, ξ are denoted α, β, γ respectively; the lengths of the sides of T opposite the vertices ζ, η, ξ are denoted a, b, c respectively; and the lengths of the altitudes of T dropped from the vertices ζ, η, ξ to their corresponding opposite sides are denoted h_a, h_b, h_c respectively. Then*

$$c(\zeta, \eta, \xi) = \frac{2\sin\alpha}{a} = \frac{2\sin\beta}{b} = \frac{2\sin\gamma}{c} = \frac{2h_a}{bc} = \frac{2h_b}{ac} = \frac{2h_c}{ab}.$$

Moreover, denoting the area of T by A, we have

$$c(\zeta, \eta, \xi) = \frac{4A}{abc} = \frac{a\cos\alpha + b\cos\beta + c\cos\gamma}{2A}.$$

Finally,

$$c^2(\zeta, \eta, \xi) = \frac{2(a\cos\alpha + b\cos\beta + c\cos\gamma)}{abc}.$$

J.J. Dudziak, *Vitushkin's Conjecture for Removable Sets*, Universitext,
DOI 10.1007/978-1-4419-6709-1_4,

We shall be cavalier about the distinction between a geometric object and its appropriate measure. Thus, for example, although a has been defined above as the length of the line segment $[\eta, \xi]$, we will frequently use it also to denote the line segment $[\eta, \xi]$ itself.

Proof Denote the circumradius and circumcenter of T by r and p, respectively. A result of elementary geometry states that the angle subtended by an arc of a circle from the center of the circle is double the angle subtended by the arc from any point of the circle not on the arc. Thus the triangle T_a determined by p and the side a has angle at p equal to 2α. The triangle T_a is also isosceles with its sides $[p, \eta]$ and $[p, \xi]$ of length r. Considering either of the two congruent right triangles into which T_a is decomposed by the perpendicular from p to the side a, we see that $r \sin\alpha = a/2$ and so $1/r = 2(\sin\alpha)/a$. Since in T we have $b \sin\alpha = h_c$ and $ch_c = 2A$, we also obtain $1/r = 2h_c/ab = 4A/abc$. Working similarly with the two triangles T_b and T_c determined by p and the two sides b and c, respectively, we have established the first displayed line and the first equality of the second displayed line of the proposition.

The triangles T_a, T_b, and T_c have areas $A_a = a(r\cos\alpha)/2$, $A_b = b(r\cos\beta)/2$, and $A_c = c(r\cos\gamma)/2$, respectively. These areas are signed areas with the signs ensuring that $A_a + A_b + A_c = A$ in all possible cases: p in, on, or outside T. Thus, adding these three areas together, we get $A = r(a\cos\alpha + b\cos\beta + c\cos\gamma)/2$ which can be rearranged to give us the second equality of the second displayed line of the proposition. The third displayed line of the proposition follows immediately from the second. □

The next proposition deals with the manner in which Menger curvature will arise in an estimate of analytic capacity that is the goal of the next section. To state this proposition however, a notational convention must first be introduced:

$$(\star)\ \sum_{\mathcal{P}erm} f(\zeta,\eta,\xi) = f(\zeta,\eta,\xi) + f(\zeta,\xi,\eta) + f(\eta,\zeta,\xi) + f(\eta,\xi,\zeta) + f(\xi,\zeta,\eta) + f(\xi,\eta,\zeta).$$

Thus the sum defined by $(\star)$ is just the sum of the six terms gotten from $f(\zeta,\eta,\xi)$ by permuting the letters ζ, η, and ξ in all possible ways.

Proposition 4.2 (Melnikov's Miracle) *Let ζ, η, and ξ be noncollinear points of the complex plane. Then*

$$\sum_{\mathcal{P}\mathrm{erm}} \frac{1}{(\zeta-\eta)\overline{(\zeta-\xi)}} = c^2(\zeta,\eta,\xi).$$

Proof Here we are considering $(\star)$ with $f(\zeta,\eta,\xi) = 1/\{(\zeta-\eta)\overline{(\zeta-\xi)}\}$. One may write $\eta - \zeta = c\, e^{i\theta}$ and $\xi - \zeta = b\, e^{i\varphi}$ where $\alpha = |\theta - \varphi|$. Then the sum of the first two terms in $(\star)$ is just

$$\frac{1}{(\zeta-\eta)(\overline{\zeta-\xi})}+\frac{1}{(\zeta-\xi)(\overline{\zeta-\eta})}=\frac{1}{bc\,e^{i(\theta-\varphi)}}+\frac{1}{bc\,e^{i(\varphi-\theta)}}$$
$$=\frac{1}{bc}\left\{e^{i|\theta-\varphi|}+e^{-i|\theta-\varphi|}\right\}=\frac{2\cos\alpha}{bc}.$$

Similarly the sum of the middle two terms in $(\star)$ is just $2(\cos\beta)/ac$ and the sum of the last two terms in $(\star)$ is just $2(\cos\gamma)/ab$. Hence

$$\sum_{\mathcal{P}\mathrm{erm}}\frac{1}{(\zeta-\eta)(\overline{\zeta-\xi})}=\frac{2\cos\alpha}{bc}+\frac{2\cos\beta}{ac}+\frac{2\cos\gamma}{ab}=\frac{2(a\cos\alpha+b\cos\beta+c\cos\gamma)}{abc}.$$

The previous proposition now completes the job. □

4.2 Melnikov's Lower Capacity Estimate

The *Melnikov curvature* of a regular positive Borel measure μ on $\mathbb{C}$, denoted $c(\mu)$, is the well-defined element of $[0,\infty]$ uniquely determined by the following equation:

$$c^2(\mu)=\iiint_{\mathbb{C}^3}c^2(\zeta,\eta,\xi)\,d\mu(\zeta)\,d\mu(\eta)\,d\mu(\xi).$$

The goal of this section is to prove the following result from [MEL] which will be essential for solving the Denjoy Conjecture.

Theorem 4.3 (Melnikov's Lower Capacity Estimate) *Let K be a compact subset of $\mathbb{C}$. Suppose that μ is a nontrivial regular positive Borel measure supported on K for which there exists a positive finite number M such that $\mu(B(c;r))\le Mr$ for all $c\in\mathbb{C}$ and all $r>0$. Then*

$$\gamma(K)\ge m\cdot\frac{M\,\mu^2(K)}{M^2\,\mu(K)+c^2(\mu)}.$$

Here m is a small positive constant: $1/375{,}000$.

Given a positive finite number M, a regular positive Borel measure μ satisfying $\mu(B(c;r))\le Mr$ for all $c\in\mathbb{C}$ and all $r>0$ is said to have *linear growth with bound M*. We say μ has *linear growth* when it has linear growth with bound M for some positive finite number M. Notice that even when we have a nontrivial regular positive Borel measure μ on K with linear growth, the theorem will yield a nontrivial lower estimate on $\gamma(K)$ only when, in addition, the measure μ has finite Melnikov curvature!

Before proving Melnikov's Theorem, we deduce a corollary from it and then establish two lemmas necessary to its proof.

Corollary 4.4 *In the situation of the previous theorem,*

$$\gamma(K) \geq \frac{m}{2} \cdot \sqrt{\frac{\mu^3(K)}{M^2\mu(K) + c^2(\mu)}}.$$

Proof Suppose, on the one hand, that $c^2(\mu) < M^2\mu(K)$. Then $\{M^2\mu(K) + c^2(\mu)\}^{1/2} < \sqrt{2}M\mu^{1/2}(K)$. Thus, applying the theorem to μ, we see that $\gamma(K)$ must be larger than m times

$$\begin{aligned}\frac{M\mu^2(K)}{M^2\mu(K)+c^2(\mu)} &= \frac{M\mu^{1/2}(K)}{\{M^2\mu(K)+c^2(\mu)\}^{1/2}} \cdot \frac{\mu^{3/2}(K)}{\{M^2\mu(K)+c^2(\mu)\}^{1/2}} \\ &> \frac{1}{\sqrt{2}} \cdot \sqrt{\frac{\mu^3(K)}{M^2\mu(K)+c^2(\mu)}}.\end{aligned}$$

Suppose, on the other hand, that $c^2(\mu) \geq M^2\mu(K)$. Then for $a = M\mu^{1/2}(K)/c(\mu)$, we have $a \leq 1$ and so the measure $a\mu$ *still* has linear growth with bound M. Thus, applying the theorem to $a\mu$, we see that $\gamma(K)$ must be larger than m times

$$\begin{aligned}\frac{Ma^2\mu^2(K)}{M^2a\mu(K)+a^3c^2(\mu)} &= \frac{Ma\mu^2(K)}{M^2\mu(K)+a^2c^2(\mu)} = \frac{M^2\mu^{5/2}(K)/c(\mu)}{M^2\mu(K)+M^2\mu(K)} \\ &= \frac{1}{2} \cdot \frac{\mu^{3/2}(K)}{c(\mu)} > \frac{1}{2} \cdot \sqrt{\frac{\mu^3(K)}{M^2\mu(K)+c^2(\mu)}}.\end{aligned}$$

Clearly we are done. □

Any $a \leq 1$ could have been used in the last proof and would result in a valid lower bound for $\gamma(K)$. The reader may wish to check that the particular choice of a made in the proof is the one that maximizes the lower bound obtained in this manner!

Lemma 4.5 *Let μ be a nontrivial regular positive Borel measure supported on K and let $\delta > 0$. Then there exist closed balls $D_j = B(c_j; \delta)$, finite in number, that satisfy*

(a) $\mu(D_j) > 0$ *for each* j,

(b) $\sum_j \mu(D_j) \geq \frac{1}{25}\mu(K)$, *and*

(c) *the distance between distinct balls D_j and D_k is at least 2δ, i.e., $|c_j - c_k| \geq 4\delta$ whenever $j \neq k$.*

Proof Tile the complex plane with closed squares whose edges are of length $\sqrt{2}\delta$. Let $\mathcal{F}$ be the family of all squares Q from this tiling for which $\mu(Q) > 0$. Since μ is compactly supported, $\mathcal{F}$ is a finite family. Given Q from the tiling, let $\mathcal{B}(Q)$ be the family of twenty-five squares from the tiling closest to Q. Thus $\mathcal{B}(Q)$ consists of Q and the twenty-four squares from the tiling that form a "belt" two squares deep around Q.

Choose a square Q_1 with maximal μ-measure from $\mathcal{F}$ and set $\mathcal{F}_1 = \mathcal{F} \cap \mathcal{B}(Q_1)$. Choose a square Q_2 with maximal μ-measure from $\mathcal{F} \setminus \mathcal{F}_1$ and set $\mathcal{F}_2 = \{\mathcal{F} \setminus \mathcal{F}_1\} \cap \mathcal{B}(Q_2)$. Choose a square Q_3 with maximal μ-measure from $\mathcal{F} \setminus (\mathcal{F}_1 \cup \mathcal{F}_2)$ and set $\mathcal{F}_3 = \{\mathcal{F} \setminus (\mathcal{F}_1 \cup \mathcal{F}_2)\} \cap \mathcal{B}(Q_3)$. Continuing in this manner, we find that after a finite number of steps, say N, we can proceed no further since $\mathcal{F} \setminus (\mathcal{F}_1 \cup \mathcal{F}_2 \cup \cdots \cup \mathcal{F}_N) = \emptyset$. The desired c_js are the centers of the N squares Q_j just generated.

(a) Since $D_j \supseteq Q_j \in \mathcal{F}$, we have $\mu(D_j) \geq \mu(Q_j) > 0$.
(b) $\mu(K) \leq \sum_{Q \in \mathcal{F}} \mu(Q) = \sum_j \sum_{Q \in \mathcal{F}_j} \mu(Q) \leq \sum_j 25\mu(Q_j) \leq 25 \sum_j \mu(D_j)$.
(c) Without loss of generality, $k > j$. Thus $Q_k \notin \mathcal{F}_1 \cup \cdots \cup \mathcal{F}_j \supseteq \mathcal{F} \cap \mathcal{B}(Q_j)$ and so $Q_k \notin \mathcal{B}(Q_j)$. But then it is clear that c_k gets closest to c_j when it is in the third "belt" of squares about c_j and either directly above, directly below, directly to the left, or directly to the right of c_k. In these situations the distance between c_j and c_k is exactly three edge lengths, i.e., $3\sqrt{2}\delta$. Since $3\sqrt{2} > 4$, we obviously have $|c_j - c_k| \geq 4\delta$ always. □

Lemma 4.6 *Let μ be a nontrivial regular positive Borel measure supported on K that has linear growth with bound 1. Given $\delta > 0$, let the closed balls $D_j = B(c_j; \delta)$ be as in the last lemma. Then the following assertions hold:*

(a) *If $\zeta \in D_j$ and $\eta \in D_k$ where $j \neq k$, then $|\zeta - \eta| \geq \frac{1}{2}|c_j - c_k|$.*
(b) *For any fixed j, $\sum_{k \neq j} \frac{\delta\mu(D_k)}{|c_j - c_k|^2} \leq \frac{7}{5}$.*
(c) *For any fixed j and $C > 0$, suppose that for each $k \neq j$ we have numbers A_k and B_k satisfying*

$$|A_k - B_k| \leq \frac{C\delta\mu(D_k)}{|c_j - c_k|^2}.$$

Then

$$\left| \sum_{k \neq j} A_k \right|^2 \leq 2 \left| \sum_{k \neq j} B_k \right|^2 + 4C^2.$$

Proof (a) By the Triangle Inequality,

$$|c_j - c_k| \leq |c_j - \zeta| + |\zeta - \eta| + |\eta - c_k| \leq |\zeta - \eta| + 2\delta \leq |\zeta - \eta| + \frac{1}{2}|c_j - c_k|$$

with the last link in this chain of inequalities following from (c) of Lemma 4.5.

(b) Set $I(l) = \{k : 4^l\delta \le |c_j - c_k| < 4^{l+1}\delta\}$. By (c) of Lemma 4.5 each k lies in $I(l)$ for a unique $l \ge 1$ and the D_js are pairwise disjoint. Thus

$$\sum_{k\neq j} \frac{\delta\mu(D_k)}{|c_j - c_k|^2} = \sum_{l\ge 1}\sum_{k\in I(l)} \frac{\delta\mu(D_k)}{|c_j - c_k|^2} \le \sum_{l\ge 1}\sum_{k\in I(l)} \frac{\delta\mu(D_k)}{\{4^l\delta\}^2}$$
$$\le \sum_{l\ge 1} \frac{\mu(B(c_j;(4^{l+1}+1)\delta))}{16^l\delta} \le \sum_{l\ge 1} \frac{4^{l+1}+1}{16^l}$$

with the last link in this chain of inequalities following from the hypothesis that the μ-measure of any closed ball is at most its radius. To finish, note that the very last term in the above is easily expressed as the sum of two geometric series which can be evaluated to deliver the trumpeted estimate of $7/5$.

(c) Using the hypothesis on $|A_k - B_k|$ and (b) just proven, we have

$$\left|\sum_{k\neq j} A_k\right| \le \left|\sum_{k\neq j} B_k\right| + \sum_{k\neq j} |A_k - B_k| \le \left|\sum_{k\neq j} B_k\right| + C\sum_{k\neq j}\frac{\delta\mu(D_k)}{|c_j - c_k|^2} \le \left|\sum_{k\neq j} B_k\right| + \frac{7}{5}C.$$

To finish, just use the trivial inequality $(x + y)^2 \le 2x^2 + 2y^2$ and note that $2(7/5)^2 < 4$. □

Proof of Melnikov's Lower Capacity Estimate (4.3) By replacing μ by μ/M, we may assume that $M = 1$.

For this paragraph, fix $\delta > 0$ and let the c_js and their associated D_js be as in the penultimate lemma. Set $r_j = \mu(D_j)$, $B_j = B(c_j; r_j)$, and $K_\delta = \cup_j B_j$. Note that $r_j \le \delta$ and so the B_js, being contained in the D_js, are pairwise disjoint. The set $\mathbb{C}^* \setminus K_\delta$ is thus a hole-punch domain and the notational conventions from the last chapter will be applied to it. In particular, letting Γ_j denote the counterclockwise boundary of B_j, the boundary of K_δ is the union of the Γ_js. Consider the following rational function f from $A_0(\mathbb{C}^* \setminus K_\delta)$:

$$f(\zeta) = \sum_j \frac{r_j}{\zeta - c_j}.$$

Clearly $f'(\infty) = \lim_{\zeta\to\infty} \zeta\{f(\zeta) - f(\infty)\} = \lim_{\zeta\to\infty} \zeta f(\zeta) = \sum_j r_j \ge \mu(K)/25$ by (b) of the penultimate lemma. The bulk of the proof will consist in showing that $\|f\|_2^2 \le 600\{\mu(K) + c^2(\mu)\}$ (the L^2 norm here is taken over the boundary of K_δ). From this, we will deduce with the crucial help of Garabedian Duality (3.1) that

$$\gamma(\{z \in \mathbb{C} : \mathrm{dist}(z, K) \le 2\delta\}) \ge \gamma(K_\delta) = \alpha_2^2(K_\delta) \ge \frac{|f'(\infty)|^2}{\|f\|_2^2}$$
$$\ge \frac{1}{600 \times 25^2} \cdot \frac{\mu^2(K)}{\mu(K) + c^2(\mu)}.$$

If we now unfix δ and let it decrease to 0, we will be done by Proposition 1.16 since $600 \times 25^2 = 375{,}000$. So to finish we must estimate $\|f\|_2^2$ appropriately.

Fixing j, we have

$$\begin{aligned}\frac{1}{2\pi}\int_{\Gamma_j}|f(\zeta)|^2\,|d\zeta| &= \frac{1}{2\pi}\int_{\Gamma_j}\left|\frac{r_j}{\zeta-c_j}+\sum_{k\neq j}\frac{r_k}{\zeta-c_k}\right|^2|d\zeta|\\ &= \frac{1}{2\pi}\int_{\Gamma_j}\frac{r_j^2}{|\zeta-c_j|^2}\,|d\zeta| + 2\mathrm{Re}\sum_{k\neq j} r_j r_k\cdot\frac{1}{2\pi}\int_{\Gamma_j}\frac{|d\zeta|}{(\overline{\zeta-c_j})(\zeta-c_k)}\\ &\quad+\frac{1}{2\pi}\int_{\Gamma_j}\left|\sum_{k\neq j}\frac{r_k}{\zeta-c_k}\right|^2|d\zeta|.\end{aligned}$$

The first term in the above is easily seen to be r_j because $|\zeta - c_j| = r_j$ for all $\zeta \in \Gamma_j$. Since $|d\zeta|/(\overline{\zeta - c_j}) = d\zeta/(ir_j)$ for all $\zeta \in \Gamma_j$, the integrals appearing in the second term above are all equal to $r_j^{-1} \cdot n(\Gamma_j; c_k) = 0$. Thus we obtain

$$(\star)\ \frac{1}{2\pi}\int_{\Gamma_j}|f(\zeta)|^2\,|d\zeta| = r_j + \frac{1}{2\pi}\int_{\Gamma_j}\left|\sum_{k\neq j}\frac{r_k}{\zeta-c_k}\right|^2|d\zeta|.$$

For $\zeta \in \Gamma_j$ and $k \neq j$, by (a) of the last lemma we have

$$\begin{aligned}\left|\frac{r_k}{\zeta-c_k}-\int_{D_k}\frac{d\mu(\eta)}{\zeta-\eta}\right| &= \left|\int_{D_k}\frac{d\mu(\eta)}{\zeta-c_k}-\int_{D_k}\frac{d\mu(\eta)}{\zeta-\eta}\right| \le \int_{D_k}\frac{|\eta-c_k|}{|\zeta-c_k||\zeta-\eta|}\,d\mu(\eta)\\ &\le \frac{4\delta\mu(D_k)}{|c_j-c_k|^2}.\end{aligned}$$

But then, by (c) of the last lemma we get

$$\left|\sum_{k\neq j}\frac{r_k}{\zeta-c_k}\right|^2 \le 2\left|\sum_{k\neq j}\int_{D_k}\frac{d\mu(\eta)}{\zeta-\eta}\right|^2 + 64.$$

Integrating both sides of this estimate with respect to $|d\zeta|/2\pi$ on Γ_j and using $(\star)$, we obtain

$$(\star\star)\ \frac{1}{2\pi}\int_{\Gamma_j}|f(\zeta)|^2\,|d\zeta| \le 65r_j + \frac{1}{\pi}\int_{\Gamma_j}\left|\sum_{k\neq j}\int_{D_k}\frac{d\mu(\eta)}{\zeta-\eta}\right|^2|d\zeta|.$$

For $\zeta, \zeta' \in D_j$ and $k \neq j$, by (a) of the last lemma we have

$$\left|\int_{D_k} \frac{d\mu(\eta)}{\zeta - \eta} - \int_{D_k} \frac{d\mu(\eta)}{\zeta' - \eta}\right| \leq \int_{D_k} \frac{|\zeta - \zeta'|}{|\zeta - \eta||\zeta' - \eta|}\, d\mu(\eta) \leq \frac{8\delta\mu(D_k)}{|c_j - c_k|^2}.$$

But then, by (c) of the last lemma we get

$$\left|\sum_{k\neq j}\int_{D_k} \frac{d\mu(\eta)}{\zeta - \eta}\right|^2 \leq 2\left|\sum_{k\neq j}\int_{D_k} \frac{d\mu(\eta)}{\zeta' - \eta}\right|^2 + 256.$$

Integrating both sides of this estimate with respect to $|d\zeta|/\pi$ on Γ_j and then with respect to $d\mu(\zeta')$ on D_j, we conclude that

$$\frac{1}{\pi}\int_{\Gamma_j}\left|\sum_{k\neq j}\int_{D_k} \frac{d\mu(\eta)}{\zeta - \eta}\right|^2 |d\zeta| \cdot r_j \leq \left\{2\int_{D_j}\left|\sum_{k\neq j}\int_{D_k} \frac{d\mu(\eta)}{\zeta' - \eta}\right|^2 d\mu(\zeta') + 256 r_j\right\}\cdot 2r_j.$$

Dividing an r_j out of this estimate, changing each ζ' into a ζ, and then using the result in $(\star\star)$, we obtain

$$(\star\star\star)\ \frac{1}{2\pi}\int_{\Gamma_j} |f(\zeta)|^2\, |d\zeta| \leq 577 r_j + 4\int_{D_j}\left|\sum_{k\neq j}\int_{D_k} \frac{d\mu(\eta)}{\zeta - \eta}\right|^2 d\mu(\zeta).$$

Now note that the integral on the right-hand side of the above can be written

$$(\star\star\star\star)\ \int_{D_j}\left\{\sum_{k\neq j}\int_{D_k} \frac{d\mu(\eta)}{\zeta - \eta}\right\}\left\{\overline{\sum_{l\neq j}\int_{D_l} \frac{d\mu(\xi)}{\zeta - \xi}}\right\} d\mu(\zeta) =$$

$$\int_{D_j}\sum_{k\neq j}\left|\int_{D_k} \frac{d\mu(\eta)}{\zeta - \eta}\right|^2 d\mu(\zeta) + \int_{D_j}\left\{\sum_{k,l\neq j\ \&\ k\neq l}\int_{D_k}\int_{D_l} \frac{d\mu(\xi)\, d\mu(\eta)}{(\zeta - \eta)(\overline{\zeta - \xi})}\right\} d\mu(\zeta).$$

For $\zeta \in D_j$, by the Cauchy–Schwarz Inequality [RUD, 3.5] and (a) of the last lemma we have

$$\left|\int_{D_k} \frac{d\mu(\eta)}{\zeta - \eta}\right|^2 \leq \left\{\int_{D_k} \frac{d\mu(\eta)}{|\zeta - \eta|^2}\right\}\left\{\int_{D_k} 1^2\, d\mu(\eta)\right\} \leq \frac{4\delta\mu(D_k)}{|c_j - c_k|^2}.$$

But then, by (b) of the last lemma we get

$$(\star\star\star\star\star)\quad \int_{D_j} \sum_{k\neq j} \left| \int_{D_k} \frac{d\mu(\eta)}{\zeta-\eta} \right|^2 d\mu(\zeta) \leq \int_{D_j} 4\left(\frac{7}{5}\right) d\mu(\zeta) = \frac{28r_j}{5}.$$

From $(\star\star\star)$, $(\star\star\star\star)$, and $(\star\star\star\star\star)$, one may immediately conclude that

$$\frac{1}{2\pi}\int_{\Gamma_j} |f(\zeta)|^2\,|d\zeta| \leq 600r_j +$$
$$4 \sum_{k,l\neq j\ \&\ k\neq l} \int_{D_j}\int_{D_k}\int_{D_l} \frac{1}{(\zeta-\eta)\overline{(\zeta-\xi)}}\, d\mu(\xi)\, d\mu(\eta)\, d\mu(\zeta).$$

Unfix j now and sum the above over all the js. Since each $r_j = \mu(B_j)$ and the B_js are pairwise disjoint, $\sum_j r_j \leq \mu(K)$. Hence we get

$$(\dagger)\ \|f\|_2^2 \leq 600\mu(K) + 4 \sum_{\neq(j,k,l)} \int_{D_j}\int_{D_k}\int_{D_l} \frac{1}{(\zeta-\eta)\overline{(\zeta-\xi)}}\, d\mu(\xi)\, d\mu(\eta)\, d\mu(\zeta)$$

where the "$\neq (j,k,l)$" under the sigma indicates that one is to sum over all *distinct* triples, i.e., over all (j,k,l) with $j\neq k$, $j\neq l$, and $k\neq l$. Further below, "$j<k<l$" under a sigma indicates that one is to sum over all *increasing* triples, i.e., over all (j,k,l) with $j<k$ and $k<l$.

Let us consider a particular distinct triple, say $(j,k,l) = (4,7,2)$, for purposes of illustration. One may use Fubini's Theorem [RUD, 8.8] to change the order of integration so that from left to right at the front end of the integral appearing in $(\dagger)$ one encounters the discs in "ascending" order, i.e., D_2 first, D_4 second, and D_7 third. Note that at the back end of the integral one then encounters the differentials in the order $d\mu(\eta)\, d\mu(\zeta)\, d\mu(\xi)$. Now relabel the variables ζ, η, and ξ so that this back end becomes $d\mu(\zeta)\, d\mu(\eta)\, d\mu(\xi)$. In the particular case under consideration one merely needs to interchange ζ with η and leave ξ as is. The result is

$$\int_{D_4}\int_{D_7}\int_{D_2} \frac{1}{(\zeta-\eta)\overline{(\zeta-\xi)}}\, d\mu(\xi)\, d\mu(\eta)\, d\mu(\zeta) =$$
$$\int_{D_2}\int_{D_4}\int_{D_7} \frac{1}{(\eta-\zeta)\overline{(\eta-\xi)}}\, d\mu(\zeta)\, d\mu(\eta)\, d\mu(\xi).$$

So our prescription for changing the distinct triple $(4,7,2)$ into its associated increasing triple $(2,4,7)$ has merely permuted the variables ζ, η, and ξ in our integrand. The same is obviously true for any other permutation of $(2,4,7)$ with all possible permutations of $(2,4,7)$ producing all possible permutations of ζ, η, and ξ in our integrand. Thus the sum of the six integrals corresponding to all possible permutations of $(2,4,7)$ becomes the integral over the common increasing triple $(2,4,7)$ of the sum of the six terms gotten by permuting the variables ζ, η, and ξ all possible ways in our integrand. Clearly nothing is special about $(2,4,7)$ here. Any distinct increasing triple could have been used. This justifies the first equality in the following chain of in/equalities:

$$
\begin{aligned}
(\ddagger)\ & \sum_{\neq(j,k,l)} \int_{D_j}\int_{D_k}\int_{D_l} \frac{1}{(\zeta-\eta)\overline{(\zeta-\xi)}}\, d\mu(\xi)\, d\mu(\eta)\, d\mu(\zeta) \\
&= \sum_{j<k<l} \int_{D_j}\int_{D_k}\int_{D_l} \sum_{\mathcal{P}erm} \frac{1}{(\zeta-\eta)\overline{(\zeta-\xi)}}\, d\mu(\zeta)\, d\mu(\eta)\, d\mu(\xi) \\
&= \sum_{j<k<l} \int_{D_j}\int_{D_k}\int_{D_l} c^2(\zeta,\eta,\xi)\, d\mu(\zeta)\, d\mu(\eta)\, d\mu(\xi) \\
&= \frac{1}{6} \sum_{\neq(j,k,l)} \int_{D_j}\int_{D_k}\int_{D_l} c^2(\zeta,\eta,\xi)\, d\mu(\zeta)\, d\mu(\eta)\, d\mu(\xi) \\
&\le \frac{1}{6} \int_{\cup_j D_j}\int_{\cup_k D_k}\int_{\cup_l D_l} c^2(\zeta,\eta,\xi)\, d\mu(\zeta)\, d\mu(\eta)\, d\mu(\xi) \\
&\le \frac{1}{6} c^2(\mu).
\end{aligned}
$$

The second equality follows from Melnikov's Miracle (4.2). The third equality's justification is similar to that of the first; it is a consequence of Fubini's Theorem [RUD, 8.8] and relabeling variables with the addition now of an appeal to the invariance of Menger curvature under all permutations of its variables. The two inequalities that follow are trivial.

From (†) and (‡), one may conclude that

$$\|f\|_2^2 \le 600\mu(K) + \frac{2}{3}c^2(\mu) \le 600\{\mu(K) + c^2(\mu)\}.$$

This finishes the proof. □

4.3 Interlude: A Fourier Transform Review

In the next section we shall prove a theorem of Melnikov and Verdera asserting the finiteness of the curvature of some measures associated with Lipschitz graphs. The proof is essentially a clever computation involving Fourier Transforms. Thus this section.

When speaking of "$L^p(E)$" where $E \subseteq \mathbb{R}$, the measure is understood to be linear Lebesgue measure $\mathcal{L}^1$ on $\mathbb{R}$ restricted to E. In what follows, rather than writing our differentials as "$d\mathcal{L}^1(t)$," "$d\mathcal{L}^1(u)$," etc., we shall more simply write "dt," "du," etc.

The following is the first of the two basic results from real analysis that will prove very useful to us.

Proposition 4.7 *If g is an absolutely continuous function on $[a, b]$, then g' exists almost everywhere on $[a, b]$, $g' \in L^1([a, b])$, and for all $t \in [a, b]$ we have*

$$g(t) - g(a) = \int_a^t g'(u)\, du.$$

Proof See [RUD, 7.20]. □

Recall that given $f \in L^1(\mathbb{R})$, the L^1 *Fourier transform* of f, denoted by $\hat{f}$, is the function well defined for all $\xi \in \mathbb{R}$ by the formula

$$\hat{f}(\xi) = \int_{-\infty}^{\infty} e^{-i\xi t} f(t)\, \frac{dt}{\sqrt{2\pi}}.$$

The following three results gather together all that we need to know about the L^1 Fourier transform.

Proposition 4.8 *For $f \in L^1(\mathbb{R})$ and $\alpha \in \mathbb{R}$, set $g(t) = f(t + \alpha)$. Then $\hat{g}(\xi) = e^{i\alpha} \hat{f}(\xi)$ for every $\xi \in \mathbb{R}$.*

Proof Simply use the definition of the transform and an appropriate change of variables. □

Proposition 4.9 *Suppose that g is absolutely continuous and compactly supported on $\mathbb{R}$. Then $\widehat{(g')}(\xi) = i\, \xi \hat{g}(\xi)$ for every $\xi \in \mathbb{R}$.*

Proof Let g be supported on $[a, b]$. Then by Proposition 4.7 and Fubini's Theorem [RUD, 8.8],

$$\begin{aligned}
i\xi\, \hat{g}(\xi) &= i\xi \int_{-\infty}^{\infty} e^{-i\xi t} g(t)\, \frac{dt}{\sqrt{2\pi}} \\
&= \frac{i\xi}{\sqrt{2\pi}} \int_a^b e^{-i\xi t} \left\{ \int_a^t g'(u)\, du \right\} dt \\
&= \frac{1}{\sqrt{2\pi}} \int_a^b \left\{ \int_u^b i\xi e^{-i\xi t}\, dt \right\} g'(u)\, du \\
&= \frac{1}{\sqrt{2\pi}} \int_a^b \left\{ e^{-i\xi u} - e^{-i\xi b} \right\} g'(u)\, du \\
&= \int_{-\infty}^{\infty} e^{-i\xi u} g'(u)\, \frac{du}{\sqrt{2\pi}} - \frac{e^{-i\xi b}}{\sqrt{2\pi}} \int_a^b g'(u)\, du \\
&= \widehat{(g')}(\xi) - \frac{e^{-i\xi b}}{\sqrt{2\pi}} \{g(b) - g(a)\} \;=\; \widehat{(g')}(\xi).
\end{aligned}$$

□

The following is the second of the two basic results from real analysis that will prove very useful to us.

Theorem 4.10 (Plancherel) *Let* $f \in L^1(\mathbb{R}) \cap L^2(\mathbb{R})$. *Then* $\hat{f} \in L^2(\mathbb{R})$ *and* $\|\hat{f}\|_2 = \|f\|_2$.

Proof See [RUD, proof of 9.13]. □

Given $f \in L^2(\mathbb{R})$, let f_n denote the function agreeing with f on the interval $[-n, n]$ and equal to 0 otherwise. Each f_n is in $L^1(\mathbb{R}) \cap L^2(\mathbb{R})$ and the sequence $\{f_n\}$ converges in $L^2(\mathbb{R})$ to f by Lebesgue's Dominated Convergence Theorem [RUD, 1.34]. Thus $L^1(\mathbb{R}) \cap L^2(\mathbb{R})$ is dense in $L^2(\mathbb{R})$. But then Plancherel (4.10) allows us to uniquely extend the map $f \in L^1(\mathbb{R}) \cap L^2(\mathbb{R}) \mapsto \hat{f} \in L^2(\mathbb{R})$ to a linear isometry Λ on all of $L^2(\mathbb{R})$ as follows: given $f \in L^2(\mathbb{R})$, define $\Lambda(f)$ to be the limit in $L^2(\mathbb{R})$ of $\{\widehat{f_n}\}$ where $\{f_n\}$ is *any* sequence from $L^1(\mathbb{R}) \cap L^2(\mathbb{R})$ converging in $L^2(\mathbb{R})$ to f. Simple Cauchy sequence arguments show that this definition is proper and are left to the reader to supply. The map Λ is known as the L^2 *Fourier transform*. We will simply write "$\hat{f}$" instead of "$\Lambda(f)$" when "f" denotes a function from $L^2(\mathbb{R})$ and depend on context to figure out whether any given "$\hat{f}$" denotes an L^1 or L^2 Fourier transform. Note that $\hat{f}$ is defined pointwise everywhere on $\mathbb{R}$ when $f \in L^1(\mathbb{R})$ while $\hat{f}$ is only defined pointwise almost everywhere on $\mathbb{R}$ when $f \in L^2(\mathbb{R})$. The following two results gather together all that we need to know about the L^2 Fourier transform.

Proposition 4.11 *For* $f \in L^2(\mathbb{R})$ *and* $\alpha \in \mathbb{R}$, *set* $g(t) = f(t + \alpha)$. *Then* $\hat{g}(\xi) = e^{i\alpha}\hat{f}(\xi)$ *for almost every* $\xi \in \mathbb{R}$.

Proof Let $\{f_n\}$ be a sequence from $L^1(\mathbb{R}) \cap L^2(\mathbb{R})$ that converges in $L^2(\mathbb{R})$ to f. Then $\hat{f}$ is the limit in $L^2(\mathbb{R})$ of the sequence $\{\widehat{f_n}\}$. Set $g_n(t) = f_n(t + \alpha)$. Then $\{g_n\}$ is a sequence from $L^1(\mathbb{R}) \cap L^2(\mathbb{R})$ that converges in $L^2(\mathbb{R})$ to g, and so $\hat{g}$ is the limit in $L^2(\mathbb{R})$ of the sequence $\{\widehat{g_n}\}$. Without loss of generality, by dropping to a subsequence and relabeling, we may assume that the sequences $\{\widehat{f_n}\}$ and $\{\widehat{g_n}\}$ converge pointwise almost everywhere to $\hat{f}$ and $\hat{g}$ respectively [RUD, 3.12]. Involking Proposition 4.8, we then see that for almost every ξ

$$\hat{g}(\xi) = \lim_{n\to\infty} \widehat{g_n}(\xi) = \lim_{n\to\infty} i\xi \widehat{f_n}(\xi) = i\xi \hat{f}(\xi).$$

□

Proposition 4.12 *The function* $E(t) = \dfrac{e^{it} - 1}{it}$ *is in* $L^2(\mathbb{R})$ *and has Fourier transform*

$$\widehat{E} = \sqrt{2\pi}\,\mathcal{X}_{(0,1)}.$$

Proof That $E \in L^2(\mathbb{R})$ is a consequence of two facts: $|E(t)| \leq 2/|t|$ for all $t \neq 0$ and $E(t) \to 1$ as $t \to 0$.

Let E_n denote the function agreeing with E on the interval $[-n, n]$ and equal to 0 otherwise. Then $\widehat{E}$ is the limit in $L^2(\mathbb{R})$ of the sequence of functions $\{\widehat{E_n}\}$. Hence to finish off the proposition, it suffices by [RUD, 3.12] to show that

$$\lim_{n\to\infty} \int_{-n}^{n} e^{-i\xi t} \cdot \frac{e^{it}-1}{it} \frac{dt}{\sqrt{2\pi}} = \begin{cases} \sqrt{2\pi} & \text{when } 0 < \xi < 1 \\ 0 & \text{when } \xi < 0 \text{ or } \xi > 1 \end{cases}.$$

What follows is very similar to [RUD, 10.44]. Consider the real variable t in the integral above to be a complex variable z. Then by Cauchy's Integral Theorem [RUD, 10.35], the left-hand side of the above is unchanged when the straight-line path from $-n$ to n is replaced by the path Γ_n which comprises the straight-line path from $-n$ to -1, followed by the lower half of the unit circle at the origin traversed from -1 to 1, and finished by the straight-line path from 1 to n. Hence, with a bit of algebraic fiddling, it in turn suffices to show that

$$\lim_{n\to\infty} \frac{1}{2\pi i} \int_{\Gamma_n} e^{-i\xi z} \cdot \frac{e^{iz}-1}{z} dz = \begin{cases} 1 & \text{when } 0 < \xi < 1 \\ 0 & \text{when } \xi < 0 \text{ or } \xi > 1 \end{cases}.$$

Setting $\varphi_n(s) = \dfrac{1}{2\pi i} \displaystyle\int_{\Gamma_n} \frac{e^{isz}}{z}\, dz$, it finally suffices to show

$$(\star)\ \lim_{n\to\infty} \varphi_n(1-\xi) - \varphi_n(-\xi) = \begin{cases} 1 & \text{when } 0 < \xi < 1 \\ 0 & \text{when } \xi < 0 \text{ or } \xi > 1 \end{cases}.$$

Consider $s > 0$ and complete Γ_n to a cycle by the semicircular path $\gamma(\theta) = n\, e^{i\theta}$, $0 < \theta < \pi$. The function e^{isz}/z is analytic everywhere except $z = 0$ where it has a simple pole with residue 1. Hence by the Residue Theorem [RUD, 10.42],

$$\varphi_n(s) + \frac{1}{2\pi} \int_0^{\pi} \exp(isn\, e^{i\theta})\, d\theta = 1.$$

Now $|\exp(isn\, e^{i\theta})| = \exp(-sn \sin\theta)$ which is less than *1* and decreases to 0 as $n \to \infty$ when $s > 0$ and $0 < \theta < \pi$. Hence letting $n \to \infty$ in the above, by Lebesgue's Dominated Convergence Theorem [RUD, 1.34] we obtain

$$(\star\star)\ \lim_{n\to\infty} \varphi_n(s) = 1 \text{ when } s > 0.$$

For $s < 0$, a similar argument involving completing Γ_n to a cycle by the semicircular path $\gamma(\theta) = n\, e^{-i\theta}$, $0 < \theta < \pi$, shows that

$$(\star\star\star)\ \lim_{n\to\infty} \varphi_n(s) = 0 \text{ when } s < 0.$$

The relation $(\star)$ follows immediately from $(\star\star)$ and $(\star\star\star)$. □

4.4 Melnikov Curvature of Some Measures on Lipschitz Graphs

To operate with maximal generality for a paragraph, suppose that (X, d_X) and (Y, d_Y) are metric spaces, E is a subset of X, f is a function whose domain contains E and whose range is contained in Y, and M is a positive finite number. Then we say that f is a *Lipschitz function on E with bound M* when $d_Y(f(u), f(v)) \leq M\, d_X(u, v)$ for all $u, v \in E$. If we make no reference to the set E, then E is assumed to be the whole of the domain of f. Thus a Lipschitz function with bound M is one which is Lipschitz on its whole domain with bound M.

To move to the more specific context of interest to us now, suppose I is a subinterval of $\mathbb{R}$, M is a positive finite number, and $f : I \mapsto \mathbb{R}$ is a Lipschitz function on I with bound M. Then we call the set of points $\Gamma = \{t + if(t) : t \in I\}$ a *Lipschitz graph in standard position over I with bound M*. More generally, given a subinterval I of $\mathbb{C}$ and a positive finite number M, we shall call a set of points Γ in $\mathbb{C}$ a *Lipschitz graph over I with bound M* when there exists an isometry $\Phi : \mathbb{C} \mapsto \mathbb{C}$ such that $\Phi(\Gamma)$ is a Lipschitz graph in standard position over $\Phi(I) \subseteq \mathbb{R}$ with bound M. We will also abuse language somewhat and apply the terminology of this paragraph to the curve $\gamma(t) = t + if(t), t \in I$, whose graph is Γ.

In all these contexts, when we refer to an object as *Lipschitz* without reference to a bound M, what is meant is that the object is Lipschitz with bound M for some positive finite number M.

The following theorem is for our purposes the main result of [MV]. We will turn to its proof only after using it to obtain a corollary dealing with the analytic capacity of compact subsets of Lipschitz graphs that will later lead to the solution of the Denjoy Conjecture.

Theorem 4.13 (Melnikov Curvature on Lipschitz Graphs) *Suppose that Γ is a Lipschitz graph over a nondegenerate closed interval $[a, b]$ of $\mathbb{C}$ with bound M. Let $\pi : \mathbb{C} \mapsto L$ denote the orthogonal projection of the complex plane onto the line L containing $[a, b]$. Define a regular positive Borel measure μ on Γ by setting*

$$\mu(E) = \mathcal{L}^1(\pi(E \cap \Gamma))$$

for any Borel subset E of $\mathbb{C}$. Then

$$c^2(\mu) \leq 2{,}048\, M^2\, |b - a|.$$

Corollary 4.14 *In the situation of the previous theorem, suppose that K is a compact subset of Γ. Then*

$$\gamma(K) \geq \frac{m}{2} \cdot \frac{\mathcal{L}^1(\pi(K))^2}{\mathcal{L}^1(\pi(K)) + 512\, M^2\, |b - a|} \text{ and } \frac{m}{4} \cdot \sqrt{\frac{\mathcal{L}^1(\pi(K))^3}{\mathcal{L}^1(\pi(K)) + 512\, M^2\, |b - a|}}.$$

The number m here is as in Melnikov's Lower Capacity Estimate (4.3): $1/375{,}000$.

Proof Consider the measure $\tilde{\mu}$ on K which is just the measure μ of the theorem restricted to K. Note that $\tilde{\mu}(B(c;r)) \le \mu(B(c;r)) \le 2r$ for all $c \in \mathbb{C}$ and all $r > 0$, so $\tilde{\mu}$ has linear growth with bound $M = 2$. Also, $\tilde{\mu}(K) = \mu(K) = \mathcal{L}^1(\pi(K))$. Finally, by the theorem (not yet proven), $c^2(\tilde{\mu}) \le c^2(\mu) \le 2{,}048\, M^2\, |b-a|$. Thus Melnikov's Lower Capacity Estimate (4.3) and its Corollary (4.4) may be applied to $\tilde{\mu}$ and doing so gives us what we want. □

The following simple consequence of Proposition 4.7 will be useful here and elsewhere to follow.

Lemma 4.15 *Suppose that $g : I \mapsto \mathbb{C}$ is Lipschitz with bound M where I is a subinterval of $\mathbb{R}$. Then g' exists and is bounded in modulus by M $\mathcal{L}^1$-a.e. on I. Moreover, fixing any point $a \in I$, we have*

$$g(t) = g(a) + \int_a^t g'(u)\, du$$

for all $t \in I$.

Proof The Lipschitz condition on g trivially implies that g is absolutely continuous on $[a,b]$ (with $\delta = \varepsilon/M$ working in the definition of absolute continuity). The Lipschitz condition on g also implies that $|g'| \le M$ wherever g' exists. Proposition 4.7 now finishes off the proof. □

Proof of Theorem 4.13 By rotating and translating appropriately, we may assume that $\Gamma = \{t + if(t) : t \in [a,b]\}$ where f is Lipschitz on $[a,b] \subseteq \mathbb{R}$ with bound M. Setting

$$g(t) = \left[f(t) - \{f(a) + \frac{f(b)-f(a)}{b-a}(t-a)\} \right] \mathcal{X}_{[a,b]}(t),$$

we get a function defined on all of $\mathbb{R}$ and supported on $[a,b]$. For all $t, u \in [a,b]$, one has

$$g(u) - g(t) = f(u) - f(t) + \frac{f(b)-f(a)}{b-a}(u-t).$$

From this, with a little work one can show two things. First, $|g(u) - g(t)| \le 2M|u-t|$ for all $t, u \in \mathbb{R}$, i.e., g is Lipschitz on $\mathbb{R}$ with bound $2M$. Hence, from the lemma and the fact that $g = 0$ off $[a,b]$, one has

$$(\star)\ |g'| \le 2M \text{ a.e. on } [a,b] \text{ and } |g'| = 0 \text{ off } [a,b].$$

Second,

$$(\star\star)\ \frac{g(u)-g(t)}{u-t} - \frac{g(v)-g(t)}{v-t} = \frac{f(u)-f(t)}{u-t} - \frac{f(v)-f(t)}{v-t} \text{ for all}$$
$t, u, v \in [a,b]$ with $t \ne u, v$.

Let $z_j = x_j + i\,y_j$ for $j = 1, 2, 3$. Using the first expression for curvature involving area from Proposition 4.1 and recalling that the area of the triangle determined by two vectors is just half of the size of their cross product, we see that

$$
\begin{aligned}
(\star\star\star)\ c(z_1, z_2, z_3) &= \frac{4\,\text{Area of Triangle}\,(z_1, z_2, z_3)}{|z_2 - z_1||z_3 - z_1||z_2 - z_3|} \\
&= \frac{2|(x_2 - x_1)(y_3 - y_1) - (x_3 - x_1)(y_2 - y_1)|}{|z_2 - z_1||z_3 - z_1||z_2 - z_3|} \\
&\le \frac{2|(x_3 - x_1)(y_2 - y_1) - (x_2 - x_1)(y_3 - y_1)|}{|x_2 - x_1||x_3 - x_1||x_2 - x_3|} \\
&= 2\left|\left\{\frac{y_2 - y_1}{x_2 - x_1} - \frac{y_3 - y_1}{x_3 - x_1}\right\}\Big/\ \{x_2 - x_3\}\right|.
\end{aligned}
$$

In what follows we will have a number of chains of in/equalities. Immediately after each chain, a justification will be given for each of its lines.

$$
\begin{aligned}
(\star\star\star\star)\ c^2(\mu) &\le 4\int_a^b\int_a^b\int_a^b \left|\left\{\frac{f(u) - f(t)}{u - t} - \frac{f(v) - f(t)}{v - t}\right\}\Big/\ \{u - v\}\right|^2 dt\,du\,dv \\
&\le 4\int_{-\infty}^{\infty}\int_{-\infty}^{\infty}\int_{-\infty}^{\infty} \left|\left\{\frac{g(u) - g(t)}{u - t} - \frac{g(v) - g(t)}{v - t}\right\}\Big/\ \{u - v\}\right|^2 dt\,du\,dv \\
&= 4\int_{-\infty}^{\infty}\int_{-\infty}^{\infty}\int_{-\infty}^{\infty} \left|\left\{\frac{g(t + h) - g(t)}{h} - \frac{g(t + k) - g(t)}{k}\right\}\Big/\ \{h - k\}\right|^2 dh\,dk\,dt \\
&= 4\int_{-\infty}^{\infty}\int_{-\infty}^{\infty}\int_{-\infty}^{\infty} \left|\left\{\frac{e^{i\xi h} - 1}{h} - \frac{e^{i\xi k} - 1}{k}\right\}\Big/\ \{h - k\}\right|^2 |\hat{g}(\xi)|^2\,d\xi\,dh\,dk \\
&= 4\int_{-\infty}^{\infty}\int_{-\infty}^{\infty}\int_{-\infty}^{\infty} \left|\left\{\frac{e^{iu} - 1}{iu} - \frac{e^{iv} - 1}{iv}\right\}\Big/\ \{u - v\}\right|^2 |i\xi\hat{g}(\xi)|^2\,du\,dv\,d\xi \\
&= 4\left\{\int_{-\infty}^{\infty}\int_{-\infty}^{\infty} \left|\frac{E(u) - E(v)}{u - v}\right|^2 du\,dv\right\}\left\{\int_{-\infty}^{\infty} |i\xi\hat{g}(\xi)|^2\,d\xi\right\}.
\end{aligned}
$$

[The first line follows from the definition of curvature, the definition of μ, and $(\star\star\star)$; the second from $(\star\star)$; the third from Fubini's Theorem [RUD, 8.8] to make the dt outermost and the changes of variables $u = t + h$ and $v = t + k$; the fourth from Fubini's Theorem [RUD, 8.8] to make the dt innermost, the isometricality of the L^2 Fourier transform applied to the resulting innermost integral, and Proposition 4.11; the fifth from Fubini's Theorem [RUD, 8.8] to make the $d\xi$ outermost and the changes of variables $u = \xi h$ and $v = \xi k$; and the last from the definition of E and a bit of rearranging.]

We now turn to estimating the double integral in the last line of $(\star\star\star\star)$:

$$\int_{-\infty}^{\infty}\int_{-\infty}^{\infty}\left|\frac{E(u)-E(v)}{u-v}\right|^2 du\,dv = \int_{-\infty}^{\infty}\int_{-\infty}^{\infty}\left|\frac{E(v+t)-E(v)}{t}\right|^2 dt\,dv$$
$$= \int_{-\infty}^{\infty}\int_{-\infty}^{\infty}\left|\frac{e^{i\xi t}-1}{t}\right|^2 |\hat{E}(\xi)|^2\, d\xi\, dt$$
$$= 2\pi\int_{-\infty}^{\infty}\int_{0}^{1}\left|\frac{e^{i\xi t}-1}{t}\right|^2 d\xi\, dt.$$

[The first line follows from the change of variables $u = v + t$; the second from Fubini's Theorem [RUD, 8.8] to make the dv innermost, the isometricality of the L^2 Fourier transform applied to the resulting innermost integral, and Proposition 4.11; and the last from Proposition 4.12.] For $|t| \le 1$ and $0 < \xi < 1$, we have

$$\left|\frac{e^{i\xi t}-1}{t}\right| = \left|\sum_{n=1}^{\infty}\frac{(i\xi)^n t^{n-1}}{n!}\right| \le \sum_{n=1}^{\infty}\frac{|(i\xi)^n t^{n-1}|}{n!} \le \sum_{n=1}^{\infty}\frac{1}{n!} = e-1 \le 2.$$

Also, $|e^{i\xi t}-1| \le 2$ always. Hence

$$(\dagger)\quad \int_{-\infty}^{\infty}\int_{-\infty}^{\infty}\left|\frac{E(u)-E(v)}{u-v}\right|^2 du\,dv \le$$
$$2\pi\left\{\int_{|t|\le 1}\int_0^1 4\, d\xi\, dt + \int_{|t|>1}\int_0^1 \frac{4}{t^2}\, d\xi\, dt\right\} = 32\pi \le 128.$$

Next we turn to estimating the single integral in the last line of $(\star\star\star\star)$:

$$(\ddagger)\quad \int_{-\infty}^{\infty} |i\xi\hat{g}(\xi)|^2\, d\xi = \int_{-\infty}^{\infty} |\widehat{(g')}(\xi)|^2\, d\xi = \int_{-\infty}^{\infty} |g'(t)|^2\, dt$$
$$\le \int_a^b (2M)^2\, dt = 4M^2(b-a).$$

[The first equality follows from Proposition 4.9 since g is compactly supported and Lipschitz on all of $\mathbb{R}$; the second from the isometricality of the L^2 Fourier transform; the only inequality from $(\star)$; while the last equality is trivial.]

Using $(\dagger)$ and $(\ddagger)$ in $(\star\star\star\star)$, we obtain $c^2(\mu) \le 2{,}048M^2(b-a)$. □

4.5 Arclength and Arclength Measure: Enough to Do the Job

In this section we define arclength measure on curves and elucidate its relation to linear Hausdorff measure enough to be able to dispose of the Denjoy Conjecture in the next section. A complete elucidation of the relation between the two notions is not attempted here since that requires us to know more of linear Hausdorff measure than just Proposition 2.1; we must also know that it is indeed a positive measure defined on a σ-algebra containing the Borel subsets of $\mathbb{C}$! Since this topic will be taken up only in the next chapter, the conclusion of the story begun here must wait till then. The section closes with a result enabling us to deal with the Lipschitz graphs introduced in the previous section.

Recall that a *curve* in $\mathbb{C}$ is simply a continuous function $\gamma : [a,b] \mapsto \mathbb{C}$. (The use of γ to denote a curve is a time-honored tradition. Since we will have no occasion to speak of analytic capacity in this section, no confusion should thereby result.) The *length* of the curve γ is by definition

$$l(\gamma) = \sup \sum_{j=1}^{n} |\gamma(t_j) - \gamma(t_{j-1})|$$

where the supremum is taken over all partitions $a = t_0 \leq t_1 \leq t_2 \leq \cdots \leq t_n = b$ of $[a,b]$ (here n is finite but arbitrary). Consideration of the trivial two-point partition $a = t_0 \leq t_1 = b$ yields the Shortest Path Property:

$$|\gamma(a) - \gamma(b)| \leq l(\gamma).$$

By the Triangle Inequality, the addition of a point to a partition can only increase the sum occurring in the definition of γ's length. From this one easily gets the Subarc Additivity Property:

$$l(\gamma) = l(\gamma|[a,c]) + l(\gamma|[c,b])$$

for any $c \in [a,b]$ (here $\gamma|[a,c]$ and $\gamma|[c,b]$ simply denote the restrictions of the function γ to the subintervals $[a,c]$ and $[c,b]$, respectively). These two properties of arclength will frequently be used without mention. Finally, we say that γ is *rectifiable* if and only if $l(\gamma) < \infty$.

Let us immediately clear up a point concerning length and complex integration theory that may have occurred to the reader already. Recall that a *path* is a piecewise continuously differentiable curve in $\mathbb{C}$ (if needed, see [RUD, 10.8] for more explication of this). Given a path $\gamma : [a,b] \mapsto \mathbb{C}$ and a function f that is continuous on $\gamma([a,b])$, by definition

$$\int_{\gamma} f(z)\, dz = \int_{a}^{b} f(\gamma(t))\gamma'(t)\, dt.$$

From this definition it trivially follows that if $|f| \leq M$ on $\gamma([a,b])$, then

$$\left| \int_\gamma f(z)\, dz \right| \leq M \int_a^b |\gamma'(t)|\, dt.$$

Of course one wants to rewrite the right-hand side of this inequality as $M\, l(\gamma)$, and for line segments and arcs of circles parametrized in the usual way the integral of the absolute value of the derivative of the parametrization does indeed end up being the expected length from elementary geometry. In many texts, e.g., [RUD, just before 10.9], this is made to hold without exception by *defining* the length of a path to be the integral of the absolute value of the derivative of the parametrization. We do not have this luxury, having already defined length in a different manner, and so we must prove the desired relation instead of making it into a definition. Hence the following.

Proposition 4.16 *If* $\gamma : [a,b] \mapsto \mathbb{C}$ *is a path, then* γ *is rectifiable and*

$$l(\gamma) = \int_a^b |\gamma'(t)|\, dt.$$

Proof Without loss of generality, we may assume that γ is continuously differentiable on $[a,b]$. Let $a = t_0 \leq t_1 \leq t_2 \leq \cdots \leq t_n = b$ be any partition of $[a,b]$. Then, by the Fundamental Theorem of Calculus,

$$\sum_{j=1}^n |\gamma(t_j) - \gamma(t_{j-1})| = \sum_{j=1}^n \left| \int_{t_{j-1}}^{t_j} \gamma'(t)\, dt \right| \leq \sum_{j=1}^n \int_{t_{j-1}}^{t_j} |\gamma'(t)|\, dt = \int_a^b |\gamma'(t)|\, dt.$$

Suping over all partitions, we see that $l(\gamma) \leq \int_a^b |\gamma'(t)|\, dt$ and so γ is rectifiable.

Since γ is continuously differentiable on the compact set $[a,b]$, γ' is uniformly continuous there. Thus given $\varepsilon > 0$, there exists a partition $a = t_0 \leq t_1 \leq t_2 \leq \cdots \leq t_n = b$ of $[a,b]$ such that $|\gamma'(t_j) - \gamma'(t)| < \varepsilon$ for any j and any $t \in [t_{j-1}, t_j]$. Hence

$$\begin{aligned}
\int_{t_{j-1}}^{t_j} |\gamma'(t)|\, dt &\leq \{|\gamma'(t_j)| + \varepsilon\}(t_j - t_{j-1}) \\
&= \left| \int_{t_{j-1}}^{t_j} \gamma'(t_j)\, dt \right| + \varepsilon(t_j - t_{j-1}) \\
&\leq \left| \int_{t_{j-1}}^{t_j} \gamma'(t)\, dt \right| + \left| \int_{t_{j-1}}^{t_j} \gamma'(t_j) - \gamma'(t)\, dt \right| + \varepsilon(t_j - t_{j-1}) \\
&\leq |\gamma(t_j) - \gamma(t_{j-1})| + 2\varepsilon(t_j - t_{j-1})
\end{aligned}$$

with the last inequality again using the Fundamental Theorem of Calculus. Summing over all j, we see that

$$\int_a^b |\gamma'(t)|\,dt \leq \sum_{j=1}^n |\gamma(t_j) - \gamma(t_{j-1})| + 2\varepsilon(b-a) \leq l(\gamma) + 2\varepsilon(b-a).$$

Letting $\varepsilon \downarrow 0$, we obtain $\int_a^b |\gamma'(t)|\,dt \leq l(\gamma)$ and so are done. □

Let us now turn to a relation between the length of a curve and the linear Hausdorff measure of its range.

Proposition 4.17 *For $\gamma : [a,b] \mapsto \mathbb{C}$ a curve, set $\Gamma = \gamma([a,b])$. Then*

$$\mathcal{H}^1(\Gamma) \leq l(\gamma).$$

Proof Since γ is uniformly continuous on $[a,b]$, given $\delta > 0$, there exists a partition $a = t_0 \leq t_1 \leq t_2 \leq \cdots \leq t_n = b$ of $[a,b]$ such that $|\gamma([t_{j-1},t_j])| < \delta$ for each j. For each j, choose $u_j, v_j \in [t_{j-1},t_j]$ with $u_j \leq v_j$ such that $|\gamma([t_{j-1},t_j])| = |\gamma(u_j) - \gamma(v_j)|$. Then from the Definition of $\mathcal{H}^1_\delta$, the Shortest Path Property, and the Subarc Additivity Property, it follows that

$$\mathcal{H}^1_\delta(\Gamma) \leq \sum_{j=1}^n |\gamma([t_{j-1},t_j])| = \sum_{j=1}^n |\gamma(u_j) - \gamma(v_j)| \leq \sum_{j=1}^n l(\gamma|[u_j,v_j]) \leq l(\gamma).$$

Letting $\delta \downarrow 0$, we are done. □

One cannot expect equality in the last result to hold without exception for the obvious reason: a curve can double back upon itself any number of times resulting in some portions of the curve contributing more than once to the total length. However, for one-to-one curves all is well and equality does obtain. For a general curve one must take into account multiplicity. All this must wait till the next chapter however! For the record we here define an *arc* to be a one-to-one curve.

Having introduced the notion of the length of a curve, we now consider how to define the arclength of an arbitrary subset of a rectifiable curve. This is most conveniently approached via the notion of arclength parametrization. We say that a rectifiable curve $\tilde{\gamma} : [0,l] \mapsto \mathbb{C}$ is *parametrized by arclength* if $l(\tilde{\gamma}|[0,s]) = s$ for all $s \in [0,l]$. (When this happens we must trivially have $l(\tilde{\gamma}) = l$.) We further call this $\tilde{\gamma}$ an *arclength parametrization* of another rectifiable curve $\gamma : [a,b] \mapsto \mathbb{C}$ if, in addition, $\tilde{\gamma}$ *reparametrizes* γ, i.e., if, in addition, there exists a continuous and increasing function $\varphi : [a,b] \mapsto [0,l]$ such that $\tilde{\gamma} \circ \varphi = \gamma$. (When this happens we must have $l = l(\gamma)$ by Lemma 4.19 below.) We now aim to show that every rectifiable curve has a *unique* arclength parametrization. As usual, some preliminary lemmas are necessary.

Lemma 4.18 *Let $\gamma : [a,b] \mapsto \mathbb{C}$ be a rectifiable curve with length l. Then the function $\varphi : [a,b] \mapsto [0,l]$ defined by $t \mapsto l(\gamma|[a,t])$ is continuous and increasing.*

Proof By the Subarc Additivity Property, for $a \le u \le v \le b$ we have $\varphi(v) = \varphi(u) + l(\gamma|[u, v]) \ge \varphi(u)$. Thus φ is increasing. Note that we have not tried to show that φ is *strictly* increasing. Indeed, any attempt to do so is doomed to failure since φ is constant on any open subinterval of $[a, b]$ where γ may happen to be constant!

We show right continuity of φ by contradiction (left continuity is similar). Given $a \le u < b$, if φ is not right continuous at u, then there exists an $\varepsilon > 0$ such that $l(\gamma|[u, v]) > \varepsilon$ for all v in a subset of $(u, b]$ which accumulates at u. Since φ is increasing, this may be immediately be strengthened to $l(\gamma|[u, v]) > \varepsilon$ for all $v \in (u, b]$.

We inductively construct a decreasing sequence $\{u_m\}$ in $(u, b]$ such that $l(\gamma|[u_{m+1}, u_m]) > \varepsilon/2$. Start by setting $u_0 = b$. Assuming u_m defined, since $l(\gamma|[u, u_m]) > \varepsilon$, there exists a partition $u = t_0 \le t_1 \le t_2 \le \cdots \le t_n = u_m$ of $[u, u_m]$ such that

$$(\star) \quad \sum_{j=1}^{n} |\gamma(t_j) - \gamma(t_{j-1})| > \varepsilon.$$

Since γ is continuous at $t_0 = u$, there exists an $u_{m+1} \in (u, t_1)$ such that $(\star\star)$ $|\gamma(u_{m+1}) - \gamma(t_0)| < \varepsilon/2$. But then consideration of the partition $u_{m+1} \le t_1 \le t_2 \le \cdots \le t_n = u_m$ of $[u_{m+1}, u_m]$, a use of the Triangle Inequality, and use of $(\star)$ as well as $(\star\star)$ show that

$$\begin{aligned} l(\gamma|[u_{m+1}, u_m]) &\ge |\gamma(t_1) - \gamma(u_{m+1})| + \sum_{j=2}^{n} |\gamma(t_j) - \gamma(t_{j-1})| \\ &= \{\, |\gamma(t_1) - \gamma(u_{m+1})| + |\gamma(u_{m+1}) - \gamma(t_0)| - |\gamma(t_1) - \gamma(t_0)| \,\} \\ &\quad - |\gamma(u_{m+1}) - \gamma(t_0)| + \sum_{j=1}^{n} |\gamma(t_j) - \gamma(t_{j-1})| \\ &> 0 - \varepsilon/2 + \varepsilon = \varepsilon/2. \end{aligned}$$

This clearly implies that $l(\gamma|[u, b])$ is arbitrarily large, and so infinite. With this contradiction to rectifiability, the proof is finished. □

Lemma 4.19 *Suppose $\varphi : [a, b] \mapsto [0, l]$ is a continuous increasing function and $\tilde{\gamma} : [0, l] \mapsto \mathbb{C}$ is a curve. Then for any $s \in [0, l]$ and any $t \in [a, b]$ such that $\varphi(t) = s$, it is the case that*

$$l(\tilde{\gamma}|[0, s]) = l((\tilde{\gamma} \circ \varphi)|[a, t]).$$

Proof On the one hand, if $a = t_0 \le t_1 \le t_2 \le \cdots \le t_n = t$ is a partition of $[a, t]$, then $0 = \varphi(t_0) \le \varphi(t_1) \le \varphi(t_2) \le \cdots \le \varphi(t_n) = s$ is a partition of $[0, s]$. On the other hand, if $0 = s_0 \le s_1 \le s_2 \le \cdots \le s_n = s$ is a partition of $[0, s]$, then one can find a partition $a = t_0 \le t_1 \le t_2 \le \cdots \le t_n = t$ of $[a, t]$ such that $\varphi(t_j) = s_j$

for each j. What is to be proved is now clear from the way the length of a curve has been defined. □

Proposition 4.20 *Every rectifiable curve has a unique arclength parametrization.*

Proof Existence. Given $\gamma : [a,b] \mapsto \mathbb{C}$ rectifiable with length l, let $\varphi : [a,b] \mapsto [0,l]$ be the function of Lemma 4.18, so $\varphi(t) = l(\gamma|[a,t])$. Given $s \in [0,l]$, define $\tilde{\gamma}(s)$ to be $\gamma(t)$ where t is the smallest number in $[a,b]$ such that $\varphi(t) = s$. By Lemma 4.18 and the Intermediate Value Theorem, $\tilde{\gamma} : [0,l] \mapsto \mathbb{C}$ is well defined. We must show that $\tilde{\gamma}$ is a curve, that $\tilde{\gamma}$ reparametrizes γ, and that $\tilde{\gamma}$ is parametrized by arclength.

Given $s_1, s_2 \in [0,l]$, without loss of generality $s_1 < s_2$. Let $t_1, t_2 \in [a,b]$ be the smallest numbers such that $\varphi(t_1) = s_1$ and $\varphi(t_2) = s_2$. Then $t_1 < t_2$. Hence

$$|\tilde{\gamma}(s_2)-\tilde{\gamma}(s_1)| = |\gamma(t_2)-\gamma(t_1)| \le l(\gamma|[t_1,t_2]) = \varphi(t_2)-\varphi(t_1) = s_2-s_1 = |s_2-s_1|.$$

We conclude that $\tilde{\gamma}$ is continuous and so a curve. As a matter of fact, $\tilde{\gamma}$ is absolutely continuous!

Given $u \in [a,b]$, let t be the smallest number in $[a,b]$ such that $\varphi(t) = \varphi(u)$. Then by the definition of $\tilde{\gamma}$, we have $\tilde{\gamma}(\varphi(u)) = \gamma(t)$. However, we also have $t \le u$ and $l(\gamma|[t,u]) = \varphi(u) - \varphi(t) = 0$. The only way this can happen is for γ to be constant on the interval $[t,u]$. In particular, $\gamma(t) = \gamma(u)$. We conclude that $\tilde{\gamma}(\varphi(u)) = \gamma(u)$, i.e., $\tilde{\gamma} \circ \varphi = \gamma$ and so $\tilde{\gamma}$ reparametrizes γ.

Given $s \in [0,l]$, let t be the smallest number in $[a,b]$ such that $\varphi(t) = s$. Then by Lemma 4.19, the fact that $\tilde{\gamma}$ reparametrizes γ via φ, and the definition of φ,

$$l(\tilde{\gamma}|[0,s]) = l((\tilde{\gamma} \circ \varphi)|[a,t]) = l(\gamma|[a,t]) = \varphi(t) = s.$$

We conclude that $\tilde{\gamma}$ is parametrized by arclength.

Uniqueness. Given a curve $\tilde{\gamma} : [0,l] \mapsto \mathbb{C}$ parametrized by arclength and a function $\varphi : [a,b] \mapsto [0,l]$ showing that $\tilde{\gamma}$ reparametrizes $\gamma : [a,b] \mapsto \mathbb{C}$, we will show that $\tilde{\gamma}$ and φ are uniquely determined by γ.

The following shows that φ is uniquely determined by γ:

$$\varphi(t) = l(\tilde{\gamma}|[0,\varphi(t)]) = l((\tilde{\gamma} \circ \varphi)|[a,t]) = l(\gamma|[a,t]).$$

The first equality above holds since $\tilde{\gamma}$ is parametrized by arclength; the second by Lemma 4.19; and the third since $\tilde{\gamma}$ is a reparametrization of γ.

Because $\tilde{\gamma}(\varphi(t)) = \gamma(t)$ and the domain of $\tilde{\gamma}$ is covered by the range of φ, we now see that $\tilde{\gamma}$ is also uniquely determined by γ. □

Let $\tilde{\gamma}$ denote the arclength parametrization of a rectifiable curve γ of length l. For a subset E of the range of γ, the *arclength measure* of E with respect to γ is defined by

$$l_\gamma(E) = \mathcal{L}^1(\tilde{\gamma}^{-1}(E)).$$

This is motivated by the fact that for all $0 \le s \le t \le l$, we have $l(\tilde{\gamma}|[s,t]) = t - s = \mathcal{L}^1([s,t])$, i.e., the map $[s,t] \mapsto \tilde{\gamma}|[s,t]$ from all subintervals of $[0,l]$ to all subarcs of $\tilde{\gamma}$ preserves lengths. However, note that as we have defined it, $l_\gamma(\tilde{\gamma}([s,t])) = \mathcal{L}^1(\tilde{\gamma}^{-1}(\tilde{\gamma}([s,t])))$, which may be *strictly* greater than $\mathcal{L}^1([s,t])$ since $\tilde{\gamma}^{-1}(\tilde{\gamma}([s,t]))$ may be a *proper* superset of $[s,t]$ due to γ covering parts of itself more than once. Thus our l_γ, which assigns "length" to subsets of $\gamma([a,b])$ as opposed to subarcs of γ, takes into account multiplicity.

Proposition 4.21 *Let E be a subset of the range of a rectifiable curve γ. Then*

$$\mathcal{H}^1(E) \le l_\gamma(E).$$

Proof If $\tilde{\gamma}$ is the arclength parametrization of γ, then from the Shortest Path Property we have

$$|\tilde{\gamma}(s_2) - \tilde{\gamma}(s_1)| \le l(\tilde{\gamma}|[s_1,s_2]) = |s_2 - s_1|.$$

Thus by Proposition 2.2 and the coincidence of $\mathcal{H}^1$ and $\mathcal{L}^1$ for linear sets (from Section 2.1), we get

$$\mathcal{H}^1(E) = \mathcal{H}^1(\tilde{\gamma}(\tilde{\gamma}^{-1}(E))) \le \mathcal{H}^1(\tilde{\gamma}^{-1}(E)) = \mathcal{L}^1(\tilde{\gamma}^{-1}(E)) = l_\gamma(E).$$

□

The next result generalizes Proposition 4.16.

Proposition 4.22 *If $\gamma : [a,b] \mapsto \mathbb{C}$ is absolutely continuous, then γ is rectifiable and*

$$l(\gamma) = \int_a^b |\gamma'(t)|\, dt.$$

Proof That $l(\gamma) \le \int_a^b |\gamma'(t)|\, dt$ and that γ is thus rectifiable follows as in the first paragraph of the proof of Proposition 4.16 with the appeal to the Fundamental Theorem of Calculus there replaced by an appeal to Proposition 4.7 here.

Let $\varphi : [a,b] \mapsto [0, l(\gamma)]$ be the function of Lemma 4.18, so $\varphi(t) = l(\gamma|[a,t])$. First note that from the definition of arclength one easily sees that a $\delta > 0$ that works for an $\varepsilon > 0$ to show γ absolutely continuous will also work for the same $\varepsilon > 0$ to show φ absolutely continuous. So Proposition 4.7 applies to φ as well as γ. Second note that from the definition of the derivative and the Shortest Path Property one easily has $|\gamma'(t)| \le \varphi'(t)$ whenever both derivatives exist, which is $\mathcal{L}^1$-a.e. Hence we get the rest of what we need:

$$\int_a^b |\gamma'(t)|\, dt \le \int_a^b \varphi'(t)\, dt = \varphi(b) - \varphi(a) = l(\gamma).$$

□

Given a positive finite number M, a *Lipschitz curve with bound M* is simply a curve that is a Lipschitz function with bound M, i.e., it is simply a curve $\gamma : [a, b] \mapsto \mathbb{C}$ such that $|\gamma(t) - \gamma(u)| \leq M|t - u|$ for all $t, u \in [a, b]$. A Lipschitz curve with bound M is trivially absolutely continuous (with $\delta = \varepsilon/M$ working in the definition of absolute continuity). Note that any Lipschitz graph with bound M, as defined at the beginning of the previous section, is a Lipschitz curve with bound $\sqrt{1 + M^2}$, as just defined. Thus we immediately obtain the following.

Corollary 4.23 *If $\gamma : [a, b] \mapsto \mathbb{C}$ is a Lipshitz curve or Lipschitz graph, then γ is rectifiable and*

$$l(\gamma) = \int_a^b |\gamma'(t)|\, dt.$$

4.6 The Denjoy Conjecture Resolved Affirmatively

In this section confusion will result if we persist in using "γ" to denote a typical curve: Is "$\gamma(E)$" the image of E under the curve γ or is it simply the analytic capacity of E? To avoid this confusion we will use "f" instead to denote a typical curve.

Lemma 4.24 *Let $f : [0, l] \mapsto \mathbb{C}$ be a rectifiable curve parametrized by arclength and suppose that $d = |f(l) - f(0)| > 0$. Let π denote the orthogonal projection of the complex plane onto the line through $f(0)$ and $f(l)$. Then given any finite number $M > l/d$, there exists a nonempty compact subset E of $[0, l]$ such that*

(a) *$f(E)$ lies on a Lipschitz graph with bound M over some interval contained in the line through $f(0)$ and $f(l)$,*
(b) *$\mathcal{L}^1(\pi(f(\widetilde{E}))) \geq \dfrac{1}{M}\mathcal{L}^1(\widetilde{E})$ for any $\mathcal{L}^1$-measurable subset $\widetilde{E}$ of E, and*
(c) *$\mathcal{L}^1(E) \geq \dfrac{M}{M+1}\left\{d - \dfrac{l}{M}\right\}$.*

Proof By translating and rotating appropriately, we may assume that $f(0) = 0$ and $f(l) = d$. Then our projection π is onto the real axis and so, writing $f(t)$ in terms of its real and imaginary parts as $x(t) + iy(t)$, we have $\pi(f(t)) = x(t)$. The nonempty, compact subset E of $[0, l]$ that we seek is easily defined:

$$E = \{t \in [0, l] : x(t) - x(s) \geq \frac{1}{M}(t - s) \text{ for all } s \in [0, t]\}.$$

(a) From a trivial inequality, the Shortest Path Property, the parametrization of our curve by arclength, and the definition of E, we see that for any $s, t \in E$ $(s \leq t)$, one has

$$|y(t) - y(s)| \leq |f(t) - f(s)| \leq l(f|[s, t]) = |t - s| \leq M|x(t) - x(s)|.$$

This implies that $K = f(E)$, clearly nonempty and compact, is also a *Lipschitz set with bound* M, i.e., a set for which one has $|y_2 - y_1| \leq M|x_2 - x_1|$ whenever $x_1 + iy_1$ and $x_2 + iy_2$ are points of the set. By a *Lipschitz set* we simply mean a set which is Lipschitz for some positive finite bound M.

Now *any* nonempty, compact Lipschitz set K is contained in a Lipschitz graph in standard position with same bound. To see this, note that given any $x \in \operatorname{Re} K$, by the Lipschitz-set property there exists only one number y such that $x + iy \in K$. Denote this unique number y by $g(x)$. Set $a = \min \operatorname{Re} K$ and $b = \max \operatorname{Re} K$. Without loss of generality, $[a, b]$ is nondegenerate, i.e., $a < b$ (otherwise K consists of a singleton for which our assertion is trivially true). Now $[a, b] \setminus \operatorname{Re} K$, being open, is the union of a countable collection $\{(a_j, b_j)\}$ of pairwise disjoint open intervals. Extend g to all of $[a, b]$ by defining it on each (a_j, b_j) to be linear on $[a_j, b_j]$. Using the Lipschitz-set property, one can easily verify that the function g so defined is Lipschitz on $[a, b]$ with the same bound and so determines a Lipschitz graph in standard position with the same bound containing K.

(b) By the way E was defined, $|x(t) - x(s)| \geq |t - s|/M$ for any $s, t \in E$. Thus the function $x|E : E \mapsto x(E)$ has an inverse $(x|E)^{-1} : x(E) \mapsto E$ satisfying

$$|(x|E)^{-1}(x_2) - (x|E)^{-1}(x_1)| \leq M|x_2 - x_1|$$

for all $x_1, x_2 \in x(E)$. Hence, by Proposition 2.2 and the coincidence of $\mathcal{H}^1$ and $\mathcal{L}^1$ for linear sets (from Section 2.1), for any $\mathcal{L}^1$-measurable $\widetilde{E} \subseteq E$, we have

$$\begin{aligned}\mathcal{L}^1(\widetilde{E}) &= \mathcal{L}^1((x|E)^{-1}(x(\widetilde{E}))) = \mathcal{H}^1((x|E)^{-1}(x(\widetilde{E}))) \\ &\leq M\mathcal{H}^1(x(\widetilde{E})) = M\mathcal{L}^1(x(\widetilde{E})).\end{aligned}$$

This is what we want since $\pi(f(\widetilde{E})))$ is just $x(\widetilde{E})$.

(c) Set $\varphi(t) = x(t) - t/M$ and note that $t \in E \Leftrightarrow \varphi(t) \geq \varphi(s)$ for all $s \in [0, t] \Leftrightarrow \varphi(t) = \max_{[0,t]} \varphi$. Since E is compact, $(0, l) \setminus E$ is an open set and so we may write

$$(\star)\ (0, l) \setminus E = \bigcup_j (c_j, d_j)$$

for a countable collection $\{(c_j, d_j)\}$ of pairwise disjoint open intervals.

First Claim. $\varphi(c_j) \geq \varphi(d_j)$ for each j. (Indeed, one actually has $\varphi(c_j) = \varphi(d_j)$ unless $d_j = l$, but we do not need that much!)

Consider $t \in (c_j, d_j)$. Then $\varphi(t) < \max_{[0,t]} \varphi$. Choose $s \in [0, t]$ such that $\varphi(s) = \max_{[0,t]} \varphi$. Then $\varphi(s) = \max_{[0,s]} \varphi$ too and so $s \in E$. Hence $s \in [0, t] \setminus (c_j, t] = [0, c_j]$. But now $\varphi(c_j) = \max_{[0,c_j]} \varphi \geq \varphi(s) > \varphi(t)$. Letting $t \uparrow d_j$, we have $\varphi(c_j) \geq \varphi(d_j)$.

Second Claim. φ is Lipschitz on $[0, l]$ with bound $(M + 1)/M$.

From a trivial inequality, the Shortest Path Property, and the parametrization of our curve by arclength, we see that for any $s, t \in [0, l]$ $(s \leq t)$, one has

$$|x(t) - x(s)| \leq |f(t) - f(s)| \leq l(f|[s,t]) \leq |t - s|.$$

Hence, by the way φ was defined, for any $s, t \in [0, l]$, one has

$$|\varphi(t) - \varphi(s)| \leq |x(t) - x(s)| + \frac{1}{M}|t - s| \leq \frac{M+1}{M}|t - s|.$$

From our two Claims, Lemma 4.15, and $(\star)$, we now deduce that

$$\begin{aligned} d - \frac{l}{M} = \varphi(l) - \varphi(0) &= \int_0^l \varphi'(t)\,dt = \int_E \varphi'(t)\,dt + \sum_j \int_{c_j}^{d_j} \varphi'(t)\,dt \\ &\leq \int_E \frac{M+1}{M}\,dt + \sum_j \{\varphi(d_j) - \varphi(c_j)\} \leq \frac{M+1}{M}\mathcal{L}^1(E) + 0 \\ &= \frac{M+1}{M}\mathcal{L}^1(E). \end{aligned}$$

□

Although in the preface we stated Denjoy's and Vitushkin's Conjectures in terms of removability, it will now prove convenient to state them in terms of nonremovability. Indeed, we shall be cavalier in the future about whether these conjectures deal with removability or nonremovability!

Theorem 4.25 (Denjoy's Conjecture Resolved: Arclength-Measure Version) *Suppose K is a compact subset of a rectifiable curve f. Then K is nonremovable if and only if $l_f(K) > 0$.*

Proof The contrapositive of the forward implication follows from Proposition 4.21 and Painlevé's Theorem (2.7).

Turn now to the backward implication which is the meat of the result. Without loss of generality, let $f : [0, l] \mapsto \mathbb{C}$ be parametrized by arclength. Then $\mathcal{L}^1(f^{-1}(K)) > 0$ by assumption. Find a partition $0 = t_0 \leq t_1 \leq t_2 \leq \cdots \leq t_n = l$ of $[0, l]$ such that

$$\sum_{j=1}^{n} |f(t_j) - f(t_{j-1})| > l - \mathcal{L}^1(f^{-1}(K)).$$

Setting $l_j = t_j - t_{j-1}$ and $d_j = |f(t_j) - f(t_{j-1})|$, we may rewrite the above as

$$\sum_{j=1}^{n} \mathcal{L}^1(f^{-1}(K) \cap [t_{j-1}, t_j]) > \sum_{j=1}^{n} \{l_j - d_j\}.$$

Clearly, for some $j \geq 1$ and some $\varepsilon > 0$, we must then have

$$(\star)\ \mathcal{L}^1(f^{-1}(K) \cap [t_{j-1}, t_j]) > l_j - d_j + \varepsilon.$$

Since $\mathcal{L}^1(f^{-1}(K)\cap[t_{j-1},t_j]) \leq l_j$, one must have $d_j > \varepsilon > 0$ here. We wish to apply the last lemma to the curve $f|[t_{j-1},t_j]$ but cannot do so since it is not *quite* parametrized by arclength! This is easily fixed up: apply the lemma to the curve $t \in [0,l_j] \mapsto f(t+t_{j-1}) \in \mathbb{C}$ and then take the nonempty, compact subset of $[0,l_j]$ produced and translate it to the right by t_{j-1} to give us a nonempty, compact subset of $[t_{j-1},t_j]$ which we will denote by E. One can easily check that the conclusions of the lemma still hold for our original f and this E. But what was our M when we applied the lemma? We are free to choose M to be anything we please greater than l_j/d_j. Choose M large enough so that from (c) of the lemma we get

$$(\star\star)\ \mathcal{L}^1(E) > d_j - \varepsilon.$$

From $(\star)$ and $(\star\star)$ we see that

$$\mathcal{L}^1(f^{-1}(K)\cap[t_{j-1},t_j]) + \mathcal{L}^1(E) > l_j = \mathcal{L}^1([t_{j-1},t_j]).$$

Since $f^{-1}(K)\cap[t_{j-1},t_j]$ and E are both subsets of $[t_{j-1},t_j]$, the only way this can happen is for their intersection to have positive measure. Thus $\mathcal{L}^1(f^{-1}(K)\cap E) > 0$. Then by (b) of the lemma, $\mathcal{L}^1(\pi(f(f^{-1}(K)\cap E))) > 0$. Here π is the orthogonal projection of the complex plane onto the line L through $f(t_{j-1})$ and $f(t_j)$. Also, by (a) of the lemma, $f(f^{-1}(K)\cap E)$ lies on a Lipschitz graph over some interval contained in L. Thus by Corollary 4.14, $\gamma(f(f^{-1}(K)\cap E)) > 0$. Since $f(f^{-1}(K)\cap E) \subseteq K$, we see that $\gamma(K) > 0$ and so are done. □

Removing the first two and last three sentences from this proof and invoking Proposition 4.21, we obtain a proof of the following result which shall come in useful in the next section.

Scholium 4.26 *Suppose K is a compact subset of a rectifiable curve f such that $l_f(K) > 0$ or $\mathcal{H}^1(K) > 0$. Then there exists a line L such that the orthogonal projection of K onto L has positive linear Lebesgue measure.*

From Proposition 4.21, the last theorem, and Painlevé's Theorem (2.7), we immediately obtain the following which we label a theorem as opposed to a corollary because of its importance.

Theorem 4.27 (The Denjoy Conjecture Resolved: Linear-Hausdorff-Measure Version) *Suppose K is a compact subset of a rectifiable curve. Then K is nonremovable if and only if $\mathcal{H}^1(K) > 0$.*

4.7 Conjecture and Refutation: The Joyce–Mörters Set

The affirmative resolution of the Denjoy Conjecture implies that any compact subset of the complex plane which intersects a rectifiable curve in a set of positive linear Hausdorff measure is nonremovable. This brings to mind the obvious conjecture that the reverse also holds ... which would imply that nonremovability is equivalent to intersecting some rectifiable curve in a set of positive linear Hausdorff measure. Alas, this turns out to be false and so life is more complicated than expected! In [JM2] Helen Joyce and Peter Mörters constructed a compact subset K of the complex plane with the following two properties: first, the Lebesgue measure of the orthogonal projection of K onto any line is zero, and second, K supports a probability measure that has finite Melnikov curvature and linear growth. This K is a counterexample to our conjecture because, on the one hand, it must intersect every rectifiable curve in a set of linear Hausdorff measure zero by the first enunciated property and Scholium 4.26, while, on the other hand, it must be nonremovable by the second enunciated property and Melnikov's Lower Capacity Estimate (4.3). The example of Joyce and Mörters also shows that a theorem of John M. Marstrand is in some sense the strongest possible. Not being interested in this aspect of their set, we are here able to present a simplified version of it and make our life a little easier.

As an aside, let us note that when we speak of a set intersecting a rectifiable curve in a set of positive measure, we are abusing language ... the set actually intersects the *range* of the rectifiable curve in a set of positive measure. In future, we shall frequently and vigorously continue to abuse language in this way without further comment!

Let us now turn to the construction of K. Define a sequence $\{\alpha_2, \alpha_3, \ldots\}$ of angles by

$$\alpha_j = \frac{\pi}{2^J} \text{ where } J \text{ is the unique integer such that } 2^J \leq j < 2^{J+1}.$$

Thus α_2 through α_3 are $\pi/2$, α_4 through α_7 are $\pi/4$, α_8 through α_{15} are $\pi/8$, etc. Define a sequence $\{N_n\}$ of integers recursively by

$$N_1 = 1 \text{ and } N_{n+1} = 4(n+1)N_n.$$

Define a sequence $\{D_n\}$ of diameters recursively by

$$D_1 = 1 \text{ and } D_{n+1} = \left\{\frac{n+1}{n}\right\}^{3/4} \cdot \frac{D_n}{4(n+1)} = \frac{D_n}{4n^{3/4}(n+1)^{1/4}}.$$

Note that

$$N_n = 4^{n-1}\{n!\} \text{ and } D_n = \frac{n^{3/4}}{4^{n-1}\{n!\}} = \frac{1}{4^{n-1}n^{1/4}\{(n-1)!\}}.$$

Clearly $N_n \uparrow \infty$ and $D_n \downarrow 0$. Also, $N_n D_n = n^{3/4}$.

Start with K_1 being any single closed ball B of diameter 1. So K_1 consists of N_1 closed balls each with diameter D_1. Inside B place eight closed balls of diameter $2^{3/4}/8$ so that their centers are on the diameter of B which makes an angle of $\theta_2 = \alpha_2$ with respect to the positive x-axis. These balls overlap but we seek to minimize their overlap: space their centers evenly along this diameter of B with the two extreme balls being internally tangent to B. The set K_2 is the union of these eight balls. So K_2 consists of N_2 closed balls each with diameter D_2.

Now supposing that we have constructed K_n as the union of N_n closed balls, called *balls of the* n^{th} *stage*, each with diameter D_n, we show how to construct K_{n+1}. Let B be a closed ball of the nth stage. Inside B place $4(n+1)$ closed balls of diameter D_{n+1} so that their centers are on the diameter of B which makes an angle of $\theta_{n+1} = \alpha_2 + \alpha_3 + \cdots + \alpha_{n+1}$ with respect to the positive x-axis. We say that these balls have been *generated* from B. They overlap but we seek to minimize their overlap: space their centers evenly along this diameter of B with the two extreme balls being internally tangent to B. The N_{n+1} closed balls of diameter D_{n+1} generated from B as B ranges over all balls of the nth stage comprise the *balls of the* $(n+1)^{\text{st}}$ *stage* and the set K_{n+1} is simply their union.

Finally, the *Joyce–Mörters set* K is just the intersection of all the K_ns so generated.

Proposition 4.28 *The orthogonal projection of the Joyce–Mörters set upon any line of the complex plane has linear Lebesgue measure zero.*

Proof Fix an angular direction $\theta \in [0, \pi)$ and an integer n. Let $J = J(n)$ be the integer satisfying

$$(\star)\ \frac{\pi}{2^J} < D_n \leq \frac{\pi}{2^{J-1}}.$$

Note that

$$(\star\star)\ \sum_{j=2}^{2^J-1} \alpha_j = \pi(J-1).$$

Now let $n_0 = n_0(n, \theta)$ be the integer greater than or equal to 2^J but strictly less than 2^{J+1} for which

$$(\star\star\star)\ \sum_{j=2^J}^{n_0-1} \alpha_j = (n_0 - 2^J)\frac{\pi}{2^J} \leq \theta < (n_0 - 2^J + 1)\frac{\pi}{2^J} = \sum_{j=2^J}^{n_0} \alpha_j.$$

(Interpret the leftmost sum in the above to be 0 when $n_0 = 2^J$.)

Let B_{n_0} be a ball of the $n_0{}^{\text{th}}$ stage and B_{n_0+n} be a ball of the $(n_0+n)^{\text{th}}$ stage *generated* from B_{n_0}. By this we mean that the short sequence $\{B_{n_0}, B_{n_0+n}\}$ can be filled in to an expanded sequence $\{B_m : m = n_0, n_0+1, \ldots, n_0+n\}$ such that each B_m is a ball of the mth stage and each B_{m+1} is generated from B_m. Let c_m denote the center of B_m. We may write $c_{m+1} - c_m = \rho_{m+1}\exp(i\theta_{m+1})$ where ρ_m can be negative as well as positive. Clearly

$$|\rho_{m+1}| \leq D_m/2.$$

From $(\star\star)$, the fact that $|\sin x| \leq |x|$, $(\star\star\star)$, and $(\star)$, we see that for $m = n_0, n_0 + 1, \ldots, n_0 + n - 1$,

$$\begin{aligned} |\sin(\theta_{m+1} - \theta)| = |\sin(\sum_{j=2^J}^{m+1} \alpha_j - \theta)| &\leq |\sum_{j=2^J}^{m+1} \alpha_j - \theta| = \sum_{j=2^J}^{m+1} \alpha_j - \theta \\ &= \sum_{j=n_0+1}^{m+1} \alpha_j + \left(\sum_{j=2^J}^{n_0} \alpha_j - \theta \right) \leq (m - n_0 + 1)\frac{\pi}{2^J} + \frac{\pi}{2^J} \\ &\leq (n+1) \cdot D_n. \end{aligned}$$

From our recursive definition of the D_ns, we see that

$$\sum_{m=n_0}^{\infty} D_m \leq \sum_{m=n_0}^{\infty} \frac{D_{n_0}}{4^{m-n_0}} = \frac{4}{3} D_{n_0} \leq 2D_{n_0}.$$

The three displayed inequalities of the last paragraph imply that the distance between c_{n_0+n} and the line through c_{n_0} which makes an angle of θ with respect to the positive x-axis is bounded by

$$\sum_{m=n_0}^{n_0+n-1} |\rho_{m+1}||\sin(\theta_{m+1} - \theta)| \leq \sum_{m=n_0}^{n_0+n-1} \frac{D_m}{2} \cdot (n+1)D_n \leq (n+1)D_n D_{n_0}.$$

Let L be any line perpendicular to the direction θ and let π denote the orthogonal projection of the complex plane onto L. From the last paragraph it follows that the orthogonal projections onto L of the centers of the balls of the $(n_0 + n)^{\text{th}}$ stage generated from a single ball B of the $n_0{}^{\text{th}}$ stage all lie within a distance $(n+1)D_n D_{n_0}$ of the orthogonal projection onto L of the center of B. Hence $\pi(K_{n_0+n} \cap B)$, for any ball B of the $n_0{}^{\text{th}}$ stage, is contained in a segment of L of length $D_{n_0+n} + 2(n+1)D_n D_{n_0}$. Since K_{n_0+n} is the union of $K_{n_0+n} \cap B$ over N_{n_0} such balls B, we conclude that

$$\mathcal{L}^1(\pi(K)) \leq \mathcal{L}^1(\pi(K_{n_0+n})) \leq N_{n_0} \cdot D_{n_0+n} + 2N_{n_0} \cdot (n+1)D_n D_{n_0}.$$

Now $N_{n_0} \cdot D_{n_0+n} \to 0$ as $n \to \infty$ since

$$\begin{aligned} N_{n_0} \cdot D_{n_0+n} &= 4^{n_0-1}(n_0!) \cdot \frac{(n_0+n)^{3/4}}{4^{n_0+n-1}\{(n_0+n)!\}} \\ &= \frac{1}{4^n} \cdot \frac{(n_0+n)^{3/4}}{(n_0+1)(n_0+2)\cdots(n_0+n)} \leq \frac{1}{4^n}. \end{aligned}$$

To show that the $N_{n_0} \cdot (n+1)D_n D_{n_0} \to 0$ as $n \to \infty$ and be finished, note that by the definition of n_0 and $(\star)$, $n_0 < 2^{J+1} \leq 4\pi/D_n$. Hence

$$N_{n_0} \cdot (n+1) D_n D_{n_0} = n_0^{3/4}(n+1)D_n < (4\pi)^{3/4}(n+1)D_n^{1/4}$$
$$= (4\pi)^{3/4}\left\{\frac{(n+1)^4}{4^{n-1}n^{1/4}\{(n-1)!\}}\right\}^{1/4}.$$

However,

$$\frac{(n+1)^4}{4^{n-1}n^{1/4}\{(n-1)!\}} \to 0 \text{ as } n \to \infty$$

by an easy use of the Ratio and Divergence Tests for infinite series. □

To construct a probability measure on K that has finite Melnikov curvature and linear growth, we need a lemma dealing with the geometry of intersection of the balls of the various stages. Each geometric assertion of the lemma below is followed by an inequality that will be proved and shown to imply it. These inequalities involve an awkward quantity, the distance between the centers of any two adjacent balls of the $(n+1)^{\text{st}}$ stage generated from the same ball of the nth stage:

$$d_{n+1} = \frac{D_n - D_{n+1}}{4n+3}.$$

Lemma 4.29 *With conventions and definitions as above, we have:*

(a) *Adjacent balls of the $(n+1)^{\text{st}}$ stage generated from the same ball of the nth stage overlap:*

$$\frac{d_{n+1}}{D_{n+1}} < 1 \text{ for } n \geq 1.$$

(b) *Nonadjacent balls of the $(n+1)^{\text{st}}$ stage generated from the same ball of the nth stage do not intersect:*

$$\frac{d_{n+1}}{D_{n+1}} > \frac{1}{2} \text{ for } n \geq 1.$$

(c) *Balls of the $(n+1)^{\text{st}}$ stage generated from the different balls of the nth stage do not intersect (and are in fact a distance at least $3D_{n+1}$ from one another):*

$$d_n \sin \alpha_{n+1} > 4D_{n+1} \text{ for } n \geq 2.$$

Proof (a) and (b) Clearly $D_n/D_{n+1} = 4n^{3/4}(n+1)^{1/4}$ and so $4n < D_n/D_{n+1} < 4(n+1)$. From our formula for d_{n+1}, it now follows that

$$\frac{4n-1}{4n+3} < \frac{d_{n+1}}{D_{n+1}} < 1.$$

To finish the proof of these two inequalities, note that for $n \geq 2$, $(4n-1)/(4n+3) \geq 7/11 > 1/2$ while for $n = 1$, $d_{n+1}/D_{n+1} = \{4(2^{1/4}) - 1\}/7 > 1/2$.

The geometric assertions now follow since two closed balls, each of diameter D and with centers c_1 and c_2, overlap if and only if $D > |c_1 - c_2|$ and do not intersect if and only if $D < |c_1 - c_2|$ (when $D = |c_1 - c_2|$, they are externally tangent to one another).

(c) Since $\sin x \geq 2\sqrt{2}x/\pi$ for $0 \leq x \leq \pi/4$, it follows that $\sin \alpha_{n+1} \geq 2\sqrt{2}/(n+1)$ for $n \geq 3$. From this, (b) just proven, and $D_n/D_{n+1} > 4n$ noted above, we have,

$$\begin{aligned} d_n \sin \alpha_{n+1} &= \sin \alpha_{n+1} \cdot \frac{d_n}{D_n} \cdot \frac{D_n}{D_{n+1}} \cdot D_{n+1} > \frac{2\sqrt{2}}{n+1} \cdot \frac{1}{2} \cdot 4n \cdot D_{n+1} \\ &= \frac{4\sqrt{2}n}{n+1} D_{n+1} > 4D_{n+1} \end{aligned}$$

for any $n \geq 3$. When $n = 2$, use $\sin \alpha_3 = \sin \pi/2 = 1$ in place of $\sin \alpha_{n+1} \geq 2/(n+1)$ in the above to deduce that $d_2 \sin \alpha_3 > 1 \cdot \frac{1}{2} \cdot 8 \cdot D_3 = 4D_3$.

Let B_{n+1} and B'_{n+1} be two balls of the $(n+1)^{\text{st}}$ stage generated from different balls B_n and B'_n of the nth stage, respectively. Extend the short sequence $\{B_n, B_{n+1}\}$ backward to an expanded sequence $\{B_m : m = 1, 2, \ldots n+1\}$ such that each B_m is a ball of the mth stage and each B_{m+1} is generated from B_m. Let $\{B'_m : m = 1, 2, \ldots n+1\}$ be similarly obtained from $\{B'_n, B'_{n+1}\}$. Define n_0 to be the smallest integer for which $B_{n_0} \neq B'_{n_0}$. Since $B_1 = B'_1$ and $B_n \neq B'_n$, $2 \leq n_0 \leq n$.

Being generated from the *same* ball of the $(n_0 - 1)^{\text{st}}$ stage, the distance between the centers of B_{n_0} and B'_{n_0} is at least d_{n_0} and the distance between the diameters of B_{n_0} and B'_{n_0} upon which the centers of B_{n_0+1} and B'_{n_0+1} lie is at least $d_{n_0} \sin \alpha_{n_0+1}$. Thus the inequality just established implies that B_{n_0+1} and B'_{n_0+1} do not intersect (and are in fact a distance at least $3D_{n_0+1}$ from one another). But $B_{n+1} \subseteq B_{n_0+1}$, $B'_{n+1} \subseteq B'_{n_0+1}$, and $D_{n+1} \leq D_{n_0+1}$, so B_{n+1} and B'_{n+1} also do not intersect (and are in fact a distance at least $3D_{n+1}$ from one another). □

Given $n \geq 1$, let μ_n denote the discrete probability measure which puts mass $1/N_n$ at each of the N_n centers of the balls of the nth stage. Applying [RUD, 11.29] just as in the proof of Frostman's Lemma (2.9), we obtain a subsequence $\{\mu_{n_k}\}$ of $\{\mu_n\}$ which converges weakly to some probability measure μ on K. The key to showing that this μ has finite Melnikov curvature and linear growth is a nonlinear growth condition that μ satisfies. To this we now turn.

From the recursive definition of the D_ns, we see that $D_n < D_{n-1}/2$ for $n \geq 2$. Hence

$$d_{n+1} = \frac{D_n - D_{n+1}}{4n+3} < \frac{D_n}{4n-1} < \frac{D_{n-1} - D_n}{4n-1} = d_n$$

for $n \geq 2$. Thus the d_ns decrease, and clearly they decrease to 0. Hence we may properly define a increasing function φ on $(0, \infty)$ by

$$\varphi(t) = \begin{cases} t/(n+1)^{3/4} & \text{when} \quad d_{n+1} \le t < d_n \\ d_2/2^{3/4} & \text{when} \quad d_2 \le t \end{cases}.$$

Lemma 4.30 *For all points $c \in \mathbb{C}$ and all $r > 0$, $\mu(B(c; r)) < 56\varphi(r)$.*

Proof First Claim. If A is the union of m adjacent balls of the $(n+1)^{\text{st}}$ stage generated from a given single ball of the nth stage, then

$$\mu(A) \le \frac{m+2}{N_{n+1}}.$$

By the last lemma, at most two more balls of the $(n+1)^{\text{st}}$ stage intersect A and the rest of the balls of the $(n+1)^{\text{st}}$ stage, whose union we will denote by B, stay a positive distance d from A. Choose a compactly supported, continuous function ψ such that $0 \le \psi \le 1$ everywhere, $\psi = 1$ on A, and $\psi = 0$ on B ($\psi(z) = \max\{1 - \text{dist}(z, A)/d, 0\}$ will do nicely). Then

$$\mu(A) \le \int \psi \, d\mu = \lim_{k\to\infty} \int \psi \, d\mu_{n_k} \le \limsup_{k\to\infty} \mu_{n_k}(K_{n+1} \setminus B) \le \frac{m+2}{N_{n+1}}.$$

Second Claim. $B(c; r)$ intersects at most $(2r/d_{n+1}) + 3$ balls of the $(n+1)^{\text{st}}$ stage generated from a given single ball of the nth stage. Moreover, these balls are adjacent.

We may assume that m, the number of balls to be estimated, is not 0. Clearly these m balls are adjacent and strung out along a diameter of the single given ball of the nth stage with the two extreme balls having centers a distance $(m-1)d_{n+1}$ apart. By the Triangle Inequality,

$$(m-1)d_{n+1} \le \frac{D_{n+1}}{2} + r + r + \frac{D_{n+1}}{2} = 2r + D_{n+1}.$$

But then $m \le (2r/d_{n+1}) + (D_{n+1}/d_{n+1}) + 1 < (2r/d_{n+1}) + 3$ by (b) of the last lemma.

Third Claim. If $r < d_n$, then $B(c; r)$ intersects at most four balls of the nth stage.

By (c) and (a) of the last lemma, $B(c; r)$ cannot intersect balls of the nth stage that come from different balls of the $(n-1)^{\text{st}}$ stage. Hence by the last Claim, the number of balls of the nth stage that intersect $B(c; r)$ is at most $(2r/d_n) + 3$, i.e., at most 4 since $r/d_n < 1$.

By (b) of the last lemma,

$$\frac{1}{N_{n+1}d_{n+1}} < \frac{2}{N_{n+1}D_{n+1}} = \frac{2}{(n+1)^{3/4}}.$$

From our three Claims and this observation, it follows that if $d_{n+1} \le r < d_n$, then

$$\mu(B(c;r)) \leq 4 \cdot \frac{1}{N_{n+1}}\left[\frac{2r}{d_{n+1}}+5\right] = \left[2+\frac{5d_{n+1}}{r}\right] \cdot \frac{4r}{N_{n+1}d_{n+1}}$$
$$< 7 \cdot \frac{8r}{(n+1)^{3/4}} = 56\varphi(r).$$

The case of $r \geq d_2$ is trivial since then $\mu(B(c;r)) \leq 1 < 4(2^{1/4}) - 1 = 56d_2/2^{3/4} = 56\varphi(r)$. □

Lemma 4.31 $\displaystyle\int_0^\infty \frac{\varphi^2(t)}{t^3}\,dt < \infty.$

Proof From (a) and (b) of Lemma 4.29, we see that for $n \geq 2$, $d_n/d_{n+1} < 2D_n/D_{n+1} = 8n^{1/4}(n+1)^{3/4} < 8(n+1)$ and so

$$\int_{d_{n+1}}^{d_n} \frac{\varphi^2(t)}{t^3}\,dt = \frac{1}{(n+1)^{3/2}}\int_{d_{n+1}}^{d_n} \frac{1}{t}\,dt = \frac{\ln(d_n/d_{n+1})}{(n+1)^{3/2}} < \frac{\ln\{8(n+1)\}}{(n+1)^{3/2}}.$$

But a simple use of the Integral Test (along with Parts and L'Hôpital) shows that

$$\sum_{n=2}^{\infty} \frac{\ln\{8(n+1)\}}{(n+1)^{3/2}} < \infty.$$

Since $\displaystyle\int_{d_2}^\infty \frac{\varphi^2(t)}{t^3}\,dt = \frac{1}{2^{5/2}} < \infty$, we are done. □

The curvature estimate that appears in the proof of the following result is from [MAT4].

Proposition 4.32 *The Joyce–Mörters set supports a probability measure with finite Melnikov curvature and linear growth.*

Proof Since $\varphi(t) \leq t$ for all $t > 0$, Lemma 4.30 implies that the probability measure μ we have constructed has linear growth with bound 56.

Consider $S_1 = \{(\zeta,\eta,\xi) \in \mathbb{C}^3 : |\zeta-\eta| \leq |\xi-\eta|\}$ and $S_2 = \{(\zeta,\eta,\xi) \in \mathbb{C}^3 : |\zeta-\eta| > |\xi-\eta|\}$. Since $\mathbb{C}^3 = S_1 \cup S_2$, to show $c^2(\mu) < \infty$ it suffices to show that

$$\iiint_{S_j} c^2(\zeta,\eta,\xi)\,d\mu(\zeta)\,d\mu(\eta)\,d\mu(\xi) < \infty$$

for $j = 1, 2$. We only show this for the case $j = 1$, the proof for $j = 2$ being similar.

In what follows we have a chain of in/equalities. Immediately after this chain, a justification will be given for each of its lines.

$$\begin{aligned}
\iiint_{S_1} c^2(\zeta,\eta,\xi)\,d\mu(\zeta)\,d\mu(\eta)\,d\mu(\xi) &\le 4\iint \frac{\mu(B(\eta;|\xi-\eta|))}{|\xi-\eta|^2}\,d\mu(\eta)\,d\mu(\xi)\\
&\le 224\iint \frac{\varphi(|\xi-\eta|)}{|\xi-\eta|^2}\,d\mu(\eta)\,d\mu(\xi)\\
&= 448\int\left\{\int\int_{|\xi-\eta|}^{\infty}\frac{\varphi(|\xi-\eta|)}{t^3}\,dt\,d\mu(\eta)\right\}d\mu(\xi)\\
&= 448\int\left\{\int_0^{\infty}\int_{B(\xi;t)}\frac{\varphi(|\xi-\eta|)}{t^3}\,d\mu(\eta)dt\right\}d\mu(\xi)\\
&\le 448\int\left\{\int_0^{\infty}\frac{\varphi(t)}{t^3}\mu(B(\xi;t))\,dt\right\}d\mu(\xi)\\
&= 25{,}088\int_0^{\infty}\frac{\varphi^2(t)}{t^3}\,dt.
\end{aligned}$$

[The first line holds since $c(\zeta,\eta,\xi) = 2\sin\alpha/a \le 2/|\xi-\eta|$ by Proposition 4.1; the second by Lemma 4.30; to see the third line just work out the dt-integral; the fourth line is Fubini's Theorem [RUD, 8.8]; the fifth line holds since φ is increasing; and the last line is just Lemma 4.30 again along with the fact that μ is a probability measure.]

By the last lemma, we are now done. □

Putting together Propositions 4.28 and 4.32 with Scholium 4.26 and Melnikov's Lower Capacity Estimate (4.3), we immediately obtain the following.

Theorem 4.33 *The Joyce–Mörters set intersects any rectifiable curve in a set of arclength measure and linear Hausdorff measure zero, yet is nonremovable.*

How big is the Joyce–Mörters set? According to the next proposition, its Hausdorff dimension is one and it is as big as such a set can possibly be!

Proposition 4.34 *The Joyce–Mörters set has Hausdorff dimension one and is of non-σ-finite linear Hausdorff measure, i.e., it cannot be written as a countable union of sets each of which has finite linear Haudsorff measure.*

Proof Let $\alpha > 0$. The collection of balls of the nth stage is a $2D_n$-cover of K. Hence

$$(\star)\ \mathcal{H}^{1+\alpha}_{2D_n}(K) \le N_n\cdot D_n^{1+\alpha} = N_nD_n\cdot D_n^{\alpha} = n^{3/4}\cdot\left\{\frac{n^{3/4}}{4^{n-1}(n!)}\right\}^{\alpha} = \left\{\frac{n^{\beta}}{4^{n-1}(n!)}\right\}^{\alpha}$$

where $\beta = (3/4)[(1/\alpha)+1]$. An easy use of the Ratio and Divergence Tests for infinite series shows that $n^{\beta}/[4^{n-1}(n!)] \to 0$. Thus one may let $n\to\infty$ in $(\star)$ and conclude that $\mathcal{H}^{1+\alpha}(K) = 0$. Since $\alpha > 0$ is otherwise arbitrary, it must be the case that $\dim_{\mathcal{H}}(K) \le 1$.

To finish, we suppose K to be written as a countable union of sets $\{E_m\}$ and show that $\mathcal{H}^1(E_m)$ must be infinite for at least one integer m. Since $\mu(K) = 1$, one must have $\mu(E_m) > 0$ for some m. Fixing N, select a d_N-cover $\{U_n\}$ of E_m such that

$$\sum_n |U_n| \le \mathcal{H}^1_{d_N}(E_m) + \frac{1}{N}.$$

Choosing points $z_n \in U_n$, use Lemma 4.30 to note that

$$\mu(U_n) \le \mu(B(z_n; |U_n|)) \le 56\varphi(|U_n|) \le \frac{56}{(N+1)^{3/4}}|U_n|.$$

Hence

$$\mu(E_m) \le \sum_n \mu(U_n) \le \frac{56}{(N+1)^{3/4}}\left\{\mathcal{H}^1_{d_N}(E_m) + \frac{1}{N}\right\}.$$

Now unfixing N and letting it get arbitrarily large, the fact that $\mu(E_m)$ is strictly positive forces $\mathcal{H}^1(E_m)$ to be infinite. □

Hidden in the last proof is the notion of the *generalized Hausdorff φ-measure*, $\mathcal{H}^\varphi(E)$, of a subset E of $\mathbb{C}$ where φ is any right continuous, increasing function defined on $[0, \infty)$ such that $\varphi(0) = 0$ and $\varphi(t) > 0$ for all $t > 0$. It is defined just as s-dimensional Hausdorff measure was only with $|U_n|^s$ replaced by $\varphi(|U_n|)$. So $\mathcal{H}^s(E)$ arises as the special case of $\mathcal{H}^\varphi(E)$ where $\varphi(t) = t^s$. For K the Joyce–Mörters set and φ the function defined just before Lemma 4.30, the reader may try his or her hand at showing that $1/56 \le \mathcal{H}^\varphi(K) \le 1$. For more on this topic see [FALC], [MAT3], or the older [ROG]. The last reference is the only one that introduces the generalized Hausdorff measure $\mathcal{H}^\varphi$ as the basic object of study from the beginning, specializing to s-dimensional Hausdorff measure $\mathcal{H}^s$ from time to time.

Chapter 5
Some Measure Theory

5.1 The Carathéodory Criterion and Metric Outer Measures

We remind the reader of some terminology. Given a set X, the *power set* of X, denoted $\mathcal{P}(X)$, is the collection of all subsets of X. An *outer measure* on X is a set function $\mu : \mathcal{P}(X) \mapsto [0,\infty]$ such that (i) $\mu(\emptyset) = 0$, (ii) $\mu(E) \leq \mu(F)$ whenever $E \subseteq F \in \mathcal{P}(X)$, and (iii) $\mu(\bigcup_{n=1}^{\infty} E_n) \leq \sum_{n=1}^{\infty} \mu(E_n)$ whenever $E_1, E_2, \ldots \in \mathcal{P}(X)$. When property (iii) just enunciated holds we say that μ is *countably subadditive* on X. A *σ-algebra* on X is a subcollection $\mathcal{M} \subseteq \mathcal{P}(X)$ that contains the empty set and is closed under complements and countable unions. A *positive measure* on X is a set function $\mu : \mathcal{M} \mapsto [0,\infty]$ such that (i) $\mathcal{M}$ is a σ-algebra on X, (ii) $\mu(\emptyset) = 0$, and (iii) $\mu(\bigcup_{n=1}^{\infty} E_n) = \sum_{n=1}^{\infty} \mu(E_n)$ whenever $E_1, E_2, \ldots \in \mathcal{M}$ are pairwise disjoint. When property (iii) just enunciated holds we say that μ is *countably additive* on $\mathcal{M}$.

Suppose now that X is a topological space and $\mu : \mathcal{M} \mapsto [0,\infty]$ is a positive measure on X. The *Borel sets* of X are the elements of the smallest σ-algebra $\mathcal{B}(X)$ on X containing the closed subsets of X. Rudin says "A [positive] measure μ defined on the σ-algebra of all Borel sets in a locally compact Hausdorff space X is called a [positive] *Borel measure* on X" [RUD, 2.15]. This is ambiguous. Does it mean $\mathcal{M} = \mathcal{B}(X)$? Or does it merely mean $\mathcal{M} \supseteq \mathcal{B}(X)$? We have opted for the latter. Thus when we say that μ is a *positive Borel measure* on X we mean that $\mathcal{M} \supseteq \mathcal{B}(X)$.

Up till now it has sufficed to know merely that $\mathcal{H}^s$ satisfies Proposition 2.1, i.e., that $\mathcal{H}^s$ is an outer measure on $\mathbb{C}$. For what follows it will be necessary to know that $\mathcal{H}^s$ restricted to an appropriate σ-algebra is a positive Borel measure on $\mathbb{C}$. Proving this fact is the goal of the present section!

In [RUD] when a positive Borel measure on a topological space is needed, one typically produces a positive linear functional on the space of compactly supported, continuous functions on the space and then invokes the Riesz Representation Theorem [RUD, 2.14]. The positive measures so obtained all assign finite mass to every compact set [RUD, 2.14(b)]. Since the Hausdorff dimension of any compact set K with nonempty interior is 2, $\mathcal{H}^s(K) = \infty$ for such a K when $s < 2$. Thus the Rudin approach will not provide a means of showing that $\mathcal{H}^s$ restricted to an appropriate σ-algebra is a positive Borel measure on $\mathbb{C}$! Instead we must invoke a two-part

J.J. Dudziak, *Vitushkin's Conjecture for Removable Sets*, Universitext,
DOI 10.1007/978-1-4419-6709-1_5,

approach due to Carathéodory: first, starting with an arbitrary outer measure on a set, a measure is obtained by simply restricting the outer measure to an appropriate σ-algebra of subsets of the set, and second, when the original set is a metric space, the measure obtained is shown to be Borel if the original outer measure is additive on pairs of subsets a positive distance apart. What we now do is standard measure theory, but not in [RUD]. Our treatment is taken from the first chapter of [FALC].

It is not at all obvious how to get a grip on the "appropriate σ-algebra" mentioned above. Thus the following definition, usually referred to as the *Carathéodory Criterion for Measurability*, is a major accomplishment: A subset E of X is called *μ-measurable* if

$$\mu(A) = \mu(A \cap E) + \mu(A \setminus E).$$

for every subset A of X. Let $\mathcal{M}_\mu$ denote the collection of all μ-measurable subsets of X.

Proposition 5.1 *Suppose μ is an outer measure on a set X. Then $\mathcal{M}_\mu$ is a σ-algebra on X on which μ is countably additive. Thus the restriction of μ to $\mathcal{M}_\mu$ is a positive measure on X.*

Proof The empty set is trivially contained in $\mathcal{M}_\mu$. Due to the symmetry of the definition of μ-measurability, it is clear that $\mathcal{M}_\mu$ is closed under complements. Hence to show that $\mathcal{M}_\mu$ is a σ-algebra it suffices to show that it is closed under countable unions.

Suppose $\{E_n\}$ is a sequence from $\mathcal{M}_\mu$. Then for any subset A of X,

$$\begin{aligned}
\mu(A) &= \mu(A \cap E_1) + \mu(A \setminus E_1) \\
&= \mu(A \cap E_1) + \mu(\{A \setminus E_1\} \cap E_2) + \mu((A \setminus \{E_1 \cup E_2\}) \\
&\ \vdots \\
&= \sum_{m=1}^{M} \mu\left(\left\{A \setminus \bigcup_{n=1}^{m-1} E_n\right\} \cap E_m\right) + \mu\left(A \setminus \bigcup_{n=1}^{M} E_n\right) \\
&\geq \sum_{m=1}^{M} \mu\left(\left\{A \setminus \bigcup_{n=1}^{m-1} E_n\right\} \cap E_m\right) + \mu\left(A \setminus \bigcup_{n=1}^{\infty} E_n\right).
\end{aligned}$$

Letting $M \to \infty$, using the subadditivity of μ, a little set algebra, and the subadditivity of μ again, we see that

$$\begin{aligned}
(\star)\ \mu(A) &\geq \sum_{m=1}^{\infty} \mu\left(\left\{A \setminus \bigcup_{n=1}^{m-1} E_n\right\} \cap E_m\right) + \mu\left(A \setminus \bigcup_{n=1}^{\infty} E_n\right) \\
&\geq \mu\left(\bigcup_{m=1}^{\infty}\left(\left\{A \setminus \bigcup_{n=1}^{m-1} E_n\right\} \cap E_m\right)\right) + \mu\left(A \setminus \bigcup_{n=1}^{\infty} E_n\right)
\end{aligned}$$

$$= \mu\left(A \cap \bigcup_{n=1}^{\infty} E_n\right) + \mu\left(A \setminus \bigcup_{n=1}^{\infty} E_n\right)$$

$$\geq \mu(A).$$

Ignoring the right half of the first line and all of the second line of ($\star$), it follows that $\bigcup_{n=1}^{\infty} E_n$ is μ-measurable and so $\mathcal{M}_\mu$ is closed under countable unions.

Now suppose our sequence $\{E_n\}$ from $\mathcal{M}_\mu$ consists of pairwise disjoint sets. Then setting $A = \bigcup_{m=1}^{\infty} E_m$ in ($\star$) and ignoring the resulting second and third lines, we get

$$\mu\left(\bigcup_{m=1}^{\infty} E_m\right) \geq \sum_{m=1}^{\infty} \mu(E_m) \geq \mu\left(\bigcup_{m=1}^{\infty} E_m\right).$$

Hence μ is countably additive on $\mathcal{M}_\mu$. □

Of course this proposition may be no victory at all if $\mathcal{M}_\mu$ is very small! What are the μ-measurable sets? A little thought convinces one that any *μ-null set*, i.e., any set with μ-measure zero, is μ-measurable. A small victory! Anything else? To go further one needs to assume more about X and μ. This leads to a second great definition of Carathéodory: Given a metric space X, call an outer measure μ on X a *metric outer measure* if

$$\mu(A \cup B) = \mu(A) + \mu(B),$$

whenever A and B are subsets of X that are a positive distance apart, i.e., whenever A and B are subsets of X such that $\operatorname{dist}(A, B) > 0$. In this situation the Borel sets are μ-measurable and the following lemma is the key to seeing this fact.

Lemma 5.2 (Carathéodory) *Let μ be a metric outer measure on X. Given $\{A_n\}$ an increasing sequence of subsets of X, set $A = \bigcup_{n=1}^{\infty} A_n$. Suppose that A_n and $A \setminus A_{n+1}$ are a positive distance apart for all $n \geq 1$. Then*

$$\mu(A) = \lim_{n\to\infty} \mu(A_n).$$

Proof It suffices to show

$$(\star)\ \mu(A) \leq \lim_{n\to\infty} \mu(A_n)$$

since the reverse inequality is trivial. Set $B_1 = A_1$ and $B_m = A_m \setminus A_{m-1}$ for $m \geq 2$. One may easily verify that B_m and B_n are a positive distance apart whenever $|m - n| \geq 2$. From the metricity hypothesis one then gets

$$\mu\left(\bigcup_{m=1}^{n} B_{2m}\right) = \sum_{m=1}^{n} \mu(B_{2m}) \text{ and } \mu\left(\bigcup_{m=1}^{n} B_{2m-1}\right) = \sum_{m=1}^{n} \mu(B_{2m-1}).$$

We may assume that both these sums remain bounded as $n \to \infty$ (otherwise $\lim_{n\to\infty} \mu(A_n) = \infty$ and $(\star)$ becomes trivially true). But then the series $\sum_{m=1}^{\infty} \mu(B_m)$ converges. Note that

$$\mu(A) = \mu\left(A_n \cup \bigcup_{m=n+1}^{\infty} B_m\right) \leq \mu(A_n) + \sum_{m=n+1}^{\infty} \mu(B_m)$$

$$\leq \lim_{n\to\infty} \mu(A_n) + \sum_{m=n+1}^{\infty} \mu(B_m).$$

Since $\sum_{m=n+1}^{\infty} \mu(B_m) \to 0$ as $n \to \infty$, $(\star)$ now follows. □

Proposition 5.3 *Let μ be a metric outer measure on X. Then the collection $\mathcal{M}_\mu$ of all μ-measurable subsets of X is a σ-algebra on X on which μ is countably additive. Moreover, $\mathcal{M}_\mu \supseteq \mathcal{B}(X)$. Thus the restriction of μ to $\mathcal{M}_\mu$ is a positive Borel measure on X.*

Proof By the last proposition, we need only show that $\mathcal{M}_\mu$ contains the Borel sets. Since the collection of Borel subsets of X is just the smallest σ-algebra on X containing the closed sets of X, it suffices to show that an arbitrary closed set E of X is μ-measurable. Given any subset A of X, set $A_n = \{z \in A : \text{dist}(z, E) \geq 1/n\}$. Since $A \cap E$ and A_n are a distance at least $1/n$ apart, the metricity hypothesis implies that

$$\mu(A \cap E) + \mu(A_n) = \mu((A \cap E) \cup A_n) \leq \mu(A).$$

Note that A_n and $A \setminus A_{n+1}$ are a distance at least $1/n - 1/(n+1)$ apart and that $\bigcup_{n=1}^{\infty} A_n = A \setminus E$ since E is closed. Hence letting $n \to \infty$ in the last displayed inequality and invoking the last lemma, we obtain

$$\mu(A \cap E) + \mu(A \setminus E) \leq \mu(A).$$

The reverse inequality being trivial, we are done. □

Lemma 5.4 *For $0 \leq s \leq 2$, $\mathcal{H}^s$ is a metric outer measure on $\mathbb{C}$.*

Proof Let A and B be subsets of $\mathbb{C}$ that are a positive distance d apart. Consider any positive δ smaller than d. Then any element of a δ-cover of $A \cup B$ can intersect at most one of A or B. Thus it is possible to write the δ-cover as the disjoint union of a δ-cover of A and a δ-cover of B. It follows that

$$\mathcal{H}^s_\delta(A \cup B) = \mathcal{H}^s_\delta(A) + \mathcal{H}^s_\delta(B).$$

Letting $\delta \downarrow 0$, we are done. □

From the last lemma and the last proposition we immediately obtain the following.

Proposition 5.5 *For $0 \leq s \leq 2$, the collection $\mathcal{M}_{\mathcal{H}^s}$ of all $\mathcal{H}^s$-measurable subsets of $\mathbb{C}$ is a σ-algebra on $\mathbb{C}$ on which $\mathcal{H}^s$ is countably additive. Moreover, $\mathcal{M}_{\mathcal{H}^s} \supseteq \mathcal{B}(\mathbb{C})$. Thus the restriction of $\mathcal{H}^s$ to $\mathcal{M}_{\mathcal{H}^s}$ is a positive Borel measure on $\mathbb{C}$.*

In the future, when we speak of $\mathcal{H}^s$ as being a measure, we really mean $\mathcal{H}^s$ restricted to the collection $\mathcal{M}_{\mathcal{H}^s}$ of all $\mathcal{H}^s$-measurable subsets of $\mathbb{C}$.

5.2 Arclength and Arclength Measure: The Rest of the Story

This section is in the nature of a mopping-up operation, completing the determination of the precise relation between arclength and arclength measure on the one hand and linear Hausdorff measure on the other that was begun in Section 4.5. That we can now give the rest of the story is due to Proposition 5.5. Our treatment in Section 4.5 and for the first two items in this section is as in Section 3.2 of [FALC] but the rest of this section goes further in that we do not restrict our attention, as [FALC] does, to arcs.

There are two distinctions operating here: arclength versus arclength measure and arc versus curve. Accordingly, we have four propositions. A lemma is needed first though. It is the linear Hausdorff measure analogue of the Shortest Path Property.

Lemma 5.6 *If K is a continuum containing points z and w, then $|z - w| \leq \mathcal{H}^1(K)$. In particular,*

$$|K| \leq \mathcal{H}^1(K).$$

Proof Assuming $z \neq w$, let L be the line containing z and w and let π be the orthogonal projection of $\mathbb{C}$ onto L. Recall that in Section 2.1 we showed that $\mathcal{H}^1$ and $\mathcal{L}^1$ coincide for linear sets. Note that $[z, w] \subseteq \pi(K)$ since K is connected. Since orthogonal projection onto a line decreases distances, we may invoke Proposition 2.2. Thus we have the first assertion of the lemma:

$$|z - w| = \mathcal{L}^1([z, w]) = \mathcal{H}^1([z, w]) \leq \mathcal{H}^1(\pi(K)) \leq \mathcal{H}^1(K).$$

Suping this inequality over all $z, w \in K$, we obtain the last assertion of the lemma. □

Proposition 5.7 *For $\gamma : [a, b] \mapsto \mathbb{C}$ an arc, set $\Gamma = \gamma([a, b])$. Then*

$$l(\gamma) = \mathcal{H}^1(\Gamma).$$

Proof Let $a = t_0 \leq t_1 \leq t_2 \leq \cdots \leq t_n = b$ be any partition of $[a, b]$. Then by the lemma, the fact that $\mathcal{H}^1$ has no atoms, the one-to-oneness of γ and Proposition 5.5, one gets

$$\sum_{j=1}^{n} |\gamma(t_j) - \gamma(t_{j-1})| \le \sum_{j=1}^{n} \mathcal{H}^1(\gamma([t_{j-1}, t_j])) \le \sum_{j=1}^{n} \mathcal{H}^1(\gamma([t_{j-1}, t_j))) \le \mathcal{H}^1(\Gamma).$$

Suping over all partitions, we see that $l(\gamma) \le \mathcal{H}^1(\Gamma)$. By Proposition 4.17, we are done. □

To handle a general curve one needs to take account of multiplicity. While the last proposition used the countable additivity of $\mathcal{H}^1$ in a simple way, the next one uses it in a more complex way since we need to integrate against $\mathcal{H}^1$ in its statement and proof.

Proposition 5.8 *For $\gamma : [a, b] \mapsto \mathbb{C}$ a curve, set $\Gamma = \gamma([a, b])$ and $\Gamma_n = \{\zeta \in \Gamma : \#\gamma^{-1}(\zeta) = n\}$ for $n = 1, 2, \dots,$ and ∞. Then*

$$l(\gamma) = \int_{\Gamma} \#\gamma^{-1}(\zeta)\, d\mathcal{H}^1(\zeta) = \sum_{1 \le n \le \infty} n \cdot \mathcal{H}^1(\Gamma_n).$$

In particular, if γ is rectifiable, then $\mathcal{H}^1(\Gamma_\infty) = 0$ and so the $n = \infty$ term in the sum above may be discarded.

Proof For this proof it will be convenient to change our conception of a partition a bit. Instead of viewing a partition $\mathcal{P}$ as an ordered set of *points* $\{t_0, t_1, t_2, \dots, t_n\}$ where $a = t_0 \le t_1 \le t_2 \le \dots \le t_n = b$, we view $\mathcal{P}$ as an ordered set of *intervals* $\{I_1, I_2, \dots, I_n\}$ with each $I_j = [t_{j-1}, t_j)$ for $1 \le j < n$ and $I_n = [t_{n-1}, t_n]$. Transfer the old notion of one partition being a refinement of another partition over to this new context. Denote the right and left endpoints, respectively, of an interval I by REP(I) and LEP(I). Using the definition of the length of a curve and the fact that the splitting of an interval of a partition into two subintervals can only increase the sum occurring in the definition of length, one can construct a sequence $\{\mathcal{P}_n\}$ of partitions of $[a, b]$ such that

1. $\sum_{I \in \mathcal{P}_n} |\gamma(\text{REP}(I)) - \gamma(\text{LEP}(I))| \to l(\gamma)$ as $n \to \infty$,
2. $\mathcal{P}_{n+1}$ is a refinement of $\mathcal{P}_n$ for each n, and
3. $\max\{|I| : I \in \mathcal{P}_n\} \to 0$ as $n \to \infty$.

By Lemma 5.6 and Proposition 4.17, for any subinterval I of $[a, b]$, we have

$$|\gamma(\text{REP}(I)) - \gamma(\text{LEP}(I))| \le \mathcal{H}^1(\gamma(I)) \le l(\gamma|I).$$

Summing over all $I \in \mathcal{P}_n$, thinking of the measure of a set as the integral of its characteristic function, and using the Subarc Additivity Property, we obtain

$$(\star)\ \sum_{I \in \mathcal{P}_n} |\gamma(\text{REP}(I)) - \gamma(\text{LEP}(I))| \leq \int_\Gamma \varphi_n(\zeta)\, d\mathcal{H}^1(\zeta) \leq l(\gamma)$$

where φ_n is the sum of the characteristic functions of the sets $\gamma(I)$ with I varying over all of $\mathcal{P}_n$. A little thought shows that $\varphi_n(\zeta)$ is just the number of intervals I from $\mathcal{P}_n$ which intersect $\gamma^{-1}(\zeta)$. Thus by items 2 and 3 above, for each $\zeta \in \mathbb{C}$ we have $\varphi_n(\zeta) \uparrow$ the number of preimages of ζ under $\gamma = \#\gamma^{-1}(\zeta)$. The Monotone Convergence Theorem [RUD, 1.26] now implies that as $n \to \infty$, $\int_\Gamma \varphi_n(\zeta)\, d\mathcal{H}^1(\zeta) \to \int_\Gamma \#\gamma^{-1}(\zeta)\, d\mathcal{H}^1(\zeta)$. Hence by $(\star)$, item 1 above, and the Squeeze Rule, we are done. □

Having disposed of arclength, we now turn to arclength measure. Recall two facts from Section 4.5: first, for a rectifiable curve γ the arclength measure of a subset E of the range Γ of γ is defined by $l_\gamma(E) = \mathcal{L}^1(\tilde{\gamma}^{-1}(E))$ where $\tilde{\gamma}$ is the arclength reparametrization of γ, and second, $l_\gamma(\Gamma) = l(\tilde{\gamma}) = l(\gamma)$.

Proposition 5.9 *Let E be a Borel subset of the range of a rectifiable arc γ. Then*

$$l_\gamma(E) = \mathcal{H}^1(E).$$

Proof By Proposition 4.21 applied twice, one has

$$(\star)\ \mathcal{H}^1(E) \leq l_\gamma(E) \text{ and } \mathcal{H}^1(\Gamma \setminus E) \leq l_\gamma(\Gamma \setminus E).$$

We have assumed that E is Borel only to ensure that both $\tilde{\gamma}^{-1}(E)$ and $\tilde{\gamma}^{-1}(\Gamma \setminus E)$ are also Borel which in turn ensures that

$$(\star\star)\ l_\gamma(E) + l_\gamma(\Gamma \setminus E) = l_\gamma(\Gamma) = l(\gamma).$$

Thus by Proposition 5.7, Proposition 2.1, $(\star)$, and $(\star\star)$, we have

$$l(\gamma) = \mathcal{H}^1(\Gamma) \leq \mathcal{H}^1(E) + \mathcal{H}^1(\Gamma \setminus E) \leq l_\gamma(E) + l_\gamma(\Gamma \setminus E) = l(\gamma).$$

The equality of the extreme terms of the above forces the two inequalities in between to be equalities. The second of these forced equalities in turn forces the inequalities of $(\star)$ to be equalities and so we are done. □

Proposition 5.10 *For E a Borel subset of the range of a rectifiable curve γ, set $E_n = \{\zeta \in E : \#\gamma^{-1}(\zeta) = n\}$ for $n = 1, 2, \ldots,$ and ∞. Then*

$$l_\gamma(E) = \int_E \#\gamma^{-1}(\zeta)\, d\mathcal{H}^1(\zeta) = \sum_{1 \leq n \leq \infty} n \cdot \mathcal{H}^1(E_n) = \sum_{1 \leq n < \infty} n \cdot \mathcal{H}^1(E_n).$$

In particular, $l_\gamma(E) = 0$ if and only if $\mathcal{H}^1(E) = 0$.

Proof We again think of partitions as we did in proving Proposition 5.8. Let $\{\mathcal{P}_n\}$ be a sequence of partitions of $[0, l]$ such that

$(\star)$ $\mathcal{P}_{n+1}$ is a refinement of $\mathcal{P}_n$ for each n and $\max\{|I| : I \in \mathcal{P}_n\} \to 0$ as $n \to \infty$.

By Proposition 2.2 and the coincidence of $\mathcal{H}^1$ and $\mathcal{L}^1$ for linear sets (from section 2.1), for any subinterval I of $[0, l]$, we have

$$\mathcal{H}^1(E \cap \tilde{\gamma}(I)) = \mathcal{H}^1(\tilde{\gamma}(\tilde{\gamma}^{-1}(E) \cap I)) \leq \mathcal{H}^1(\tilde{\gamma}^{-1}(E) \cap I) = \mathcal{L}^1(\tilde{\gamma}^{-1}(E) \cap I).$$

Summing over all $I \in \mathcal{P}_n$ and thinking of the measure of a set as the integral of its characteristic function, we obtain

$$(\star\star) \quad \int_E \varphi_n(\zeta)\, d\mathcal{H}^1(\zeta) \leq \sum_{I \in \mathcal{P}_n} \mathcal{L}^1(\tilde{\gamma}^{-1}(E) \cap I) = \mathcal{L}^1(\tilde{\gamma}^{-1}(E)) = l_\gamma(E)$$

where φ_n is the sum of the characteristic functions of the sets $\tilde{\gamma}(I)$ with I varying over all of $\mathcal{P}_n$. A little thought shows that $\varphi_n(\zeta)$ is just the number of intervals I from $\mathcal{P}_n$ which intersect $\tilde{\gamma}^{-1}(\zeta)$. Thus by $(\star)$, for each $\zeta \in \mathbb{C}$ we have $\varphi_n(\zeta) \uparrow$ the number of preimages of ζ under $\tilde{\gamma} = \#\tilde{\gamma}^{-1}(\zeta)$. The Monotone Convergence Theorem [RUD, 1.26] now implies that as $n \to \infty$, $\int_E \varphi_n(\zeta)\, d\mathcal{H}^1(\zeta) \to \int_E \#\tilde{\gamma}^{-1}(\zeta)\, d\mathcal{H}^1(\zeta)$. Hence from $(\star\star)$, we deduce

$$(\dagger) \quad \int_E \#\tilde{\gamma}^{-1}(\zeta)\, d\mathcal{H}^1(\zeta) \leq l_\gamma(E).$$

We may similarly obtain

$$(\ddagger) \quad \int_{\Gamma \setminus E} \#\tilde{\gamma}^{-1}(\zeta)\, d\mathcal{H}^1(\zeta) \leq l_\gamma(\Gamma \setminus E).$$

Adding $(\dagger)$ and $(\ddagger)$, it follows that

$$\begin{aligned}\int_\Gamma \#\tilde{\gamma}^{-1}(\zeta)\, d\mathcal{H}^1(\zeta) &= \int_E \#\tilde{\gamma}^{-1}(\zeta)\, d\mathcal{H}^1(\zeta) + \int_{\Gamma \setminus E} \#\tilde{\gamma}^{-1}(\zeta)\, d\mathcal{H}^1(\zeta) \\ &\leq l_\gamma(E) + l_\gamma(\Gamma \setminus E).\end{aligned}$$

On the one hand, by Proposition 5.8, the leftmost term in the above is just $l(\tilde{\gamma})$, while, on the other hand, the rightmost term in the above is trivially just $l_\gamma(\Gamma) = l(\tilde{\gamma})$. Thus we have equality of the extreme terms in the above. This forces the single inequality in the above to be an equality which in turn forces $(\dagger)$ to be an equality. This is what we want *except* for the presence of a $\tilde{\gamma}$ where we want a γ.

So to finish, it suffices to show that $\#\tilde{\gamma}^{-1}(\zeta) = \#\gamma^{-1}(\zeta)$ for $\mathcal{H}^1$-a.e. $\zeta \in \Gamma$. Recall that $\tilde{\gamma} \circ \varphi = \gamma$, where $\varphi : [a, b] \mapsto [0, l(\gamma)]$ defined by $\varphi(t) = l(\gamma|[a, t])$ is as in Lemma 4.18. Thus for $s \in [0, l(\gamma)]$, $\varphi^{-1}(s)$ is always either a singleton or a nontrivial interval, and so except for countably many exceptional points $s \in [0, l(\gamma)]$, $\varphi^{-1}(s)$ is a singleton. Since $\varphi(\gamma^{-1}(\zeta)) = \tilde{\gamma}^{-1}(\zeta)$ for all ζ, it follows that φ is a bijection between $\gamma^{-1}(\zeta)$ and $\tilde{\gamma}^{-1}(\zeta)$ whenever $\tilde{\gamma}^{-1}(\zeta)$ contains none of these countably many exceptional points s. Pushing these exceptional points s forward to Γ using $\tilde{\gamma}$, we conclude that $\#\tilde{\gamma}^{-1}(\zeta) = \#\gamma^{-1}(\zeta)$ for all but countably many $\zeta \in \Gamma$. Since $\mathcal{H}^1$ has no atoms, we are done. □

5.3 Vitali's Covering Lemma and Planar Lebesgue Measure

The main goal of this section is to show the coincidence of two-dimensional Hausdorff measure with a multiple of planar Lebesgue measure, the multiple being $4/\pi$. One half of this follows from Vitali's Covering Lemma for Hausdorff measure; the other half from a geometric result, the Isodiametric Inequality, which is a consequence of the polar coordinates change of variable formula for planar Lebesgue integrals. We have used planar Lebesgue measure and this change of variable result uncritically till now, e.g., Lemma 1.20, and will do so in the future, e.g., Lemma 6.8 and Lemma P.3, treating both as part of our unproblematic background knowledge. However, during this section the author has elected to get very picky, wiping the slate clean and starting from ground zero: anything needed to reach our main goal will be proved from the definitions of planar Lebesgue measure and the number π which we now give.

The *planar Lebesgue measure* of a subset E of $\mathbb{C}$ is defined by

$$\mathcal{L}^2(E) = \inf\left\{\sum_n \text{Area}(R_n) : \{R_n\} \text{ is a countable cover of } E \text{ by coordinate rectangles}\right\}$$

where a *coordinate rectangle* is simply the Cartesian product of two compact intervals and its *area* is simply the product of the lengths of the two intervals. For convenience, let us take the empty set to be a coordinate rectangle with zero area.

We take π to be the number $\mathcal{L}^2(B(0; 1))$. Since the map $z \in \mathbb{C} \mapsto rz \in \mathbb{C}$, $r > 0$, takes any coordinate rectangle into another coordinate rectangle with area r^2 times that of the old, we see that $\mathcal{L}^2(B(0; r)) = \pi r^2$. Since the map $z \in \mathbb{C} \mapsto z + c \in \mathbb{C}$, $c \in \mathbb{C}$, takes any coordinate rectangle into another coordinate rectangle with the same area, we see that $\mathcal{L}^2(B(c; r)) = \pi r^2$. Thus for any closed ball B, $\mathcal{L}^2(B) = (\pi/4)|B|^2$.

The following may appear obvious but nevertheless is in need of proof!

Proposition 5.11 $\mathcal{L}^2(R) = Area(R)$ *for any coordinate rectangle* R.

Proof Say that a collection $\mathcal{C}$ of coordinate rectangles *nicely decomposes* a coordinate rectangle R if there exist sequences $a_0 \leq a_1 \leq \cdots \leq a_l$ and $b_0 \leq b_1 \leq \cdots \leq b_m$ such that $R = [a_0, a_l] \times [b_0, b_m]$ and $\mathcal{C} = \{[a_{j-1}, a_j] \times [b_{k-1}, b_k] : 1 \leq j \leq l \text{ and } 1 \leq k \leq m\}$. In this situation, we shall also say that $\mathcal{C}$ *arises* from $a_0 \leq a_1 \leq \cdots \leq a_l$ and $b_0 \leq b_1 \leq \cdots \leq b_m$. Then

$$(\star)\ \text{Area}(R) = \sum_{Q \in C} \text{Area}(Q).$$

To prove this note that the left-hand side of $(\star)$ is just $(a_l - a_0) \times (b_m - b_0)$, write $a_l - a_0$ and $b_m - b_0$ as $\sum_{j=1}^{l}(a_j - a_{-1})$ and $\sum_{k=1}^{m}(b_k - b_{k-1})$, respectively, use the distributive law twice, and then recognize the resulting double sum as the right-hand side of $(\star)$.

Considering the trivial cover of R by itself, we see that $\mathcal{L}^2(R) \leq \text{Area}(R)$. Given $\varepsilon > 0$, choose a countable cover $\{R_n\}$ of R by coordinate rectangles such that $\sum_n \text{Area}(R_n) < \mathcal{L}^2(R) + \varepsilon$. By expanding each R_n slightly and using the compactness of R, we may also assume that the R_ns are finite in number. So it suffices to show that $\text{Area}(R) \leq \sum_n \text{Area}(R_n)$ whenever $\{R_n\}$ is a finite collection of coordinate rectangles covering R. Replacing R_n by $R_n \cap R$, we may also assume that each $R_n \subseteq R$. Then the x-coordinates of the vertical sides of the R_ns are finite in number and so can be labeled $a_0 \leq a_1 \leq \cdots \leq a_l$. Similarly, the y-coordinates of the horizontal sides of the R_ns are finite in number and so can be labeled $b_0 \leq b_1 \leq \cdots \leq b_m$. Let C arise from $a_0 \leq a_1 \leq \cdots \leq a_l$ and $b_0 \leq b_1 \leq \cdots \leq b_m$. Then C nicely decomposes R. Also, setting $C_n = \{Q \in C : Q \subseteq R_n\}$, we see that C_n nicely decomposes R_n. Since the R_ns cover R, each $Q \in C$ must be in some C_n. Thus, applying $(\star)$ $1 + n$ times,

$$\text{Area}(R) = \sum_{Q \in C} \text{Area}(Q) \leq \sum_n \sum_{Q \in C_n} \text{Area}(Q) = \sum_n \text{Area}(R_n) < \mathcal{L}^2(R) + \varepsilon.$$

Letting $\varepsilon \downarrow 0$, we are done. □

Proposition 5.12 *The collection $\mathcal{M}_{\mathcal{L}^2}$ of all $\mathcal{L}^2$-measurable subsets of $\mathbb{C}$ is a σ-algebra on $\mathbb{C}$ on which $\mathcal{L}^2$ is countably additive. Moreover, $\mathcal{M}_{\mathcal{L}^2} \supseteq \mathcal{B}(\mathbb{C})$. Thus the restriction of $\mathcal{L}^2$ to $\mathcal{M}_{\mathcal{L}^2}$ is a positive Borel measure on $\mathbb{C}$.*

Proof Clearly $\mathcal{L}^2$ is an outer measure on $\mathbb{C}$. Given $\delta > 0$ and E a subset of $\mathbb{C}$, define $\mathcal{L}^2_\delta(E)$ as $\mathcal{L}^2(E)$ was defined only with the word "cover" replaced by "δ-cover." By $(\star)$ from the previous proof, $\mathcal{L}^2(E) = \mathcal{L}^2_\delta(E)$. Then the proof of Lemma 5.4 with $\mathcal{H}^s_\delta$ replaced by $\mathcal{L}^2_\delta$ shows that $\mathcal{L}^2$ is a metric outer measure. By Proposition 5.3, we are now done. □

The following lemma is a useful fact about s-dimensional Hausdorff measure and will be applied immediately to give us part (b) of our Vitali Covering Lemma for s-dimensional Hausdorff measure.

Lemma 5.13 *Let $s \geq 0$, E be any subset of $\mathbb{C}$ for which $\mathcal{H}^s(E) < \infty$, and $\varepsilon > 0$. Then there exists a $\rho > 0$ such that*

$$\mathcal{H}^s\left(E \cap \bigcup_n U_n\right) < \sum_n |U_n|^s + \varepsilon$$

whenever $\{U_n\}$ *is a sequence of subsets of* $\mathbb{C}$ *with* $\bigcup_n U_n$ $\mathcal{H}^s$ *measurable and* $0 < |U_n| < \rho$ *always.*

Proof From the definition of $\mathcal{H}^s$, there exists a $\rho > 0$ such that

$$(\star)\ \mathcal{H}^s_\rho(E) > \mathcal{H}^s(E) - \varepsilon/2.$$

From the definiton of $\mathcal{H}^s_\rho$, there exists a ρ-cover $\{V_n\}$ of $E \setminus \bigcup_n U_n$ such that

$$(\star\star)\ \sum_n |V_n|^s < \mathcal{H}^s_\rho\left(E \setminus \bigcup_n U_n\right) + \varepsilon/2 \leq \mathcal{H}^s\left(E \setminus \bigcup_n U_n\right) + \varepsilon/2.$$

Then $\{U_n\} \cup \{V_n\}$ is a ρ-cover of E and so from $(\star)$ we have

$$(\star\star\star)\ \mathcal{H}^s(E) < \mathcal{H}^s_\rho(E) + \varepsilon/2 \leq \sum_n |U_n|^s + \sum_n |V_n|^s + \varepsilon/2.$$

Also, by the assumed $\mathcal{H}^s$ measurability of $\bigcup_n U_n$, we know that

$$\mathcal{H}^s\left(E \cap \bigcup_n U_n\right) = \mathcal{H}^s(E) - \mathcal{H}^s\left(E \setminus \bigcup_n U_n\right).$$

Using $(\star\star)$ and $(\star\star\star)$ in this last equality finishes the proof. □

A collection $\mathcal{V}$ of subsets of $\mathbb{C}$ is called a *Vitali class* for a subset E of $\mathbb{C}$ if for every $z \in E$ and every $\delta > 0$, there exists a $U \in \mathcal{V}$ such that $z \in U$ and $0 < |U| < \delta$.

Lemma 5.14 (Vitali's Covering) (a) *Let* E *be a subset of* $\mathbb{C}$ *and* $\mathcal{V}$ *be a Vitali class of closed sets for* E. *Then there exists a countable pairwise disjoint subcollection* $\{U_n\}$ *of* $\mathcal{V}$ *such that either* $\mathcal{H}^s(E \setminus \bigcup_n U_n) = 0$ *or* $\sum_n |U_n|^s = \infty$.

(b) *If* $\mathcal{H}^s(E) < \infty$, *then, given* $\varepsilon > 0$, *we may also require that*

$$\mathcal{H}^s(E) < \sum_n |U_n|^s + \varepsilon.$$

Proof (a) Without loss of generality, $0 < |U| < 1$ for every $U \in \mathcal{V}$. Choose $U_1 \in \mathcal{V}$ arbitrarily. If $E \subseteq U_1$, we are done. Otherwise, set $d_1 = \sup\{|U| : U \in \mathcal{V} \text{ and } U \cap U_1 = \emptyset\}$ and note that $d_1 > 0$. Choose $U_2 \in \mathcal{V}$ disjoint from U_1 such that $|U_2| > d_1/2$. If $E \subseteq U_1 \cup U_2$, we are done. Otherwise, set $d_2 = \sup\{|U| : U \in \mathcal{V} \text{ and } U \cap (U_1 \cup U_2) = \emptyset\}$ and note that $d_2 > 0$. Choose $U_3 \in \mathcal{V}$ disjoint from $U_1 \cup U_2$ such that $|U_3| > d_2/2$. Continue in this manner. If the process ever stops, then we have generated a *finite* pairwise disjoint subcollection $\{U_n\}$ of $\mathcal{V}$ for which $E \setminus \bigcup_n U_n = \emptyset$. Otherwise, the process generates a *countably infinite* pairwise disjoint subcollection $\{U_n\}$ of $\mathcal{V}$. In this case we shall assume that $\sum_n |U_n|^s < \infty$ and finish by showing that $\mathcal{H}^s(E \setminus \bigcup_n U_n) = 0$.

For each n, let B_n be a closed ball centered on U_n of radius $3|U_n|$. Fix N for the moment and suppose $z \in E \setminus \bigcup_{n \le N} U_n$. Choose $U \in \mathcal{V}$ containing z such that $U \cap \bigcup_{n \le N} U_n = \emptyset$. Suppose that U is disjoint from all the U_ns. Then $d_n \ge |U| > 0$ for all $n > N$. However, by the assumed convergence of $\sum_n |U_n|^s$, we have $|U_n| \to 0$ and thus $d_n \to 0$ also. This contradiction shows that there is a smallest integer m such that U intersects U_m. Clearly $m > N$ and $|U| \le d_{m-1} < 2|U_m|$. Hence $z \in U \subseteq B_m \subseteq \bigcup_{n>N} B_n$.

We have thus shown that $E \setminus \bigcup_{n \le N} U_n \subseteq \bigcup_{n>N} B_n$ for all N. Given $\delta > 0$, consider those N large enough so that $6|U_n| < \delta$ for all $n > N$. Then for such N we have

$$\mathcal{H}^s_\delta\left(E \setminus \bigcup_n U_n\right) \le \mathcal{H}^s_\delta\left(E \setminus \bigcup_{n \le N} U_n\right) \le \sum_{n>N} |B_n|^s = 6^s \sum_{n>N} |U_n|^s.$$

Letting $N \to \infty$ and then $\delta \downarrow 0$, we obtain $\mathcal{H}^s(E \setminus \bigcup_n U_n) = 0$ by the assumed convergence of $\sum_n |U_n|^s$.

(b) Given $\varepsilon > 0$, let $\rho > 0$ be as in the last lemma for our set E and our ε. Now rerun the proof of (a) just given only with the phrase "$0 < |U| < 1$" of the first sentence replaced by "$0 < |U| < \rho$" to obtain $\{U_n\}$ as before. If $\sum_n |U_n|^s = \infty$, then (b) holds trivially. Otherwise, by (a) and the last lemma,

$$\mathcal{H}^s(E) \le \mathcal{H}^s\left(E \cap \bigcup_n U_n\right) + \mathcal{H}^s\left(E \setminus \bigcup_n U_n\right) = \mathcal{H}^s\left(E \cap \bigcup_n U_n\right) < \sum_n |U_n|^s + \varepsilon.$$

□

We now know enough to prove half of the main result of the section.

Proposition 5.15 *For any subset E of $\mathbb{C}$,*

$$\mathcal{H}^2(E) \le \frac{4}{\pi}\mathcal{L}^2(E).$$

Proof Given $\varepsilon > 0$, let $\{R_n\}$ be a cover of E by coordinate rectangles such that $\sum_n \text{Area}(R_n) < \mathcal{L}^2(E) + \varepsilon$. By expanding the coordinate rectangles a bit, we may also assume their interiors cover E. Given $\delta > 0$, fix n for the moment. The collection of all closed balls contained in int R_n and of diameter less than δ is a Vitali class for int R_n. Hence (a) of Vitali's Covering Lemma (5.14) produces, for each n, a collection of pairwise disjoint, closed balls $\{B_{n,j}\}_j$ contained in int R_n and with diameters less than δ such that $\mathcal{H}^2(\text{int } R_n \setminus \bigcup_j B_{n,j}) = 0$ or $\sum_j |B_{n,j}|^2 = \infty$. We now eliminate the second alternative. Within each closed ball $B_{n,j}$ one can inscribe a coordinate square $Q_{n,j}$ whose edges have length $|B_{n,j}|/\sqrt{2}$. Thus, by Propositions 5.11 and 5.12,

$$\sum_j |B_{n,j}|^2 = 2\sum_j \text{Area}(Q_{n,j}) = 2\sum_j \mathcal{L}^2(Q_{n,j}) = 2\mathcal{L}^2\left(\bigcup_j Q_{n,j}\right)$$
$$\leq 2\mathcal{L}^2(R_n) = 2\text{Area}(R_n) < \infty.$$

Unfixing n now, since $\mathcal{H}^2_\delta$ is countably subadditive we see that

$$\mathcal{H}^2_\delta(E) \leq \sum_n \mathcal{H}^2_\delta(\text{int}\, R_n) \leq \sum_n\sum_j \mathcal{H}^2_\delta(B_{n,j}) + \sum_n \mathcal{H}^2_\delta\left(\text{int}\, R_n \setminus \bigcup_j B_{n,j}\right)$$
$$\leq \sum_n\sum_j |B_{n,j}|^2 + 0 = \frac{4}{\pi}\sum_n\sum_j \mathcal{L}^2(B_{n,j}) = \frac{4}{\pi}\sum_n \mathcal{L}^2\left(\bigcup_j B_{n,j}\right)$$
$$\leq \frac{4}{\pi}\sum_n \mathcal{L}^2(R_n) = \frac{4}{\pi}\sum_n \text{Area}(R_n) < \frac{4}{\pi}\{\mathcal{L}^2(E)+\varepsilon\}.$$

Letting first $\delta \downarrow 0$ and then $\varepsilon \downarrow 0$, we are done. □

We now make a series of elementary observations with some comments as to why some of them are true (the reader is left to fill in details):

(a) Call a collection of sets *nonoverlapping* if the pairwise intersection of each two distinct sets from the collection has planar Lebesgue measure zero. Then the planar Lebesgue measure of a union of countably many nonoverlapping $\mathcal{L}^2$-measurable sets is just the sum of the planar Lebesgue measures of the sets.
(b) Any point or line segment has planar Lebesgue measure zero.
(c) Consider the reflection map of $\mathbb{C}$ through any fixed point of $\mathbb{C}$. It must preserve the planar Lebesgue measure of any set since it takes any coordinate rectangle into another coordinate rectangle with the same area.
(d) For us a *triangle* will be the closed convex hull of any three noncollinear points and its area will be half the product of the length of a side times the length of the corresponding altitude (that the resulting number does not depend on which side is chosen is a similar triangles argument).
(e) A *coordinate triangle* is a right triangle whose legs are parallel to the coordinate axes. The planar Lebesgue measure of any coordinate triangle is its area (reflect the triangle through the midpoint of its hypotenuse and consider the coordinate rectangle that is the union of the two nonoverlapping coordinate triangles that result).
(f) The planar Lebesgue measure of any triangle with a horizontal side is just its area (consider the vertical line passing through the vertex of the triangle not lying on the horizontal side – two cases result depending on whether the intersection point of this line with the line containing the horizontal side is between the other two vertices or not).

(g) The planar Lebesgue measure of any triangle is just its area (consider the three horizontal lines passing through the three vertices; one of these lines lies between the others and splits the triangle up in a nonoverlapping way into two triangles with a horizontal side).

(h) Just to make it official, define the area of any rectangle, noncoordinate as well as coordinate, to be the product of the lengths of any two of its nonparallel sides. The planar Lebesgue measure of any rectangle is then just its area (consider how a diagonal of the rectangle splits it into two nonoverlapping triangles).

(i) Consider the reflection map of $\mathbb{C}$ about any fixed line of $\mathbb{C}$. It must preserve the planar Lebesgue measure of any set (given a covering of a set by coordinate rectangles, use the subadditivity of $\mathcal{L}^2$ on the collection of reflected rectangles, the last item, the clear invariance of the area of a rectangle under any isometry, and the definition of $\mathcal{L}^2$ to deduce that $\mathcal{L}^2$ of the reflected set is smaller than $\mathcal{L}^2$ of the set; equality now follows as soon as one realizes that reflections are idempotent).

(j) The items amassed to this point justify the ancient method of approximating π by trapping it between the areas of inscribed and circumscribed regular polygons for a ball of unit radius. Thus, for example, using inscribed and circumscribed squares one easily sees that $2 \leq \pi \leq 4$. With more effort, using inscribed and circumscribed octagons one obtains:

$$2.8284\cdots = 2\sqrt{2} \leq \pi \leq 8\sqrt{3 - 2\sqrt{2}} = 3.3137\cdots .$$

(k) Any two perpendicular diameters of a closed ball of radius r determine four nonoverlapping closed quarters of the ball with each closed quarter being mapped onto any other via either a point or line reflection. Thus each of these closed quarters, which we shall call 0th *stage pie-shaped pieces*, has planar Lebesgue measure $\pi r^2/4$.

The angle bisector of any of these closed quarters determines two nonoverlapping closed eighths of the ball mapped onto one another by reflection through the angle bisector. Thus each of the closed eighths arising from the closed quarters, which we shall call 1st *stage pie-shaped pieces*, has planar Lebesgue measure $\pi r^2/8$.

The angle bisector of any of these closed eighths determines two nonoverlapping closed sixteenths of the ball mapped onto one another by reflection through the angle bisector. Thus each of the closed sixteenths arising from the closed eighths, which we shall call 2nd *stage pie-shaped pieces*, has planar Lebesgue measure $\pi r^2/16$.

Continuing in this manner, we see that a closed ball is a nonoverlapping union of nth *stage pie-shaped pieces*, 4×2^n in number, each having planar Lebesgue measure $\pi r^2/(4 \times 2^n)$.

Lemma 5.16 *Let A and B be two n^{th} stage pie-shaped pieces from some closed ball centered at the origin. Suppose that the angle bisectors of A and B are perpendicular. Then for any $a \in A$ and any $b \in B$,*

$$|a|^2 + |b|^2 \leq |a - b|^2 + \frac{4|a||b|}{\sqrt{2^{n-1}}}.$$

Proof Without loss of generality, the angle bisectors of A and B are the positive x- and y-axes, respectively. Write a and b as $a_1 + ia_2$ and $b_1 + ib_2$, respectively. As a matter of pure algebra we have $|a - b|^2 = |a|^2 + |b|^2 - 2(a_1b_1 + a_2b_2)$ and so

$$(\star)\ |a|^2 + |b|^2 \leq |a - b|^2 + 2(|a||b_1| + |a_2||b|).$$

Since the closed right triangle T with vertices the origin, a, and a_1 is contained in the nth stage pie-shaped piece $A \cap B(0; |a|)$, we have $(1/2)|a_1||a_2| = \mathcal{L}^2(T) \leq \mathcal{L}^2(A \cap B(0; |a|)) = \pi|a|^2/(4 \cdot 2^n) \leq |a|^2/2^n$. Also, since the angle at the vertex of A is right ($n = 0$) or acute ($n > 0$), $|a_1| \geq |a_2|$. Thus $|a_2|^2 \leq |a_1||a_2| \leq |a|^2/2^{n-1}$. Similarly, $|b_1|^2 \leq |b|^2/2^{n-1}$. Thus

$$(\star\star)\ |a_2| \leq \frac{|a|}{\sqrt{2^{n-1}}} \text{ and } |b_1| \leq \frac{|b|}{\sqrt{2^{n-1}}}.$$

From the two starred relations our conclusion now follows. □

Proposition 5.17 (Isodiametric Inequality) *For any subset U of $\mathbb{C}$,*

$$\mathcal{L}^2(U) \leq \frac{\pi}{4}|U|^2.$$

The following proof is really just Exercise 1.6 on page 19 of [FALC] with its use of integrals and polar coordinates avoided ... or rather hidden ... due to the author's self-imposed restrictions mentioned at the beginning of this section!

Proof Without loss of generality we may assume that U is closed and that $|U| < \infty$. Thus U is compact. Then, by translating U appropriately, we may assume that $0 \in U \subseteq \{\text{Re } z \geq 0\}$. Cut the southeast quarter of the ball $B(0; |U|)$ up into n^{th} stage pie-shaped pieces A_j, $j = 1, \ldots, 2^n$. Cut the northeast quarter of the ball $B(0; |U|)$ up into n^{th} stage pie-shaped pieces B_j, $j = 1, \ldots, 2^n$. Relabeling if necessary, we may assume the A_js and B_js are paired up so that for each j the angle bisector of A_j is perpendicular to the angle bisector of B_j. Select $a_j \in A_j \cap U$ such that $|a_j| = \max\{|z| : z \in A_j \cap U\}$. Similarly select $b_j \in B_j \cap U$. Note that $|a_j|$, $|b_j|$, and $|a_j - b_j|$ are all less than or equal to $|U|$. Thus by the last lemma,

$$|a_j|^2 + |b_j|^2 \leq |a_j - b_j|^2 + \frac{4|a_j||b_j|}{\sqrt{2^{n-1}}} \leq |U|^2\left\{1 + \frac{4}{\sqrt{2^{n-1}}}\right\}.$$

Therefore

$$\begin{aligned}\mathcal{L}^2(U) &\le \sum_{j=1}^{2^n}\{\mathcal{L}^2(A_j\cap U)+\mathcal{L}^2(B_j\cap U)\}\\ &\le \sum_{j=1}^{2^n}\{\mathcal{L}^2(A_j\cap B(0;|a_j|))+\mathcal{L}^2(B_j\cap B(0;|b_j|))\}\\ &= \sum_{j=1}^{2^n}\frac{\pi}{4\cdot 2^n}\{|a_j|^2+|b_j|^2\}\\ &\le \frac{\pi}{4}|U|^2\left\{1+\frac{4}{\sqrt{2^{n-1}}}\right\}.\end{aligned}$$

Letting $n\to\infty$, we are done. □

We now know enough to prove the rest of the main result.

Proposition 5.18 *For any subset E of $\mathbb{C}$,*

$$\mathcal{H}^2(E)=\frac{4}{\pi}\mathcal{L}^2(E).$$

Proof Given $\varepsilon>0$, let $\{U_n\}$ be a cover of E by subsets of $\mathbb{C}$ such that $\sum_n |U_n|^2 < \mathcal{H}^2(E)+\varepsilon$. Then by the previous proposition,

$$\mathcal{L}^2(E)\le\sum_n \mathcal{L}^2(U_n)\le\frac{\pi}{4}\sum_n|U_n|^2<\frac{\pi}{4}\{\mathcal{H}^2(E)+\varepsilon\}.$$

Letting $\varepsilon\downarrow 0$, we see that $\mathcal{L}^2(E)\le(\pi/4)\mathcal{H}^2(E)$. By Proposition 5.15, we are done. □

5.4 Regularity Properties of Hausdorff Measures

Suppose that $\mu:\mathcal{M}\mapsto[0,\infty]$ is a positive Borel measure defined on a topological space X. Recall from the first section of this chapter that our use of the word "Borel" here means that $\mathcal{M}\supseteq\mathcal{B}(X)$. We shall say that a set $E\in\mathcal{M}$ is *outer regular* for μ if

$$\mu(E)=\inf\{\mu(U): E\subseteq U \text{ and } U \text{ is open}\}$$

and that E is *inner regular* for μ if

$$\mu(E)=\sup\{\mu(K): K\subseteq E \text{ and } K \text{ is compact}\}.$$

If E is both outer and inner regular for μ, then we say that E is *regular* for μ. (Of course, these definitions make sense for any subset E of X if μ is an outer measure on X instead of a positive Borel measure as specified above.)

When every $E \in \mathcal{B}(X)$ is regular for μ, we say, in agreement with [RUD], that μ is a *regular positive Borel measure* on X. It is worth noting that any positive Borel measure on $\mathbb{C}$ which assigns finite mass to each compact set is automatically a regular positive Borel measure [RUD, 2.18]. When every $E \in \mathcal{M}$ is regular for μ, we say that μ is a *totally regular positive Borel measure* on X. This terminology is the author's own – called forth by the ambiguity in [RUD], mentioned in the second paragraph of this chapter, of the term "positive Borel measure." Following usual practice, when we assert that a complex measure ν satisfies a certain property that has only been defined for positive measures what is really meant is that the property in question actually holds for $|\nu|$, the total variation measure of ν. So the reader now knows what a regular or totally regular complex Borel measure is.

Since any nonempty open subset U of $\mathbb{C}$ contains a nontrivial coordinate rectangle, we see, with the help of Propositions 5.18 and 5.11, that $\mathcal{H}^2(U) > 0$ for such U. It follows that $\mathcal{H}^s(U) = \infty$ for any nonempty open subset U of $\mathbb{C}$ and any $s < 2$ (see the definition of Hausdorff dimension in the paragraph just after Corollary 2.3). Thus, given $s < 2$, any nonempty subset of $\mathbb{C}$ with finite $\mathcal{H}^s$-measure is not outer regular for $\mathcal{H}^s$. However, we do have the following different type of outer regularity.

Proposition 5.19 (G_δ-Outer Regularity of $\mathcal{H}^s$) *For any $s \in [0, 2]$ and any subset E of $\mathbb{C}$, there exists a decreasing sequence $\{U_n\}$ of open supersets of E such that $\mathcal{H}^s\left(\bigcap_n U_n\right) = \mathcal{H}^s(E)$. Hence, if E is $\mathcal{H}^s$ measurable with $\mathcal{H}^s(E) < \infty$, then one may write E as $G \setminus N$ where G is a G_δ subset of $\mathbb{C}$ and N is an $\mathcal{H}^s$-null set.*

Proof We may assume that $\mathcal{H}^s(E) < \infty$ (otherwise, taking all the U_ns to be $\mathbb{C}$ works). Let $\{\delta_n\}$ be a sequence decreasing to 0. Then by definition, for each n, we can find an *open* δ_n-cover $\{V_{n,m}\}_m$ of E such that

$$\sum_m |V_{n,m}|^s < \mathcal{H}^s_{\delta_n}(E) + \delta_n.$$

Setting $V_n = \bigcup_m V_{n,m}$, since $E \subseteq \bigcap_n V_n$ and $\{V_{n,m}\}_m$ is a δ_n cover of $\bigcap_n V_n$, we see that

$$\mathcal{H}^s_{\delta_n}(E) \leq \mathcal{H}^s_{\delta_n}\left(\bigcap_n V_n\right) < \mathcal{H}^s_{\delta_n}(E) + \delta_n.$$

Letting $n \to \infty$, we obtain $\mathcal{H}^s\left(\bigcap_n V_n\right) = \mathcal{H}^s(E)$. Setting $U_n = V_1 \cap \cdots \cap V_n$, we are done. □

This G_δ-outer regularity has a curious consequence which we cannot resist stating and proving: the measurability assumptions on the sets occurring in the usual Upper Continuity Property of Measures [RUD, 1.19(d)] can be dispensed with for the *outer measure* $\mathcal{H}^s$!

Proposition 5.20 (Upper Continuity of $\mathcal{H}^s$) *Let E be the union of an increasing sequence $\{E_n\}$ of subsets of $\mathbb{C}$. Then for any $s \in [0, 2]$,*

$$\lim_{n\to\infty} \mathcal{H}^s(E_n) = \mathcal{H}^s(E).$$

Proof By the last proposition, for each n there exists a G_δ-subset G_n of $\mathbb{C}$ containing E_n such that $\mathcal{H}^s(G_n) = \mathcal{H}^s(E_n)$. Setting $H_n = G_n \cap G_{n+1} \cap G_{n+2} \cap \cdots$, we see that $E_n \subseteq H_n \subseteq H_{n+1}$, $\mathcal{H}^s(E_n) = \mathcal{H}^s(H_n)$, and each H_n is Borel and so $\mathcal{H}^s$ measurable (Proposition 5.5). Hence by the usual Upper Continuity Property of Measures [RUD, 1.19(d)],

$$\lim_{n\to\infty} \mathcal{H}^s(E_n) = \lim_{n\to\infty} \mathcal{H}^s(H_n) = \mathcal{H}^s\left(\bigcup_n H_n\right) \geq \mathcal{H}^s(E).$$

Since the reverse inequality is trivial, we are done. □

The situation with regard to inner regularity for Hausdorff measure is in some ways better and in other ways worse than the situation with regard to outer regularity.

Proposition 5.21 (Inner Regularity of $\mathcal{H}^s$ for Some Sets) *For any $s \in [0, 2]$, any $\mathcal{H}^s$-measurable subset E of $\mathbb{C}$ with $\mathcal{H}^s(E) < \infty$, and any $\varepsilon > 0$, there exists a compact subset K of E such that $\mathcal{H}^s(K) > \mathcal{H}^s(E) - \varepsilon$. Hence one may write such an E as $F \cup N$ where F is an F_σ subset of $\mathbb{C}$ and N is an $\mathcal{H}^s$-null set.*

Proof Invoking the G_δ-Outer Regularity of $\mathcal{H}^s$ (5.19), get a decreasing sequence $\{U_n\}$ of open supersets of E such that $\mathcal{H}^s(\bigcap_n U_n) = \mathcal{H}^s(E)$. Set

$$F_{n,m} = \{z \in U_n : \text{dist}(z, \mathbb{C} \setminus U_n) \geq 1/m \text{ and } |z| \leq m\}.$$

For each fixed n, $\{F_{n,m}\}_m$ is an increasing sequence of compact sets whose union is U_n. Hence $\lim_{m\to\infty} \mathcal{H}^s(E \cap F_{n,m}) = \mathcal{H}^s(E)$ by the last proposition. Since $\mathcal{H}^s(E) < \infty$, for every n we may thus choose an index m_n such that $\mathcal{H}^s(E \setminus F_{n,m_n}) < \varepsilon/2^n$. Setting $F = \bigcap_n F_{n,m_n}$, note that F is compact and that

$$\mathcal{H}^s(F) \geq \mathcal{H}^s(E \cap F) = \mathcal{H}^s(E) - \mathcal{H}^s(E \setminus F) \geq \mathcal{H}^s(E) - \sum_n \mathcal{H}^s(E \setminus F_{n,m_n}) > \mathcal{H}^s(E) - \varepsilon.$$

Unfortunately, F is *not* necessarily contained in E and so cannot serve as our K. However, F is contained in $\bigcap_n U_n$. Thus $\mathcal{H}^s(F \setminus E) = 0$ (it is to get this that we need the $\mathcal{H}^s$ measurability of E ... as well as the finiteness of $\mathcal{H}^s(E)$ already used). Invoking the G_δ-Outer Regularity of $\mathcal{H}^s$ (5.19) again, get a decreasing sequence $\{V_n\}$ of open supersets of $F \setminus E$ such that $\mathcal{H}^s(\bigcap_n V_n) = 0$. By the last proposition, $\lim_{n\to\infty} \mathcal{H}^s(F \setminus V_n) = \mathcal{H}^s(F \setminus \bigcap_n V_n) = \mathcal{H}^s(F)$. Thus for all n large enough we have $\mathcal{H}^s(F \setminus V_n) > \mathcal{H}^s(E) - \varepsilon$. Clearly we may take $K = F \setminus V_n$ for any such n. □

The necessity of the finiteness of $\mathcal{H}^s(E)$ for this proposition's truth is shown in [BES4].

We now formally introduce a notion that we have informally used from time to time already. Given a positive or complex measure μ with domain of definition a σ-algebra $\mathcal{M}$ and a set $A \in \mathcal{M}$, the *restriction* of μ to A is the measure μ_A on $\mathcal{M}$ defined by

$$\mu_A(E) = \mu(A \cap E)$$

for every $E \in \mathcal{M}$. It turns out that all regularity problems vanish when one restricts Hausdorff measure to a measurable subset of $\mathbb{C}$ with finite measure!

Proposition 5.22 *Let* $0 \leq s \leq 2$ *and let* A *be an* $\mathcal{H}^s$*- measurable subset of* $\mathbb{C}$ *with* $\mathcal{H}^s(A) < \infty$*. Then* $\mathcal{H}^s_A$ *is a totally regular finite positive Borel measure on* $\mathbb{C}$*.*

Proof Only total regularity is not immediate. So let E be $\mathcal{H}^s$ measurable and let $\varepsilon > 0$.

Inner Regularity of E. By the Inner Regularity of $\mathcal{H}^s$ for Some Sets (5.21), there exists a compact subset K of $A \cap E$ such that $\mathcal{H}^s(K) > \mathcal{H}^s(A \cap E) - \varepsilon$. Clearly K is a compact subset of E such that $\mathcal{H}^s_A(K) > \mathcal{H}^s_A(E) - \varepsilon$.

Outer Regularity of E. By the Inner Regularity of $\mathcal{H}^s$ for Some Sets (5.21), there exists a compact subset K of $A \setminus E$ such that $\mathcal{H}^s(K) > \mathcal{H}^s(A \setminus E) - \varepsilon$. Clearly $U = \mathbb{C} \setminus K$ is an open superset of E such that $\mathcal{H}^s_A(U) < \mathcal{H}^s_A(E) + \varepsilon$. □

The next proposition, which will be needed at a crucial point in the next section and is sufficient for our purpose there, may actually follow from [RUD, 3.14] but the author is not quite sure of this since although it is not clear that a measure as in the proposition must satisfy all of the assumptions of [RUD, 3.14], namely (b) and (e) of [RUD, 2.14], these assumptions may not actually be necessary to the proof of the result as given in [RUD]. In any case, the proposition has a proof that clearly avoids these assumptions on the measure and is rather brief.

Proposition 5.23 *Let* $\mu : \mathcal{M} \to [0, \infty]$ *be a totally regular positive Borel measure on a locally compact Hausdorff space* X*. Then* $C_c(X)$*, the space of compactly supported continuous functions on* X*, is dense in* $L^p(\mu)$ *for* $1 \leq p < \infty$*.*

Proof The class of all μ-measurable simple functions s on X such that

$$\mu(\{x \in X : s(x) \neq 0\}) < \infty$$

is dense in $L^p(\mu)$ for $1 \leq p < \infty$ [RUD, 3.13] (this result involves no subtleties between measure and topology; it is an easy piece of pure measure theory). Thus it suffices, given any $E \in \mathcal{M}$ such that $\mu(E) < \infty$, to be able to find compactly supported continuous functions on X that are arbitrarily close to $\mathcal{X}_E$ in the norm of $L^p(\mu)$.

Since μ is totally regular, given any $\varepsilon > 0$, we may find an open superset U of E such that $\mu(U) < \mu(E) + \varepsilon$ and a compact subset K of E such that $\mu(K) >$

$\mu(E) - \varepsilon$. By Urysohn's Lemma [RUD, 2.12], there exists a function $\varphi \in C_c(X)$ such that $0 \leq \varphi \leq 1$, $K \subseteq \{\varphi = 1\}$, and $\operatorname{spt}(\varphi) \subseteq U$. Then

$$\|\mathcal{X}_E - \varphi\|^p_{L^p(\mu)} = \int |\mathcal{X}_E - \varphi|^p d\mu \leq 2^p \mu(U \setminus K) < 2^{p+1}\varepsilon$$

and so we are done. □

One last subtlety, we have assumed μ *totally* regular in the last proposition since we are viewing $L^p(\mu)$ as the space of equivalence classes, in the standard manner where we identify two functions that are equal μ-a.e., of those $\mathcal{M}$-measurable, not just Borel measurable, functions f for which $\int |f|^p d\mu < \infty$. It is left as an exercise for the reader to show that in the situation of the last proposition, any element of $L^p(\mu)$ has a Borel measurable representative (use the last proposition, [RUD, 3.12], and the assumed total regularity).

5.5 Besicovitch's Covering Lemma and Lebesgue Points

In this section we prove a standard piece of machinery needed in the next chapter: the plentitude of Lebesgue points for functions from $L^p(\mu)$ where μ is a fairly arbitrary measure. The generality of the measure here requires the use of the difficult Besicovitch Covering Lemma (5.26 below) as opposed to the easier Vitali Covering Lemma (5.14). In addition, Besicovitch's Covering Lemma will end up being used many times in Chapter 8. It is true in $\mathbb{R}^n$ generally and not just $\mathbb{R}^2 = \mathbb{C}$ if one replaces the constants 125 and 2,001 occurring in our enunciation of it by other constants dependent on n. The proof given is taken from [MAT3] and requires two preliminary lemmas dealing with the geometry of balls in $\mathbb{R}^2 = \mathbb{C}$.

Lemma 5.24 *Suppose that m closed balls in $\mathbb{C}$ have a point in common and do not contain each others centers. Then $m \leq 5$.*

Proof Without loss of generality, the point common to our balls is 0. Let the center of the nth ball be $c_n = |c_n|e^{i\theta_n}$ where $0 \leq \theta_n < 2\pi$. Relabeling our points, we may assume $0 \leq \theta_1 \leq \theta_2 \leq \ldots \leq \theta_m < 2\pi$. For convenience, set $c_{m+1} = c_1$ and $\theta_{m+1} = \theta_1 + 2\pi$. Notice that our second hypothesis on the balls says that in each triangle with vertices 0, c_n, and c_{n+1}, the side $[c_n, c_{n+1}]$ is greater than the sides $[0, c_n]$ and $[0, c_{n+1}]$. In any triangle the greater side subtends the greater angle (Euclid I.18) and the three interior angles of a triangle are equal to two right angles (Euclid I.32)! In consequence, each $\theta_{n+1} - \theta_n > \pi/3$, and so

$$m \cdot \frac{\pi}{3} < \sum_{n=1}^{m} (\theta_{n+1} - \theta_n) = 2\pi.$$

Hence $m < 6$. Since m is an integer, we must have $m \leq 5$. □

Given a closed ball B and $\alpha > 0$, let αB denote the closed ball concentric with B and of radius α times that of B.

Lemma 5.25 *Suppose that B and B' are intersecting closed balls in $\mathbb{C}$ with the radius of B being at most twice the radius of B'. Then there exists a closed ball B'' contained in $(2B) \cap B'$ and of radius half that of B.*

Proof By translating and rotating appropriately, we may assume that $B = B(0; r)$ and $B' = B(c'; r')$ with $c' \geq 0$. By hypothesis, $c' \leq r' + r$ and $r \leq 2r'$. We must specify c'' so that $B'' = B(c''; r/2)$ is contained in $B(0; 2r) \cap B(c'; r')$.

Case One ($c' \geq 3r/2$). Here $c'' = 3r/2$ works. Clearly, $B(c''; r/2) = B(3r/2; r/2) \subseteq B(0; 2r)$. Also, $|c' - c''| = |c' - 3r/2| = c' - 3r/2 \leq r' - r/2$. Hence $B(c''; r/2) \subseteq B(c'; |c' - c''| + r/2) \subseteq B(c'; r')$.

Case Two ($0 \leq c' < 3r/2$). Here $c'' = c'$ works. Clearly, $B(c''; r/2) = B(c'; r/2) \subseteq B(c'; r')$. Also, $|c''| = |c'| < 3r/2$. Hence $B(c''; r/2) \subseteq B(0; |c''| + r/2) \subseteq B(0; 2r)$. □

Lemma 5.26 (Besicovitch's Covering) *Suppose E is a bounded subset of $\mathbb{C}$ and $\mathcal{B}$ is a collection of nontrivial closed balls such that each point of E is the center of some ball from $\mathcal{B}$. Then:*

(a) *There exists a countable subcollection $\{B_n\}$ of $\mathcal{B}$ that covers E such that each point of $\mathbb{C}$ lies in at most 125 balls of $\{B_n\}$, i.e.,*

$$\mathcal{X}_E \leq \sum_n \mathcal{X}_{B_n} \leq 125.$$

(b) *There exist countable subcollections $\mathcal{B}_1, \ldots, \mathcal{B}_{2001}$ of $\mathcal{B}$ that collectively cover E with each subcollection consisting of pairwise disjoint balls, i.e.,*

$$E \subseteq \bigcup_{n=1}^{2001} \bigcup \mathcal{B}_n \text{ and } B \cap B' = \emptyset \text{ for all distinct } B, B' \in \mathcal{B}_n \text{ and all } n = 1, \ldots, 2{,}001.$$

In (b) we allow $\mathcal{B}_n = \mathcal{B}_m$ for some $n \neq m$ to get the number of subcollections up to 2,001 exactly! If one insists on distinct subcollections, then one should talk of $\mathcal{B}_1, \ldots, \mathcal{B}_N$ where $N \leq 2{,}001$ instead of $\mathcal{B}_1, \ldots, \mathcal{B}_{2001}$. We will only use (b) once in this book—in the proof of Lemma 8.80; for all other applications of Lemma 5.26 that we will make, (a) suffices.

Proof (a) For each $z \in E$, we may select one $B(z; r(z)) \in \mathcal{B}$ and assume $\mathcal{B} = \{B(z; r(z)) : z \in E\}$. We may also assume that

$$M_1 = \sup\{r(z) : z \in E\} < \infty$$

(for otherwise, our bounded set E could be covered by a single ball of $\mathcal{B}$). Pick $z_1 \in E$ such that $r(z_1) > M_1/2$ and set $B_1 = B(z_1; r(z_1))$. Pick $z_2 \in E \setminus B_1$ such that $r(z_2) > M_1/2$ and set $B_2 = B(z_2; r(z_2))$. Pick $z_3 \in E \setminus (B_1 \cup B_2)$ such that $r(z_3) > M_1/2$ and set $B_3 = B(z_3; r(z_3))$. Continuing in this manner, the process must eventually terminate since E is bounded, thus producing the balls of the *first tranche*: $B_n = B(z_n; r(z_n)), n = 1, 2, \ldots, n_1$.

Set

$$M_2 = \sup \left\{ r(z) : z \in E \setminus \bigcup_{n=1}^{n_1} B_n \right\}$$

and pick $z_{n_1+1} \in E \setminus \bigcup_{n=1}^{n_1} B_n$ such that $r(z_{n_1+1}) > M_2/2$. Pick $z_{n_1+2} \in E \setminus \bigcup_{n=1}^{n_1+1} B_n$ such that $r(z_{n_1+2}) > M_2/2$. Continuing in this manner, the process must eventually terminate since E is bounded, thus producing the balls of the *second tranche*: $B_n = B(z_n; r(z_n)), n = n_1 + 1, n_1 + 2, \ldots, n_2$.

Continuing in this manner, two alternatives present themselves: either the process terminates producing only finitely many tranches or the process never terminates producing countably many tranches. We complete the argument for the second alternative only, the argument for the first alternative being easier. Our process produces a sequence of decreasing positive numbers $\{M_k\}$, a sequence of balls $\{B_n\} = \{B(z_n; r(z_n))\}$, and a sequence of increasing integers $\{n_k\}$ determining the tranches into which the balls fall, so that the kth tranche of balls is $\{B_n : n \in \mathcal{T}_k\}$ where $\mathcal{T}_k = \{n_{k-1}+1, n_{k-1}+2, \ldots, n_k\}$ (set $n_0 = 0$). These satisfy the following:

1. $M_{k+1} \leq M_k/2$ for all k,
2. $z_{n+1} \in E \setminus \bigcup_{m=1}^{n} B_m$ for all n,
3. $M_k/2 < r(z_n) \leq M_k$ whenever B_n comes from the kth tranche, and
4. $z_m \notin B_n$ and $z_n \notin B_m$ whenever B_m and B_n come from different tranches.

Items 1, 2, and 3 are immediate from our construction. To verify item 4, suppose that $m < n$ with B_m from the jth tranche and B_n from the kth tranche. Of course we must have $j < k$. By item 2, $z_n \notin B_m$. But then, using items 1 and 3, $|z_m - z_n| > r(z_m) > M_j/2 \geq M_{j+1} \geq M_k \geq r(z_n)$, so $z_m \notin B_n$ also.

The following three claims will easily finish the proof of (a).

First Claim. E is covered by the sequence of balls $\{B_n\}$.

For if one had $z \in E \setminus \bigcup_n B_n$, then one would have $r(z) \leq M_k$ for all k. This is a contradiction since $r(z) > 0$ but $M_k \downarrow 0$ by item 1.

Second Claim. For any $z \in \mathbb{C}$, the balls B_n that contain z come from at most five different tranches.

Because of item 4, this follows immediately from the penultimate Lemma 5.24.

Third Claim. For any $z \in \mathbb{C}$, at most 25 balls B_n from a given tranche can contain z.

Supposing that z is contained in balls $B_{n_1}, B_{n_2}, \ldots, B_{n_m}$ from a single tranche, say the kth one, we will show that $m \leq 25$ (thus these indices $n_1, n_2, \ldots$ are not those appearing earlier which marked out the tranches). Considering any two indices n_i and n_j, one is smaller – say n_i. Then by items 2 and 3 we have $|z_{n_i} - z_{n_j}| >$

$r(z_{n_i}) > M_k/2$ and so conclude that the balls $B(z_{n_1}; M_k/4)$, $B(z_{n_2}; M_k/4)$, ..., $B(z_{n_m}; M_k/4)$ are pairwise disjoint. But $|z - z_{n_j}| \leq r(z_{n_j}) \leq M_k$ by item 3, so all of these balls are contained in $B(z; 5M_k/4)$. Hence

$$m \cdot \pi(M_k/4)^2 = \mathcal{L}^2\left(\bigcup_{j=1}^{m} B(z_{n_j}; M_k/4)\right) \leq \mathcal{L}^2(B(z; 5M_k/4)) = \pi(5M_k/4)^2.$$

It now easily follows that $m \leq 25$.

(b) By items 1 and 3, $r(z_n) \to 0$ and so we may, and do, reorder the B_ns so that the sequence $\{r(z_n)\}$ is decreasing. Set $B_{1,1} = B_1$. Set $B_{1,2} = B_n$ where n is the smallest index such that $B_n \cap B_{1,1} = \emptyset$. Set $B_{1,3} = B_n$ where n is the smallest index such that $B_n \cap (B_{1,1} \cup B_{1,2}) = \emptyset$. Continuing in this manner, we generate a finite or countably infinite subcollection $\mathcal{B}_1 = \{B_{1,1}, B_{1,2}, B_{1,3}, \ldots\}$ of $\{B_n\}$ consisting of pairwise disjoint balls.

If E is not covered by $\mathcal{B}_1$, then start all over and set $B_{2,1} = B_n$ where n is the smallest index such that $B_n \notin \mathcal{B}_1$. Set $B_{2,2} = B_n$ where n is the smallest index such that $B_n \notin \mathcal{B}_1$ and $B_n \cap B_{2,1} = \emptyset$. Set $B_{2,3} = B_n$ where n is the smallest index such that $B_n \notin \mathcal{B}_1$ and $B_n \cap (B_{2,1} \cup B_{2,2}) = \emptyset$. Continuing in this manner, we generate a finite or countably infinite subcollection $\mathcal{B}_2 = \{B_{2,1}, B_{2,2}, B_{2,3}, \ldots\}$ of $\{B_n\}$ consisting of pairwise disjoint balls.

If E is not covered by $\mathcal{B}_1 \cup \mathcal{B}_2$, then start all over and set $B_{3,1} = B_n$ where n is the smallest index such that $B_n \notin \mathcal{B}_1 \cup \mathcal{B}_2$. Set $B_{3,2} = B_n$ where n is the smallest index such that $B_n \notin \mathcal{B}_1 \cup \mathcal{B}_2$ and $B_n \cap B_{3,1} = \emptyset$. Set $B_{3,3} = B_n$ where n is the smallest index such that $B_n \notin \mathcal{B}_1 \cup \mathcal{B}_2$ and $B_n \cap (B_{3,1} \cup B_{3,2}) = \emptyset$. Continuing in this manner, we generate a finite or countably infinite subcollection $\mathcal{B}_3 = \{B_{3,1}, B_{3,2}, B_{3,3}, \ldots\}$ of $\{B_n\}$ consisting of pairwise disjoint balls.

Continue generating subcollections in this manner. To finish it suffices to show that if E is not covered by $\mathcal{B}_1 \cup \cdots \cup \mathcal{B}_m$, then $m \leq 2{,}000$.

So suppose that $z \in E \setminus (\mathcal{B}_1 \cup \cdots \cup \mathcal{B}_m)$. Since E is covered by $\{B_n\}$, z is in some B_n. Clearly $B_n \notin \mathcal{B}_j$ for $j = 1, 2, \ldots, m$, and so by construction there must exist indices $n_j \leq n$ such that B_n intersects $B_{n_j} \in \mathcal{B}_j$ (thus these indices $n_1, n_2, \ldots$ are not those appearing earlier which marked out the tranches nor those involved in establishing the Third Claim). By the way we reordered our balls, each $r(z_{n_j}) \geq r(z_n)$. To simplify our notation somewhat, write B_n, B_{n_j}, and $r(z_n)$ as B, B'_j, and r, respectively. Then we may apply the last lemma to each pair of balls B and B'_j to get balls B''_j contained in $(2B) \cap B'_j$ and of radius $r/2$. Note that each point of $\mathbb{C}$ is contained in at most 125 of the balls B''_j since the same is true of the balls B'_j. Hence

$$m \cdot \pi(r/2)^2 = \int \sum_{j=1}^{m} \mathcal{X}_{B''_j} \, d\mathcal{L}^2 \leq \int 125 \mathcal{X}_{2B} \, d\mathcal{L}^2 = 125 \cdot \pi(2r)^2.$$

It now easily follows that $m \leq 2{,}000$. □

Let μ and ν be two positive Borel measures on $\mathbb{C}$ with μ being *locally finite*, i.e., with $\mu(K) < \infty$ for every compact subset K of $\mathbb{C}$. The *maximal function* of ν with respect to μ is the function $\mathcal{M}_\mu \nu$ with domain the support of μ and range contained in $[0, \infty]$ defined by

$$(\mathcal{M}_\mu \nu)(z) = \sup_{r>0} \frac{\nu(B(z; r))}{\mu(B(z; r))}.$$

Proposition 5.27 (Maximal Function Inequality) *Let μ and ν be two positive Borel measures on $\mathbb{C}$ with μ being locally finite. Then for $\lambda > 0$,*

$$E_\lambda = \{z \in \operatorname{spt}(\mu) : (\mathcal{M}_\mu \nu)(z) > \lambda\}$$

is a Borel set and

$$\mu(E_\lambda) \le 125 \frac{\nu(\mathbb{C})}{\lambda}.$$

Of course the second conclusion only has content when ν is a finite measure!

Proof We will show that E_λ is Borel by showing that it is relatively open in $\operatorname{spt}(\mu)$. So we let z be a point of E_λ and show there exists a $\delta > 0$ such that $B(z; \delta) \cap \operatorname{spt}(\mu) \subseteq E_\lambda$. By the definition of E_λ, there exists an $r > 0$ such that $\nu(B(z; r)) > \lambda\mu(B(z; r))$. Of course there then exists an $\varepsilon > 0$ such that

$$(\star)\ \nu(B(z; r)) > (1 + \varepsilon)\lambda\mu(B(z; r)).$$

By the local finiteness assumption and the Lower Continuity Property of Measures [RUD, 1.19(e)], there exists a $\delta > 0$ such that

$$(\star\star)\ \mu(B(z; r + 2\delta)) < (1 + \varepsilon)\mu(B(z; r)).$$

Then for any $w \in B(z; \delta) \cap \operatorname{spt}(\mu)$, we have

$$(\star\star\star)\ \nu(B(w; r + \delta)) \ge \nu(B(z; r)) \text{ and } \mu(B(z; r + 2\delta)) \ge \mu(B(w; r + \delta)).$$

From $(\star)$, $(\star\star)$, and $(\star\star\star)$ we deduce that $\nu(B(w; r + \delta)) > \lambda\mu(B(w; r + \delta))$ and so $w \in E_\lambda$.

For each $z \in E_\lambda$, choose a closed ball B_z centered at z such that $\nu(B_z) > \lambda\mu(B_z)$. Then for any $R > 0$, we may apply (a) of Besicovitch's Covering Lemma (5.26) to the bounded set $E_\lambda \cap B(0; R)$ and the collection $\mathcal{B} = \{B_z : z \in E_\lambda\}$ to obtain a countable subcollection $\{B_n\}$ of $\mathcal{B}$ that covers $E_\lambda \cap B(0; R)$ and has a bounded overlap of at most 125. Thus

$$\mu(E_\lambda \cap B(0; R)) \le \sum_n \mu(B_n) \le \frac{1}{\lambda} \sum_n \nu(B_n) = \frac{1}{\lambda} \int \sum_n \mathcal{X}_{B_n}\, d\nu \le 125 \frac{\nu(\mathbb{C})}{\lambda}.$$

Letting $R \uparrow \infty$ and invoking the Upper Continuity Property of Measures [RUD, 1.19(d)], we are done. □

The following result will be needed in proving Proposition 6.13 of the next chapter.

Proposition 5.28 (Existence of Lebesgue Points) *Let $f \in L^p(\mu)$ where μ is a totally regular locally finite positive Borel measure on $\mathbb{C}$ and $1 \leq p < \infty$. Then for μ-a.e. $z \in \mathbb{C}$,*

$$\lim_{r \downarrow 0} \frac{1}{\mu(B(z;r))} \int_{B(z;r)} |f(\zeta) - f(z)|^p \, d\mu(\zeta) = 0.$$

A point z as in the conclusion of this theorem is called a *Lebesgue point* for f.

Proof Defining

$$f^*(z) = \limsup_{r \downarrow 0} \left\{ \frac{1}{\mu(B(z;r))} \int_{B(z;r)} |f(\zeta) - f(z)|^p \, d\mu(\zeta) \right\}^{1/p}$$

for $z \in \operatorname{spt}(\mu)$, it suffices to show that $f^*(z) = 0$ for μ-a.e. $z \in \operatorname{spt}(\mu)$.

Given $\varepsilon > 0$, use Proposition 5.23 to choose g continuous and compactly supported on $\mathbb{C}$ such that $\|f - g\|_{L^p(\mu)} < \varepsilon$ and set $h = f - g$. Trivially, $g^* = 0$ by the continuity of g. Thus, using Minkowski's Inequality [RUD, 3.5] twice,

$$f^* = (g+h)^* \leq g^* + h^* = h^* \leq \{\mathcal{M}_\mu(|h|^p d\mu)\}^{1/p} + |h|.$$

Hence for any $\lambda > 0$,

$$\begin{aligned} (\star)\ & \mu(\{z \in \operatorname{spt}(\mu) : f^*(z) > \lambda\}) \\ & \leq \mu(\{z \in \operatorname{spt}(\mu) : \{\mathcal{M}_\mu(|h|^p d\mu)\}^{1/p}(z) > \lambda/2\}) \leftarrow \text{(I)!} \\ & + \mu(\{z \in \operatorname{spt}(\mu) : |h(z)| > \lambda/2\}). \leftarrow \text{(II)!} \end{aligned}$$

Now by the last Proposition,

$$\begin{aligned} (\star\star)\ \text{(I)} &= \mu(\{z \in \operatorname{spt}(\mu) : \{\mathcal{M}_\mu(|h|^p d\mu)\}(z) > (\lambda/2)^p\}) \\ &\leq 125 \frac{\int |h|^p \, d\mu}{(\lambda/2)^p} < 125 \frac{(2\varepsilon)^p}{\lambda^p}. \end{aligned}$$

We also clearly have

$$(\star\star\star)\ \text{(II)} \leq \frac{\int |h|^p \, d\mu}{(\lambda/2)^p} < \frac{(2\varepsilon)^p}{\lambda^p}.$$

The relations $(\star)$, $(\star\star)$, and $(\star\star\star)$ imply that $\mu(\{z \in \operatorname{spt}(\mu) : f^*(z) > \lambda\}) < 126(2\varepsilon)^p/\lambda^p$. Letting first $\varepsilon \downarrow 0$ and then $\lambda \downarrow 0$, we see that $\mu(\{z \in \operatorname{spt}(\mu) : f^*(z) > 0\}) = 0$. Since $f^* \geq 0$, we are done. □

Chapter 6
A Solution to Vitushkin's Conjecture Modulo Two Difficult Results

6.1 Statement of the Conjecture and a Reduction

From two chapters ago we learned that hitting some rectifiable curve in a set of positive linear Hausdorff measure implies nonremovability (an easy consequence of Theorem 4.27). However, the Joyce–Mörters set that concluded that chapter showed that the converse of this result fails in general (Theorem 4.33). We were in a similar situation before in Chapter 2 when we learned that the converse of Painlevé's Theorem (2.7) fails. There we asked if the converse becomes true when the compact set in question is restricted in some suitable way – in that case, restricted to be a subset of a rectifiable curve. The restriction to make in the case now facing us is clear because the Joyce–Mörters set, although of Hausdorff dimension one, has *non-σ-finite* linear Hausdorff measure (Proposition 4.34). Thus what has come to be known as *Vitushkin's Conjecture* is simply the restriction of our converse to compact sets of *finite* linear Hausdorff measure: For every compact subset K of $\mathbb{C}$,

> if K is nonremovable *and* $\mathcal{H}^1(K) < \infty$, then $\mathcal{H}^1(K \cap \Gamma) > 0$ for some rectifiable curve Γ.

This chapter is devoted to proving this conjecture assuming two difficult results: a $T(b)$ theorem due to Nazarov, Treil, and Volberg and a curvature theorem due to David and Léger. The two chapters immediately following this one will then establish these results.

In proving Vitushkin's Conjecture, it will be very convenient to be able to assume that the restriction measure $\mathcal{H}^1_K$ has linear growth (recall the definition in the paragraph after the statement of Melnikov's Lower Capacity Estimate (4.3)). That this convenience is justified is the point of the rest of this section which is devoted to showing that the full Vitushkin Conjecture follows from the validity of a stronger conjecture for compact sets K whose associated restriction measure $\mathcal{H}^1_K$ has linear growth!

Proposition 6.1 *Suppose E is a compact subset of $\mathbb{C}$ such that $\mathcal{H}^1(E) < \infty$. Then, given $\varepsilon > 0$, there exists a compact subset K of $\mathbb{C}$ such that*

J.J. Dudziak, *Vitushkin's Conjecture for Removable Sets*, Universitext,
DOI 10.1007/978-1-4419-6709-1_6, © Springer Science+Business Media, LLC 2010

(a) $\gamma(K) \geq \gamma(E)$,
(b) $\mathcal{H}^1(K \vartriangle E) < \varepsilon$, *and*
(c) $\mathcal{H}^1_K$ *has linear growth.*

Here $K \vartriangle E$ denotes the *symmetric difference* of the sets K and E, i.e., the set $(K \setminus E) \cup (E \setminus K)$.

If one takes the set K produced by this proposition, considers the set $E \cap K$, and then rechristens it K, one obtains the following corollary which will prove useful in Chapter 8.

Corollary 6.2 *Suppose E is a compact subset of $\mathbb{C}$ such that $\mathcal{H}^1(E) < \infty$. Then, given $\varepsilon > 0$, there exists a compact subset K of E such that $\mathcal{H}^1(E \setminus K) < \varepsilon$ and $\mathcal{H}^1_K$ has linear growth.*

Before proving the proposition, let us see that it indeed delivers the goods in conjunction with the following.

Reduction 6.3 *Let K be a compact subset of $\mathbb{C}$ such that $\mathcal{H}^1_K$ has linear growth. Then there exists a countable family of rectifiable curves whose union Γ satisfies*

$$\mathcal{H}^1_K(\Gamma) \geq \gamma(K).$$

Note that the conclusion here is $\mathcal{H}^1_K(\Gamma) \geq \gamma(K)$ and not merely $\mathcal{H}^1_K(\Gamma) > 0$. Thus the desired *qualitative* conclusion of Vitushkin's Conjecture has been replaced by the stronger *quantitative* conclusion of the reduction. This is the price to be paid for the linear growth that the reduction assumes and the proposition provides!

Theorem 6.4 *Let E be a compact subset of $\mathbb{C}$ such that $\mathcal{H}^1(E) < \infty$. Then there exists a countable family of rectifiable curves whose union Γ satisfies*

$$\mathcal{H}^1_E(\Gamma) \geq \gamma(E).$$

Proof of Theorem 6.4 *from Proposition* 6.1 *and Reduction* 6.3 Let $\{\varepsilon_n\}$ be a sequence decreasing to 0. Let K_n be gotten by applying Proposition 6.1 to E and ε_n and then let Γ_n be gotten by applying Reduction 6.3 to K_n. Clearly $\Gamma = \bigcup_n \Gamma_n$ is the union of a countable family of rectifiable curves. Moreover,

$$\begin{aligned}
\mathcal{H}^1_E(\Gamma) &\geq \mathcal{H}^1(E \cap K_n \cap \Gamma_n) \\
&= \mathcal{H}^1(K_n \cap \Gamma_n) - \mathcal{H}^1(\{K_n \cap \Gamma_n\} \setminus E) \\
&\geq \gamma(K_n) - \mathcal{H}^1(K_n \vartriangle E) \\
&> \gamma(E) - \varepsilon_n.
\end{aligned}$$

Letting $n \to \infty$, we are done. □

Clearly this last theorem and Theorem 4.27 immediately yield the following which we label a theorem as opposed to a corollary due to its importance ... after all it is the whole point of this book!

Theorem 6.5 (Vitushkin's Conjecture Resolved) *Suppose K is a compact subset of $\mathbb{C}$ such that $\mathcal{H}^1(K) < \infty$. Then K is nonremovable if and only if $\mathcal{H}^1(K \cap \Gamma) > 0$ for some rectifiable curve Γ.*

The rest of this section is devoted to proving Proposition 6.1. Reduction 6.3 will then be verified in the remaining sections of this chapter ... and the remaining chapters of this book!

Proof of Proposition 6.1 By Lemma 5.13, there exists a number $\rho > 0$ such that

$$(\star)\quad \mathcal{H}^1(E \cap \bigcup_n U_n) < \sum_n |U_n| + \frac{\varepsilon}{2}$$

for any countable collection of Borel sets $\{U_n\}$ such $0 < |U_n| < \rho$ always. Choose $M_0 > \pi$ so that

$$(\star\star)\quad \frac{\mathcal{H}^1(E)}{M_0 - \pi} < \rho \text{ and } \frac{(\pi + 1)\mathcal{H}^1(E)}{M_0 - \pi} < \frac{\varepsilon}{2}.$$

Set $K_0 = E$ and consider the collection $\mathcal{F}_1$ of all closed balls B centered on K_0 such that $\mathcal{H}^1(K_0 \cap B) > M_0|B|$. If $\mathcal{F}_1 = \emptyset$, then $K = K_0$ works. So we may as well assume that $\mathcal{F}_1 \neq \emptyset$. Choose $B_1 \in \mathcal{F}_1$ such that $|B_1| > \frac{1}{2}\sup\{|B| : B \in \mathcal{F}_1\}$. Set $K_1 = \{E \cup B_1\} \setminus \operatorname{int} B_1$ and consider the collection $\mathcal{F}_2$ of all balls B centered on K_1 such that $\mathcal{H}^1(K_1 \cap B) > M_0|B|$. If $\mathcal{F}_2 \neq \emptyset$, choose $B_2 \in \mathcal{F}_2$ such that $|B_2| > \frac{1}{2}\sup\{|B| : B \in \mathcal{F}_2\}$. Set $K_2 = \{E \cup (B_1 \cup B_2)\} \setminus \operatorname{int}(B_1 \cup B_2)$ and consider the collection $\mathcal{F}_3$ of all balls B centered on K_2 such that $\mathcal{H}^1(K_2 \cap B) > M_0|B|$. Proceeding in this manner, we may suppose that the process continues through step N and produces

$$K_N = \left\{E \cup \bigcup_{n \le N} B_n\right\} \setminus \operatorname{int} \bigcup_{n \le N} B_n.$$

Since $K_n = (K_n \setminus B_n) \cup (K_n \cap B_n) \subseteq (K_{n-1} \setminus B_n) \cup \partial B_n$, we see that

$$\mathcal{H}^1(K_n) \le \mathcal{H}^1(K_{n-1}) - \mathcal{H}^1(K_{n-1} \cap B_n) + \mathcal{H}^1(\partial B_n) < \mathcal{H}^1(K_{n-1}) - M_0|B_n| + \pi|B_n|,$$

and so

$$(M_0 - \pi)|B_n| < \mathcal{H}^1(K_{n-1}) - \mathcal{H}^1(K_n).$$

Summing this from $n = 1$ to N, noting the obvious inequality on the telescoping side, and then dividing by $M_0 - \pi$, we obtain

$$(\star\star\star)\quad \sum_{n \le N} |B_n| < \frac{\mathcal{H}^1(E)}{M_0 - \pi}.$$

Case One. Suppose the process terminates, i.e., suppose $\mathcal{F}_{N+1} = \emptyset$ for some $N \geq 1$.
We claim that $K = K_N$ works.

(a) By Proposition 1.10,

$$\gamma(K) = \gamma(\widehat{K}) \geq \gamma\left(E \cup \bigcup_{n \leq N} B_n\right) \geq \gamma(E).$$

(b) Note that

$$K \bigtriangleup E = (K \setminus E) \cup (E \setminus K) \subseteq \left\{\bigcup_{n \leq N} \partial B_n\right\} \cup \left\{E \cap \bigcup_{n \leq N} B_n\right\},$$

so by $(\star)$, $(\star\star)$, and $(\star\star\star)$ we obtain

$$\begin{aligned}
\mathcal{H}^1(K \bigtriangleup E) &\leq \sum_{n \leq N} \mathcal{H}^1(\partial B_n) + \mathcal{H}^1\left(E \cap \bigcup_{n \leq N} B_n\right) \\
&< \sum_{n \leq N} \pi |B_n| + \sum_{n \leq N} |B_n| + \frac{\varepsilon}{2} \\
&= (\pi + 1) \sum_{n \leq N} |B_n| + \frac{\varepsilon}{2} \\
&< \frac{(\pi + 1)\mathcal{H}^1(E)}{M_0 - \pi} + \frac{\varepsilon}{2} \quad < \quad \varepsilon.
\end{aligned}$$

(c) Since $\mathcal{F}_{N+1} = \emptyset$, $\mathcal{H}^1(K \cap B(c; r)) \leq 2M_0 r$ for all $c \in K$ and $r > 0$. But what about $c \notin K$? Clearly we need only consider $r \geq \operatorname{dist}(c, K)$ in this case. Choose $\tilde{c} \in K$ such that $|\tilde{c} - c| = \operatorname{dist}(c, K)$. Then by the Triangle Inequality, $B(c; r) \subseteq B(\tilde{c}; 2r)$, so $\mathcal{H}^1(K \cap B(c; r)) \leq \mathcal{H}^1(K \cap B(\tilde{c}; 2r)) \leq 4M_0 r$. We conclude that $\mathcal{H}^1_K$ has linear growth with bound $M = 4M_0$.

Case Two. Suppose the process does not terminate, i.e., suppose $\mathcal{F}_{N+1} \neq \emptyset$ for all N.

We claim that $K = \left\{E \cup \operatorname{cl} \bigcup_n B_n\right\} \setminus \operatorname{int} \bigcup_n B_n$ works.

(a) By Proposition 1.10,

$$\gamma(K) = \gamma(\widehat{K}) \geq \gamma\left(E \cup \mathrm{cl}\bigcup_n B_n\right) \geq \gamma(E).$$

(b) Setting $H = \left\{\mathrm{cl}\bigcup_n B_n\right\} \setminus \left\{E \cup \bigcup_n B_n\right\}$, note that

$$K \bigtriangleup E = (K \setminus E) \cup (E \setminus K) \subseteq H \cup \left\{\bigcup_n \partial B_n\right\} \cup \left\{E \cap \bigcup_n B_n\right\},$$

so by ($\star$), ($\star\star$), and ($\star\star\star$) we obtain

$$\begin{aligned}
\mathcal{H}^1(K \bigtriangleup E) &\leq \mathcal{H}^1(H) + \sum_n \mathcal{H}^1(\partial B_n) + \mathcal{H}^1\left(E \cap \bigcup_n B_n\right) \\
&< \mathcal{H}^1(H) + \sum_n \pi |B_n| + \sum_n |B_n| + \frac{\varepsilon}{2} \\
&= \mathcal{H}^1(H) + (\pi + 1) \sum_n |B_n| + \frac{\varepsilon}{2} \\
&\leq \mathcal{H}^1(H) + \frac{(\pi + 1)\mathcal{H}^1(E)}{M_0 - \pi} + \frac{\varepsilon}{2} \quad < \quad \mathcal{H}^1(H) + \varepsilon.
\end{aligned}$$

Thus to finish verifying (b), it suffices to show that $\mathcal{H}^1(H) = 0$, and for this, it in turn suffices to show that H can be covered by families of balls whose radii sum to be arbitrarily small.

To this end, fix N. The ball B_{N+1} is centered on K_N, so create a "pile" for it. The ball B_{N+2} is centered on K_{N+1}, and so either on K_N or on ∂B_{N+1}. If on ∂B_{N+1}, then throw B_{N+2} into the pile containing B_{N+1}; otherwise, create a new pile for B_{N+2}. The ball B_{N+3} is centered on K_{N+2}, and so either on K_N, on ∂B_{N+1}, or on ∂B_{N+2}. If on ∂B_{N+2}, then throw B_{N+3} into the pile containing B_{N+2}; if not on ∂B_{N+2} but on ∂B_{N+1}, then throw B_{N+3} into the pile containing B_{N+1}; otherwise, create a new pile for B_{N+3}. Proceeding in this manner, we see that $\{B_n : n > N\}$ can be written as $\bigcup_m \mathcal{B}_m$, a disjoint union of families $\mathcal{B}_1$ $\mathcal{B}_2$, $\mathcal{B}_3$, ...of balls, such that each family $\mathcal{B}_m$ consists of balls from $\{B_n : n > N\}$ which may be "chained back" to K_N. More precisely, each $\mathcal{B}_m$ may be written $\{\Delta_1^{(m)}, \Delta_2^{(m)}, \Delta_3^{(m)}, \ldots\}$ with $\Delta_1^{(m)}$ centered on K_N and each $\Delta_{i+1}^{(m)}$ centered on some $\partial\Delta_j^{(m)}$, $j \leq i$. Let c_m denote the center of $\Delta_1^{(m)}$.

Fix $z \in H$. Then $z \notin K_N$ and so $r = \mathrm{dist}(z, K_N) > 0$. Since $z \in \left(\mathrm{cl}\bigcup_{n>N} B_n\right) \setminus \left(\bigcup_{n>N} B_n\right)$ and $|B_n| \to 0$ as $n \to \infty$ by ($\star\star\star$), $B(z; r/2)$ contains some B_n, $n > N$. This $B_n = \Delta_i^{(m)}$ for some m and i. Since $\Delta_i^{(m)}$, with center, say, $c \in B(z; r/2)$, is "chained back" to $\Delta_1^{(m)}$, with center $c_m \in K_N$ and so $\notin \operatorname{int} B(z;r)$, we clearly have

$$\frac{r}{2} \leq |c - c_m| \leq \sum_i \frac{|\Delta_i^{(m)}|}{2}.$$

Thus $|z - c_m| \leq |z - c| + |c - c_m| \leq r/2 + |c - c_m| \leq \sum_i |\Delta_i^{(m)}|$, i.e., $z \in B\big(c_m; \sum_i |\Delta_i^{(m)}|\big)$.

It follows that H is covered by the collection of balls $\big\{B\big(c_m; \sum_i |\Delta_i^{(m)}|\big)\big\}$, whose radii sum to $\sum_m \sum_i |\Delta_i^{(m)}| = \sum_{n>N} |B_n|$. Since $\sum_{n>N} |B_n| \to 0$ as $N \to \infty$ by $(\star\star\star)$, we have shown that $\mathcal{H}^1(H) = 0$.

(c) Suppose that $\mathcal{H}^1(K \cap B(c;r)) > 12M_0 r$ for some $c \in K$ and $r > 0$. Slice $B(c;r)$ up into six $60°$ pie-shaped pieces (distribute their boundaries as you please so that they are pairwise disjoint). Clearly, $\mathcal{H}^1(K \cap \Pi) > 2M_0 r$ for at least one of these pieces Π. For the set H introduced in the proof of (b), we have

$$(\star\star\star\star)\ K = \left(\left\{E \cup \bigcup_n B_n\right\} \setminus \operatorname{int} \bigcup_n B_n\right) \cup H \text{ with } \mathcal{H}^1(H) = 0.$$

Thus we may choose a point p from the set

$$\left(\left\{E \cup \bigcup_n B_n\right\} \setminus \operatorname{int} \bigcup_n B_n\right) \cap \Pi.$$

Clearly, $p \in K_N$ for all N large enough and $\mathcal{H}^1(K \cap B(p;r)) \geq \mathcal{H}^1(K \cap \Pi) > 2M_0 r = M_0 |B(p;r)|$. Also, by Proposition 5.20 and $(\star\star\star\star)$,

$$\begin{aligned}
\mathcal{H}^1(K_N \cap B(p;r)) &\geq \mathcal{H}^1\left(\left(\left\{E \cup \bigcup_{n \leq N} B_n\right\} \setminus \operatorname{int} \bigcup_n B_n\right) \cap B(p;r)\right) \\
&\nearrow \mathcal{H}^1\left(\left(\left\{E \cup \bigcup_n B_n\right\} \setminus \operatorname{int} \bigcup_n B_n\right) \cap B(p;r)\right) \\
&= \mathcal{H}^1(K \cap B(p;r))
\end{aligned}$$

as $N \to \infty$. We conclude that $\mathcal{H}^1(K_N \cap B(p;r)) > M_0|B(p;r)|$ for all N large enough. In consequence, $B(p;r) \in \mathcal{F}_N$ for all N large enough and so $|B_N| > \frac{1}{2}\sup\{|B| : B \in \mathcal{F}_N\} \geq r$ for these N. This is a contradiction since $|B_N| \to 0$ by $(\star\star\star)$.

So we must have $\mathcal{H}^1(K \cap B(c;r)) \le 12M_0 r$ for all $c \in K$ and $r > 0$. We handle the case $c \notin K$ as before and conclude that $\mathcal{H}^1_K$ has linear growth with bound $M = 24M_0$. □

The next three sections will be devoted to assembling the pieces needed to prove Reduction 6.3 from two difficult results, one due to Nazarov, Treil, and Volberg, the other due to David and Léger, which, in the spirit of Just-In-Time Mathematics, will be stated only at their point of application!

6.2 Cauchy Integral Representation

Let K be a compact subset of $\mathbb{C}$ such that $\mathcal{H}^1(K) < \infty$. In this section, we seek to represent any function bounded and analytic off K which vanishes at ∞ as the Cauchy transform of a measure of the form $h\, d\mathcal{H}^1_K$ where $h \in L^\infty(\mathcal{H}^1_K)$. Such a Cauchy transform will be referred to as a *Cauchy integral* in what follows.

Given a subset E of $\mathbb{C}$, the *s-dimensional spherical Hausdorff measure* of E, denoted $\mathcal{S}^s(E)$, is defined just as the s-dimensional Hausdorff measure only with the δ-coverings not being arbitrary but restricted to collections of balls. Proposition 5.5 holds when s-dimensional spherical Hausdorff measure replaces s-dimensional Hausdorff measure. The two measures are related as follows:

$$(\star)\ \mathcal{H}^s(E) \le \mathcal{S}^s(E) \le 2^s \mathcal{H}^s(E)$$

for any subset E of $\mathbb{C}$. The first inequality is totally trivial, while the second follows easily from the observation that any subset U of $\mathbb{C}$ is contained in a ball of radius $|U|$. If one invokes Jung's Theorem instead (recall the paragraph after the proof of Proposition 1.18), then the factor of 2^s in the above may be replaced by $(2/\sqrt{3})^s$, an improvement. Besicovitch has an example for the case $s = 1$ which shows that this is the best one can do (see Section 53 of [BES1]). As an exercise, after finishing this chapter, the reader may want to check that if one uses this Jung-improved version of $(\star)$ in the proof of the next result instead of $(\star)$ itself, then the conclusion of Reduction 6.3 may be improved to $\mathcal{H}^1_K(\Gamma) \ge \sqrt{3}\gamma(K)$!

Proposition 6.6 *Suppose that K is a compact subset of $\mathbb{C}$ such that $\mathcal{H}^1(K) < \infty$ and that f is a nonconstant element of $H^\infty_0(\mathbb{C}^* \setminus K)$. Then there exists a function $h \in L^\infty(\mathcal{H}^1_K)$ such that for all $z \in \mathbb{C} \setminus K$,*

$$f(z) = \int \frac{h(\zeta)}{\zeta - z}\, d\mathcal{H}^1_K(\zeta).$$

Moreover, $\|h\|_\infty \le \|f\|_\infty$ and $\mathcal{H}^1_K(\{h \ne 0\}) \ge |f'(\infty)|/\|f\|_\infty$.

In particular, when f is the Ahlfors function of K the associated h satisfies

$$\|h\|_\infty \le 1 \text{ and } \mathcal{H}^1_K(\{h \ne 0\}) \ge \gamma(K).$$

Proof Fix a closed ball B big enough to contain K and all points within a distance 1 of K. Given ε between 0 and 1, there exists a family $\{B_{\varepsilon,n}\}_n$ of balls of diameter less than ε whose interiors cover K such that

$$(\star\star) \quad \sum_n |B_{\varepsilon,n}| < \mathcal{S}^1(K) + \varepsilon.$$

Since K is compact, we may assume that for each ε there are on only finitely many $B_{\varepsilon,n}$ and that each $B_{\varepsilon,n}$ intersects K. Let Γ_ε be the boundary of $\bigcup_n B_{\varepsilon,n}$ oriented to have winding number 1 about each point of K and consider the continuous linear functional Λ_ε on $C(B)$ defined by

$$\Lambda_\varepsilon(\varphi) = \frac{-1}{2\pi i}\int_{\Gamma_\varepsilon} \varphi(\zeta) f(\zeta)\, d\zeta.$$

From $(\star\star)$ we see that

$$\|\Lambda_\varepsilon\| \le \frac{1}{2\pi}\|f\|_\infty \sum_n \pi |B_{\varepsilon,n}| < \frac{1}{2}\|f\|_\infty \left\{\mathcal{S}^1(K) + 1\right\}.$$

Thus we may apply [RUD, 11.29] and [RUD, 6.19] as in the proof of Frostman's Lemma (2.9) to conclude that there exists a sequence $\{\varepsilon_n\}$ decreasing to 0 such that the sequence $\{\Lambda_{\varepsilon_n}\}$ converges weakly to a continuous linear functional Λ on $C(B)$ represented by a regular complex Borel measure μ on B. This means that for any $\varphi \in C(B)$,

$$(\star\star\star) \quad \lim_{n\to\infty} \frac{-1}{2\pi i}\int_{\Gamma_{\varepsilon_n}} \varphi(\zeta) f(\zeta)\, d\zeta = \lim_{n\to\infty} \Lambda_{\varepsilon_n}(\varphi) = \Lambda(\varphi) = \int \varphi(\zeta)\, d\mu(\zeta).$$

Since each Λ_ε annihilates those continuous functions whose support is a distance more that ε from K, it follows from $(\star\star\star)$ that Λ annihilates those continuous functions compactly supported in $B \setminus K$. Thus by [RUD, 6.19 (2) and 3.17], $|\mu|(B \setminus K) = \|\Lambda | C_c(B \setminus K)\| = 0$, i.e., μ is supported on K.

Given $z \in \mathbb{C}\setminus K$, use Tietze's Extension Theorem [RUD, 20.4] to get a continuous function φ such that $\varphi(\zeta) = 1/(\zeta - z)$ for all ζ whose distance to K is at most half the distance of z to K. By Lemma 2.13, for every $\varepsilon \in (0, 1)$ which is less than half the distance of z to K, we then have

$$f(z) = \frac{-1}{2\pi i}\int_{\Gamma_\varepsilon} \frac{f(\zeta)}{\zeta - z}\, d\zeta = \frac{-1}{2\pi i}\int_{\Gamma_\varepsilon} \varphi(\zeta) f(\zeta)\, d\zeta.$$

Thus from $(\star\star\star)$ we deduce that

$$f(z) = \int \varphi(\zeta)\, d\mu(\zeta) = \int \frac{1}{\zeta - z}\, d\mu(\zeta).$$

Next we must show that $d\mu = h\, d\mathcal{H}^1_K$ for an appropriate h. To this end, let U be any open subset of $\mathbb{C}$ and φ be any continuous function on $\mathbb{C}$ compactly supported in U with $\|\varphi\|_\infty \leq 1$. Consider only those $\varepsilon \in (0, 1)$ which are less than the distance between the support of φ and the compliment of U. Let I_ε consist of those indices n such that $B_{\varepsilon,n}$ intersects the support of φ and let J_ε consist of all the remaining indices. Then $K \setminus U$ must be covered by $\{B_{\varepsilon,n} : n \in J_\varepsilon\}$ and so $\sum_{n\in J_\varepsilon} |B_{\varepsilon,n}| \geq \mathcal{S}^1_\varepsilon(K \setminus U)$. Consequently,

$$\begin{aligned}
|\Lambda_\varepsilon(\varphi)| &= \left| \frac{-1}{2\pi i} \int_{\Gamma_\varepsilon} \varphi(\zeta) f(\zeta)\, d\zeta \right| \\
&\leq \frac{1}{2\pi} \|f\|_\infty \sum\nolimits_{n\in I_\varepsilon} \pi |B_{\varepsilon,n}| \\
&\leq \frac{1}{2} \|f\|_\infty \left\{ \sum_n |B_{\varepsilon,n}| - \sum\nolimits_{n\in J_\varepsilon} |B_{\varepsilon,n}| \right\} \\
&< \frac{1}{2} \|f\|_\infty \left\{ \mathcal{S}^1(K) + \varepsilon - \mathcal{S}^1_\varepsilon(K \setminus U) \right\}
\end{aligned}$$

where $(\star\star)$ helps justify the last inequality. Thus from $(\star\star\star)$ and $(\star)$ we deduce that

$$|\Lambda(\varphi)| \leq \frac{1}{2} \|f\|_\infty \left\{ \mathcal{S}^1(K) - \mathcal{S}^1(K \setminus U) \right\} = \frac{1}{2} \|f\|_\infty \mathcal{S}^1(K \cap U) \leq \|f\|_\infty \mathcal{H}^1(K \cap U).$$

Suping over all our φ's, we conclude that $\|\Lambda | C_c(U)\| \leq \|f\|_\infty \mathcal{H}^1_K(U)$. From [RUD, 6.19 (2) and 3.17] it now follows that $|\mu|(U) \leq \|f\|_\infty \mathcal{H}^1_K(U)$. Since U is an arbitrary open set, Proposition 5.22 now implies that

$$(\star\star\star\star)\ |\mu|(E) \leq \|f\|_\infty \mathcal{H}^1_K(E)$$

for any Borel set E. The Radon–Nikodym Theorem [RUD, 6.10] now implies that $d\mu = h\, d\mathcal{H}^1_K$ for some $h \in L^1(\mathcal{H}^1_K)$. Given $\alpha > 0$, set $E_\alpha = \{z : |h(z)| \geq \|f\|_\infty + \alpha\}$. Invoking Proposition 5.22 once again, we note that E_α is the union of a Borel set and a set of measure zero for $\mathcal{H}^1_K$. Since $d|\mu| = |h|\, d\mathcal{H}^1_K$ [RUD, 6.13], we see that $(\star\star\star\star)$ now implies that

$$(\|f\|_\infty + \alpha)\mathcal{H}^1_K(E_\alpha) \leq \int_{E_\alpha} |h|\, d\mathcal{H}^1_K = |\mu|(E_\alpha) \leq \|f\|_\infty \mathcal{H}^1_K(E_\alpha).$$

In consequence, $\mathcal{H}^1_K(E_\alpha) = 0$ for all $\alpha > 0$. Thus h is an element of $L^\infty(\mathcal{H}^1_K)$ with $\|h\|_\infty \leq \|f\|_\infty$.

Finally, recall that the derivative at infinity of the Cauchy transform of a measure is negative the mass of the measure. We have shown that f is the Cauchy transform of $h\, d\mathcal{H}^1_K$. Hence

$$|f'(\infty)| = \left| -\int h\, d\mathcal{H}^1_K \right| \le \|h\|_\infty \mathcal{H}^1_K(\{h \ne 0\}) \le \|f\|_\infty \mathcal{H}^1_K(\{h \ne 0\}).$$

□

6.3 Estimates of Truncated Cauchy Integrals

The Cauchy integral produced in the last section is bounded off K. One would like to extend it to be defined and bounded on K, but it may not make sense there since it may diverge absolutely. Standing at the threshold of the fearsome theory of singular integral operators, we opt for the simple expedient of modifying the kernel $\zeta \mapsto 1/(\zeta - z)$ to be 0 in a small disc of radius $\varepsilon > 0$ about its singularity z. The Cauchy integral arising from this "truncated" kernel, appropriately called a *truncated Cauchy integral*, now converges absolutely and boundedly on the whole complex plane. However, as payment for this gift one must obtain a bound on the truncated Cauchy integral that is independent of the parameter ε. Accomplishing this is the point of the proposition with which we end this section. As usual, some preliminary lemmas are necessary.

The first lemma is a trivial consequence of Proposition 5.18, but we avoid appeal to this difficult result with an easy ad hoc argument.

Lemma 6.7 *Suppose E is a subset of $\mathbb{C}$ such that $\mathcal{H}^1(E) < \infty$. Then $\mathcal{L}^2(E) = 0$.*

Proof Given $\delta > 0$, there exists a δ-cover $\{U_n\}$ of E such that $\sum_n |U_n| < \mathcal{H}^1(E) + 1$. Then, since each U_n is contained in a ball of radius $|U_n|$, we must have

$$\mathcal{L}^2(E) \le \sum_n \mathcal{L}^2(U_n) \le \sum_n \pi |U_n|^2 \le \pi\delta \sum_n |U_n| \le \pi\delta\{\mathcal{H}^1(E) + 1\}.$$

Since $\delta > 0$ is otherwise arbitrary, we are done. □

Note the similarity of the following lemma to Lemma 1.20. Also note the similarity of its proof to that of Lemma 1.20: both involve moving the mass of part of the set in question to form a ball in such a way that that the integrand can only get bigger in the process, and then evaluating the integral over the resulting ball via polar coordinates!

Lemma 6.8 *Let E be an $\mathcal{L}^2$-measurable subset of $\mathbb{C}$. Then for any point z of $\mathbb{C}$ we have*

$$\iint_E \frac{1}{|\zeta - z|}\, d\mathcal{L}^2(\zeta) \le 2\sqrt{\pi \mathcal{L}^2(E)}.$$

Proof Without loss of generality, $0 < \mathcal{L}^2(E) < \infty$. Choose $a > 0$ such that $\pi a^2 = \mathcal{L}^2(E)$. Set $B = B(z; a)$. Then $\mathcal{L}^2(E \setminus B) = \mathcal{L}^2(B \setminus E)$. Also, $|\zeta - z| > a$ for $\zeta \in E \setminus B$ and $|\zeta - z| \leq a$ for $\zeta \in B \setminus E$. Thus

$$\iint_{E\setminus B} \frac{1}{|\zeta - z|}\, d\mathcal{L}^2(\zeta) \leq \frac{\mathcal{L}^2(E \setminus B)}{a} = \frac{\mathcal{L}^2(B \setminus E)}{a} \leq \iint_{B\setminus E} \frac{1}{|\zeta - z|}\, d\mathcal{L}^2(\zeta)$$

and so, adding the integral of $1/|\zeta - z|$ over $E \cap B$ to both sides of this inequality, we get

$$\iint_E \frac{1}{|\zeta - z|}\, d\mathcal{L}^2(\zeta) \leq \iint_B \frac{1}{|\zeta - z|}\, d\mathcal{L}^2(\zeta).$$

Finally, using polar coordinates centered at the point z, we have

$$\iint_B \frac{1}{|\zeta - z|}\, d\mathcal{L}^2(\zeta) = \int_0^{2\pi} \int_0^a \frac{1}{r}\, r\,dr\,d\theta = 2\pi a = 2\sqrt{\pi \mathcal{L}^2(E)}.$$

□

Because of the this lemma, for any $z \in \mathbb{C}$ the function $\zeta \in \mathbb{C} \mapsto 1/(\zeta - z)$ is in $L^1(\mathcal{L}^2_E)$ for any $\mathcal{L}^2$-measurable E with $\mathcal{L}^2(E) < \infty$. Thus in Lemma 1.20 we did not actually need to restrict our attention to a *compact* K or to a point z lying *off* K; we could have stated it for *any* $\mathcal{L}^2$-measurable subset E of $\mathbb{C}$ such that $\mathcal{L}^2(E) < \infty$ and *any* point z of $\mathbb{C}$.

Proposition 6.9 *Let K be a compact subset of $\mathbb{C}$ such that $\mathcal{H}^1_K$ has linear growth with bound M. Suppose h is an element of $L^\infty(\mathcal{H}^1_K)$ with $\|h\|_\infty \leq 1$ such that*

$$\left| \int \frac{h(\zeta)}{\zeta - z}\, d\mathcal{H}^1_K(\zeta) \right| \leq 1$$

for every $z \in \mathbb{C} \setminus K$. Then

$$\left| \int_{\mathbb{C}\setminus B(z;\varepsilon)} \frac{h(\zeta)}{\zeta - z}\, d\mathcal{H}^1_K(\zeta) \right| \leq 6M + 1$$

for every $z \in \mathbb{C}$ and $\varepsilon > 0$.

Proof We dominate the absolute value of our integral by the sum of three pieces, each of which will be estimated differently:

$$\left|\int_{\mathbb{C}\setminus B(z;\varepsilon)} \frac{h(\zeta)}{\zeta - z}\, d\mathcal{H}_K^1(\zeta)\right| \leq$$

$$\left|\int_{\mathbb{C}\setminus B(z;\varepsilon)} \frac{h(\zeta)}{\zeta - z}\, d\mathcal{H}_K^1(\zeta) - \frac{4}{\pi\varepsilon^2}\int_{B(z;\varepsilon/2)} \left\{\int_{\mathbb{C}\setminus B(z;\varepsilon)} \frac{h(\zeta)}{\zeta - \xi}\, d\mathcal{H}_K^1(\zeta)\right\} d\mathcal{L}^2(\xi)\right| \leftarrow \text{(I)!}$$

$$+ \left|\frac{4}{\pi\varepsilon^2}\int_{B(z;\varepsilon/2)} \left\{\int \frac{h(\zeta)}{\zeta - \xi}\, d\mathcal{H}_K^1(\zeta)\right\} d\mathcal{L}^2(\xi)\right| \leftarrow \text{(II)!}$$

$$+ \left|-\frac{4}{\pi\varepsilon^2}\int_{B(z;\varepsilon/2)} \left\{\int_{B(z;\varepsilon)} \frac{h(\zeta)}{\zeta - \xi}\, d\mathcal{H}_K^1(\zeta)\right\} d\mathcal{L}^2(\xi)\right|. \leftarrow \text{(III)!}$$

For $\zeta \in \mathbb{C} \setminus B(z;\varepsilon)$ and $\xi \in B(z;\varepsilon/2)$, we have $|\zeta - \xi| \geq |\zeta - z| - |\xi - z| \geq \varepsilon - \varepsilon/2 = \varepsilon/2$. Hence $|\zeta - z| \leq |\zeta - \xi| + |\xi - z| \leq |\zeta - \xi| + \varepsilon/2 \leq 2|\zeta - \xi|$ and so

$$\left|\frac{1}{\zeta - z} - \frac{1}{\zeta - \xi}\right| = \frac{|\xi - z|}{|\zeta - z||\zeta - \xi|} \leq \frac{2|\xi - z|}{|\zeta - z|^2} \leq \frac{\varepsilon}{|\zeta - z|^2}.$$

Hence, using this, the hypothesis $\|h\|_\infty \leq 1$, Fubini's Theorem [RUD, 8.8], and the linear growth of $\mathcal{H}_K^1$, we see that

$$\begin{aligned}
\text{(I)} &= \left|\frac{4}{\pi\varepsilon^2}\int_{B(z;\varepsilon/2)}\int_{\mathbb{C}\setminus B(z;\varepsilon)} \left\{\frac{1}{\zeta - z} - \frac{1}{\zeta - \xi}\right\} h(\zeta)\, d\mathcal{H}_K^1(\zeta)\, d\mathcal{L}^2(\xi)\right| \\
&\leq \frac{4}{\pi\varepsilon^2}\int_{B(z;\varepsilon/2)}\int_{\mathbb{C}\setminus B(z;\varepsilon)} \frac{\varepsilon}{|\zeta - z|^2}\, |h(\zeta)|\, d\mathcal{H}_K^1(\zeta)\, d\mathcal{L}^2(\xi) \\
&\leq \int_{\mathbb{C}\setminus B(z;\varepsilon)} \frac{\varepsilon}{|\zeta - z|^2}\, d\mathcal{H}_K^1(\zeta) \;=\; \varepsilon\int_{\mathbb{C}\setminus B(z;\varepsilon)} \left\{\int_{|\zeta - z|}^{\infty} \frac{2}{t^3}\, dt\right\} d\mathcal{H}_K^1(\zeta) \\
&= 2\varepsilon\int_{\varepsilon}^{\infty} \frac{\mathcal{H}_K^1(B(z;t) \setminus B(z;\varepsilon))}{t^3}\, dt \;\leq\; 2\varepsilon\int_{\varepsilon}^{\infty} \frac{Mt}{t^3}\, dt = 2M.
\end{aligned}$$

By assumption, our Cauchy integral is bounded in absolute value by 1 off K and so $\mathcal{L}^2$-almost everywhere on $\mathbb{C}$ by the first lemma of this section. Hence

$$\text{(II)} \leq \frac{4}{\pi\varepsilon^2}\int_{B(z;\varepsilon/2)} \left|\int \frac{h(\zeta)}{\zeta - \xi}\, d\mathcal{H}_K^1(\zeta)\right| d\mathcal{L}^2(\xi) \leq \frac{4}{\pi\varepsilon^2}\mathcal{L}^2(B(z;\varepsilon/2)) = 1.$$

Finally, by the hypothesis $\|h\|_\infty \leq 1$, Fubini's Theorem [RUD, 8.8], the second lemma of this section, and the linear growth of $\mathcal{H}_K^1$, we see that

$$\begin{aligned}
(\mathrm{III}) &\le \frac{4}{\pi\varepsilon^2}\int_{B(z;\varepsilon/2)}\int_{B(z;\varepsilon)}\frac{|h(\zeta)|}{|\zeta-\xi|}\,d\mathcal{H}^1_K(\zeta)\,d\mathcal{L}^2(\xi)\\
&\le \frac{4}{\pi\varepsilon^2}\int_{B(z;\varepsilon)}\int_{B(z;\varepsilon/2)}\frac{1}{|\zeta-\xi|}\,d\mathcal{L}^2(\xi)\,d\mathcal{H}^1_K(\zeta)\\
&\le \frac{4}{\pi\varepsilon^2}\int_{B(z;\varepsilon)} 2\sqrt{\pi\mathcal{L}^2(B(z;\varepsilon/2))}\,d\mathcal{H}^1_K(\zeta)\\
&= \frac{4}{\varepsilon}\mathcal{H}^1_K(B(z;\varepsilon)) \;\le\; \frac{4}{\varepsilon}M\varepsilon = 4M.
\end{aligned}$$

Adding our estimates together, we get $6M+1$ and are done. □

6.4 Estimates of Truncated Suppressed Cauchy Integrals

When we say that Φ is a *Lip*(1)*-function*, we shall mean that Φ is a function with domain $\mathbb{C}$ and range contained in $[0,\infty)$ that is Lipschitz with bound 1, i.e., $\Phi : \mathbb{C} \mapsto [0,\infty)$ is such that $|\Phi(z)-\Phi(w)| \le |z-w|$ for all $z, w \in \mathbb{C}$. The *suppressed Cauchy kernel* k_Φ associated with such a Φ is defined for $\zeta \ne z$ by the equation

$$k_\Phi(\zeta,z) = \frac{\overline{\zeta - z}}{|\zeta-z|^2+\Phi(\zeta)\Phi(z)}.$$

Proposition 6.10 *For k_Φ as above and $\zeta \ne z$, we have the following*:

(a) $k_\Phi(\zeta,z) = -k_\Phi(z,\zeta)$.

(b) $k_\Phi(\zeta,z) = \dfrac{1}{\zeta - z}$ *whenever* $\Phi(\zeta)$ *or* $\Phi(z)=0$.

(c) $|k_\Phi(\zeta,z)| \le \dfrac{1}{|\zeta-z|}$.

(d) $|k_\Phi(\zeta,z)| \le \dfrac{1}{\Phi(\zeta)}$ *whenever* $\Phi(\zeta)\ne 0$ *and* $\le \dfrac{1}{\Phi(z)}$ *whenever* $\Phi(z)\ne 0$.

(e) $\left|k_\Phi(\zeta,z) - \dfrac{1}{\zeta-z}\right| \le \dfrac{\Phi(z)}{|\zeta-z|^2} + \dfrac{\Phi^2(z)}{|\zeta-z|^3}$.

(f) $|k_\Phi(\zeta,z) - k_\Phi(\zeta',z)| \le \dfrac{4|\zeta'-\zeta|}{|\zeta'-z|^2}$ *whenever* $|\zeta'-\zeta| \le \dfrac{1}{2}|\zeta'-z|$.

Proof (a), (b), and (c): Clear.

(d) Since Φ is Lip(1), $|\Phi(\zeta)-\Phi(z)| \le |\zeta - z|$ and so $\Phi(\zeta) - |\zeta - z| \le \Phi(z)$. Hence

$$\Phi(\zeta)|\overline{\zeta - z}| \leq \Phi(\zeta)|\zeta - z| + \{\Phi(\zeta) - |\zeta - z|\}^2$$
$$= |\zeta - z|^2 + \Phi(\zeta)\{\Phi(\zeta) - |\zeta - z|\} \leq |\zeta - z|^2 + \Phi(\zeta)\Phi(z).$$

Rearranging, we get $|k_\Phi(\zeta, z)| \leq 1/\Phi(\zeta)$. That $|k_\Phi(\zeta, z)| \leq 1/\Phi(z)$ now follows from (a).

(e) Since Φ is Lip(1), $|\Phi(\zeta) - \Phi(z)| \leq |\zeta - z|$ and so $\Phi(\zeta) \leq |\zeta - z| + \Phi(z)$. Hence

$$\begin{aligned}\left|k_\Phi(\zeta, z) - \frac{1}{\zeta - z}\right| &= \left|\frac{\overline{\zeta - z}}{|\zeta - z|^2 + \Phi(\zeta)\Phi(z)} - \frac{\overline{\zeta - z}}{|\zeta - z|^2}\right| \\ &= \left|\frac{-(\overline{\zeta - z})\Phi(\zeta)\Phi(z)}{\{|\zeta - z|^2 + \Phi(\zeta)\Phi(z)\}|\zeta - z|^2}\right| \\ &\leq \frac{\Phi(\zeta)\Phi(z)}{|\zeta - z|^3} \\ &\leq \frac{\{|\zeta - z| + \Phi(z)\}\Phi(z)}{|\zeta - z|^3} \\ &= \frac{\Phi(z)}{|\zeta - z|^2} + \frac{\Phi^2(z)}{|\zeta - z|^3}.\end{aligned}$$

(f) Use Den(a, b) to denote $|a - b|^2 + \Phi(a)\Phi(b)$ and note that Den$(a, b) \geq |a - b|^2$ and $\Phi(a)\Phi(b)$ always (these inequalities will be used many times in what follows). Plugging in the definition of k_Φ, forming a common denominator, and then segregating terms in the numerator into those not involving Φ and those that do, we obtain

$$(\star)\ k_\Phi(\zeta, z) - k_\Phi(\zeta', z) = \frac{\text{Num I}}{\text{Den}(\zeta, z)\text{Den}(\zeta', z)} + \frac{\text{Num II}}{\text{Den}(\zeta, z)\text{Den}(\zeta', z)}$$

where Num I $= (\overline{\zeta - z})|\zeta' - z|^2 - (\overline{\zeta' - z})|\zeta - z|^2$ and Num II $= (\overline{\zeta - z})\Phi(\zeta')\Phi(z) - (\overline{\zeta' - z})\Phi(\zeta)\Phi(z)$.

To handle the first quotient, note that Num I $= (\overline{\zeta - z})(\overline{\zeta' - z})(\zeta' - \zeta)$ (write each of the two absolute values squared in Num I as the conjugate times the number, pull out two common factors, and simplify what remains). Then note that if $|\zeta' - \zeta| \leq |\zeta' - z|/2$, then $|\zeta - z| \geq |\zeta' - z| - |\zeta' - \zeta| \geq |\zeta' - z|/2$. Thus

$$(\star\star)\ \frac{|\text{Num I}|}{\text{Den}(\zeta, z)\text{Den}(\zeta', z)} \leq \frac{|\zeta - z||\zeta' - z||\zeta' - \zeta|}{|\zeta - z|^2|\zeta' - z|^2} = \frac{|\zeta' - \zeta|}{|\zeta - z||\zeta' - z|} \leq \frac{2|\zeta' - \zeta|}{|\zeta' - z|^2}.$$

To handle the second quotient, note that Num II $= (\overline{\zeta - z})\{\Phi(\zeta') - \Phi(\zeta)\}\Phi(z) - (\overline{\zeta' - \zeta})\Phi(\zeta)\Phi(z)$ (subtract and add $(\overline{\zeta - z})\Phi(\zeta)\Phi(z)$ and tidy up the second term). Note also that $|\Phi(\zeta') - \Phi(\zeta)| \leq |\zeta' - \zeta|$. Thus

$$(\star\star\star)\quad \frac{|\text{Num II}|}{\text{Den}(\zeta,z)\text{Den}(\zeta',z)} \leq \frac{|\zeta-z||\zeta'-\zeta|\Phi(z)}{\text{Den}(\zeta,z)\text{Den}(\zeta',z)} + \frac{|\zeta'-\zeta|\Phi(\zeta)\Phi(z)}{\text{Den}(\zeta,z)\text{Den}(\zeta',z)}$$

$$= \frac{|\zeta'-\zeta|}{\text{Den}(\zeta',z)}\left\{\frac{|\zeta-z|}{\text{Den}(\zeta,z)}\cdot\Phi(z) + \frac{\Phi(\zeta)\Phi(z)}{\text{Den}(\zeta,z)}\right\}$$

$$\leq \frac{|\zeta'-\zeta|}{|\zeta'-z|^2}\{1+1\}.$$

The first 1 in the last inequality is justified by (d) since $|\zeta - z|/\text{Den}(\zeta,z) = |k_\Phi(\zeta,z)|$!

Considering $(\star)$, $(\star\star)$, and $(\star\star\star)$, we are done. □

The integral whose absolute value is estimated in the conclusion of the next proposition could be called a *truncated suppressed Cauchy integral*. We will meet the *suppressed Cauchy integral* that it is the truncation of in the next section!

Proposition 6.11 *Let K be a compact subset of $\mathbb{C}$ such that $\mathcal{H}^1_K$ has linear growth with bound M. Suppose h is an element of $L^\infty(\mathcal{H}^1_K)$ with $\|h\|_\infty \leq 1$ such that*

$$\left|\int \frac{h(\zeta)}{\zeta - z}\, d\mathcal{H}^1_K(\zeta)\right| \leq 1$$

for every $z \in \mathbb{C}\setminus K$. Then for any Lip(1)-function Φ, we have

$$\left|\int_{\mathbb{C}\setminus B(z;\varepsilon)} k_\Phi(\zeta,z)h(\zeta)\, d\mathcal{H}^1_K(\zeta)\right| \leq 11M + 1$$

for every $z \in \mathbb{C}$ and $\varepsilon > 0$.

Proof By (b) of the last proposition and Proposition 6.9, we may as well assume that $\Phi(z) > 0$. Set $R = \max\{\varepsilon, \Phi(z)\}$. We split our integral into three pieces, each of which will be estimated differently:

$$\left|\int_{\mathbb{C}\setminus B(z;\varepsilon)} k_\Phi(\zeta,z)h(\zeta)\, d\mathcal{H}^1_K(\zeta)\right| \leq \left|\int_{B(z;R)\setminus B(z;\varepsilon)} k_\Phi(\zeta,z)h(\zeta)\, d\mathcal{H}^1_K(\zeta)\right| \leftarrow \text{(I)!}$$

$$+\left|\int_{\mathbb{C}\setminus B(z;R)} \frac{h(\zeta)}{\zeta-\xi}\, d\mathcal{H}^1_K(\zeta)\right| \leftarrow \text{(II)!}$$

$$+\left|\int_{\mathbb{C}\setminus B(z;R)} \left\{k_\Phi(\zeta,z) - \frac{1}{\zeta - z}\right\} h(\zeta)\, d\mathcal{H}^1_K(\zeta)\right|. \leftarrow \text{(III)!}$$

If $R = \varepsilon$, then (I) $= 0 \leq M$ trivially. Otherwise, $R = \Phi(z)$ and so by (d) of the last proposition and the linear growth of $\mathcal{H}^1_K$, we have

$$\text{(I)} \le \frac{\mathcal{H}^1_K(B(z;R))}{\Phi(z)} \le \frac{MR}{\Phi(z)} = M.$$

By Proposition 6.9, (II) $\le 6M + 1$.

By (e) of the last proposition,

$$\text{(III)} \le \Phi(z) \int_{\mathbb{C}\setminus B(z;R)} \frac{1}{|\zeta - z|^2}\, d\mathcal{H}^1_K(\zeta) + \Phi^2(z) \int_{\mathbb{C}\setminus B(z;R)} \frac{1}{|\zeta - z|^3}\, d\mathcal{H}^1_K(\zeta).$$

Denote the first and second summands on the right-hand side of the above inequality by (IIIa) and (IIIb), respectively. Then, using Fubini [RUD, 8.8] and the linear growth of $\mathcal{H}^1_K$,

$$\begin{aligned}
\text{(IIIa)} &= \Phi(z) \int_{\mathbb{C}\setminus B(z;R)} \left\{ \int_{|\zeta - z|}^{\infty} \frac{2}{t^3}\, dt \right\} d\mathcal{H}^1_K(\zeta) \\
&= 2\Phi(z) \int_R^{\infty} \frac{\mathcal{H}^1_K(B(z;t)\setminus B(z;R))}{t^3}\, dt \\
&\le 2\Phi(z) \int_R^{\infty} \frac{Mt}{t^3}\, dt \; = \; \frac{2\Phi(z)M}{R} \; \le \; 2M
\end{aligned}$$

since $R \ge \Phi(z)$ always. One can similarly show that (IIIb) $\le 3M/2$ ... which we dominate by $2M$ to avoid fractions!

Adding our estimates together, we get $11M + 1$ and are done. □

6.5 Vitushkin's Conjecture Resolved Affirmatively Modulo Two Difficult Results

In this section we resolve Vitushkin's Conjecture affirmatively by verifying Reduction 6.3 using two difficult results: a $T(b)$ theorem due to Nazarov, Treil, and Volberg and a curvature theorem due to David and Léger. These two difficult results will then be proved in following two chapters.

So suppose K is a compact subset of $\mathbb{C}$ such that $\mathcal{H}^1_K$ has linear growth with bound M for some $M < \infty$. Note that this trivially implies that $\mathcal{H}^1(K) < \infty$. Also, we may as well assume that K is nonremovable (otherwise the conclusion of Reduction 6.3 holds trivially). Thus we may apply Proposition 6.6 to the Ahlfors function of K to obtain a function $h \in L^\infty(\mathcal{H}^1_K)$ with $\|h\|_\infty \le 1$ such that $\mathcal{H}^1_K(\{h \ne 0\}) \ge \gamma(K)$ and for all $z \in \mathbb{C} \setminus K$,

$$\left| \int \frac{h(\zeta)}{\zeta - z}\, d\mathcal{H}^1_K(\zeta) \right| \le 1.$$

For the remainder of this section, let K, M, and h be fixed as in the last paragraph. To verify Reduction 6.3, it suffices to find a countable family of rectifiable curves whose union Γ satisfies

$$\mathcal{H}^1_K(\Gamma) \geq \gamma(K).$$

Given $\delta > 0$, a closed ball $B(c; r)$ of $\mathbb{C}$ is called *δ-good* (for K and h) if

1. $0 < r < 1/8$,
2. $\mathcal{H}^1_K(\partial B(c; r)) = 0$,
3. $\mathcal{H}^1_K(B(c; r)) > r/2$, and
4. $\displaystyle\int_{B(c;r)} |h - h(c)|^2 \, d\mathcal{H}^1_K < \frac{\delta^4}{2} |h(c)|^2 \mathcal{H}^1_K(B(c; r))$.

We wish to split up $\{h \neq 0\}$ nicely with δ-good balls. For that the next lemma lets us deal with item 3 of the above definition. The lemma is essentially a special case of a general upper density result for s-dimensional Hausdorff measures (see Corollary 2.6 of [FALC]).

Lemma 6.12 *Let E be a subset $\mathbb{C}$ such that $\mathcal{H}^1(E) < \infty$ and let $0 < \alpha < 1$. Then for $\mathcal{H}^1$-a.e. $z \in E$, $\mathcal{H}^1(E \cap B(z; r)) > \alpha r$ for arbitrarily small $r > 0$.*

Proof Given $\delta > 0$, set

$$E_\delta = \{z \in E : \mathcal{H}^1(E \cap B(z; r)) \leq \alpha r \text{ for all } 0 < r < \delta\}.$$

Since the set of points of E for which the conclusion of the lemma fails is equal to the union of the E_δs over any sequence of δs decreasing to 0, it suffices to show that $\mathcal{H}^1(E_\delta) = 0$ for a typical fixed δ.

Given $\varepsilon > 0$, let $\{U_n\}$ be a δ-cover of E_δ such that $\sum_n |U_n| < \mathcal{H}^1(E_\delta) + \varepsilon$. Without loss of generality, each $E_\delta \cap U_n$ is nonempty, so we may choose $z_n \in E_\delta \cap U_n$. Then

$$\mathcal{H}^1(E_\delta) \leq \sum_n \mathcal{H}^1(E_\delta \cap B(z_n; |U_n|)) \leq \sum_n \alpha |U_n| < \alpha\{\mathcal{H}^1(E_\delta) + \varepsilon\}.$$

Letting $\varepsilon \downarrow 0$, we get $\mathcal{H}^1(E_\delta) \leq \alpha \mathcal{H}^1(E_\delta)$. Since $\mathcal{H}^s(E_\delta) < \infty$ and $0 < \alpha < 1$, this forces $\mathcal{H}^1(E_\delta) = 0$ and so we are done. □

Proposition 6.13 *For any $\delta > 0$, there exists a countable pairwise disjoint collection $\{B_n\}$ of δ-good balls such that*

$$\mathcal{H}^1_K\left(\{h \neq 0\} \setminus \bigcup_n B_n\right) = 0.$$

Proof By the last lemma (with $E = K, \alpha = 1/2$), Proposition 5.22, and Proposition 5.28 (with $f = h, \mu = \mathcal{H}^1_K, p = 2$), there exists a subset H of $\{h \neq 0\}$ with $\mathcal{H}^1_K(\{h \neq 0\} \setminus H) = 0$ such that every point $c \in H$ satisfies item 3 of the definition of δ-good ball for arbitrarily small $r > 0$ and item 4 for all sufficiently small $r > 0$. So given $c \in H$, construct a sequence $\{r_n\}$ decreasing to 0 with each r_n satisfying item 3 for c. Item 2 for c may fail for at most countably many $r > 0$. Unfortunately all of our r_ns may be among these countably many $r > 0$! But consider that if r_n satisfies item 3 for c, then any r between r_n and $\min\{2\mathcal{H}^1_K(B(c; r_n)), r_{n-1}\}$ also satisfies item 3 for c. Since the open interval with endpoints r_n and $\min\{2\mathcal{H}^1_K(B(c; r_n)), r_{n-1}\}$ is an uncountable set, we may obviously choose an r'_n from it satisfying items 2 and 3 for c. Then $\{r'_n\}$ is a sequence decreasing to 0 and r'_n satisfies items 1, 2, 3, and 4 for c for all sufficiently large n.

We have thus shown that every $c \in H$ is the center of δ-good balls with arbitrarily small radii, i.e., the collection of δ-good balls is a Vitali class for H. Applying (a) of Vitali's Covering Lemma (5.14), there exists a countable pairwise disjoint collection $\{B_n\}$ of δ-good balls such that either $\mathcal{H}^1_K(H \setminus \bigcup_n B_n) = 0$ or $\sum_n |B_n| = \infty$. If we can rule out the latter alternative, we are done. This is easy due to item 3 of the definition of a δ-good ball and the pairwise disjointness of the B_ns:

$$\sum_n |B_n| < \sum 4\mathcal{H}^1_K(B_n) \leq 4\mathcal{H}^1(K) < \infty.$$

□

It is at this point that we wish to apply the first difficult result, whose proof will be postponed to the next chapter, the $T(b)$ theorem of Nazarov, Treil, and Volberg. To state this result, however, a definition is needed:

Given a positive Borel measure μ on $\mathbb{C}$, the *suppressed Cauchy integral* $\mathcal{K}_\Phi\varphi$ of a function $\varphi \in L^2(\mu)$ with respect to a Lip(1)-function Φ is defined by the equation

$$(\mathcal{K}_\Phi\varphi)(z) = \int_{\mathbb{C}} k_\Phi(\zeta, z)\varphi(\zeta)\, d\mu(\zeta).$$

This definition has a problem that needs going into: For $z \in \mathrm{spt}(\mu)$, the integral defining $(\mathcal{K}_\Phi\varphi)(z)$ may not converge absolutely and so may not be well defined. To circumvent this annoyance, we introduce a parameter $\lambda > 0$ and consider $\mathcal{K}_{\Phi+\lambda}\varphi$ in place of $\mathcal{K}_\Phi\varphi$. The integral defining $\mathcal{K}_{\Phi+\lambda}\varphi$ is then absolutely convergent and so unproblematic. For, using the Cauchy–Schwarz Inequality [RUD, 3.5] and (d) of Proposition 6.10, we have

$$\begin{aligned} \int_{\mathbb{C}} |k_{\Phi+\lambda}(\zeta, z)\varphi(\zeta)|\, d\mu(\zeta) &\leq \left\{\int_{\mathbb{C}} |k_{\Phi+\lambda}(\zeta, z)|^2 d\mu(\zeta)\right\}^{1/2} \|\varphi\|_{L^2(\mu)} \\ &\leq \left\{\frac{\mu(\mathbb{C})}{\lambda^2}\right\}^{1/2} \cdot \|\varphi\|_{L^2(\mu)}. \end{aligned}$$

Moreover, from this it follows that

$$\|\mathcal{K}_{\Phi+\lambda}\,\varphi\|^2_{L^2(\mu)} = \int_{\mathbb{C}} \left| \int_{\mathbb{C}} k_{\Phi+\lambda}(\zeta, z)\varphi(\zeta)\, d\mu(\zeta) \right|^2 d\mu(z)$$

$$\leq \int_{\mathbb{C}} \frac{\mu(\mathbb{C})}{\lambda^2} \cdot \|\varphi\|^2_{L^2(\mu)}\, d\mu(z) \;=\; \frac{\mu^2(\mathbb{C})}{\lambda^2} \cdot \|\varphi\|^2_{L^2(\mu)}$$

and so

$$\|\mathcal{K}_{\Phi+\lambda}\|_{L^2(\mu)\mapsto L^2(\mu)} \leq \frac{\mu(\mathbb{C})}{\lambda} < \infty.$$

One of the major points of the $T(b)$ theorem of Nazarov, Treil, and Volberg is that, given a few more hypotheses, there is a *finite* bound on $\|\mathcal{K}_{\Phi+\lambda}\|_{L^2(\mu)\mapsto L^2(\mu)}$ that is *independent* of $\lambda > 0$ (unlike the bound just obtained). This will allow us to let $\lambda \downarrow 0$ and obtain information about $\mathcal{K}_\Phi$. So in future when we speak of $\mathcal{K}_\Phi$ as being bounded from $L^2(\mu)$ to $L^2(\mu)$ what we really mean is that $\|\mathcal{K}_{\Phi+\lambda}\|_{L^2(\mu)\mapsto L^2(\mu)}$ has a finite bound that is independent of $\lambda > 0$.

There is another major point to their theorem, but to properly go into that we first need to state the result!

Theorem 6.14 (Nazarov, Treil, and Volberg) *Let* $0 < \delta \leq 1/27{,}000$, $0 < M < \infty$, $0 < \widetilde{M} < \infty$, *and* $c \in \mathbb{C}$. *Suppose that* μ *is a totally regular positive Borel measure on* $\mathbb{C}$, h *is an element of* $L^\infty(\mu)$, *and* $\widetilde{\Phi}$ *is a Lip*(1)*-function that together satisfy*

(1) $\mathrm{spt}(\mu) \subseteq B(c; 1/8)$,

(2) μ *has linear growth with bound* M,

(3) $\displaystyle\int_{\mathbb{C}} |h - h(c)|^2\, d\mu < \frac{\delta^4}{2}|h(c)|^2\mu(\mathbb{C})$,

(4) $\mu(\{\widetilde{\Phi} > 0\}) \leq \dfrac{\delta^2}{2}\mu(\mathbb{C})$, *and*

(5) *for all points* $z \in \mathbb{C}$, *numbers* $\varepsilon > 0$, *and Lip*(1)*-functions* $\Theta \geq \delta\widetilde{\Phi}$,

$$\left| \int_{\mathbb{C}\setminus B(z;\varepsilon)} k_\Theta(\zeta, z)h(\zeta)\, d\mu(\zeta) \right| \leq \widetilde{M}|h(c)|.$$

Then there exists a Lip(1)*-function* Φ *such that*

(a) $\mu(\{\Phi > 0\}) \leq \delta\mu(\mathbb{C})$ *and*

(b) $\displaystyle\sup_{\lambda>0} \|\mathcal{K}_{\Phi+\lambda}\|_{L^2(\mu)\mapsto L^2(\mu)} < \infty$.

Note that (1) and (2) imply that μ is compactly supported and finite. (The reader may be wondering why the bound of 1/27,000 on the number δ in the enunciation of this theorem. If so, see the Proof of Theorem 7.1 from Reduction 7.21 in Section 7.10!)

The ideal result here, which is unobtainable, would be that the Cauchy integral operator

$$(\mathcal{K}_0\varphi)(z) = \int_{\mathbb{C}} \frac{\varphi(\zeta)}{\zeta - z}\, d\mu(\zeta)$$

is always bounded from $L^2(\mu)$ to $L^2(\mu)$. What the theorem instead asserts, or rather implies, is that if one throws away μ_U, the restriction of μ to a small "bad" set U, then the Cauchy integral operator $\mathcal{K}_0$ is bounded on L^2 of what is left, μ_V, the restriction of μ to the complementary "good" set V. To see this, we first note that the purpose of the Lip(1)-function Φ of the theorem's conclusion is to single out these "bad" and "good" sets: $U = \{\Phi > 0\}$ and $V = \{\Phi = 0\}$. Next, we note that $k_\Phi(\zeta, z) = 1/(\zeta - z)$ on V and so $\mathcal{K}_\Phi\varphi = \mathcal{K}_0\varphi$ for all $\varphi \in L^2(\mu_V)$. This in conjunction with (b) of the theorem's conclusion allow us to conclude that $\mathcal{K}_0$ is bounded from $L^2(\mu_V)$ to $L^2(\mu_V)$. Finally, that U is indeed "small" is the import of (a) of the theorem's conclusion.

Proposition 6.15 *Let* $0 < \delta < 1/27{,}000$. *If B is a δ-good ball, then there exists an open subset U of $\mathbb{C}$ such that*

$$\mathcal{H}^1_K(B \cap U) \le \delta \mathcal{H}^1_K(B) \text{ and } c^2\big(\mathcal{H}^1_{K\cap(B\setminus U)}\big) < \infty.$$

Proof Letting $B = B(c; r)$ and setting $\mu = \mathcal{H}^1_{K\cap B}$, note that hypotheses (1), (2), and (3) of Theorem 6.14 hold. We shall produce a Φ and $\widetilde{M}$ that make hypotheses (4) and (5) hold too.

Since $\mathcal{H}^1_K(\partial B) = 0$, we may choose ρ close enough to but less than r so that

$$\mathcal{H}^1_K(B \setminus B(c;\rho)) \le \frac{\delta^2}{2}\mathcal{H}^1_K(B)$$

by the Lower Continuity Property of Measures [RUD, 1.19(e)]. Then $\widetilde{\Phi}(z) = \operatorname{dist}(z, B(c;\rho))$ is a Lip(1)-function that satisfies hypothesis (4). Given z, ε, and Θ as in hypothesis (5), note that

$$\left|\int_{\mathbb{C}\setminus B(z;\varepsilon)} k_\Theta(\zeta,z)h(\zeta)\,d\mu(\zeta)\right| \le \left|\int_{\mathbb{C}\setminus B(z;\varepsilon)} k_\Theta(\zeta,z)h(\zeta)\,d\mathcal{H}^1_K(\zeta)\right| \leftarrow \text{(I)!}$$
$$+ \left|\int_{\mathbb{C}\setminus B(z;\varepsilon)} k_\Theta(\zeta,z)h(\zeta)\,d\mathcal{H}^1_{K\setminus B}(\zeta)\right|. \leftarrow \text{(II)!}$$

By Proposition 6.11, (I) $\le 11M + 1$, and by (d) of Proposition 6.10,

$$\text{(II)} \le \int_{\mathbb{C}} \frac{1}{\Theta(\zeta)}\, d\mathcal{H}^1_{K\setminus B}(\zeta) \le \frac{\mathcal{H}^1(K)}{\delta(r-\rho)}.$$

Then hypothesis (5) will hold if we set

$$\widetilde{M} = \frac{1}{h(c)}\left\{(11M+1) + \frac{\mathcal{H}^1(K)}{\delta(r-\rho)}\right\}.$$

We may thus apply Theorem 6.14 here to get a Lip(1)-function Φ as in its conclusion. Setting $U = \{\Phi > 0\}$, part (a) of the theorem's conclusion immediately translates into $\mathcal{H}^1_K(B \cap U) \le \delta \mathcal{H}^1_K(B)$. Denoting the supremum in part (b) of the theorem's conclusion by N, we see that

$$(\star)\ \|\mathcal{K}_{\Phi+\lambda}\, \mathcal{X}_{\mathbb{C}\setminus U}\|^2_{L^2(\mu)} \le N^2 \|\mathcal{X}_{\mathbb{C}\setminus U}\|^2_{L^2(\mu)} \le N^2 \mu(\mathbb{C}).$$

We shall now develop the leftmost term of $(\star)$ into a triple integral from which the desired curvature will emerge. The arguments will be similar to those of Section 4.1. Indeed, the reader should now reacquaint himself with the notation and results of that section for they will be called upon. It is for this reason that we shall suddenly start to use ξ where up to now we have been using z! Since $\Phi = 0$ off of U, we have

$$\begin{aligned}
&(\star\star)\ \|\mathcal{K}_{\Phi+\lambda}\, \mathcal{X}_{\mathbb{C}\setminus U}\|^2_{L^2(\mu)} \\
&= \int_{\mathbb{C}} \left| \int_{\mathbb{C}} k_{\Phi+\lambda}(\zeta,\xi)\, \mathcal{X}_{\mathbb{C}\setminus U}(\zeta)\, d\mu(\zeta) \right|^2 d\mu(\xi) \\
&\ge \int_{\mathbb{C}\setminus U} \left| \int_{\mathbb{C}\setminus U} \frac{\overline{\zeta-\xi}}{|\zeta-\xi|^2+\lambda^2}\, d\mu(\zeta) \right|^2 d\mu(\xi) \\
&= \iiint_{(\mathbb{C}\setminus U)^3} \frac{(\overline{\zeta-\xi})(\eta-\xi)}{(|\zeta-\xi|^2+\lambda^2)(|\eta-\xi|^2+\lambda^2)}\, d\mu(\zeta)\, d\mu(\eta)\, d\mu(\xi) \\
&= \frac{1}{6} \iiint_{(\mathbb{C}\setminus U)^3} \sum_{\mathcal{P}erm} \frac{(\overline{\zeta-\xi})(\eta-\xi)}{(|\zeta-\xi|^2+\lambda^2)(|\eta-\xi|^2+\lambda^2)}\, d\mu(\zeta)\, d\mu(\eta)\, d\mu(\xi).
\end{aligned}$$

[When one permutes the variables in the second to last integral in $(\star\star)$ in any particular way obviously it does not change, but Fubini [RUD, 8.8] always allows us to undo the resulting permuting of the $d\mu$s. The upshot is that any permuting of the variables in the integrand alone does not change the value of the second to last integral in $(\star\star)$. The last equality of $(\star\star)$ follows in consequence.]

One may write $\zeta - \xi = be^{i\theta}$, $\eta - \xi = ae^{i\varphi}$, and $\gamma = |\theta - \varphi|$. Then two terms of the six-term sum making up the integrand of the last integral in $(\star\star)$ sum to

$$\frac{\overline{(\zeta-\xi)}(\eta-\xi)}{(|\zeta-\xi|^2+\lambda^2)(|\eta-\xi|^2+\lambda^2)}+\frac{\overline{(\eta-\xi)}(\zeta-\xi)}{(|\eta-\xi|^2+\lambda^2)(|\zeta-\xi|^2+\lambda^2)}$$
$$=\frac{bae^{i(\varphi-\theta)}}{(b^2+\lambda^2)(a^2+\lambda^2)}+\frac{abe^{i(\theta-\varphi)}}{(a^2+\lambda^2)(b^2+\lambda^2)}=\frac{2ab\cos\gamma}{(a^2+\lambda^2)(b^2+\lambda^2)}.$$

Two similar expressions may be gotten when the remaining four terms are appropriately paired off. Adding the three resulting expressions, we see that the integrand of the last integral in $(\star\star)$ may be written as the sum of

$$\text{(I)}=\frac{2abc(a\cos\alpha+b\cos\beta+c\cos\gamma)}{(a^2+\lambda^2)(b^2+\lambda^2)(c^2+\lambda^2)}$$

and

$$\text{(II)}=\frac{\lambda^2(2bc\cos\alpha+2ac\cos\beta+2ab\cos\gamma)}{(a^2+\lambda^2)(b^2+\lambda^2)(c^2+\lambda^2)}.$$

One the one hand, from Proposition 4.1 we see that

$$\text{(I)}=\frac{2(a\cos\alpha+b\cos\beta+c\cos\gamma)}{abc}\cdot\frac{a^2b^2c^2}{(a^2+\lambda^2)(b^2+\lambda^2)(c^2+\lambda^2)}$$
$$=c^2(\zeta,\eta,\xi)\cdot\frac{a^2b^2c^2}{(a^2+\lambda^2)(b^2+\lambda^2)(c^2+\lambda^2)}.$$

On the other hand, from the Law of Cosines we see that

$$\text{(II)}=\frac{\lambda^2\big([b^2+c^2-a^2]+[a^2+c^2-b^2]+[a^2+b^2-c^2]\big)}{(a^2+\lambda^2)(b^2+\lambda^2)(c^2+\lambda^2)}$$
$$=\frac{\lambda^2(a^2+b^2+c^2)}{(a^2+\lambda^2)(b^2+\lambda^2)(c^2+\lambda^2)}\geq 0.$$

Thus this whole paragraph has shown the following:

$(\star\star\star)$ The integrand of the last integral in $(\star\star)\geq$
$$c^2(\zeta,\eta,\xi)\frac{a^2b^2c^2}{(a^2+\lambda^2)(b^2+\lambda^2)(c^2+\lambda^2)}.$$

From $(\star)$, $(\star\star)$, and $(\star\star\star)$, it now follows that

$$\iiint_{(\mathbb{C}\setminus U)^3}c^2(\zeta,\eta,\xi)\cdot\frac{a^2b^2c^2}{(a^2+\lambda^2)(b^2+\lambda^2)(c^2+\lambda^2)}d\mu(\zeta)\,d\mu(\eta)\,d\mu(\xi)\leq 6N^2\mu(\mathbb{C}).$$

Letting $\lambda \downarrow 0$ and invoking the Monotone Convergence Theorem [RUD, 1.26], we deduce that

$$c^2(\mu_{\mathbb{C}\setminus U}) = \iiint_{(\mathbb{C}\setminus U)^3} c^2(\zeta,\eta,\xi)\, d\mu(\zeta)\, d\mu(\eta)\, d\mu(\xi) \le 6N^2\mu(\mathbb{C}) < \infty.$$

Since $\mu_{\mathbb{C}\setminus U} = \mathcal{H}^1_{K\cap(B\setminus U)}$, we are done. □

At this point we need to apply the second difficult result, whose proof will be postponed to the chapter after the next, the curvature theorem of David and Léger.

Theorem 6.16 (David & Léger) *Suppose E is an $\mathcal{H}^1$-measurable subset of $\mathbb{C}$ such that $\mathcal{H}^1(E) < \infty$ and $c^2(\mathcal{H}^1_E) < \infty$. Then there exists a countable family of rectifiable curves whose union Γ satisfies*

$$\mathcal{H}^1(E \setminus \Gamma) = 0.$$

From this result and our last proposition we immediately get the following.

Proposition 6.17 *Let $0 < \delta < 1/27{,}000$. If B is a δ-good ball, then there exists an open subset U of $\mathbb{C}$ and a set Γ that is a union of a countable family of rectifiable curves which together satisfy*

$$\mathcal{H}^1_K(B \cap U) \le \delta\mathcal{H}^1_K(B) \text{ and } \mathcal{H}^1_K(\{B \setminus U\} \setminus \Gamma) = 0.$$

Completion of the Proof of Reduction 6.3 For the moment, fix $0 < \delta < 1/27{,}000$. Let $\{B_n\}$ be as in Proposition 6.13. Then, applying the last proposition to each B_n, get open sets U_n and sets Γ_n, each a union of a countable family of rectifiable curves, that satisfy

$$\mathcal{H}^1_K(B_n \cap U_n) \le \delta\mathcal{H}^1_K(B_n) \text{ and } \mathcal{H}^1_K(\{B_n \setminus U_n\} \setminus \Gamma_n) = 0$$

for each n. Set $\Gamma_\delta = \bigcup_n \Gamma_n$. Then

$$\begin{aligned}
\mathcal{H}^1(K \cap \Gamma_\delta) &\ge \sum_n \mathcal{H}^1_K(\{B_n \setminus U_n\} \cap \Gamma_n) \\
&= \sum_n \mathcal{H}^1_K(B_n \setminus U_n) \\
&= \sum_n \{\mathcal{H}^1_K(B_n) - \mathcal{H}^1_K(B_n \cap U_n)\} \\
&\ge \sum_n (1-\delta)\mathcal{H}^1_K(B_n) \\
&\ge (1-\delta)\mathcal{H}^1_K(\{h \neq 0\}) \ \ge \ (1-\delta)\gamma(K).
\end{aligned}$$

To finish, simply take a sequence $\{\delta_n\}$ decreasing to 0 and consider $\Gamma = \bigcup_n \Gamma_{\delta_n}$. □

We close this section with a few comments on the history of our solution to Vitushkin's Conjecture. Michael Christ came up with the idea of using an appropriate $T(b)$ theorem to obtain information on the L^2 boundedness of the Cauchy transform and thus too on analytic capacity. He applied it to *Ahlfors–David regular sets* in [CHRIST]. These are linearly measurable subsets E for which there exists a positive finite number M such that for all $z \in E$ and all positive $r < |E|$,

$$M^{-1}r \leq \mathcal{H}^1(E \cap B(z; r)) \leq Mr.$$

When a linearly measurable E satisfies the first inequality of this definition, we call it *lower Ahlfors–David regular*. When it satisfies the second inequality, we call it *upper Ahlfors–David regular*. It is an easy exercise to see that E is upper Ahlfors–David regular if and only if $\mathcal{H}^1_E$ has linear growth. Shortly after the introduction of Melnikov curvature in [MEL], Pertti Mattila, Mark Melnikov, and Joan Verdera showed in [MMV] that Vitushkin's Conjecture was correct for Ahlfors–David regular compact sets. As seen in the opening section of this chapter, upper Ahlfors–David regularity may be assumed when verifying the general conjecture. Not so however for lower Ahlfors–David regularity! Indeed, getting around lack of lower Ahlfors–David regularity was the main obstacle to resolving the full conjecture. Guy David obtained an appropriate $T(b)$ theorem in [DAV1] with a truly heroic and difficult proof that avoided lower Ahlfors–David regularity. This work built on an earlier paper [DM], a joint effort with Pertti Mattila. Shortly thereafter or at about the same time, Fedor Nazarov, Sergei Treil, and Alexander Volberg obtained another $T(b)$ theorem in [NTV] which did the job of David's $T(b)$ theorem and had a simpler and smoother proof which is more easily adapted to other situations. Thus we have replaced David's $T(b)$ theorem with theirs in our solution of Vitushkin's Conjecture. To finish off Vitushkin's Conjecture, David also obtained an appropriate curvature result in the absence of lower Ahlfors–David regularity. However, he did not publish this last proof since Jean-Christophe Léger, a student of his, managed in [LÉG] to prove the same result in a smoother fashion that, unlike David's original proof, easily generalizes from $\mathbb{C} = \mathbb{R}^2$ to $\mathbb{R}^n$. Recently, in [TOL5], Xavier Tolsa obtained another proof of this curvature theorem. His proof leans heavily on Besicovitch's theory of 1-sets which we describe in the next section. For this and other reasons, we have elected to stick with the earlier proof of Léger.

6.6 Postlude: Vitushkin's Original Conjecture

The conjecture which has just been resolved affirmatively in the last section and has come to bear Anatoli Vitushkin's name grew out of an earlier conjecture of his which was resolved negatively. We here provide some background on this earlier conjecture.

Given $\theta \in [0, \pi)$, let π_θ denote the orthogonal projection of the complex plane onto the line that passes through the origin and makes a counterclockwise angle of θ with the positive x-axis. The *Crofton measure* (also known as the *Buffon needle probability* or the *Favard length*) of a subset E of $\mathbb{C}$ is

$$\mathcal{CR}(E) = \frac{1}{\pi}\int_0^\pi \mathcal{L}^1(\pi_\theta(E))\,d\theta.$$

Thus $\mathcal{CR}(E) = 0$ if and only if the set E orthogonally projects in almost every direction onto a set of linear Lebesgue measure zero.

There is a result of John M. Marstrand, Theorem 1 of [MARST] coupled with the penultimate paragraph of Section 5 of the same work, from which it follows that a compact subset of $\mathbb{C}$ with Hausdorff dimension greater than 1 has positive Crofton measure. By contrast, it is an easy consequence of Proposition 2.2 that a compact subset of $\mathbb{C}$ with Hausdorff dimension less than 1 has zero Crofton measure. This contrast appears again when one turns to removability: a compact subset of $\mathbb{C}$ is nonremovable or removable depending on whether its Hausdorff dimension is greater or less than 1 respectively (Corollaries 2.12 and 2.8). This naturally leads to *Vitushkin's Original Conjecture* (VOC): For every compact subset K of $\mathbb{C}$,

$$(\star)\ K \text{ is removable if and only if } \mathcal{CR}(K) = 0.$$

To the author's knowledge, this was first stated in Section 3 of chapter I of [VIT2]. The fate of this conjecture clearly hinges upon the compact subsets of $\mathbb{C}$ of Hausdorff dimension one!

Of course, the alert reader has probably realized that the Joyce–Mörters set K that concluded Chapter 4 refutes VOC: K is nonremovable, yet $\mathcal{CR}(K) = 0$ (Theorem 4.33 and Proposition 4.28). However, this is only the most recent and simplest counterexample. The first counterexample, in 1986, was the whole point of Pertti Mattila's paper [MAT2] and upon it hangs an amusing tale. Being removable is conformally invariant: if K is removable and f is a conformal map defined on a neighborhood of K, then $f(K)$ is also removable. Mattila proved that given any C^2 diffeomorphism f which does not always map line segments to line segments, there exists a compact subset K of the domain of f such that $\mathcal{CR}(K) = 0$ but $\mathcal{CR}(f(K)) > 0$. Applying this result to the map $f(z) = z^2$ on the open right half-plane, we realize that either the K or the $f(K)$ produced must be a counterexample to VOC. The amusing bit of the tale is that since we do not know whether K and $f(K)$ are both removable or both nonremovable, we do not know which set is the counterexample! Thus Matilla's example gives us no idea which implication of VOC fails! Later on, in 1988, Peter Jones and Takafumi Murai produced, by very complicated means, a nonremovable compact set with zero Crofton measure in [JM1]. The simpler Joyce–Mörters example from 2000 had to wait for the invention of Melnikov curvature in 1995. At present it is still not known if there exists a removable compact set with positive Crofton measure!

Recall that the Joyce–Mörters set, although of Hausdorff dimension one, has infinite, indeed non-σ-finite, linear Hausdorff measure (Proposition 4.34). This suggests what we will call the Finite Case of Vitushkin's Original Conjecture (FCVOC): For every compact subset K of $\mathbb{C}$ with $\mathcal{H}^1(K) < \infty$,

$$(\star)\ K \text{ is removable if and only if } \mathcal{CR}(K) = 0.$$

What compact sets with finite linear Hausdorff measure can we come up with to test this out? For the rest of this section, call a set *trivial* if it has zero linear Hausdorff measure. Then $(\star)$ holds for trivial compact sets (Painlevé's Theorem (2.7)). Another class to consider are the nontrivial compact sets that are contained in a rectifiable curve. These all have finite linear Hausdorff measure and are nonremovable (Proposition 4.21 and Theorem 4.27). Since it can be shown that the Crofton measure of such sets is positive (see Theorem 6.10 of [FALC]), we see that $(\star)$ holds again. The only other nontrivial compact set with finite linear Hausdorff measure that we have encountered is the planar Cantor quarter set which is removable (Theorem 2.19). If $(\star)$ is to hold here, then we must check that its Crofton measure is zero. At first sight this appears unlikely since we showed this set to be nontrivial at the beginning of Section 2.4 by observing that its orthogonal projection onto the line $y = x/2$ was an interval of length $3\sqrt{5}/5$. Of course the same is true of the eight lines gotten by applying the symmetries of $[0, 1] \times [0, 1]$ to the line $y = x/2$. It is left to the reader to show that in all directions except the eight determined by these lines the orthogonal projection of the set has linear Lebesgue measure zero! Thus the planar Cantor quarter set satisfies $(\star)$. Indeed, for a long time all test sets that were "computable" turned out to satisfy $(\star)$, thus giving strong support to the FCVOC.

To gain better perspective here, let us turn to an old and fascinating piece of mathematics. A subset E of $\mathbb{C}$ is called a *1-set* if E is $\mathcal{H}^1$-measurable and $\mathcal{H}^1(E) < \infty$. In a series of papers, [BES1], [BES2], and [BES3], Abram Samoilovitch Besicovitch uncovered and developed a beautiful and deep theory of 1-sets. We now describe his results.

The *lower* and *upper densities* of a 1-set E at a point $z \in \mathbb{C}$ are defined by

$$\underline{D}(E, z) = \liminf_{r \downarrow 0} \frac{\mathcal{H}^1(E \cap B(z; r))}{2r} \text{ and } \overline{D}(E, z) = \limsup_{r \downarrow 0} \frac{\mathcal{H}^1(E \cap B(z; r))}{2r}$$

respectively. When $\underline{D}(E, z) = \overline{D}(E, z)$, we write $D(E, z)$ for the common value and call it the *density* of E at z. A point $z \in E$ is called a *regular point* of E if $D(E, z)$ exists and equals 1; otherwise, z is called an *irregular point* of E. The set of regular points of E is called the *regular part* of E and denoted E_{reg}; the set of irregular points of E is called the *irregular part* of E and denoted E_{irr}. We say that E is a *regular set* if $\mathcal{H}^1(E \setminus E_{\text{reg}}) = 0$ and an *irregular set* if $\mathcal{H}^1(E \setminus E_{\text{irr}}) = 0$. Besicovitch proved, among other things, the following results:

1. The regular part of a 1-set is a regular 1-set; the irregular part of a 1-set is an irregular 1-set.
2. Any $\mathcal{H}^1$-measureable subset of a regular 1-set is a regular 1-set; any $\mathcal{H}^1$-measureable subset of an irregular 1-set is an irregular 1-set.

3. The density of an irregular 1-set fails to exist at $\mathcal{H}^1$-almost all of its points. Thus a 1-set will never be irregular because it has a subset of positive $\mathcal{H}^1$-measure at each point of which the density exists but happens to take a value other than 1.
4. A regular 1-set is, modulo a set of $\mathcal{H}^1$-measure zero, contained in the union of countably many rectifiable curves; an irregular 1-set intersects every rectifiable curve in a set of $\mathcal{H}^1$-measure zero. Each of these statements is reversible, thus giving alternative characterizations of regularity and irregularity for 1-sets.
5. One can give a measure-theoretic definition of the notion of a *tangent line* to a set at a given point of the set. Using this measure-theoretic notion of tangent line, one has the following: a regular 1-set has a tangent line at $\mathcal{H}^1$-almost all of its points; an irregular 1-set fails to have a tangent line at $\mathcal{H}^1$-almost all of its points. Each of these statements is reversible, thus giving alternative characterizations of regularity and irregularity for 1-sets.
6. A regular nontrivial 1-set has positive Crofton measure; an irregular nontrivial 1-set has Crofton measure zero. Each of these statements is reversible, thus giving alternative characterizations of regularity and irregularity for nontrivial 1-sets.

(Aside from the original references mentioned above, an excellent modern exposition of Besicovitch's results, as well as of the theorem of Marstrand quoted earlier, is the concise and elegant book [FALC]. Be aware that what [FALC] calls a 1-set is what is here called a nontrivial 1-set.)

Note that in terms of our earlier test sets, a nontrivial compact subset of a rectifiable curve is a canonical example of a regular set while the planar Cantor quarter set is a canonical example of an irregular set. (A comment in passing: one can show the planar Cantor quarter set to be irregular straight from the definition in terms of densities – consideration of the radii $r_n = \sqrt{2}/4^n$ leads to the conclusion that the lower density of the set at any of its points is at most $1/2$.)

From item 6 we see that for a compact 1-set K, $\mathcal{CR}(K) = 0$ is equivalent to K being irregular. But then from item 4 we see that the FCVOC can be equivalently restated as follows: For every compact set subset K of $\mathbb{C}$ with $\mathcal{H}^1(K) < \infty$,

($\star\star$) K is removable if and only if $\mathcal{H}^1(K \cap \Gamma) = 0$ for every rectifiable curve Γ.

or, contrapositively,

($\star\star\star$) K is nonremovable if and only if $\mathcal{H}^1(K \cap \Gamma) > 0$ for some rectifiable curve Γ.

As noted before, the backward implication of ($\star\star\star$) follows easily from Theorem 4.27. So the FCVOC is reduced to the forward implication of ($\star\star\star$): For every compact subset K of $\mathbb{C}$,

if K is nonremovable and $\mathcal{H}^1(K) < \infty$, then $\mathcal{H}^1(K \cap \Gamma) > 0$ for some rectifiable curve Γ.

But this is just what we have been calling Vitushkin's Conjecture!

Chapter 7
The *T(b)* Theorem of Nazarov, Treil, and Volberg

7.1 Restatement of the Result

The goal of this long chapter is to prove Theorem 6.14, the first of the two difficult results needed to complete the resolution of Vitushkin's Conjecture. Our treatment here, and in various parts of the previous chapter, is from [NTV]. Via the change of variable $\zeta \mapsto \zeta - c$ and the replacement of h by $h/h(c)$, we may assume that $c = 0$ and $h(c) = 1$ in Theorem 6.14. So it suffices to prove the following.

Theorem 7.1 (Nazarov, Treil, and Volberg) *Let* $0 < \delta \leq 1/27{,}000$, $0 < M < \infty$, *and* $0 < \widetilde{M} < \infty$. *Suppose that* μ *is a totally regular positive Borel measure on* $\mathbb{C}$, h *is an element of* $L^\infty(\mu)$, *and* $\widetilde{\Phi}$ *is a Lip*(1)*-function that together satisfy*

(1) $\operatorname{spt}(\mu) \subseteq B(0; 1/8)$,

(2) μ *has linear growth with bound* M,

(3) $\displaystyle\int_{\mathbb{C}} |h - 1|^2 \, d\mu < \frac{\delta^4}{2}\mu(\mathbb{C})$,

(4) $\mu(\{\widetilde{\Phi} > 0\}) \leq \dfrac{\delta^2}{2}\mu(\mathbb{C})$, *and*

(5) *for all points* $z \in \mathbb{C}$, *numbers* $\varepsilon > 0$, *and Lip*(1)*-functions* $\Theta \geq \delta\widetilde{\Phi}$,

$$\left| \int_{\mathbb{C} \setminus B(z;\varepsilon)} k_\Theta(\zeta, z) h(\zeta) \, d\mu(\zeta) \right| \leq \widetilde{M}.$$

Then there exists a Lip(1)*-function* Φ *such that*

(a) $\mu(\{\Phi > 0\}) \leq \delta\mu(\mathbb{C})$ *and*

(b) $\sup_{\lambda > 0} \|\mathcal{K}_{\Phi+\lambda}\|_{L^2(\mu) \mapsto L^2(\mu)} < \infty.$

J.J. Dudziak, *Vitushkin's Conjecture for Removable Sets*, Universitext,
DOI 10.1007/978-1-4419-6709-1_7,

At this point the reader may wish to go back and reread the paragraphs before the statement of Theorem 6.14 for the definition of and information about $\mathcal{K}_{\Phi \underset{\sim}{+} \lambda}$.

A standing assumption for the rest of this chapter is that δ, M, $\widetilde{M}$, μ, h, and $\underset{\sim}{\Phi}$ are as in Theorem 7.1. During the course of this chapter many inequalities concerning δ will be stated and used without comment. Their validity follows from the fact that δ is positive and less than $1/27{,}000$. Finally *and most importantly*, g will always denote the function $h - 1$!

7.2 Random Dyadic Lattice Construction

It will be convenient in this chapter to require of all squares that they be half-open as in the proof of Frostman's Lemma (2.9). Indeed, we carry forward to the present context the terminology concerning "children" and other genealogical terms introduced in the first paragraph of the proof of that lemma. New to this context is the following: call any square Q *transit* if

$$\int_Q |g|^2 \, d\mu < \delta^2 \mu(Q)$$

and *terminal* otherwise. Clearly $\mu(Q) > 0$ whenever Q is transit, or equivalently, Q is terminal whenever $\mu(Q) = 0$.

Set $\Omega = [-1/4, 1/4) \times [-1/4, 1/4)$ and make it into a probability space via the measure $\mathcal{P} = 4\mathcal{L}^2_{\Omega}$, four times planar Lebesgue measure restricted to Ω. Given $\omega \in \Omega$, set $Q^0_\omega = \omega + [-1/2, 1/2) \times [-1/2, 1/2)$. Note that by hypotheses (1) and (3) of Theorem 7.1,

$$\int_{Q^0_\omega} |g|^2 \, d\mu = \int_{\mathbb{C}} |g|^2 \, d\mu < \frac{\delta^4}{2} \mu(\mathbb{C}) < \delta^2 \mu(\mathbb{C}) = \delta^2 \mu\left(Q^0_\omega\right),$$

so Q^0_ω is always transit.

We now construct a *random dyadic lattice* $\mathcal{D}_\omega$ for each $\omega \in \Omega$ as follows. We always place Q^0_ω in $\mathcal{D}_\omega$. Since Q^0_ω is transit, we place its four children in $\mathcal{D}_\omega$ too. Those of these four children which happen to be transit will have their four children placed in $\mathcal{D}_\omega$ in turn, while those of these four children which happen to be terminal will have none of their descendants placed in $\mathcal{D}_\omega$. The lattice $\mathcal{D}_\omega$ is gotten by continuing in this manner: always placing the four children of a transit square of $\mathcal{D}_\omega$ into $\mathcal{D}_\omega$ in turn and always barring the descendants of a terminal square of $\mathcal{D}_\omega$ from $\mathcal{D}_\omega$.

Let $\mathcal{D}^{\mathrm{tran}}_\omega$ and $\mathcal{D}^{\mathrm{term}}_\omega$ denote the subcollections of transit and terminal squares of $\mathcal{D}_\omega$ respectively. Note that a descendant of Q^0_ω does not become a member of $\mathcal{D}^{\mathrm{tran}}_\omega$ simply by being transit; it only becomes a member if it and all its ancestors up to Q^0_ω are transit. Similarly, a descendant of Q^0_ω does not become a member of $\mathcal{D}^{\mathrm{term}}_\omega$ simply by being terminal; it only becomes a member if it is terminal and all its ancestors up to Q^0_ω are transit. Thus $\mathcal{D}_\omega$ does not contain all the descendants of Q^0_ω;

any descendant of Q^0_ω with even a single terminal ancestor does not make it into $\mathcal{D}_\omega$. Lastly, note that the squares of $\mathcal{D}^{\text{term}}_\omega$ are pairwise disjoint.

7.3 Lip(1)-Functions Attached to Random Dyadic Lattices

To construct the Lip(1)-function Φ of Theorem 7.1, we need to construct Lip(1)-functions Φ_ω for each $\mathcal{D}_\omega$.

Proposition 7.2 *For any $\omega \in \Omega$, set $\Phi_\omega = \max\{\widetilde{\Phi}, \widetilde{\Phi}_\omega\}$ where*

$$\widetilde{\Phi}_\omega(z) = \begin{cases} \operatorname{dist}(z, \partial Q) & \text{when } z \in Q \text{ for some } \mathcal{D}^{\text{term}}_\omega \\ 0 & \text{when } z \in \mathbb{C} \setminus \bigcup \mathcal{D}^{\text{term}}_\omega \end{cases}.$$

Then

(a) Φ_ω *is a Lip(1)-function,*

(b) $\Phi_\omega \geq \widetilde{\Phi}$ *on* $\mathbb{C}$,

(c) $\Phi_\omega \geq \operatorname{dist}(z, \partial Q)$ *whenever* $z \in Q \in \mathcal{D}^{\text{term}}_\omega$, *and*

(d) $\mu(\{\Phi_\omega > 0\}) \leq \delta^2 \mu(\mathbb{C})$.

Proof (a) Note that we could also have defined $\widetilde{\Phi}_\omega(z)$ to be $\sup\{\operatorname{dist}(z, \mathbb{C} \setminus Q) : Q \in \mathcal{D}^{\text{term}}_\omega\}$. Thus (a) follows from two easily verified facts: first, for any nonempty subset E of $\mathbb{C}$, the function $z \in \mathbb{C} \mapsto \operatorname{dist}(z, E) \in [0, \infty)$ is Lip(1), and second, the supremum of any number of Lip(1)-functions is again a Lip(1)-function as soon as it is finite at even one point of $\mathbb{C}$.

(b) and (c) Immediate.

(d) From our various definitions, hypotheses (3) and (4) of Theorem 7.1, and the pairwise disjointness of the squares from $\mathcal{D}^{\text{term}}_\omega$, we see that

$$\begin{aligned}
\mu(\{\Phi_\omega > 0\}) &\leq \mu(\{\widetilde{\Phi} > 0\}) + \sum_{Q \in \mathcal{D}^{\text{term}}_\omega} \mu(Q) \\
&\leq \frac{\delta^2}{2}\mu(\mathbb{C}) + \sum_{Q \in \mathcal{D}^{\text{term}}_\omega} \frac{1}{\delta^2} \int_Q |g|^2 \, d\mu \\
&\leq \frac{\delta^2}{2}\mu(\mathbb{C}) + \frac{1}{\delta^2} \int_{\mathbb{C}} |g|^2 \, d\mu \\
&< \frac{\delta^2}{2}\mu(\mathbb{C}) + \frac{1}{\delta^2} \left\{ \frac{\delta^4}{2}\mu(\mathbb{C}) \right\} = \delta^2 \mu(\mathbb{C}).
\end{aligned}$$

□

7.4 Construction of the Lip(1)-Function of the Theorem

We call the function Φ of the next proposition the *truncated expectation* of the Φ_ωs.

Proposition 7.3 *For any $z \in \mathbb{C}$, set* $\Phi(z) = \inf\left\{\int_{\Omega\setminus\widetilde{\Omega}} \Phi_\omega(z)\,d\mathcal{P}(\omega) : \mathcal{P}(\widetilde{\Omega}) \le \delta\right\}$.
Then

(a) Φ *is a Lip*(1)*-function and*

(b) $\mu(\{\Phi > 0\}) \le \delta\mu(\mathbb{C})$.

Proof (a) Given $z \in \mathbb{C}$ and $\varepsilon > 0$, choose $\widetilde{\Omega}$ with $\mathcal{P}(\widetilde{\Omega}) \le \delta$ such that

$$\int_{\Omega\setminus\widetilde{\Omega}} \Phi_\omega(z)\,d\mathcal{P}(\omega) < \Phi(z) + \varepsilon.$$

Then for any $\zeta \in \mathbb{C}$,

$$\begin{aligned}\Phi(\zeta) - \Phi(z) &< \Phi(\zeta) - \int_{\Omega\setminus\widetilde{\Omega}} \Phi_\omega(z)\,d\mathcal{P}(\omega) + \varepsilon\\ &\le \int_{\Omega\setminus\widetilde{\Omega}} \{\Phi_\omega(\zeta) - \Phi_\omega(z)\}\,d\mathcal{P}(\omega) + \varepsilon \le |\zeta - z| + \varepsilon.\end{aligned}$$

Thus, letting $\varepsilon \downarrow 0$, we obtain $\Phi(\zeta) - \Phi(z) \le |\zeta - z|$. Similarly, $\Phi(z) - \Phi(\zeta) \le |\zeta - z|$ and so we are done.

(b) Set $E = \{z \in \mathbb{C} : \mathcal{P}(\{\omega \in \Omega : \Phi_\omega(z) > 0\}) > \delta\}$. Note that for any $z \in \mathbb{C}\setminus E$, the set $\widetilde{\Omega} = \{\omega \in \Omega : \Phi_\omega(z) > 0\}$ has $\mathcal{P}(\widetilde{\Omega}) \le \delta$ and so by the definition of Φ,

$$0 \le \Phi(z) \le \int_{\Omega\setminus\widetilde{\Omega}} \Phi_\omega(z)\,d\mathcal{P}(\omega) = 0.$$

It follows that $\{\Phi > 0\} \subseteq E$ and so to finish it suffices to show that $\mu(E) \le \delta\mu(\mathbb{C})$. Using (d) of Proposition 7.2, this is an easy consequence of Fubini's Theorem [RUD, 8.8]:

$$\begin{aligned}\delta\mu(E) &\le \int_{\mathbb{C}} \mathcal{P}(\{\omega \in \Omega : \Phi_\omega(z) > 0\})\,d\mu(z)\\ &= \int_{\Omega} \mu(\{z \in \mathbb{C} : \Phi_\omega(z) > 0\})\,d\mathcal{P}(\omega) \le \delta^2\mu(\mathbb{C}).\end{aligned}$$

Dividing both sides by δ, we are done. □

The alert reader will have noticed that the use of Fubini in the last proof calls for a verification of product measurability which in this case is not immediately obvious. Thus the following ...

Proposition 7.4 *The function* $(z, \omega) \in \mathbb{C} \times \Omega \mapsto \Phi_\omega(z) \in [0, \infty)$ *is* $\mu \times \mathcal{P}$*-measurable.*

Proof It suffices to show that $(z, \omega) \in \mathbb{C} \times \Omega \mapsto \widetilde{\Phi}_\omega(z) \in [0, \infty)$ is $\mu \times \mathcal{P}$-measurable.

For this we need to introduce some notation. First, let $\mathcal{F}$ denote the *full dyadic lattice* consisting of the square $[-1/2, 1/2) \times [-1/2, 1/2)$ along with all its descendants, transit or terminal. Note that $\mathcal{F}$ is countable. Second, given $Q \in \mathcal{F}$, let $\mathcal{A}(Q)$ denote the collection consisting of all of the ancestors of Q from $\mathcal{F}$ (not including Q itself). Note that $\mathcal{A}(Q)$ is finite. Third, given $Q \in \mathcal{F}$, define a function F_Q on Ω by

$$F_Q(\omega) = \begin{cases} 1 \text{ when } \omega + Q \text{ is transit} \\ 0 \text{ when } \omega + Q \text{ is terminal.} \end{cases}$$

Then a little thought shows that

$$\widetilde{\Phi}_\omega(z) = \sup_{Q \in \mathcal{F}} \left\{ \operatorname{dist}(z, \mathbb{C} \setminus \{\omega + Q\}) \cdot (1 - F_Q(\omega)) \cdot \prod_{R \in \mathcal{A}(Q)} F_R(\omega) \right\}.$$

Now for any $Q \in \mathcal{F}$ the function $(z, \omega) \in \mathbb{C} \times \Omega \mapsto \operatorname{dist}(z, \mathbb{C} \setminus \{\omega + Q\}) \in [0, \infty)$ is continuous. Indeed, for any $z, \tilde{z} \in \mathbb{C}$ and $\omega, \tilde{\omega} \in \Omega$ we have

$$|\operatorname{dist}(z, \mathbb{C} \setminus \{\omega + Q\}) - \operatorname{dist}(\tilde{z}, \mathbb{C} \setminus \{\tilde{\omega} + Q\}) \leq |z - \tilde{z}| + |\omega - \tilde{\omega}|.$$

Thus, fixing $Q \in \mathcal{F}$, we simply need to show that F_Q is $\mathcal{P}$-measurable, i.e., that the set

$$\{\omega \in \Omega : \int_{\omega+Q} |g|^2 \, d\mu < \int_{\omega+Q} \delta^2 \, d\mu\}$$

is $\mathcal{P}$-measurable.

This follows if we show that for any nonnegative function $f \in L^1(\mu)$, the function

$$\omega \in \Omega \mapsto \int_{\omega+Q} f \, d\mu = \int f(z) \mathcal{X}_Q(z - \omega) \, d\mu(z) \in [0, \infty)$$

is $\mathcal{P}$-measurable (above we need the two cases $f = |g|^2$ and $f = \delta^2$). By Fubini [RUD, 8.8] this in turn follows from the $\mu \times \mathcal{P}$-measurability of the function

$$(z, \omega) \in \mathbb{C} \times \Omega \mapsto f(z)\mathcal{X}_Q(z - \omega) \in [0, \infty).$$

But $(z, \omega) \in \mathbb{C} \times \Omega \mapsto f(z) \in [0, \infty)$ is $\mu \times \mathcal{P}$-measurable since f is μ-measurable and $(z, \omega) \in \mathbb{C} \times \Omega \mapsto \mathcal{X}_Q(z - \omega) \in [0, \infty)$, being the characteristic function of a half-open skewed "strip", is Borel measurable. Thus we are done. □

7.5 The Standard Martingale Decomposition

Having produced the function Φ of Theorem 7.1 and verified that it satisfies (a), it remains to show that it satisfies (b). We now sketch, in a vague and impressionistic way, how this will be done over the course of the rest of this chapter. Instead of getting "appropriate" estimates on the norms $\|\mathcal{K}_{\Phi+\lambda}\,\varphi\|_{L^2(\mu)}$ where $\varphi \in L^2(\mu)$, we shall get "appropriate" estimates on the inner products $\langle \mathcal{K}_{\Phi+\lambda}\,\varphi, \psi\rangle$ where $\varphi, \psi \in L^2(\mu)$. In this and the next two sections we shall show how a function $\varphi \in L^2(\mu)$ can be decomposed in a "nice" way over a random dyadic lattice $\mathcal{D}_\omega$:

$$\varphi = \Lambda\,\varphi + \sum_{Q \in \mathcal{D}_\omega^{\text{tran}}} \Delta_Q\,\varphi.$$

Here $\Lambda\,\varphi$ is just the multiple of h with the same average value over $\mathbb{C}$ as φ and each $\Delta_Q\,\varphi$ is a function constructed from h that lives on Q. For our purposes we shall be able to assume that $\Lambda\,\varphi$ and $\Lambda\,\psi$ are both 0. Thus we can decompose our inner product as follows:

$$\langle \mathcal{K}_{\Phi+\lambda}\,\varphi, \psi\rangle = \sum_{(Q,R) \in \mathcal{D}_{\omega_1}^{\text{tran}} \times \mathcal{D}_{\omega_2}^{\text{tran}}} \langle \mathcal{K}_{\Phi+\lambda}\,\Delta_Q\,\varphi, \Delta_R\,\psi\rangle.$$

The inner products in the above double sum will be grouped into three categories that depend on the relative sizes and relative positions of the squares Q and R. The sums of inner products corresponding to the different categories will then be estimated with analyses appropriate to each category. Everyone before Nazarov, Treil, and Volberg worked with nonrandom dyadic lattices and completion of the argument required them to assume that the measure μ satisfied a doubling condition or lower Ahlfors regularity in order to get around certain "bad" situations that could arise. Nazarov, Treil, and Volberg introduced randomness in the dyadic lattices being used and showed by means of a probabilistic argument that for any pair (φ, ψ) of functions from $L^2(\mu)$, it is always possible to choose a pair of points (ω_1, ω_2) from Ω such that these "bad" situations do not occur, thus obviating any assumption of a doubling condition or lower Ahlfors regularity! (David's proof proceeds quite differently, accepting "bad" situations as they arise and modifying the dyadic decomposition to deal with them.)

The martingale decomposition we desire will be adapted to take advantage of our function h and will be put forward two sections hence. Just now in this section we introduce the standard martingale decomposition. In the next section we shall prove the Dyadic Carleson Imbedding Inequality which will then allow us to deduce properties of the adapted decomposition from the corresponding properties of the standard decomposition. During this section and the next two we shall be considering a fixed dyadic lattice $\mathcal{D}_\omega$ and so will drop reference to ω. Thus we shall write "$\mathcal{D}$" instead of "$\mathcal{D}_\omega$," "Q^0" instead of "Q^0_ω," etc.

Define the *average value* $\langle\varphi\rangle_E$ of a function $\varphi \in L^2(\mu)$ over a μ-measurable set E by

$$\langle\varphi\rangle_E = \begin{cases} \dfrac{1}{\mu(E)} \displaystyle\int_E \varphi \, d\mu & \text{when} \quad \mu(E) > 0 \\ 0 & \text{when} \quad \mu(E) = 0. \end{cases}$$

Note that under either clause of the definition one has $\int_E \varphi \, d\mu = \langle\varphi\rangle_E \cdot \mu(E)$. Also, for any $Q \in \mathcal{D}^{\text{tran}}$, we have $\mu(Q) > 0$ and so $\langle\varphi\rangle_Q$ is defined via the first clause of the definition. We require two lemmas about these average values.

Lemma 7.5 *Let $\mathcal{D}^{\text{tran}}_{n+1}$ denote the set of all $Q \in \mathcal{D}^{\text{tran}}$ with $l(Q) = 2^{-(n+1)}$. Then*

$$\lim_{n\to\infty} \sum_{Q\in\mathcal{D}^{\text{tran}}_{n+1}} \int_Q |\varphi - \langle\varphi\rangle_Q|^2 \, d\mu = 0$$

for any $\varphi \in L^2(\mu)$.

Proof Given $\varepsilon > 0$, choose a compactly supported continuous function f on $\mathbb{C}$ such that $\int_{\mathbb{C}} |\varphi - f|^2 \, d\mu < \varepsilon/12$ via Proposition 5.23. From the trivial inequality $(a+b+c)^2 \le 4a^2 + 4b^2 + 4c^2$ we then obtain

$$(\star) \sum_{Q\in\mathcal{D}^{\text{tran}}_{n+1}} \int_Q |\varphi - \langle\varphi\rangle_Q|^2 \, d\mu \le \sum_{Q\in\mathcal{D}^{\text{tran}}_{n+1}} \int_Q 4|\varphi - f|^2 \, d\mu \leftarrow \text{(I)!}$$

$$+ \sum_{Q\in\mathcal{D}^{\text{tran}}_{n+1}} \int_Q 4|f - \langle f\rangle_Q|^2 \, d\mu \leftarrow \text{(II)!}$$

$$+ \sum_{Q\in\mathcal{D}^{\text{tran}}_{n+1}} \int_Q 4|\langle f\rangle_Q - \langle\varphi\rangle_Q|^2 \, d\mu. \leftarrow \text{(III)!}$$

Now the squares of $\mathcal{D}^{\text{tran}}$ are not in general pairwise disjoint, but the ones of a given fixed size are, so (I) $\le 4\int_{\mathbb{C}} |\varphi - f|^2 \, d\mu < \varepsilon/3$.

Since f is uniformly continuous on $\mathbb{C}$, there exists an integer N such that

$$\operatorname{osc}\big(f, \mathbb{C}; 2^{-N}\big) < \sqrt{\frac{\varepsilon}{12\mu(\mathbb{C})}}.$$

Thus given any $n \geq N$, any $Q \in \mathcal{D}_{n+1}^{\text{tran}}$, and any $z \in Q$, one has

$$|f(z) - \langle f \rangle_Q| \leq \left| \frac{1}{\mu(Q)} \int_Q \{f(z) - f(\zeta)\} \, d\mu(\zeta) \right| \leq \sqrt{\frac{\varepsilon}{12\mu(\mathbb{C})}}.$$

Hence for any $n \geq N$, using the pairwise disjointness of $\mathcal{D}_{n+1}^{\text{tran}}$ again, we see that

$$(\text{II}) \leq \sum_{Q \in \mathcal{D}_{n+1}^{\text{tran}}} 4 \cdot \frac{\varepsilon}{12\mu(\mathbb{C})} \cdot \mu(Q) \leq \frac{\varepsilon}{3}.$$

Lastly, using the Cauchy–Schwarz Inequality [RUD, 3.5] and the pairwise disjointness of $\mathcal{D}_{n+1}^{\text{tran}}$ yet again, we see that

$$\begin{aligned}
(\text{III}) &= \sum_{Q \in \mathcal{D}_{n+1}^{\text{tran}}} 4 \left| \frac{1}{\mu(Q)} \int_Q \{f - \varphi\} \, d\mu \right|^2 \mu(Q) \\
&\leq \sum_{Q \in \mathcal{D}_{n+1}^{\text{tran}}} 4 \left\{ \frac{1}{\mu(Q)} \int_Q |f - \varphi|^2 \, d\mu \right\} \left\{ \frac{1}{\mu(Q)} \int_Q 1^2 \, d\mu \right\} \mu(Q) \\
&\leq 4 \int_{\mathbb{C}} |f - \varphi|^2 \, d\mu \quad < \quad \frac{\varepsilon}{3}.
\end{aligned}$$

The lemma now follows from ($\star$) and our estimates on (I), (II), and (III). □

Lemma 7.6 *Let E be a μ-measurable set and $\varphi \in L^2(\mu)$. Then*

(a) $\displaystyle\int_E |\varphi|^2 \, d\mu = \int_E |\langle \varphi \rangle_E|^2 \, d\mu + \int_E |\varphi - \langle \varphi \rangle_E|^2 \, d\mu,$

(b) $\langle |\varphi|^2 \rangle_E = |\langle \varphi \rangle_E|^2 + \langle |\varphi - \langle \varphi \rangle_E|^2 \rangle_E$, *and*

(c) $|\langle \varphi \rangle_E|^2 \leq \langle |\varphi|^2 \rangle_E$ *and* $\langle |\varphi - \langle \varphi \rangle_E|^2 \rangle_E \leq \langle |\varphi|^2 \rangle_E$.

Proof To prove (a) it suffices to show that the left-hand side of its conclusion minus its right-hand side is 0. But the trivial equality $|a+b|^2 = |a|^2 + |b|^2 + 2\text{Re}\,\bar{a}b$ shows that this difference is

$$\begin{aligned}
\int_E 2\text{Re}\, \overline{\langle \varphi \rangle_E} \{\varphi - \langle \varphi \rangle_E\} \, d\mu &= 2\text{Re}\, \overline{\langle \varphi \rangle_E} \int_E \{\varphi - \langle \varphi \rangle_E\} \, d\mu \\
&= 2\text{Re}\, \overline{\langle \varphi \rangle_E} \left\{ \int_E \varphi \, d\mu - \langle \varphi \rangle_E \cdot \mu(E) \right\} = 0.
\end{aligned}$$

Note that (b) follows immediately from (a) and then (c) follows immediately from (b). □

We are now ready to define the *standard martingale decomposition* of a function $\varphi \in L^2(\mu)$. Set $\widetilde{\Lambda}\,\varphi = \langle \varphi \rangle_{\mathbb{C}}$. Since all of the mass of μ is concentrated on Q^0, we also have $\widetilde{\Lambda}\,\varphi = \langle \varphi \rangle_{Q^0}$. Given $Q \in \mathcal{D}^{\text{tran}}$, let Q_j, $j = 1, 2, 3, 4$, denote the four children of Q labeled starting with the northeast one and proceeding counterclockwise for the sake of definiteness. Set

$$(\widetilde{\Delta}_Q\,\varphi)(z) = \begin{cases} 0 & \text{when} \quad z \in \mathbb{C} \setminus Q \\ \langle \varphi \rangle_{Q_j} - \langle \varphi \rangle_Q & \text{when} \quad z \in Q_j \in \mathcal{D}^{\text{tran}} \\ \varphi - \langle \varphi \rangle_Q & \text{when} \quad z \in Q_j \in \mathcal{D}^{\text{term}}. \end{cases}$$

Let $\mathcal{D}^{\text{tran}}_{\leq n}$ denote the set of all $Q \in \mathcal{D}^{\text{tran}}$ with $l(Q) \geq 2^{-n}$. An expression of the form

$$\sum_{Q \in \mathcal{D}^{\text{tran}}} f(Q)$$

is to be interpreted as the limit as $n \to \infty$ of the sum of $f(Q)$ over all $Q \in \mathcal{D}^{\text{tran}}_{\leq n}$.

Proposition 7.7 (Standard Martingale Decomposition) *For any* $\varphi \in L^2(\mu)$,

$$\varphi = \widetilde{\Lambda}\,\varphi + \sum_{Q \in \mathcal{D}^{\text{tran}}} \widetilde{\Delta}_Q\,\varphi$$

with the convergence being in $L^2(\mu)$. *Moreover,*

$$\|\varphi\|^2_{L^2(\mu)} = \|\widetilde{\Lambda}\,\varphi\|^2_{L^2(\mu)} + \sum_{Q \in \mathcal{D}^{\text{tran}}} \|\widetilde{\Delta}_Q\,\varphi\|^2_{L^2(\mu)}.$$

Proof Since μ is supported on Q^0, $\mathcal{X}_{Q^0} = 1$ everywhere as far as μ is concerned. Thus

$$(\star)\ \varphi - \widetilde{\Lambda}\,\varphi = \{\varphi - \langle \varphi \rangle_{Q^0}\}\mathcal{X}_{Q^0} = \sum_{Q \in \mathcal{D}^{\text{tran}}_{\leq 0}} \{\varphi - \langle \varphi \rangle_Q\}\mathcal{X}_Q.$$

If $Q \in \mathcal{D}^{\text{tran}}$ and Q_j, $j = 1, 2, 3, 4$, are as in the definition of $\widetilde{\Delta}_Q$, then the reader may verify

$$(\star\star)\ \{\varphi - \langle \varphi \rangle_Q\}\mathcal{X}_Q - \widetilde{\Delta}_Q\,\varphi = \sum_{Q_j \in \mathcal{D}^{\text{tran}}} \{\varphi - \langle \varphi \rangle_{Q_j}\}\mathcal{X}_{Q_j}.$$

Starting with $(\star)$ and applying $(\star\star)$ over and over again, we see that

$$\varphi - \left\{ \widetilde{\Lambda}\,\varphi + \sum_{Q \in \mathcal{D}^{\text{tran}}_{\leq n}} \widetilde{\Delta}_Q\,\varphi \right\} = \sum_{Q \in \mathcal{D}^{\text{tran}}_{n+1}} \{\varphi - \langle \varphi \rangle_Q\} \mathcal{X}_Q.$$

Hence the pairwise disjointness of $\mathcal{D}^{\text{tran}}_{n+1}$ implies that

$$\|\varphi - \{\widetilde{\Lambda}\,\varphi + \sum_{Q \in \mathcal{D}^{\text{tran}}_{\leq n}} \widetilde{\Delta}_Q\,\varphi\}\|^2_{L^2(\mu)} = \sum_{Q \in \mathcal{D}^{\text{tran}}_{n+1}} \int_Q |\varphi - \langle \varphi \rangle_Q|^2\, d\mu.$$

The desired L^2 convergence now follows from Lemma 7.5.

By (a) of the last lemma,

$$(\star\star\star)\ \|\varphi\|^2_{L^2(\mu)} = \int_{Q^0} |\varphi|^2\, d\mu \ = \ \int_{Q^0} |\langle \varphi \rangle_{Q^0}|^2\, d\mu + \int_{Q^0} |\varphi - \langle \varphi \rangle_{Q^0}|^2\, d\mu$$
$$= \|\widetilde{\Lambda}\,\varphi\|^2_{L^2(\mu)} + \sum_{Q \in \mathcal{D}^{\text{tran}}_{\leq 0}} \int_Q |\varphi - \langle \varphi \rangle_Q|^2\, d\mu.$$

Also, letting Q and the Q_js be as before, we may use (a) of the last lemma again to obtain

$$(\star\star\star\star)\ \int_Q |\varphi - \langle \varphi \rangle_Q|^2\, d\mu = \sum_{Q_j \in \mathcal{D}^{\text{term}}} \int_{Q_j} |\varphi - \langle \varphi \rangle_Q|^2\, d\mu$$
$$+ \sum_{Q_j \in \mathcal{D}^{\text{tran}}} \int_{Q_j} |\varphi - \langle \varphi \rangle_Q|^2\, d\mu$$
$$= \sum_{Q_j \in \mathcal{D}^{\text{term}}} \int_{Q_j} |\widetilde{\Delta}_Q\,\varphi|^2\, d\mu$$
$$+ \sum_{Q_j \in \mathcal{D}^{\text{tran}}} \left\{ \int_{Q_j} |\langle \varphi \rangle_{Q_j} - \langle \varphi \rangle_Q|^2\, d\mu + \int_{Q_j} |\varphi - \langle \varphi \rangle_{Q_j}|^2\, d\mu \right\}$$
$$= \sum_{Q_j \in \mathcal{D}^{\text{term}}} \int_{Q_j} |\widetilde{\Delta}_Q\,\varphi|^2\, d\mu + \sum_{Q_j \in \mathcal{D}^{\text{tran}}} \int_{Q_j} |\widetilde{\Delta}_Q\,\varphi|^2\, d\mu$$
$$+ \sum_{Q_j \in \mathcal{D}^{\text{tran}}} \int_{Q_j} |\varphi - \langle \varphi \rangle_{Q_j}|^2\, d\mu$$
$$= \|\widetilde{\Delta}_Q\,\varphi\|^2_{L^2(\mu)} + \sum_{Q_j \in \mathcal{D}^{\text{tran}}} \int_{Q_j} |\varphi - \langle \varphi \rangle_{Q_j}|^2\, d\mu.$$

Starting with $(\star\star\star)$ and applying $(\star\star\star\star)$ over and over again, we see that

$$\|\varphi\|_{L^2(\mu)}^2 = \|\widetilde{\Lambda}\,\varphi\|_{L^2(\mu)}^2 + \sum_{Q\in\mathcal{D}_{\le n}^{\text{tran}}} \|\widetilde{\Delta}_Q\,\varphi\|_{L^2(\mu)}^2 + \sum_{Q\in\mathcal{D}_{n+1}^{\text{tran}}} \int |\varphi - \langle\varphi\rangle_Q|^2\, d\mu.$$

The desired L^2 norm equality now follows by Lemma 7.5. □

7.6 Interlude: The Dyadic Carleson Imbedding Inequality

We let $\mathcal{F}$ denote the *full dyadic lattice* generated by Q^0. Thus $\mathcal{F}$ consists of the square Q^0 along with all its dyadic descendants. (Note that in the proof of Proposition 7.4 we encountered the $\mathcal{F}$ corresponding to $Q^0 = [-1/2, 1/2) \times [-1/2, 1/2)$.) Given a square $Q \in \mathcal{F}$, let $\mathcal{F}(Q)$ denote the set of squares $R \in \mathcal{F}$ such that $R \subseteq Q$.

Proposition 7.8 (Dyadic Carleson Imbedding Inequality) *Suppose the indexed numbers $\{a_Q\}_{Q\in\mathcal{F}}$ are nonnegative and there exists a positive number $A < \infty$ such that for every $Q \in \mathcal{F}$,*

$$\sum_{R\in\mathcal{F}(Q)} a_R \le A\mu(Q).$$

Then for any $\varphi \in L^2(\mu)$,

$$\sum_{Q\in\mathcal{F}} a_Q |\langle\varphi\rangle_Q|^2 \le 4A\|\varphi\|_{L^2(\mu)}^2.$$

The proof of this proposition will follow easily from the next lemma which introduces a magic function f out of thin air. This magic function is known as a *Bellman function* associated with the Dyadic Carleson Imbedding Inequality. Many nontrivial inequalities of analysis have Bellman functions associated with them which once found make their proofs practically trivial. To learn about the general theory of Bellman functions, what they are, how to find them, and how to use them to get relatively easy proofs of otherwise difficult inequalities, see [NT] of which this section gives an incomplete glimpse.

Lemma 7.9 *Consider $f(x, y, z) = 4\left\{x - \dfrac{y^2}{1+z}\right\}$ defined on $\mathbb{R}^3$ minus the plane $z = -1$.*

(a) *For x and y arbitrary and z and Δz satisfying $0 \le z \le z + \Delta z \le 1$, we have*

$$y^2\Delta z \le f(x, y, z + \Delta z) - f(x, y, z).$$

(b) *Let x_j and y_j be arbitrary and z_j and λ_j be nonnegative for $1 \le j \le N$. Suppose also that $\sum_j \lambda_j = 1$. Set $\bar{x} = \sum_j \lambda_j x_j$, $\bar{y} = \sum_j \lambda_j y_j$, and $\bar{z} = \sum_j \lambda_j z_j$. Then*

$$f(\bar{x}, \bar{y}, \bar{z}) \geq \sum_j \lambda_j f(x_j, y_j, z_j).$$

Here all sums are from $j = 1$ *to* $j = N$.

(c) *For* x, y, *and* z *nonnegative with* $y \leq \sqrt{x}$, *we have* $f(x, y, z) \geq 0$.

(d) *For* x *and* y *arbitrary and* z *nonnegative, we have* $f(x, y, z) \leq 4x$.

Proof (a) As a matter of algebra we have

$$f(x, y, z + \Delta z) - f(x, y, z) = \frac{4y^2 \Delta z}{(1+z)(1+z+\Delta z)}.$$

This does it since by our hypotheses on z and Δz, we have $(1+z)(1+z+\Delta z) \leq 4$.

(b) It suffices to prove this for $N = 2$. Clearly $[(1+z_2)y_1 - (1+z_1)y_2]^2 \geq 0$. Squaring this out and rearranging, we obtain

$$2y_1 y_2 \leq \frac{y_1^2}{1+z_1}(1+z_2) + \frac{y_2^2}{1+z_2}(1+z_1).$$

Hence

$$\begin{aligned}
\bar{y}^2 &= \lambda_1^2 y_1^2 + \lambda_1\lambda_2(2y_1y_2) + \lambda_2^2 y_2^2 \\
&\leq \lambda_1^2 y_1^2 + \lambda_1\lambda_2 \left\{ \frac{y_1^2}{1+z_1}(1+z_2) + \frac{y_2^2}{1+z_2}(1+z_1) \right\} + \lambda_2^2 y_2^2 \\
&= \lambda_1 \frac{y_1^2}{1+z_1}\{\lambda_1(1+z_1) + \lambda_2(1+z_2)\} + \lambda_2 \frac{y_2^2}{1+z_2}\{\lambda_1(1+z_1) + \lambda_2(1+z_2)\} \\
&= \lambda_1 \frac{y_1^2}{1+z_1}\{1+\bar{z}\} + \lambda_2 \frac{y_2^2}{1+z_2}\{1+\bar{z}\}.
\end{aligned}$$

Dividing by $1 + \bar{z}$ yields

$$\frac{\bar{y}^2}{1+\bar{z}} \leq \lambda_1 \frac{y_1^2}{1+z_1} + \lambda_2 \frac{y_2^2}{1+z_2}.$$

This last inequality easily yields what we want.

(c) and (d) Clear. □

Proof of the Dyadic Carleson Imbedding Inequality (7.8) Replacing each a_Q by a_Q/A, we may assume that $A = 1$. Since $|\langle\varphi\rangle_Q| \leq \langle|\varphi|\rangle_Q$ and $\|\varphi\|^2_{L^2(\mu)} = \|\,|\varphi|\,\|^2_{L^2(\mu)}$, we may also assume that $\varphi \geq 0$. As a notational convenience, given $Q \in \mathcal{F}$ set

$$\langle a\rangle_Q = \begin{cases} \dfrac{1}{\mu(Q)} \displaystyle\sum_{R\in\mathcal{F}(Q)} a_R & \text{when } \mu(Q) > 0 \\[2ex] 0 & \text{when } \mu(Q) = 0. \end{cases}$$

Then by hypothesis, $\langle a\rangle_Q \le 1$ for every $Q \in \mathcal{F}$.

For the moment fix $Q \in \mathcal{F}$ and assume $\mu(Q) > 0$. Let Q_j, $j = 1, 2, 3, 4$, denote the four children of Q. Set $\lambda_j = \mu(Q_j)/\mu(Q)$, $x_j = \langle\varphi^2\rangle_{Q_j}$, $y_j = \langle\varphi\rangle_{Q_j}$, $z_j = \langle a\rangle_{Q_j}$, and $\Delta z = a_Q/\mu(Q)$. Note that all these numbers are nonnegative and that $\sum_j \lambda_j = 1$. Simple computations show that $\bar{x} = \langle\varphi^2\rangle_Q$, $\bar{y} = \langle\varphi\rangle_Q$, and $\bar{z} + \Delta z = \langle a\rangle_Q$. Clearly $0 \le \bar{z} \le \bar{z} + \Delta z \le 1$. Thus from (a) and (b) of the last lemma we get

$$\begin{aligned}(\star)\ a_Q\langle\varphi\rangle_Q^2 &= \mu(Q)\bar{y}^2\Delta z \\ &\le \mu(Q)\big[f(\bar{x},\bar{y},\bar{z}+\Delta z) - f(\bar{x},\bar{y},\bar{z})\big] \\ &\le \mu(Q)\Big[f(\bar{x},\bar{y},\bar{z}+\Delta z) - \sum_j \lambda_j f(x_j, y_j, z_j)\Big] \\ &= \mu(Q)f\big(\langle\varphi^2\rangle_Q, \langle\varphi\rangle_Q, \langle a\rangle_Q\big) - \sum_j \mu(Q_j) f\big(\langle\varphi^2\rangle_{Q_j}, \langle\varphi\rangle_{Q_j}, \langle a\rangle_{Q_j}\big).\end{aligned}$$

By our conventions $(\star)$ is also true when $\mu(Q) = 0$. Thus $(\star)$ holds for every $Q \in \mathcal{F}$.

Let $\mathcal{F}_{\le n}$ and $\mathcal{F}_{n+1}$ denote the squares $Q \in \mathcal{F}$ with $l(Q) \ge 2^{-n}$ and $l(Q) = 2^{-(n+1)}$ respectively. Applying $(\star)$ to all squares $Q \in \mathcal{F}_{\le n}$ and then adding the results together, an orgy of telescoping cancelling occurs on the right-hand side and we obtain

$$\begin{aligned}(\star\star)\ \sum_{Q\in\mathcal{F}_{\le n}} a_Q\langle\varphi\rangle_Q^2 &\le \mu(Q^0)f\big(\langle\varphi^2\rangle_{Q^0}, \langle\varphi\rangle_{Q^0}, \langle a\rangle_{Q^0}\big) \\ &\quad - \sum_{Q\in\mathcal{F}_{n+1}} \mu(Q)f\big(\langle\varphi^2\rangle_Q, \langle\varphi\rangle_Q, \langle a\rangle_Q\big).\end{aligned}$$

By (c) of Lemma 7.6, $\langle\varphi\rangle_Q \le \sqrt{\langle\varphi^2\rangle_Q}$ and so (c) of the last lemma implies that each term of the sum on the right-hand side of $(\star\star)$ is nonnegative. Also, (d) of the last lemma applies to the first term on right-hand side of $(\star\star)$. Hence

$$\sum_{Q\in\mathcal{F}_{\le n}} a_Q\langle\varphi\rangle_Q^2 \le \mu(Q^0)f\big(\langle\varphi^2\rangle_{Q^0}, \langle\varphi\rangle_{Q^0}, \langle a\rangle_{Q^0}\big) \le \mu(Q^0)\cdot 4\langle\varphi^2\rangle_{Q^0} = 4\|\varphi\|_{L^2(\mu)}^2.$$

Letting $n \to \infty$, we are done. □

7.7 The Adapted Martingale Decomposition

We shall be wanting to divide various quantities by $\langle h\rangle_Q$ where $Q \in \mathcal{D}^{\text{tran}}$ and h is the function of Theorem 7.1. For this we need to know that $\langle h\rangle_Q \neq 0$. The following result guarantees this and will also prove useful in establishing desired norm estimates.

Lemma 7.10 *Suppose* $Q \in \mathcal{D}^{\text{tran}}$. *Then*

(a) $1 - \delta < |\langle h\rangle_Q| < 1 + \delta$ *and*

(b) $1 \le \dfrac{\langle |h|^2\rangle_Q}{|\langle h\rangle_Q|^2} < 1 + \delta.$

Proof We make a useful observation: a square Q is transit if and only if $\mu(Q) > 0$ and $\langle |g|^2\rangle_Q < \delta^2$ where $g = h - 1$.

From this observation and (c) of Lemma 7.6 we get $|\langle h\rangle_Q - 1| = |\langle g\rangle_Q| \le \sqrt{\langle |g|^2\rangle_Q} < \delta$ and so all of (a) follows.

Since (a) implies that $|\langle h\rangle_Q| \neq 0$ (recall that $0 < \delta < 1/27{,}000$), the first inequality of (b) is now an immediate consequence of (c) of Lemma 7.6.

Using (b) and (c) of Lemma 7.6 and our observation, we see that

$$\begin{aligned}\langle |h|^2\rangle_Q &= |\langle h\rangle_Q|^2 + \langle |h - \langle h\rangle_Q|^2\rangle_Q \\ &= |\langle h\rangle_Q|^2 + \langle |g - \langle g\rangle_Q|^2\rangle_Q \le |\langle h\rangle_Q|^2 + \langle |g|^2\rangle_Q < |\langle h\rangle_Q|^2 + \delta^2.\end{aligned}$$

Dividing through by $|\langle h\rangle_Q|^2$ and using the first inequality of (a), we obtain the second inequality of (b) since $\delta/(1-\delta)^2 \le 1$. □

We are now ready to define the *adapted martingale decomposition* of a function $\varphi \in L^2(\mu)$. Set

$$\Lambda\,\varphi = \frac{\langle\varphi\rangle_{\mathbb{C}}}{\langle h\rangle_{\mathbb{C}}} h = \frac{\langle\varphi\rangle_{Q^0}}{\langle h\rangle_{Q^0}} h.$$

The last equality in the above holds since all of the mass of μ is concentrated on Q^0. Given $Q \in \mathcal{D}^{\text{tran}}$, let Q_j, $j = 1, 2, 3, 4$, denote the four children of Q labeled starting with the northeast one and proceeding counterclockwise for the sake of definiteness. Set

$$(\Delta_Q\,\varphi)(z) = \begin{cases} 0 & \text{when} \quad z \in \mathbb{C} \setminus Q \\ \left\{ \dfrac{\langle\varphi\rangle_{Q_j}}{\langle h\rangle_{Q_j}} - \dfrac{\langle\varphi\rangle_Q}{\langle h\rangle_Q} \right\} h & \text{when} \quad z \in Q_j \in \mathcal{D}^{\text{tran}} \\ \varphi - \dfrac{\langle\varphi\rangle_Q}{\langle h\rangle_Q} h & \text{when} \quad z \in Q_j \in \mathcal{D}^{\text{term}}. \end{cases}$$

Note that the adapted martingale decomposition becomes the standard martingale decomposition when h is the constant one function. The first part of our basic proposition concerning the adapted decomposition will consist of a convergence statement whose proof will parallel the proof of the corresponding convergence statement of the standard decomposition and require one new lemma to replace Lemma 7.5. The second part will consist of two norm inequalities which will be gotten from the norm equality of the standard decomposition via two new lemmas and a corollary.

Lemma 7.11 *For any* $\varphi \in L^2(\mu)$, $\lim_{n\to\infty} \sum_{Q\in\mathcal{D}^{\text{tran}}_{n+1}} \int_Q \left|\varphi - \frac{\langle\varphi\rangle_Q}{\langle h\rangle_Q} h\right|^2 d\mu = 0.$

Proof Let $\varepsilon > 0$. Given $N > 0$, set $\mathcal{A}^{(N)}_{n+1} = \{Q \in \mathcal{D}^{\text{tran}}_{n+1} : \langle|\varphi|^2\rangle_Q \le N^2\}$, and $\mathcal{B}^{(N)}_{n+1} = \{Q \in \mathcal{D}^{\text{tran}}_{n+1} : \langle|\varphi|^2\rangle_Q > N^2\}$. Using the trivial inequality $|a+b|^2 \le 2|a|^2 + 2|b|^2$, (a) of Lemma 7.10, and (c) of Lemma 7.6, we have

$$
\begin{aligned}
\sum_{Q\in\mathcal{D}^{\text{tran}}_{n+1}} \int_Q |\varphi - \frac{\langle\varphi\rangle_Q}{\langle h\rangle_Q} h|^2\, d\mu &= 2 \sum_{Q\in\mathcal{D}^{\text{tran}}_{n+1}} \int_Q |\varphi - \langle\varphi\rangle_Q|^2\, d\mu \\
&\quad + 2 \sum_{Q\in\mathcal{D}^{\text{tran}}_{n+1}} \int_Q |\langle\varphi\rangle_Q - \frac{\langle\varphi\rangle_Q}{\langle h\rangle_Q} h|^2\, d\mu \\
&\le 2 \sum_{Q\in\mathcal{D}^{\text{tran}}_{n+1}} \int_Q |\varphi - \langle\varphi\rangle_Q|^2\, d\mu \leftarrow \text{(I)!} \\
&\quad + \frac{2}{(1-\delta)^2} \sum_{Q\in\mathcal{A}^{(N)}_{n+1}} \langle|\varphi|^2\rangle_Q \int_Q |h - \langle h\rangle_Q|^2\, d\mu \leftarrow \text{(II)!} \\
&\quad + \frac{2}{(1-\delta)^2} \sum_{Q\in\mathcal{B}^{(N)}_{n+1}} \langle|\varphi|^2\rangle_Q \int_Q |h - \langle h\rangle_Q|^2\, d\mu. \leftarrow \text{(III)!}
\end{aligned}
$$

We look at (III) first, it being the hardest to handle. Setting $B^{(N)}_{n+1} = \bigcup \mathcal{B}^{(N)}_{n+1}$, note that

$$
N^2\mu(B^{(N)}_{n+1}) = \sum_{Q\in\mathcal{B}^{(N)}_{n+1}} N^2\mu(Q) \le \sum_{Q\in\mathcal{B}^{(N)}_{n+1}} \int_Q |\varphi|^2 d\mu = \int_{B^{(N)}_{n+1}} |\varphi|^2 d\mu \le \|\varphi\|^2_{L^2(\mu)}.
$$

Thus, setting $E_N = \{|\varphi|^2 > N\}$, we see that

$$
\int_{B^{(N)}_{n+1}} |\varphi|^2\, d\mu \le \int_{B^{(N)}_{n+1}\setminus E_N} |\varphi|^2\, d\mu + \int_{E_N} |\varphi|^2\, d\mu
$$

$$\leq N\mu(B_{n+1}^{(N)}) + \int_{E_N} |\varphi|^2 \, d\mu \leq \frac{\|\varphi\|_{L^2(\mu)}^2}{N} + \int_{E_N} |\varphi|^2 \, d\mu.$$

Note that it is a simple consequence of Lebesgue's Dominated Convergence Theorem [RUD, 1.34] that $\int_{E_N} |\varphi|^2 \, d\mu \to 0$ as $N \to \infty$. Hence we can and do fix N big enough so that for any n,

$$\frac{2\delta^2}{(1-\delta)^2} \int_{B_{n+1}^{(N)}} |\varphi|^2 \, d\mu < \varepsilon/3.$$

For any $Q \in \mathcal{D}_{n+1}^{\text{tran}}$, by (c) of Lemma 7.6

$$\begin{aligned}\int_Q |h - \langle h \rangle_Q|^2 \, d\mu &= \int_Q |g - \langle g \rangle_Q|^2 \, d\mu \\ &= \langle |g - \langle g \rangle_Q|^2 \rangle_Q \cdot \mu(Q) \leq \langle |g|^2 \rangle_Q \cdot \mu(Q) < \delta^2 \mu(Q).\end{aligned}$$

Thus independently of n we have

$$\begin{aligned}\text{(III)} &\leq \frac{2\delta^2}{(1-\delta)^2} \sum_{Q \in \mathcal{B}_{n+1}^{(N)}} \langle |\varphi|^2 \rangle_Q \cdot \mu(Q) \\ &= \frac{2\delta^2}{(1-\delta)^2} \sum_{Q \in \mathcal{B}_{n+1}^{(N)}} \int_Q |\varphi|^2 \, d\mu = \frac{2\delta^2}{(1-\delta)^2} \int_{B_{n+1}^{(N)}} |\varphi|^2 \, d\mu < \varepsilon/3.\end{aligned}$$

Clearly

$$\text{(II)} \leq \frac{2N^2}{(1-\delta)^2} \sum_{Q \in \mathcal{A}_{n+1}^{(N)}} \int_Q |h - \langle h \rangle_Q|^2 \, d\mu \leq \frac{2N^2}{(1-\delta)^2} \sum_{Q \in \mathcal{D}_{n+1}^{\text{tran}}} \int_Q |h - \langle h \rangle_Q|^2 \, d\mu.$$

It thus follows from Lemma 7.5 that both (I) and (II) can be made less than $\varepsilon/3$ by taking n sufficiently large and so we are done. □

Lemma 7.12 *For* $a, b \in \mathbb{C}$, $\frac{2}{3}|a|^2 - 2|b|^2 \leq |a+b|^2 \leq \frac{3}{2}|a|^2 + 3|b|^2$.

Proof Clearly, $0 \leq (|a| - 2|b|)^2 = |a|^2 - 4|a||b| + 4|b|^2$, and so $2|a||b| \leq |a|^2/2 + 2|b|^2$. From this we obtain the desired right-hand inequality:

$$|a+b|^2 \leq (|a| + |b|)^2 = |a|^2 + 2|a||b| + |b|^2 \leq \frac{3}{2}|a|^2 + 3|b|^2.$$

From this right-hand inequality just proven we deduce that

$$|a|^2 = |(a+b) + (-b)|^2 \leq \frac{3}{2}|a+b|^2 + 3|-b|^2.$$

Upon rearrangement this becomes the desired left-hand inequality and so we are done. □

Corollary 7.13 *For E a μ-measurable subset of $\mathbb{C}$ and $f, \tilde{f} \in L^2(\mu)$,*

$$\frac{2}{3}\int_E |f|^2\, d\mu - 2\int_E |\tilde{f}|^2\, d\mu \leq \int_E |f+\tilde{f}|^2\, d\mu \leq \frac{3}{2}\int_E |f|^2\, d\mu + 3\int_E |\tilde{f}|^2\, d\mu.$$

Lemma 7.14 *For any $\varphi \in L^2(\mu)$ and any $Q \in \mathcal{D}_{n+1}^{\mathrm{tran}}$,*

(a) $\|\widetilde{\Lambda}\,\varphi\|^2_{L^2(\mu)} \leq \|\Lambda\,\varphi\|^2_{L^2(\mu)} \leq (1+\delta)\;\|\widetilde{\Lambda}\,\varphi\|^2_{L^2(\mu)}$ *and*

(b) $\frac{2}{3}\;\|\widetilde{\Delta}_Q\,\varphi\|^2_{L^2(\mu)} - 2\frac{(1+\delta)}{(1-\delta)^2}\;\|\widetilde{\Delta}_Q\, g\|^2_{L^2(\mu)}\;|\langle\varphi\rangle_Q|^2 \leq \|\Delta_Q\,\varphi\|^2_{L^2(\mu)}$
$\leq \frac{3}{2}(1+\delta)\;\|\widetilde{\Delta}_Q\,\varphi\|^2_{L^2(\mu)} + 3\frac{(1+\delta)}{(1-\delta)^2}\;\|\widetilde{\Delta}_Q\, g\|^2_{L^2(\mu)}\;|\langle\varphi\rangle_Q|^2.$

Proof (a) This follows from (b) of Lemma 7.10 since

$$\|\Lambda\varphi\|^2_{L^2(\mu)} = \frac{|\langle\varphi\rangle_{Q^0}|^2}{|\langle h\rangle_{Q^0}|^2}\int_{Q^0}|h|^2 d\mu = \frac{\langle|h|^2\rangle_{Q^0}}{|\langle h\rangle_{Q^0}|^2}\;|\langle\varphi\rangle_{Q^0}|^2\;\mu(Q^0) = \frac{\langle|h|^2\rangle_{Q^0}}{|\langle h\rangle_{Q^0}|^2}\;\|\widetilde{\Lambda}\varphi\|^2_{L^2(\mu)}.$$

(b) Let Q_j, $j = 1, 2, 3, 4$, be as in the definitions of Δ_Q and $\widetilde{\Delta}_Q$. Then simple summing shows that it suffices for each of these j to prove

$$(\star)\;\frac{2}{3}\int_{Q_j}|\widetilde{\Delta}_Q\,\varphi|^2\, d\mu - 2\frac{(1+\delta)}{(1-\delta)^2}\cdot\int_{Q_j}|\widetilde{\Delta}_Q\, g|^2\, d\mu\cdot|\langle\varphi\rangle_Q|^2 \leq \int_{Q_j}|\Delta_Q\,\varphi|^2\, d\mu$$

$$\leq \frac{3}{2}(1+\delta)\int_{Q_j}|\widetilde{\Delta}_Q\,\varphi|^2\, d\mu + 3\frac{(1+\delta)}{(1-\delta)^2}\cdot\int_{Q_j}|\widetilde{\Delta}_Q\, g|^2\, d\mu\cdot|\langle\varphi\rangle_Q|^2.$$

Case One (Q_j transit). On Q_j we may write $\Delta_Q\,\varphi = f + \tilde{f}$ where

$$f = \left\{\frac{\langle\varphi\rangle_{Q_j}}{\langle h\rangle_{Q_j}} - \frac{\langle\varphi\rangle_Q}{\langle h\rangle_{Q_j}}\right\} h \text{ and } \tilde{f} = \left\{\frac{\langle\varphi\rangle_Q}{\langle h\rangle_{Q_j}} - \frac{\langle\varphi\rangle_Q}{\langle h\rangle_Q}\right\} h.$$

Since

$$\int_{Q_j}|f|^2\, d\mu = \frac{|\langle\varphi\rangle_{Q_j} - \langle\varphi\rangle_Q|^2}{|\langle h\rangle_{Q_j}|^2}\int_{Q_j}|h|^2\, d\mu$$
$$= \frac{\langle|h|^2\rangle_{Q_j}}{|\langle h\rangle_{Q_j}|^2}\cdot|\langle\varphi\rangle_{Q_j} - \langle\varphi\rangle_Q|^2\;\mu(Q_j)$$

$$= \frac{\langle |h|^2 \rangle_{Q_j}}{|\langle h \rangle_{Q_j}|^2} \int_{Q_j} |\widetilde{\Delta}_Q \varphi|^2 \, d\mu,$$

applying (b) of Lemma 7.10 we deduce

$$(\star\star) \quad \int_{Q_j} |\widetilde{\Delta}_Q \varphi|^2 \, d\mu \leq \int_{Q_j} |f|^2 \, d\mu \leq (1+\delta) \int_{Q_j} |\widetilde{\Delta}_Q \varphi|^2 \, d\mu.$$

A similar computation using both (a) and (b) of Lemma 7.10 follows:

$$\begin{aligned}
(\star\star\star) \quad \int_{Q_j} |\tilde{f}|^2 \, d\mu &= \frac{|\langle h \rangle_Q - \langle h \rangle_{Q_j}|^2 \, |\langle \varphi \rangle_Q|^2}{|\langle h \rangle_{Q_j}|^2 \, |\langle h \rangle_Q|^2} \int_{Q_j} |h|^2 \, d\mu \\
&= \frac{\langle |h|^2 \rangle_{Q_j}}{|\langle h \rangle_{Q_j}|^2} \, \frac{1}{|\langle h \rangle_Q|^2} \cdot |\langle g \rangle_Q - \langle g \rangle_{Q_j}|^2 \, \mu(Q_j) \cdot |\langle \varphi \rangle_Q|^2 \\
&\leq \frac{(1+\delta)}{(1-\delta)^2} \cdot \int_{Q_j} |\widetilde{\Delta}_Q \, g|^2 \, d\mu \cdot |\langle \varphi \rangle_Q|^2.
\end{aligned}$$

Applying the last corollary to our decomposition of $\Delta_Q \varphi$ as $f + \tilde{f}$ and using the bounds of $(\star\star)$ and $(\star\star\star)$, we obtain $(\star)$ for the case of Q_j transit.

Case Two (Q_j is terminal). On Q_j we may write $\Delta_Q \varphi = f + \tilde{f}$ where

$$f = \varphi - \langle \varphi \rangle_Q \text{ and } \tilde{f} = \langle \varphi \rangle_Q - \frac{\langle \varphi \rangle_Q}{\langle h \rangle_Q} h.$$

Clearly

$$(\dagger) \quad \int_{Q_j} |f|^2 \, d\mu = \int_{Q_j} |\widetilde{\Delta}_Q \varphi|^2 \, d\mu.$$

Also, applying (a) of Lemma 7.10 we obtain

$$\begin{aligned}
(\ddagger) \quad \int_{Q_j} |\tilde{f}|^2 \, d\mu &= \frac{1}{|\langle h \rangle_Q|^2} \cdot \int_{Q_j} |h - \langle h \rangle_Q|^2 \, d\mu \cdot |\langle \varphi \rangle_Q|^2 \\
&\leq \frac{1}{(1-\delta)^2} \cdot \int_{Q_j} |g - \langle g \rangle_Q|^2 \, d\mu \cdot |\langle \varphi \rangle_Q|^2 \\
&= \frac{1}{(1-\delta)^2} \cdot \int_{Q_j} |\widetilde{\Delta}_Q \, g|^2 \, d\mu \cdot |\langle \varphi \rangle_Q|^2.
\end{aligned}$$

Applying the last corollary to our decomposition of $\Delta_Q \varphi$ as $f + \tilde{f}$ and using the bounds of $(\dagger)$ and $(\ddagger)$, we again obtain $(\star)$ but now for the case of Q_j terminal. □

Proposition 7.15 (Adapted Martingale Decomposition) *For any $\varphi \in L^2(\mu)$,*

$$\varphi = \Lambda\,\varphi + \sum_{Q \in \mathcal{D}^{\rm tran}} \Delta_Q\,\varphi$$

with the convergence being in $L^2(\mu)$. Moreover,

$$\frac{1}{2}\|\varphi\|^2_{L^2(\mu)} \leq \|\Lambda\,\varphi\|^2_{L^2(\mu)} + \sum_{Q \in \mathcal{D}^{\rm tran}} \|\Delta_Q\,\varphi\|^2_{L^2(\mu)} \leq 2\|\varphi\|^2_{L^2(\mu)}.$$

Proof One may argue as we previously did in proving the L^2 convergence of the Standard Martingale Decomposition (7.7) to arrive at

$$\|\varphi - \{\Lambda\,\varphi + \sum_{Q \in \mathcal{D}^{\rm tran}_{\leq n}} \Delta_Q\,\varphi\}\|^2_{L^2(\mu)} = \sum_{Q \in \mathcal{D}^{\rm tran}_{n+1}} \int_Q |\varphi - \frac{\langle\varphi\rangle_Q}{\langle h\rangle_Q} h|^2\,d\mu.$$

The desired L^2 convergence now follows from Lemma 7.11.

Turning to the norm inequalities, if we sum the inequality (a) of Lemma 7.14 along with the inequalities that result when (b) of Lemma 7.14 is applied to all $Q \in \mathcal{D}^{\rm tran}$ and then use the norm equality of the Standard Martingale Decomposition (7.7), we obtain

$$(\star)\ \frac{2}{3}\,\|\varphi\|^2_{L^2(\mu)} - 2\frac{(1+\delta)}{(1-\delta)^2} \sum_{Q \in \mathcal{D}^{\rm tran}} \|\widetilde{\Delta}_Q\,g\|^2_{L^2(\mu)}\,|\langle\varphi\rangle_Q|^2$$
$$\leq \|\Lambda\,\varphi\|^2_{L^2(\mu)} + \sum_{Q \in \mathcal{D}^{\rm tran}} \|\Delta_Q\,\varphi\|^2_{L^2(\mu)}$$
$$\leq \frac{3}{2}(1+\delta)\,\|\varphi\|^2_{L^2(\mu)} + 3\frac{(1+\delta)}{(1-\delta)^2} \sum_{Q \in \mathcal{D}^{\rm tran}} \|\widetilde{\Delta}_Q\,g\|^2_{L^2(\mu)}\,|\langle\varphi\rangle_Q|^2.$$

We now wish to use the Dyadic Carleson Imbedding Inequality (7.8) with

$$a_Q = \begin{cases} \|\widetilde{\Delta}_Q\,g\|^2_{L^2(\mu)} & \text{when} \quad Q \in \mathcal{D}^{\rm tran} \\ 0 & \text{when} \quad Q \in \mathcal{F} \setminus \mathcal{D}^{\rm tran} \end{cases}$$

and $A = \delta^2$. Setting $\mathcal{D}^{\rm tran}(Q) = \{R \in \mathcal{D}^{\rm tran} : R \subseteq Q\}$, note that for $Q \in \mathcal{D}^{\rm tran}$,

$$\sum_{R \in \mathcal{F}(Q)} a_R = \sum_{R \in \mathcal{D}^{\rm tran}(Q)} \|\widetilde{\Delta}_R\,(\mathcal{X}_Q g)\|^2_{L^2(\mu)} \leq \|\mathcal{X}_Q g\|^2_{L^2(\mu)} = \int_Q |g|^2 d\mu \leq \delta^2\mu(Q)$$

with the first "$\leq$" in the above chain being a consequence of the norm equality of the Standard Martingale Decomposition (7.7). For $Q \in \mathcal{F}\setminus\mathcal{D}^{\rm tran}$, the same inequality is trivial since then $\mathcal{D}^{\text{tran}}(Q) = \emptyset$. Having verified its hypothesis, the Dyadic Carleson Imbedding Inequality (7.8) now implies

$$(\star\star) \quad \sum_{Q \in \mathcal{D}^{\text{tran}}} \|\widetilde{\Delta}_Q\, g\|^2_{L^2(\mu)}\, |\langle \varphi \rangle_Q|^2 \leq 4\delta^2 \|\varphi\|^2_{L^2(\mu)}.$$

From $(\star)$ and $(\star\star)$ we obtain

$$\left\{ \frac{2}{3} - \frac{8\delta^2(1+\delta)}{(1-\delta)^2} \right\} \|\varphi\|^2_{L^2(\mu)} \leq \|\Lambda\, \varphi\|^2_{L^2(\mu)} + \sum_{Q \in \mathcal{D}^{\text{tran}}} \|\Delta_Q\, \varphi\|^2_{L^2(\mu)}$$

$$\leq \left\{ \frac{3}{2}(1+\delta) + \frac{12\delta^2(1+\delta)}{(1-\delta)^2} \right\} \|\varphi\|^2_{L^2(\mu)}.$$

Since the δs under consideration satisfy

$$\frac{8\delta^2(1+\delta)}{(1-\delta)^2} \leq \frac{1}{6} \quad \text{and} \quad \frac{3}{2}\delta + \frac{12\delta^2(1+\delta)}{(1-\delta)^2} \leq \frac{1}{2},$$

we are done. □

The following lemma establishes the elementary properties of the pieces, Λ and Δ_Q, of our adapted martingale decomposition.

Lemma 7.16 *Suppose $\varphi \in L^2(\mu)$ and $Q \in \mathcal{D}^{\text{tran}}$. Then*

(a) $\int_Q \Delta_Q\, \varphi\, d\mu = 0$ *with* $\Delta_Q\, \varphi = 0$ *off* Q,

(b) $\Delta_Q\, \Delta_Q\, \varphi = \Delta_Q\, \varphi$,

(c) $\Delta_Q\, \Delta_R\, \varphi = 0$ *for* $R \in \mathcal{D}^{\text{tran}}$ *distinct from* Q,

(d) $\Lambda\, \Delta_Q\, \varphi = \Delta_Q\, \Lambda\, \varphi = 0$, *and*

(e) $\Lambda\, \Lambda\, \varphi = \Lambda\, \varphi$.

Proof Let Q_j, $j = 1, 2, 3, 4$, be as in the definition of Δ_Q. Then one may easily check that whether Q_j is terminal or transit, one has

$$(\star) \quad \int_{Q_j} \Delta_Q\, \varphi\, d\mu = \int_{Q_j} \varphi\, d\mu - \frac{\langle \varphi \rangle_Q}{\langle h \rangle_Q} \int_{Q_j} h\, d\mu.$$

(a) Setting $j = 1, 2, 3, 4$ in $(\star)$ and adding the results together yield

$$\int_Q \Delta_Q\, \varphi\, d\mu = \int_Q \varphi\, d\mu - \frac{\langle \varphi \rangle_Q}{\langle h \rangle_Q} \int_Q h\, d\mu = 0.$$

That $\Delta_Q\, \varphi = 0$ off Q is simply a matter of definition.

(b) We show that $\Delta_Q \Delta_Q \varphi = \Delta_Q \varphi$ on each Q_j whether Q_j is terminal or transit. From (a) it follows that $\langle \Delta_Q \varphi \rangle_Q = 0$. Hence for the terminal case one has

$$\Delta_Q \Delta_Q \varphi = \Delta_Q \varphi - \frac{\langle \Delta_Q \varphi \rangle_Q}{\langle h \rangle_Q} h = \Delta_Q \varphi.$$

From $(\star)$ it follows that

$$\langle \Delta_Q \varphi \rangle_{Q_j} = \langle \varphi \rangle_{Q_j} - \frac{\langle \varphi \rangle_Q}{\langle h \rangle_Q} \langle h \rangle_{Q_j}$$

whenever $\mu(Q_j) \neq 0$. Hence for the transit case one has

$$\begin{aligned} \Delta_Q \Delta_Q \varphi &= \left\{ \frac{\langle \Delta_Q \varphi \rangle_{Q_j}}{\langle h \rangle_{Q_j}} - \frac{\langle \Delta_Q \varphi \rangle_Q}{\langle h \rangle_Q} \right\} h = \frac{\langle \Delta_Q \varphi \rangle_{Q_j}}{\langle h \rangle_{Q_j}} h \\ &= \left\{ \frac{\langle \varphi \rangle_{Q_j}}{\langle h \rangle_{Q_j}} - \frac{\langle \varphi \rangle_Q}{\langle h \rangle_Q} \right\} h = \Delta_Q \varphi. \end{aligned}$$

(c) Either $Q \cap R = \emptyset$, $Q \subset R$, or $R \subset Q$.

In the first case $\Delta_R \varphi = 0$ off R and so on Q. But Δ_Q of any function vanishing on Q is 0, so we are done.

In the second case we must have $Q \subseteq R_j$ for some single $j = 1, 2, 3, 4$ and this R_j must be transit. But then $\Delta_R \varphi$ is a constant multiple of h on this transit R_j and so on Q. But Δ_Q of any function that is a multiple of h on Q is 0, so we are done again.

In the third case we must have $R \subseteq Q_j$ for some single $j = 1, 2, 3, 4$ and this Q_j must be transit. By (a), $\int_R \Delta_R \varphi \, d\mu = 0$ with $\Delta_R \varphi = 0$ off R. This implies that the average value of $\Delta_R \varphi$ over both Q and this transit Q_j are 0. But then $\Delta_Q \Delta_R \varphi = 0$ on this transit Q_j. Note that $\Delta_R \varphi = 0$ on the three other Q_js. But Δ_Q of any function which has average value 0 over Q and vanishes on a Q_j is 0 on the Q_j, so we are done yet again.

(d) To show that $\Lambda \, \Delta_Q \varphi = 0$ it suffices to show $\langle \Delta_Q \varphi \rangle_{\mathbb{C}} = 0$. This follows easily from (a):

$$\langle \Delta_Q \varphi \rangle_{\mathbb{C}} = \frac{1}{\mu(\mathbb{C})} \int_{\mathbb{C}} \Delta_Q \varphi \, d\mu = \frac{1}{\mu(\mathbb{C})} \int_Q \Delta_Q \varphi \, d\mu = 0.$$

To show that $\Delta_Q \Lambda \varphi = 0$ simply note that $\Lambda \varphi$ is a constant multiple of h and that Δ_Q of any constant multiple of h is 0.

(e) Clearly $\Lambda \, h = h$. Since $\Lambda \varphi = c \, h$ for some constant c, it then follows that $\Lambda \, \Lambda \varphi = \Lambda \, c \, h = c \, \Lambda \, h = c \, h = \Lambda \varphi$. □

The next result lets us multiply the pieces of the adapted martingale decomposition of an L^2 function by a bounded "sequence" and then sum the results to obtain a function still in L^2.

Proposition 7.17 *Suppose* $\varphi \in L^2(\mu)$, $c \in \mathbb{C}$, *and* $Q \in \mathcal{D}^{\text{tran}} \mapsto c_Q \in \mathbb{C}$ *is a bounded function. Then the sum*

$$\psi = c\, \Lambda\, \varphi + \sum_{Q \in \mathcal{D}^{\text{tran}}} c_Q\, \Delta_Q\, \varphi$$

converges in $L^2(\mu)$. *Moreover,*

$$\frac{1}{2} \|\psi\|^2_{L^2(\mu)} \leq |c|^2 \|\Lambda\, \varphi\|^2_{L^2(\mu)} + \sum_{Q \in \mathcal{D}^{\text{tran}}} |c_Q|^2 \|\Delta_Q\, \varphi\|^2_{L^2(\mu)} \leq 2 \|\psi\|^2_{L^2(\mu)}.$$

Proof Case One (*Only finitely many of the* c_Q*s are nonzero*). Then clearly there is no problem with the convergence. By the last lemma, $\Lambda\, \psi = c\, \Lambda\, \varphi$ and $\Delta_Q\, \psi = c_Q\, \Delta_Q\, \varphi$ for each $Q \in \mathcal{D}^{\text{tran}}$. The desired norm inequalities now follow from the norm inequalities of the Adapted Martingale Decomposition (7.15) applied to ψ.

Case Two (*Infinitely many of the* c_Q*s are nonzero*). Set

$$\psi_n = c\, \Lambda\, \varphi + \sum_{Q \in \mathcal{D}^{\text{tran}}_{\leq n}} c_Q\, \Delta_Q\, \varphi.$$

By assumption, $C = \sup\{|c_Q| : Q \in \mathcal{D}^{\text{tran}}\} < \infty$. Given $m > n$, let $\sum_n^m$ denote a sum over all $Q \in \mathcal{D}^{\text{tran}}$ such that $2^{-m} \leq l(Q) < 2^{-n}$. Then by Case One just proven,

$$\begin{aligned} \|\psi_m - \psi_n\|^2_{L^2(\mu)} &= \|\sum_n^m c_Q\, \Delta_Q\, \varphi\|^2_{L^2(\mu)} \leq 2 \sum_n^m |c_Q|^2 \|\Delta_Q\, \varphi\|^2_{L^2(\mu)} \\ &\leq 2C^2 \sum_n^m \|\Delta_Q\, \varphi\|^2_{L^2(\mu)}. \end{aligned}$$

But by the Adapted Martingale Decomposition (7.15),

$$\sum_{Q \in \mathcal{D}^{\text{tran}}} \|\Delta_Q\, \varphi\|^2_{L^2(\mu)} \leq 2 \|\varphi\|^2_{L^2(\mu)} < \infty.$$

Thus the sequence $\{\psi_n\}$ is Cauchy, and so convergent, in $L^2(\mu)$.

Applying the desired norm inequalities to the ψ_ns (already proven in Case One) and letting $n \to \infty$, we get the desired norm inequalities for ψ. □

7.8 Bad Squares and Their Rarity

We now restore the ω that has been repressed in the last three sections. Thus we shall write "$\mathcal{D}_\omega$" instead of "$\mathcal{D}$," "Q^0_ω" instead of Q^0, etc.

For any $\omega \in \Omega$, the family of squares

$$\mathcal{E}_\omega = \left\{\omega + \left[\frac{j}{2^n}, \frac{j+1}{2^n}\right) \times \left[\frac{k}{2^n}, \frac{k+1}{2^n}\right) : n = 0, 1, 2, \ldots \text{ and } j, k = 0, \pm 1, \pm 2, \ldots\right\}$$

is called the *extended full dyadic lattice* based at ω. Call a square *dyadic* if it is an element of $\mathcal{E}_\omega$ for some $\omega \in \Omega$. Denoting $\mathcal{E}_0$ more simply by $\mathcal{E}$, note that we always have $\mathcal{E}_\omega = \{\omega + Q : Q \in \mathcal{E}\}$.

Given a positive finite number M', a subset E of $\mathbb{C}$ is called *M'-negligible* whenever $\mu(\{z \in \mathbb{C} : \mathrm{dist}(z, E) \le r\}) \le M'r$ for every $r > 0$. We now make an observation that will be of importance a few paragraphs down. Suppose E is not M'-negligible, so $\varepsilon = \mu(\{z \in \mathbb{C} : \mathrm{dist}(z, E) \le r\}) - M'r > 0$ for some $r > 0$. Then, since $\{z \in \mathbb{C} : \mathrm{dist}(z, E) \le r\} \subseteq \{z \in \mathbb{C} : \mathrm{dist}(z, w + E) \le r + |w|\}$, consideration of $r + |w|$ in place of r now shows that $w + E$ is also not M'-negligible whenever $|w| < \varepsilon/M'$.

Recall that $l(Q)$ denotes the edge length of a square Q. Let αQ denote the square concentric with Q whose edge length is α times that of Q. What follows is the most important definition of this chapter: given a dyadic square Q and a point $\omega \in \Omega$, we shall say that Q is *ω-bad* if there exists an $R \in \mathcal{E}_\omega$ such that either

1. $l(R) \ge 2^m\, l(Q)$ and $\mathrm{dist}(Q, \partial R) < 16\, l(Q)^{1/4} l(R)^{3/4}$ or
2. $l(R) \ge l(Q)/2^{m+1}$, $R \subseteq \mathrm{int}\{4 \cdot 2^m + 1\}Q$, and ∂R is not M'-negligible.

Note that our definition of "ω-bad" involves two parameters m and M' that are presently unspecified. We are free to choose these at will and shall now exploit this freedom of choice to arrange for bad squares to be "rare." Before attending to this however, we make three comments which will allow the reader to dispose of measurability questions that will arise in this and the next two sections.

First, for this section and the next, note that $\{\omega \in \Omega : Q$ is ω-bad according to clause i of the definition$\}$ is $\mathcal{P}$-measurable for any given dyadic square Q and $i = 1, 2$. The definition of ω-bad and our observation from a few paragraphs before show this and much more, namely that these two sets are open in Ω!

Second, for two sections hence, note that $\{(\omega_1, \omega_2) \in \Omega \times \Omega : \omega_1 + Q$ is ω_2-bad$\}$ is $(\mathcal{P} \times \mathcal{P})$-measurable for any given $Q \in \mathcal{E}$. Again, the definition of ω-bad and our observation from a few paragraphs before show this and much more, namely that this set is open in $\Omega \times \Omega$!

Third, again for two sections hence, recall Proposition 7.4!

Lemma 7.18 *It is possible to choose the parameter m so that for every dyadic square Q of $\mathbb{C}$,*

$$\mathcal{P}(\{\omega \in \Omega : Q \textit{ is } \omega\textit{-bad according to the first clause of the definition}\}) < \delta/2.$$

Proof Choose m large enough so that $1{,}700(2^{-1/4})^m/(1-2^{-1/4}) < \delta/2$. Suppose a dyadic square Q is ω-bad according to the first clause of the definition. Thus there exists an $R \in \mathcal{E}_\omega$ such that $l(R) \geq 2^m\, l(Q)$ and $\operatorname{dist}(Q, \partial R) < 16\, l(Q)^{1/4} l(R)^{3/4}$. Clearly $l(R) = 2^k\, l(Q)$ for some integer $k \geq m$. Setting

$$G(k) = \{x + iy : x \text{ or } y \text{ is of the form } j \cdot 2^k\, l(Q) \text{ for some } j = 0, \pm 1, \pm 2, \ldots\},$$

note that $\partial R \subseteq \omega - G(k)$. Thus, letting z_Q denote the center of Q, we have

$$\begin{aligned}\operatorname{dist}(\omega, z_Q + G(k)) &= \operatorname{dist}(z_Q, \omega - G(k)) \leq \operatorname{dist}(z_Q, \partial R) \\ &\leq l(Q) + \operatorname{dist}(Q, \partial R) < 17\, l(Q)^{1/4} l(R)^{3/4} = 17(2^{3k/4}) l(Q).\end{aligned}$$

Setting $\widetilde{G}(k) = \{\omega \in \mathbb{C} : \operatorname{dist}(\omega, z_Q + G(k)) \leq 17(2^{3k/4}) l(Q)\}$, we obtain

$$(\star)\ \mathcal{P}(\{\omega \in \Omega : Q \text{ is } \omega\text{-bad according to the first clause of the definition}\}) \leq \sum_{k \geq m} 4\mathcal{L}^2_\Omega(\widetilde{G}(k)).$$

Now $G(k)$ is a grid with $\mathcal{L}^2$-measure zero whose complement in $\mathbb{C}$ can be written as a union of a countably infinite pairwise disjoint collection $\mathcal{S}$ of open squares of edge length $l = 2^k\, l(Q)$. Since $\widetilde{G}(k)$ is just $G(k)$ "thickened up" by $17(2^{3k/4}) l(Q)$ on either side of its constituent segments, it is clear that for any square $S \in \mathcal{S}$,

$$\mathcal{L}^2(\widetilde{G}(k) \cap S) \leq 4 \cdot 17(2^{3k/4}) l(Q) \cdot l(S) = 68(2^{-k/4}) l^2.$$

Let $\mathcal{S}'$ denote the collection of squares of $\mathcal{S}$ that intersect Ω. The cardinality of $\mathcal{S}'$ is at most $\{l(\Omega)/l + 2\}^2$. Since $l(\Omega) = 1/2$ and $l \leq 1$, it follows that $l^2 \cdot \{l(\Omega)/l + 2\}^2 = \{l(\Omega) + 2l\}^2 \leq 25/4$ and so

$$(\star\star)\ \mathcal{L}^2_\Omega(\widetilde{G}(k)) = \sum_{S \in \mathcal{S}'} \mathcal{L}^2_\Omega(\widetilde{G}(k) \cap S) \leq 68(2^{-k/4}) l^2 \cdot \{l(\Omega)/l + 2\}^2 \leq 425\big(2^{-k/4}\big).$$

From $(\star)$ and $(\star\star)$ we see that the probability to be estimated is dominated by $1{,}700 \sum_{k \geq m} 2^{-k/4}$. Using the geometric series formula and recalling our choice of m, we are done. □

Proposition 7.19 *It is possible to choose the parameters m and M' so that for every dyadic square Q of $\mathbb{C}$,*

$$\mathcal{P}(\{\omega \in \Omega : Q \text{ is } \omega\text{-bad}\}) < \delta.$$

Proof Let m be as in the previous lemma. Setting

$$M'' = \left[\{4 \cdot 2^m + 1\} \cdot 2^{m+1} \cdot \frac{12}{\delta} + 1 \right] M,$$

choose $M' > M''$ large enough so that $500M\{4 \cdot 2^m + 2\}/M' < (\delta/12) \cdot (1/2^{m+1})$. Suppose a dyadic square Q is ω-bad according to the second clause of the definition. Thus there exists an $R \in \mathcal{E}_\omega$ such that $l(R) \geq l(Q)/2^{m+1}$, $R \subseteq \text{int}\{4 \cdot 2^m + 1\}Q$, and ∂R is not M'-negligible. But then $\mu(\{z \in \mathbb{C} : \text{dist}(z, \partial R) \leq r\}) > M'r$ for some $r > 0$. Note that

$$\{z \in \mathbb{C} : \text{dist}(z, \partial R) \leq r\} \subseteq B(z_Q; \{4 \cdot 2^m + 1\}l(Q) + r).$$

Suppose for the moment that $r \geq (\delta/12) \cdot \big(l(Q)/2^{m+1}\big)$. One would then have $\{4 \cdot 2^m + 1\}l(Q) + r \leq (M''/M)r$ and so

$$\mu(\{z \in \mathbb{C} : \text{dist}(z, \partial R) \leq r\}) \leq \mu(B(z_Q; (M''/M)r)) \leq M''r < M'r$$

by the linear growth of μ and our choice of M'. But this contradicts the way r arose! Hence

$$(\star)\ r < (\delta/12) \cdot \big(l(Q)/2^{m+1}\big).$$

Clearly then $r < l(Q)$ and so

$$\{z \in \mathbb{C} : \text{dist}(z, \partial R) \leq r\} \subseteq B(z_Q; \{4 \cdot 2^m + 2\}l(Q)).$$

Thus, denoting the restriction of the measure μ to $B(z_Q; \{4 \cdot 2^m + 2\}l(Q))$ by $\tilde{\mu}$ and invoking the linear growth of μ again, we have

$$(\star\star)\ \tilde{\mu}(\{z \in \mathbb{C} : \text{dist}(z, \partial R) \leq r\}) > M'r \text{ with } \tilde{\mu}(\mathbb{C}) \leq M\{4 \cdot 2^m + 2\}l(Q).$$

Set

$$\text{VL} = \left\{ x + iy : x = j \cdot \frac{l(Q)}{2^{m+1}} \quad \text{for some} \quad j = 0, \pm 1, \pm 2, \ldots \right\}$$

and

$$\text{HL} = \left\{ x + iy : y = k \cdot \frac{l(Q)}{2^{m+1}} \quad \text{for some} \quad k = 0, \pm 1, \pm 2, \ldots \right\}.$$

Since $\partial R \subseteq (\omega + \text{VL}) \cup (\omega + HL)$, it follows from $(\star\star)$ that we must either have

$$(\dagger)\ \tilde{\mu}(\{z \in \mathbb{C} : \text{dist}(z, \omega + \text{VL}) \leq r\}) > M'r/2 \quad \text{or}$$
$$(\ddagger)\ \tilde{\mu}(\{z \in \mathbb{C} : \text{dist}(z, \omega + HL) \leq r\}) > M'r/2.$$

Supposing that (†) is the case, it suffices to prove that $\mathcal{P}(\{\omega \in \Omega : \omega \text{ satisfies } (\dagger)\}) < \delta/4$. For then this and the corresponding result for (‡) yield

$$\mathcal{P}(\{\omega \in \Omega : Q \text{ is } \omega\text{-bad according to the second clause of the definition}\}) < \delta/2$$

and we shall be done by the last lemma.

Recall that $\Omega = [-1/4, 1/4) \times [-1/4, 1/4)$. The condition (†) is invariant under translation of ω by $\pm l(Q)/2^{m+1}$ and actually only depends on the real part of ω. Thus $\{\omega \in \Omega : \omega \text{ satisfies } (\dagger)\}$ is the union of the sets $j \cdot l(Q)/2^{m+1} + \widetilde{E} \times [-1/4, 1/4)$ where

$$\widetilde{E} = \{x \in [0, l(Q)/2^{m+1}) : (\dagger) \text{ holds with } \omega \text{ replaced by } x\}$$

and $j = 0, \pm 1, \pm 2, \ldots$ ranges only over those values for which $j \cdot l(Q)/2^{m+1} + \widetilde{E} \times [-1/4, 1/4) \subseteq \Omega$. Now recall that $\mathcal{P}$ is just planar Lebesgue measure on Ω normalized to have mass 1. It follows that $\mathcal{P}(\{\omega \in \Omega : \omega \text{ satisfies } (\dagger)\}) = \mathcal{L}^1(\widetilde{E})/\{l(Q)/2^{m+1}\}$, so it suffices to show that $\mathcal{L}^1(\widetilde{E}) < (\delta/4) \cdot (l(Q)/2^{m+1})$.

To this end, let ν be the positive measure defined for an arbitrary Borel subset E of $\mathbb{C}$ by the equation

$$\nu(E) = \tilde{\mu}\left(\left\{E \cap \left[0, \frac{l(Q)}{2^{m+1}}\right)\right\} + \mathrm{VL}\right).$$

If $x \in \widetilde{E}$ lies a distance at least $(\delta/12) \cdot (l(Q)/2^{m+1})$ from the points 0 and $l(Q)/2^{m+1}$, then $B(x; r) \cap [0, l(Q)/2^{m+1}) = [x - r, x + r]$ by (⋆). Hence

$$\nu(B(x; r)) = \tilde{\mu}([x - r, x + r] + \mathrm{VL}) = \tilde{\mu}(\{z \in \mathbb{C} : \mathrm{dist}(z, x + \mathrm{VL}) \leq r\}) > M'r/2$$

and so $(\mathcal{M}_{\mathcal{L}^1}\nu)(x) \geq \nu(B(x; r))/2r > M'/4$. Recall that $\mathcal{M}_{\mathcal{L}^1}\nu$ is the maximal function of ν with respect to $\mathcal{L}^1$ as defined just before Proposition 5.27. It now follows that

$$\mathcal{L}^1(\widetilde{E}) \leq 2(\delta/12) \cdot \left(l(Q)/2^{m+1}\right) + \mathcal{L}^1(\{\mathcal{M}_{\mathcal{L}^1}\nu > M'/4\}).$$

Invoking Proposition 5.27 and (⋆⋆), we see that

$$\begin{aligned}\mathcal{L}^1(\{\mathcal{M}_{\mathcal{L}^1}\nu > M'/4\}) &\leq 125\frac{\nu(\mathbb{C})}{M'/4} \leq 500\frac{\tilde{\mu}(\mathbb{C})}{M'} \leq 500\frac{M\{4 \cdot 2^m + 2\}l(Q)}{M'} \\ &< (\delta/12) \cdot \left(l(Q)/2^{m+1}\right)\end{aligned}$$

by our choice of M'. From the last two displayed equations we have $\mathcal{L}^1(\widetilde{E}) < (\delta/4) \cdot (l(Q)/2^{m+1})$ and so are done. □

The use of Proposition 5.27 above, whose proof relies on the difficult Besicovitch Covering Lemma (5.26), can be avoided by using [RUD, 7.4], whose proof relies on an almost trivial covering lemma [RUD, 7.3]. If one does this, the 125 gets replaced by 3!

For the rest of this chapter we assume m and M' to have been chosen as in Proposition 7.19. Also, to make the obvious official, we say that a dyadic square is *ω-good* if it is not ω-bad.

7.9 The Good/Bad-Function Decomposition

Supposing $\varphi \in L^2(\mu)$ and $(\omega_1, \omega_2) \in \Omega \times \Omega$, we now define a decomposition with respect to (ω_1, ω_2) of φ into a *good part* $(\varphi)_{\text{good}}$ and a *bad part* $(\varphi)_{\text{bad}}$ as follows:

$$(\varphi)_{\text{good}} = \Lambda\,\varphi + \sum_{Q \in \mathcal{D}^{\text{tran}}_{\omega_1} \text{ is } \omega_2-\text{good}} \Delta_Q\,\varphi$$

and

$$(\varphi)_{\text{bad}} = \sum_{Q \in \mathcal{D}^{\text{tran}}_{\omega_1} \text{ is } \omega_2-\text{bad}} \Delta_Q\,\varphi.$$

But how are we to interpret the two sums just written down? Via Proposition 7.17! The good part $(\varphi)_{\text{good}}$ is a sum as in Proposition 7.17 with $c = 1$, $c_Q = 1$ when $Q \in \mathcal{D}^{\text{tran}}_{\omega_1}$ is ω_2-good, and $c_Q = 0$ otherwise. The bad part $(\varphi)_{\text{bad}}$ is a sum as in Proposition 7.17 with $c = 0$, $c_Q = 1$ when $Q \in \mathcal{D}^{\text{tran}}_{\omega_1}$ is ω_2-bad, and $c_Q = 0$ otherwise. Thus our sums are partially reassembled adapted martingale decompositions. The point ω_1 determines the adapted martingale decomposition we are working with, while the point ω_2 determines which pieces of this decomposition go into each partial reassembly. If we were hyperfinicky, we would write, for example, "$(\varphi)_{(\omega_1,\omega_2)-\text{good}}$" instead of just "$(\varphi)_{\text{good}}$" and "$c_Q(\omega_2)$" instead of just "$c_Q$." We shall never do the former and only once the latter (in the proof of (c) of the following assertion since it then helps our understanding).

Proposition 7.20 *Let $(\varphi)_{\text{good}}$ and $(\varphi)_{\text{bad}}$ be the good and bad parts of $\varphi \in L^2(\mu)$ with respect to $(\omega_1, \omega_2) \in \Omega \times \Omega$. Then*

(a) $\varphi = (\varphi)_{\text{good}} + (\varphi)_{\text{bad}}$,
(b) $\|(\varphi)_{\text{good}}\|_{L^2(\mu)}$, $\|(\varphi)_{\text{bad}}\|_{L^2(\mu)} \le 2\|\varphi\|_{L^2(\mu)}$, *and*
(c) *viewing $\omega_1 \in \Omega$ as fixed and $\omega_2 \in \Omega$ as variable,*

$$\mathcal{P}\big(\big\{\omega_2 \in \Omega : \|(\varphi)_{\text{bad}}\|_{L^2(\mu)} > \delta^{1/3}\big\}\big) \le 4\delta^{1/3}\|\varphi\|^2_{L^2(\mu)}.$$

Proof (a) By Proposition 7.17 and the Adapted Martingale Decomposition (7.15),

$$(\varphi)_{\text{good}} + (\varphi)_{\text{bad}} = \Lambda\,\varphi + \sum_{Q \in \mathcal{D}^{\text{tran}}_{\omega_1}} \Delta_Q\,\varphi \;= \varphi.$$

(b) By Proposition 7.17 and the Adapted Martingale Decomposition (7.15),

$$\begin{aligned}\|(\varphi)_{\text{good}}\|^2_{L^2(\mu)} &\le 2\{\|\Lambda\,\varphi\|^2_{L^2(\mu)} + \sum_{Q\in\mathcal{D}^{\text{tran}}_{\omega_1}\text{ is }\omega_2-\text{good}} \|\Delta_Q\,\varphi\|^2_{L^2(\mu)}\} \\ &\le 2\{\|\Lambda\,\varphi\|^2_{L^2(\mu)} + \sum_{Q\in\mathcal{D}^{\text{tran}}_{\omega_1}} \|\Delta_Q\,\varphi\|^2_{L^2(\mu)}\} \;\le\; 4\|\varphi\|^2_{L^2(\mu)}.\end{aligned}$$

The estimate on $\|(\varphi)_{\text{bad}}\|_{L^2(\mu)}$ is proven similarly.

(c) Denote the set $\{\omega_2 \in \Omega : \|(\varphi)_{\text{bad}}\|_{L^2(\mu)} > \delta^{1/3}\}$ by E. Set $c_Q(\omega_2) = 1$ when Q is ω_2-bad and 0 otherwise. Then by Proposition 7.17, Proposition 7.19, and the Adapted Martingale Decomposition (7.15),

$$\begin{aligned}\delta^{2/3}\mathcal{P}(E) &\le \int_E \|(\varphi)_{\text{bad}}\|^2_{L^2(\mu)}\, d\mathcal{P}(\omega_2) \\ &\le \int_\Omega 2 \sum_{Q\in\mathcal{D}^{\text{tran}}_{\omega_1}} |c_Q(\omega_2)|^2\|\Delta_Q\,\varphi\|^2_{L^2(\mu)}\, d\mathcal{P}(\omega_2) \\ &= 2\sum_{Q\in\mathcal{D}^{\text{tran}}_{\omega_1}} \left\{\int_\Omega |c_Q(\omega_2)|^2\, d\mathcal{P}(\omega_2)\right\} \|\Delta_Q\,\varphi\|^2_{L^2(\mu)} \\ &= 2\sum_{Q\in\mathcal{D}^{\text{tran}}_{\omega_1}} \mathcal{P}(\{\omega_2\in\Omega : Q \text{ is } \omega_2\text{-bad}\})\|\Delta_Q\,\varphi\|^2_{L^2(\mu)} \\ &\le 2\delta \sum_{Q\in\mathcal{D}^{\text{tran}}_{\omega_1}} \|\Delta_Q\,\varphi\|^2_{L^2(\mu)} \;\le\; 4\delta\|\varphi\|^2_{L^2(\mu)}.\end{aligned}$$

Dividing both sides by $\delta^{2/3}$ now gives us what we want. □

7.10 Reduction to the Good Function Estimate

Given $\varphi \in L^2(\mu)$ and $(\omega_1, \omega_2) \in \Omega\times\Omega$, call φ *good* with respect to (ω_1, ω_2) if $Q \in \mathcal{D}^{\text{tran}}_{\omega_1}$ is ω_2-good whenever $\Delta_Q\,\varphi \neq 0$, or contrapositively, if $\Delta_Q\,\varphi = 0$ whenever $Q \in \mathcal{D}^{\text{tran}}_{\omega_1}$ is ω_2-bad. We shall need to know that $(\varphi)_{\text{good}}$, the good part of φ with respect to (ω_1, ω_2), is always good with respect to (ω_1, ω_2) according to this definition. This follows easily from (c) and (d) of Lemma 7.16. It is also the case that φ is good if and only if $\varphi = (\varphi)_{\text{good}}$, but we shall have no use for this result. The point of this section is to show how to complete the proof of Theorem 7.1 from the following.

Reduction 7.21 (The Good Function Estimate) *There exists a positive finite number N such that for all $(\omega_1, \omega_2) \in \Omega\times\Omega$, for all Lip(1)-functions Θ such that*

$\inf_{\mathbb{C}} \Theta > 0$ *and* $\Theta \geq \delta \max\{\Phi_{\omega_1}, \Phi_{\omega_2}\}$, *and for all* $\varphi_1, \varphi_2 \in L^2(\mu)$ *such that* φ_1 *is good with respect to* (ω_1, ω_2) *and* φ_2 *is good with respect to* (ω_2, ω_1), *we have*

$$|\langle \mathcal{K}_\Theta \varphi_1, \varphi_2 \rangle| \leq N \, \|\varphi_1\|_{L^2(\mu)} \, \|\varphi_2\|_{L^2(\mu)}.$$

In the above we have used, and will continue to use, $\langle \cdot, \cdot \rangle$ to denote the inner product in $L^2(\mu)$. Also, the condition $\inf_{\mathbb{C}} \Theta > 0$ imposed above is simply to make the integral defining $\mathcal{K}_\Theta \varphi$ converge absolutely and so be well defined.

For the rest of this section we shall frequently denote (ω_1, ω_2) more concisely by $\tilde{\omega}$.

Lemma 7.22 *Suppose* $\varphi \in L^2(\mu)$ *and* $\tilde{\omega} = (\omega_1, \omega_2) \in \Omega \times \Omega$.

(a) *Let* $S_{\tilde{\omega}} = \{z \in \mathbb{C} : \delta \max\{\Phi_{\omega_1}(z), \Phi_{\omega_2}(z)\} > \Phi(z)\}$. *Then*

$$(\mathcal{P} \times \mathcal{P})\big(\{\tilde{\omega} \in \Omega \times \Omega : \|\varphi \mathcal{X}_{S_{\tilde{\omega}}}\|_{L^2(\mu)} > \delta^{1/3}\}\big) \leq 4\delta^{1/3} \|\varphi\|^2_{L^2(\mu)}.$$

(b) *Let* $(\varphi)_{\text{bad}}$ *be the bad part of* φ *with respect to* (ω_1, ω_2). *Then*

$$(\mathcal{P} \times \mathcal{P})\big(\{\tilde{\omega} \in \Omega \times \Omega : \|(\varphi)_{\text{bad}}\|_{L^2(\mu)} > \delta^{1/3}\}\big) \leq 4\delta^{1/3} \|\varphi\|^2_{L^2(\mu)}.$$

(c) *Let* $(\varphi)_{\text{bad}}$ *be the bad part of* φ *with respect to* (ω_2, ω_1). *Note the switch in indices! Then*

$$(\mathcal{P} \times \mathcal{P})\big(\{\tilde{\omega} \in \Omega \times \Omega : \|(\varphi)_{\text{bad}}\|_{L^2(\mu)} > \delta^{1/3}\}\big) \leq 4\delta^{1/3} \|\varphi\|^2_{L^2(\mu)}.$$

Proof (a) First, given $z \in \mathbb{C}$, set $S^z = \{\omega \in \Omega : \delta \Phi_\omega(z) > \Phi(z)\}$ and suppose that $\mathcal{P}(S^z) > 2\delta$. Then, by the Upper Continuity Property of Measures [RUD, 1.19(d)], we also have $\mathcal{P}(S^z_\varepsilon) > 2\delta$ for some $\varepsilon > 0$ small enough where $S^z_\varepsilon = \{\omega \in \Omega : \delta \Phi_\omega(z) > \Phi(z) + \varepsilon\}$. Thus, given any $\widetilde{\Omega} \subseteq \Omega$ such that $\mathcal{P}(\widetilde{\Omega}) \leq \delta$, we have $\mathcal{P}(S^z_\varepsilon \setminus \widetilde{\Omega}) > \delta$ and so

$$\int_{\Omega \setminus \widetilde{\Omega}} \Phi_\omega(z) \, d\mathcal{P}(\omega) \geq \int_{S^z_\varepsilon \setminus \widetilde{\Omega}} \Phi_\omega(z) \, d\mathcal{P}(\omega) \geq \frac{\Phi(z) + \varepsilon}{\delta} \mathcal{P}(S^z_\varepsilon \setminus \widetilde{\Omega}) > \Phi(z) + \varepsilon.$$

Infing over all such $\widetilde{\Omega}$ and recalling the definition of Φ from Section 7.4, we obtain the contradiction $\Phi(z) \geq \Phi(z) + \varepsilon$. It follows that $\mathcal{P}(S^z) \leq 2\delta$ for all $z \in \mathbb{C}$.

Second, noting that $\tilde{\omega} \in (S^z \times \Omega) \cup (\Omega \times S^z)$ whenever $z \in S_{\tilde{\omega}}$, it immediately follows from the last paragraph that $(\mathcal{P} \times \mathcal{P})(\{\tilde{\omega} \in \Omega \times \Omega : z \in S_{\tilde{\omega}}\}) \leq 4\delta$ for all $z \in \mathbb{C}$.

Finally, from Fubini [RUD, 8.8] and this last estimate we obtain

$$\delta^{2/3} (\mathcal{P} \times \mathcal{P})\big(\{\tilde{\omega} \in \Omega \times \Omega : \|\varphi \mathcal{X}_{S_{\tilde{\omega}}}\|_{L^2(\mu)} > \delta^{1/3}\}\big)$$

$$\leq \int_{\Omega \times \Omega} \|\varphi \mathcal{X}_{S_{\tilde{\omega}}}\|^2_{L^2(\mu)} \, d(\mathcal{P} \times \mathcal{P})(\tilde{\omega})$$

$$= \int_{\Omega\times\Omega}\int |\varphi(z)|^2 |\mathcal{X}_{S_{\tilde{\omega}}}(z)|^2 \, d\mu(z)\, d(\mathcal{P}\times\mathcal{P})(\tilde{\omega})$$

$$= \int |\varphi(z)|^2 (\mathcal{P}\times\mathcal{P})(\{\tilde{\omega}\in\Omega\times\Omega : z\in S_{\tilde{\omega}}\})\, d\mu(z) \;\le\; 4\delta\|\varphi\|^2_{L^2(\mu)}.$$

Dividing both sides by $\delta^{2/3}$ yields what we want.

(b) This is a simple consequence of Fubini [RUD, 8.8] and (c) of Proposition 7.20:

$$(\mathcal{P}\times\mathcal{P})\big(\{\tilde{\omega}\in\Omega\times\Omega : \|(\varphi)_{\rm bad}\|_{L^2(\mu)} > \delta^{1/3}\}\big)$$

$$= \int_{\Omega} \mathcal{P}\big(\{\omega_2\in\Omega : \|(\varphi)_{\rm bad}\|_{L^2(\mu)} > \delta^{1/3}\}\big)\, d\mathcal{P}(\omega_1)$$

$$\le \int_{\Omega} 4\delta^{1/3}\|\varphi\|^2_{L^2(\mu)}\, d\mathcal{P}(\omega_1) \;=\; 4\delta^{1/3}\|\varphi\|^2_{L^2(\mu)}.$$

(c) Similar. □

Proof of Theorem 7.1 *from Reduction* 7.21 We must show that $\sup_{\lambda>0}\|\mathcal{K}_{\Phi+\lambda}\|_{L^2(\mu)\mapsto L^2(\mu)} < \infty$. We shall actually show more, namely that $\sup_{\lambda>0} N_\lambda \le 10N < \infty$ where

$$N_\lambda = \sup\{\|\mathcal{K}_\Theta\|_{L^2(\mu)\mapsto L^2(\mu)} : \Theta \text{ is Lip}(1) \text{ and } \Theta \ge \Phi+\lambda\}.$$

The reasoning given in the paragraphs after the proof of Proposition 6.13 and before the statement of Theorem 6.14 is easily adapted to show that $N_\lambda \le \mu(\mathbb{C})/\lambda$ and so each N_λ is a finite number. Hence, given $\lambda > 0$, we may choose a Lip(1)-function $\Theta \ge \Phi+\lambda$ and functions $\varphi_1, \varphi_2 \in L^2(\mu)$ with L^2 norm at most 1 such that

$$(\star)\; |\langle \mathcal{K}_\Theta\varphi_1, \varphi_2\rangle| \ge \frac{9}{10} N_\lambda.$$

Consider the following sets:

$$E_a = \{\tilde{\omega}\in\Omega\times\Omega : \|\varphi_1\mathcal{X}_{S_{\tilde{\omega}}}\|_{L^2(\mu)} > \delta^{1/3} \text{ or } \|\varphi_2\mathcal{X}_{S_{\tilde{\omega}}}\|_{L^2(\mu)} > \delta^{1/3}\},$$

$$E_b = \{\tilde{\omega}\in\Omega\times\Omega : \|(\varphi_1)_{\rm bad}\|_{L^2(\mu)} > \delta^{1/3}\}$$

where $(\varphi_1)_{\rm bad}$ is the bad part of φ_1 with respect to (ω_1, ω_2), and

$$E_c = \{\tilde{\omega}\in\Omega\times\Omega : \|(\varphi_2)_{\rm bad}\|_{L^2(\mu)} > \delta^{1/3}\}$$

where $(\varphi_2)_{\text{bad}}$ is the bad part of φ_2 with respect to (ω_2, ω_1). Then by the last lemma we have $(\mathcal{P} \times \mathcal{P})(E_a \cup E_b \cup E_c) \le 16\delta^{1/3}$.

Since $16\delta^{1/3} < 1$, we are thus ensured of the existence of an $\tilde{\omega} \in \Omega \times \Omega \setminus \{E_a \cup E_b \cup E_c\}$. Fixing such an $\tilde{\omega}$, we have $\|\varphi_1 \mathcal{X}_{S_{\tilde{\omega}}}\|_{L^2(\mu)}$, $\|\varphi_2 \mathcal{X}_{S_{\tilde{\omega}}}\|_{L^2(\mu)}$, $\|(\varphi_1)_{\text{bad}}\|_{L^2(\mu)}$, and $\|(\varphi_2)_{\text{bad}}\|_{L^2(\mu)} \le \delta^{1/3}$. Thus by (b) of Proposition 7.20,

$$\begin{aligned}\|(\varphi_1 \mathcal{X}_{\mathbb{C}\setminus S_{\tilde{\omega}}})_{\text{bad}}\|_{L^2(\mu)} &\le \|(\varphi_1)_{\text{bad}}\|_{L^2(\mu)} + \|(\varphi_1 \mathcal{X}_{S_{\tilde{\omega}}})_{\text{bad}}\|_{L^2(\mu)} \\ &\le \delta^{1/3} + 2\|\varphi_1 \mathcal{X}_{S_{\tilde{\omega}}}\|_{L^2(\mu)} \le 3\delta^{1/3}.\end{aligned}$$

Similarly $\|(\varphi_2 \mathcal{X}_{\mathbb{C}\setminus S_{\tilde{\omega}}})_{\text{bad}}\|_{L^2(\mu)} \le 3\delta^{1/3}$. Since $3\delta^{1/3} \le 1/10$ (here is where the number 1/27,000 in the hypothesis of Theorem 7.1, and Theorem 6.14, comes from), we obtain

$$\begin{aligned}&(\star\star)\ \|\varphi_1 \mathcal{X}_{S_{\tilde{\omega}}}\|_{L^2(\mu)},\ \|\varphi_2 \mathcal{X}_{S_{\tilde{\omega}}}\|_{L^2(\mu)},\ \|(\varphi_1 \mathcal{X}_{\mathbb{C}\setminus S_{\tilde{\omega}}})_{\text{bad}}\|_{L^2(\mu)},\ \text{and} \\ &\quad \|(\varphi_2 \mathcal{X}_{\mathbb{C}\setminus S_{\tilde{\omega}}})_{\text{bad}}\|_{L^2(\mu)} \le \frac{1}{10}.\end{aligned}$$

It follows from $(\star)$ and the first two estimates of $(\star\star)$ that

$$\begin{aligned}\frac{9}{10}N_\lambda &\le |\langle \mathcal{K}_\Theta \varphi_1 \mathcal{X}_{\mathbb{C}\setminus S_{\tilde{\omega}}}, \varphi_2 \mathcal{X}_{\mathbb{C}\setminus S_{\tilde{\omega}}}\rangle| + |\langle \mathcal{K}_\Theta \varphi_1 \mathcal{X}_{S_{\tilde{\omega}}}, \varphi_2 \mathcal{X}_{\mathbb{C}\setminus S_{\tilde{\omega}}}\rangle| + |\langle \mathcal{K}_\Theta \varphi_1, \varphi_2 \mathcal{X}_{S_{\tilde{\omega}}}\rangle| \\ &\le |\langle \mathcal{K}_\Theta \varphi_1 \mathcal{X}_{\mathbb{C}\setminus S_{\tilde{\omega}}}, \varphi_2 \mathcal{X}_{\mathbb{C}\setminus S_{\tilde{\omega}}}\rangle| + \frac{N_\lambda}{10} + \frac{N_\lambda}{10},\end{aligned}$$

i.e.,

$$(\star\star\star)\ \frac{7}{10}N_\lambda \le |\langle \mathcal{K}_\Theta \varphi_1 \mathcal{X}_{\mathbb{C}\setminus S_{\tilde{\omega}}}, \varphi_2 \mathcal{X}_{\mathbb{C}\setminus S_{\tilde{\omega}}}\rangle|.$$

Define $\widetilde{\Theta} = \max\{\Theta, \delta\Phi_{\omega_1}, \delta\Phi_{\omega_2}\}$. Suppose $w \in \mathbb{C}$ is such that $\widetilde{\Theta}(w) \ne \Theta(w)$. Then $\Phi(w) < \Phi(w) + \lambda \le \Theta(w) < \delta \max\{\Phi_{\omega_1}(w), \Phi_{\omega_2}(w)\}$, i.e., $w \in S_{\tilde{\omega}}$. Hence $k_{\widetilde{\Theta}}(\zeta, z) = k_\Theta(\zeta, z)$ for all $\zeta, z \in \mathbb{C} \setminus S_{\tilde{\omega}}$ and so

$$(\star\star\star\star)\ \langle \mathcal{K}_{\widetilde{\Theta}} \varphi_1 \mathcal{X}_{\mathbb{C}\setminus S_{\tilde{\omega}}}, \varphi_2 \mathcal{X}_{\mathbb{C}\setminus S_{\tilde{\omega}}}\rangle = \langle \mathcal{K}_\Theta \varphi_1 \mathcal{X}_{\mathbb{C}\setminus S_{\tilde{\omega}}}, \varphi_2 \mathcal{X}_{\mathbb{C}\setminus S_{\tilde{\omega}}}\rangle.$$

It follows from $(\star\star\star)$, $(\star\star\star\star)$, the last two estimates of $(\star\star)$, and (b) of Proposition 7.20 that

$$\begin{aligned}\frac{7}{10}N_\lambda &\le |\langle \mathcal{K}_{\widetilde{\Theta}} \varphi_1 \mathcal{X}_{\mathbb{C}\setminus S_{\tilde{\omega}}}, \varphi_2 \mathcal{X}_{\mathbb{C}\setminus S_{\tilde{\omega}}}\rangle| \\ &\le |\langle \mathcal{K}_{\widetilde{\Theta}} (\varphi_1 \mathcal{X}_{\mathbb{C}\setminus S_{\tilde{\omega}}})_{\text{good}}, (\varphi_2 \mathcal{X}_{\mathbb{C}\setminus S_{\tilde{\omega}}})_{\text{good}}\rangle| \\ &+ |\langle \mathcal{K}_{\widetilde{\Theta}} (\varphi_1 \mathcal{X}_{\mathbb{C}\setminus S_{\tilde{\omega}}})_{\text{bad}}, (\varphi_2 \mathcal{X}_{\mathbb{C}\setminus S_{\tilde{\omega}}})_{\text{good}}\rangle| \\ &+ |\langle \mathcal{K}_{\widetilde{\Theta}} \varphi_1 \mathcal{X}_{\mathbb{C}\setminus S_{\tilde{\omega}}}, (\varphi_2 \mathcal{X}_{\mathbb{C}\setminus S_{\tilde{\omega}}})_{\text{bad}}\rangle| \\ &\le |\langle \mathcal{K}_{\widetilde{\Theta}} (\varphi_1 \mathcal{X}_{\mathbb{C}\setminus S_{\tilde{\omega}}})_{\text{good}}, (\varphi_2 \mathcal{X}_{\mathbb{C}\setminus S_{\tilde{\omega}}})_{\text{good}}\rangle| + \frac{2N_\lambda}{10} + \frac{N_\lambda}{10},\end{aligned}$$

i.e.,

$$(\star\star\star\star\star)\ \frac{4N_\lambda}{10} \le |\langle \mathcal{K}_{\widetilde{\Theta}}(\varphi_1 \mathcal{X}_{\mathbb{C}\setminus S_{\tilde{\omega}}})_{\text{good}}, (\varphi_2 \mathcal{X}_{\mathbb{C}\setminus S_{\tilde{\omega}}})_{\text{good}}\rangle|.$$

Clearly $\widetilde{\Theta}$ is a Lip(1)-function and $\widetilde{\Theta} \ge \delta \max\{\Phi_{\omega_1}, \Phi_{\omega_2}\}$. Moreover, $\widetilde{\Theta} \ge \Theta \ge \Phi + \lambda \ge \lambda$, so $\inf_{\mathbb{C}} \widetilde{\Theta} > 0$. We may thus apply Reduction 7.21 to the right-hand side of $(\star\star\star\star\star)$ and then use (b) of Proposition 7.20 twice to get

$$\frac{4N_\lambda}{10} \le N\ \|(\varphi_1 \mathcal{X}_{\mathbb{C}\setminus S_{\tilde{\omega}}})_{\text{good}}\|_{L^2(\mu)}\ \|(\varphi_2 \mathcal{X}_{\mathbb{C}\setminus S_{\tilde{\omega}}})_{\text{good}}\|_{L^2(\mu)} \le 4N,$$

i.e., $N_\lambda \le 10N$ as claimed! □

7.11 A Sticky Point, More Reductions, and Course Setting

We remind the reader only once, right now, that the Good Function Estimate is Reduction 7.21!

The Sticky Point. We are clearly dealing with complex-valued L^2 spaces here. Thus our $L^2(\mu)$ is the set of all measurable functions $\varphi : \mathbb{C} \mapsto \mathbb{C}$ such that $\int |\varphi|^2\, d\mu < \infty$. Is the inner product we are using in the Good Function Estimate the "real" one or the "complex" one, i.e., is our inner product given by $\langle \varphi, \psi\rangle_{\text{Real}} = \int \varphi\psi\, d\mu$ or $\langle \varphi, \psi\rangle_{\text{Com}} = \int \varphi\overline{\psi}\, d\mu$? Ones first impulse is to opt for the "complex" one since our L^2 space is complex-valued. However, all that is required of our choice is that we be able to deduce the norm estimate $\|A\| \le N$ for a bounded linear operator $A : L^2(\mu) \mapsto L^2(\mu)$ from the inner product estimates $|\langle A\varphi, \psi\rangle| \le N\ \|\varphi\|\ \|\psi\|$ for all $\varphi, \psi \in L^2(\mu)$. Now since $\langle A\varphi, \psi\rangle_{\text{Real}} = \langle A\varphi, \overline{\psi}\rangle_{\text{Com}}$, either inner product will do for that! Moreover, it turns out that under the "real" inner product the antisymmetry property of our kernels k_Θ given in (a) of Proposition 6.10 carries over to a nice antisymmetry property of the corresponding operators $\mathcal{K}_\Theta$:

$$\begin{aligned}
\langle \mathcal{K}_\Theta\varphi, \psi\rangle_{\text{Real}} &= \int_{\mathbb{C}} (\mathcal{K}_\Theta\varphi)(z)\, \psi(z)\, d\mu(z) \\
&= \int_{\mathbb{C}} \left\{ \int_{\mathbb{C}} k_\Theta(\zeta, z)\, \varphi(\zeta)\, d\mu(\zeta) \right\} \psi(z)\, d\mu(z) \\
&= -\int_{\mathbb{C}} \varphi(\zeta) \left\{ \int_{\mathbb{C}} k_\Theta(z, \zeta)\, \psi(z)\, d\mu(z) \right\} d\mu(\zeta) \\
&= -\int_{\mathbb{C}} \varphi(\zeta)\, (\mathcal{K}_\Theta\psi)(\zeta)\, d\mu(\zeta) \\
&= -\langle \varphi, \mathcal{K}_\Theta\psi\rangle_{\text{Real}}.
\end{aligned}$$

So that settles it: We shall use the "real" inner product in verifying the Good Function Estimate! Note that this choice of inner product is always symmetric in its two variables. Thus a trivial consequence of the antisymmetry in $\mathcal{K}_\Theta$ just demonstrated is that we always have $\langle \mathcal{K}_\Theta \varphi, \varphi \rangle = 0$.

First Reduction. To prove the Good Function Estimate it suffices to show that for some positive finite number $\widetilde{N}$ we have

$$(\star)\ |\langle \mathcal{K}_\Theta \varphi_1, \varphi_2 \rangle| \le 2\widetilde{N}\ \|\varphi_1\|_{L^2(\mu)}\ \|\varphi_2\|_{L^2(\mu)}$$

for all ω_1, ω_2, Θ, φ_1, and φ_2 as in the Good Function Estimate with φ_1 and φ_2 additionally satisfying: $\Lambda\, \varphi_1 = 0$ and $\Lambda\, \varphi_2 = 0$.

Proof Note that for any $z \in \mathbb{C}$ and any $\varepsilon > 0$, the function

$$\zeta \in \mathbb{C} \mapsto k_\Theta(\zeta, z)\, h(\zeta)\, \mathcal{X}_{\mathbb{C}\setminus B(z;\varepsilon)}(\zeta) \in \mathbb{C}$$

is dominated in absolute value by the function $\zeta \in \mathbb{C} \mapsto |h(\zeta)|/\inf_{\mathbb{C}} \Theta \in [0, \infty)$ because of (d) of Proposition 6.10. Since this last function is in $L^1(\mu)$ ($L^\infty(\mu)$ even), we may invoke Lebesgue's Dominated Convergence Theorem [RUD, 1.34] and hypothesis (5) of Theorem 7.1 (recall that each $\Phi_\omega \ge \widetilde{\Phi}$ by construction) to conclude that for any $z \in \mathbb{C}$,

$$|(\mathcal{K}_\Theta h)(z)| = \left| \int_{\mathbb{C}} k_\Theta(\zeta, z) h(\zeta)\, d\mu(\zeta) \right| = \lim_{\varepsilon \downarrow 0} \left| \int_{\mathbb{C}\setminus B(z:\varepsilon)} k_\Theta(\zeta, z) h(\zeta)\, d\mu(\zeta) \right| \le \widetilde{M},$$

i.e., $\|\mathcal{K}_\Theta h\|_{L^\infty(\mu)} \le \widetilde{M}$. It follows that $\|\mathcal{K}_\Theta h\|_{L^2(\mu)} \le \|\mathcal{K}_\Theta h\|_{L^\infty(\mu)} \sqrt{\mu(\mathbb{C})} \le \widetilde{M} \sqrt{\mu(\mathbb{C})}$ and so for any $\varphi \in L^2(\mu)$,

$$\begin{aligned}(\dagger)\ \|\mathcal{K}_\Theta \Lambda\, \varphi\|_{L^2(\mu)} &= \frac{|\langle \varphi \rangle_{\mathbb{C}}|}{|\langle h \rangle_{\mathbb{C}}|} \|\mathcal{K}_\Theta h\|_{L^2(\mu)} \\ &\le \frac{1}{1-\delta} \cdot \frac{\|\varphi\|_{L^2(\mu)} \sqrt{\mu(\mathbb{C})}}{\mu(\mathbb{C})} \cdot \widetilde{M} \sqrt{\mu(\mathbb{C})} \le 2\widetilde{M}\ \|\varphi\|_{L^2(\mu)}.\end{aligned}$$

In the above we have also used (a) of Lemma 7.10, the fact that $0 < \delta < 1/2$, and the Cauchy–Schwarz Inequality [RUD, 3.5].

Now given φ_1 and φ_2 as in the Good Function Estimate, clearly we have

$$\begin{aligned}(\ddagger)\ |\langle \mathcal{K}_\Theta \varphi_1, \varphi_2 \rangle| \le\ & |\langle \mathcal{K}_\Theta \Lambda\, \varphi_1, \varphi_2 \rangle| \leftarrow \text{(I)!} \\ & + |\langle \mathcal{K}_\Theta(\varphi_1 - \Lambda\, \varphi_1), \Lambda\, \varphi_2 \rangle| \leftarrow \text{(II)!} \\ & + |\langle \mathcal{K}_\Theta(\varphi_1 - \Lambda\, \varphi_1), \varphi_2 - \Lambda\, \varphi_2 \rangle|. \leftarrow \text{(III)!}\end{aligned}$$

From (†) we immediately see that

$$\text{(I)} \leq \|\mathcal{K}_\Theta \Lambda\, \varphi_1\|_{L^2(\mu)}\, \|\varphi_2\|_{L^2(\mu)} \leq 2\widetilde{M}\, \|\varphi_1\|_{L^2(\mu)}\, \|\varphi_2\|_{L^2(\mu)}.$$

From the antisymmetry of $\mathcal{K}_\Theta$, the Adapted Martingale Decomposition (7.15), and (†), we see that

$$\begin{aligned}\text{(II)} &\leq \|\varphi_1 - \Lambda\, \varphi_1\|_{L^2(\mu)}\, \|\mathcal{K}_\Theta \Lambda\, \varphi_2\|_{L^2(\mu)}\\ &\leq (1+\sqrt{2})\|\varphi_1\|_{L^2(\mu)}\, 2\widetilde{M}\|\varphi_2\|_{L^2(\mu)}\\ &\leq 5\widetilde{M}\, \|\varphi_1\|_{L^2(\mu)}\, \|\varphi_2\|_{L^2(\mu)}.\end{aligned}$$

By (e) of Lemma 7.16, we have $\Lambda(\varphi_1 - \Lambda\, \varphi_1) = \Lambda(\varphi_2 - \Lambda\, \varphi_2) = 0$. Thus we may apply (⋆) to $\varphi_1 - \Lambda\, \varphi_1$ and $\varphi_2 - \Lambda\, \varphi_2$, followed by the Adapted Martingale Decomposition (7.15), to see that

$$\begin{aligned}\text{(III)} &\leq 2\widetilde{N}\, \|\varphi_1 - \Lambda\, \varphi_1\|_{L^2(\mu)}\, \|\varphi_2 - \Lambda\, \varphi_2\|_{L^2(\mu)}\\ &\leq 2\widetilde{N}\, (1+\sqrt{2})\|\varphi_1\|_{L^2(\mu)}\, (1+\sqrt{2})\|\varphi_2\|_{L^2(\mu)}\\ &\leq 12\widetilde{N}\, \|\varphi_1\|_{L^2(\mu)}\, \|\varphi_2\|_{L^2(\mu)}.\end{aligned}$$

Replacing (I), (II), and (III) in (‡) by our estimates on them, we see that the Good Function Estimate holds with $N = 7\widetilde{M} + 12\widetilde{N}$. □

Second Reduction. Set $\mathcal{D} = \mathcal{D}_{\omega_1}^{\text{tran}} \times \mathcal{D}_{\omega_2}^{\text{tran}}$. To prove that (⋆) of the First Reduction holds for a positive finite number $\widetilde{N}$, it suffices to show

$$(\star\star) \quad \left|\sum\nolimits_{\mathcal{D}} \langle \mathcal{K}_\Theta \Delta_{Q_1} \varphi_1, \Delta_{Q_2} \varphi_2\rangle\right| \leq 2\widetilde{N}\, \|\varphi_1\|_{L^2(\mu)}\, \|\varphi_2\|_{L^2(\mu)}$$

for all ω_1, ω_2, Θ, φ_1, and φ_2 as in the First Reduction with φ_1 and φ_2 additionally satisfying: $\Delta_{Q_1} \varphi_1 \neq 0$ for only finitely many $Q_1 \in \mathcal{D}_{\omega_1}^{\text{tran}}$ and $\Delta_{Q_2} \varphi_2 \neq 0$ for only finitely many $Q_2 \in \mathcal{D}_{\omega_2}^{\text{tran}}$.

Proof Set $\mathcal{D}_1(n) = \{Q_1 \in \mathcal{D}_{\omega_1}^{\text{tran}} : l(Q_1) \geq 2^{-n}\}$ and $\mathcal{D}_2(n) = \{Q_2 \in \mathcal{D}_{\omega_2}^{\text{tran}} : l(Q_2) \geq 2^{-n}\}$. Given φ_1 and φ_2 as in the First Reduction, set

$$\varphi_{1,n} = \sum\nolimits_{\mathcal{D}_1(n)} \Delta_{Q_1} \varphi_1 \quad \text{and} \quad \varphi_{2,n} = \sum\nolimits_{\mathcal{D}_2(n)} \Delta_{Q_2} \varphi_2.$$

By Lemma 7.16 we have $\Lambda\, \varphi_{1,n} = 0$ and $\Delta_{Q_1} \varphi_{1,n} = \Delta_{Q_1} \varphi_1$ or 0 according as $Q_1 \in \mathcal{D}_1(n)$ or not. Similar formulas hold for $\varphi_{2,n}$, so we may apply (⋆⋆) to $\varphi_{1,n}$

and $\varphi_{2,n}$. Hence, by the linearity of $\mathcal{K}_\Theta$, these formulas for $\Delta_{Q_1}\varphi_{1,n}$ and $\Delta_{Q_1}\varphi_{2,n}$, and this application of $(\star\star)$, we have

$$(\dagger)\ |\langle \mathcal{K}_\Theta \varphi_{1,n}, \varphi_{2,n}\rangle| = \left|\sum_{\mathcal{D}_1(n)\times\mathcal{D}_2(n)} \langle \mathcal{K}_\Theta \Delta_{Q_1}\varphi_1, \Delta_{Q_2}\varphi_2\rangle\right|$$

$$= \left|\sum_{\mathcal{D}} \langle \mathcal{K}_\Theta \Delta_{Q_1}\varphi_{1,n}, \Delta_{Q_2}\varphi_{2,n}\rangle\right|$$

$$\leq 2\widetilde{N}\ \|\varphi_{1,n}\|_{L^2(\mu)}\ \|\varphi_{2,n}\|_{L^2(\mu)}.$$

Since $\Lambda\,\varphi_1 = \Lambda\,\varphi_2 = 0$, we have that $\varphi_{1,n} \to \varphi_1$ and $\varphi_{2,n} \to \varphi_2$ in the norm of $L^2(\mu)$ by the Adapted Martingale Decomposition (7.15). Thus to finish we need to merely let $n \to \infty$ in $(\dagger)$. □

Remark. Since the sum appearing in $(\star\star)$ is actually a finite sum, its terms may be rearranged and regrouped at will. The same is of course true of any subsums of the sum in $(\star\star)$. These facts will prove very useful and will be used without mention in what follows!

Third Reduction. Set $\mathcal{G} = \{(Q_1, Q_2) \in \mathcal{D} : l(Q_1) \leq l(Q_2)\}$ and $\mathcal{G}^* = \{(Q_1, Q_2) \in \mathcal{D} : l(Q_1) < l(Q_2)\}$. To prove that $(\star\star)$ of the Second Reduction holds for a positive finite number $\widetilde{N}$, it suffices to show

$$(\star\star\star)\ \left|\sum_{\mathcal{G}} \langle \mathcal{K}_\Theta \Delta_{Q_1}\varphi_1, \Delta_{Q_2}\varphi_2\rangle\right| \leq \widetilde{N}\ \|\varphi_1\|_{L^2(\mu)}\ \|\varphi_2\|_{L^2(\mu)}$$

and

$$(\star\star\star\star)\ \left|\sum_{\mathcal{G}^*} \langle \mathcal{K}_\Theta \Delta_{Q_1}\varphi_1, \Delta_{Q_2}\varphi_2\rangle\right| \leq \widetilde{N}\ \|\varphi_1\|_{L^2(\mu)}\ \|\varphi_2\|_{L^2(\mu)}$$

for all ω_1, ω_2, Θ, φ_1, and φ_2 as in the Second Reduction.

Proof Clearly we may dominate the absolute value of our sum over $\mathcal{D}$ in $(\star\star)$ by the sum of the absolute value of two sums, the first over $\mathcal{G}$ and the second over $\mathcal{D}\setminus\mathcal{G}$:

$$\left|\sum_{\mathcal{D}} \langle \mathcal{K}_\Theta \Delta_{Q_1}\varphi_1, \Delta_{Q_2}\varphi_2\rangle\right| \leq \left|\sum_{\mathcal{G}} \langle \mathcal{K}_\Theta \Delta_{Q_1}\varphi_1, \Delta_{Q_2}\varphi_2\rangle\right| \leftarrow \text{(I)!}$$

$$+ \left|\sum_{\mathcal{D}\setminus\mathcal{G}} \langle \mathcal{K}_\Theta \Delta_{Q_1}\varphi_1, \Delta_{Q_2}\varphi_2\rangle\right|. \leftarrow \text{(II)!}$$

The first term (I) is taken care of by $(\star\star\star)$ immediately. Thus to finish we need only show how the second term (II) will be taken care of by $(\star\star\star\star)$.

Our notation for $\mathcal{D}$, $\mathcal{G}$, and $\mathcal{G}^*$ does not reflect the dependence of these objects on (ω_1, ω_2). This dependence is important now, so we must draw attention to it! The mapping $(Q_1, Q_2) \mapsto (Q_2, Q_1)$ takes the set $\mathcal{D}\setminus\mathcal{G}$ corresponding to (ω_1, ω_2) in a one-to-one manner onto the set $\mathcal{G}^*$ corresponding to (ω_2, ω_1). Moreover, the antisymmetry of our operator $\mathcal{K}_\Theta$ and the symmetry of our inner product $\langle \cdot\,, \cdot \rangle$ imply that

$$\sum\nolimits_{\mathcal{D}\setminus\mathcal{G}} \langle \mathcal{K}_\Theta \Delta_{Q_1} \varphi_1, \Delta_{Q_2} \varphi_2 \rangle = -\sum\nolimits_{\mathcal{D}\setminus\mathcal{G}} \langle \mathcal{K}_\Theta \Delta_{Q_2} \varphi_2, \Delta_{Q_1} \varphi_1 \rangle.$$

Thus our second term (II) is actually a term of the type estimated in $(\star\star\star\star)$ with the set $\mathcal{G}^*$ corresponding to (ω_2, ω_1) rather than (ω_1, ω_2). □

Remark. The Third Reduction has just been a matter of convenience ... it will cut six cases down to three. But note that it has broken the symmetry between φ_1 and φ_2. To emphasize this, and to avoid subscripts, let us replace φ_1, φ_2, Q_1, and Q_2 by φ, ψ, Q, and R respectively (we leave ω_1 and ω_2 as is). The import of all our work and reductions up to now, along with one more soon-to-be-justified reduction, is summarized in the following assertion: To prove Theorem 7.1 it suffices to show the following.

Reduction 7.23 (The Final Good Function Estimate) *There exists a positive finite number $\widetilde{N}$ such that*

$$\left|\sum\nolimits_{\mathcal{G}} \langle \mathcal{K}_\Theta \Delta_Q \varphi, \Delta_R \psi \rangle\right| \leq \widetilde{N}\, \|\varphi\|_{L^2(\mu)}\, \|\psi\|_{L^2(\mu)}$$

whenever

(1) $\mathcal{G} = \{(Q, R) \in \mathcal{D}^{\text{tran}}_{\omega_1} \times \mathcal{D}^{\text{tran}}_{\omega_2} : l(Q) \leq l(R)\}$ *with* $(\omega_1, \omega_2) \in \Omega \times \Omega$,

(2) Θ *is a Lip(1)-function satisfying* $\inf_{\mathbb{C}} \Theta > 0$ *and* $\Theta \geq \delta \max\{\Phi_{\omega_1}, \Phi_{\omega_2}\}$,

(3) $\varphi \in L^2(\mu)$ *with* $\Lambda\, \varphi = 0$ *and* $\psi \in L^2(\mu)$ *with* $\Lambda\, \psi = 0$,

(4) $\Delta_Q\, \varphi \neq 0$ *for only finitely many* $Q \in \mathcal{D}^{\text{tran}}_{\omega_1}$ *with each such* Q *being* ω_2*-good, and*

(5) $\Delta_R\, \psi \neq 0$ *for only finitely many* $R \in \mathcal{D}^{\text{tran}}_{\omega_2}$ *with each such* R *being* ω_1*-good.*

Remark. Except for the change in notation, this reduction differs from the Third Reduction above only in dropping the estimate $(\star\star\star\star)$. Although $(\star\star\star\star)$ does not follow directly from $(\star\star\star)$, it will be the case that the way we prove $(\star\star\star)$ will also prove $(\star\star\star\star)$. To see this, and to see how the rest of the proof will proceed, we now turn to the last order of business in this section.

Course Setting. Split $\mathcal{G}$ up disjointly into three pieces $\mathcal{G}_1$, $\mathcal{G}_2$, and $\mathcal{G}_3$ as follows:

$$\mathcal{G}_1 = \{(Q, R) \in \mathcal{G} : l(Q) \geq 2^{-m}\, l(R) \text{ and } \operatorname{dist}(Q, R) < l(R)\},$$

$$\mathcal{G}_2 = \{(Q, R) \in \mathcal{G} : l(Q) \geq 2^{-m}\, l(R) \text{ and } \operatorname{dist}(Q, R) \geq l(R)\} \cup$$

$$\{(Q, R) \in \mathcal{G} : l(Q) < 2^{-m}\, l(R) \text{ and } Q \cap R = \emptyset\}, \text{ and}$$

$$\mathcal{G}_3 = \{(Q, R) \in \mathcal{G} : l(Q) < 2^{-m}\, l(R) \text{ and } Q \cap R \neq \emptyset\}.$$

Our proof of the Final Good Function Estimate will start with the obvious inequality

$$(\dagger) \quad \left|\sum\nolimits_{\mathcal{G}} \langle \mathcal{K}_\Theta \Delta_Q\, \varphi, \Delta_R\, \psi\rangle\right| \leq \sum\nolimits_{\mathcal{G}_1} |\langle \mathcal{K}_\Theta \Delta_Q\, \varphi, \Delta_R\, \psi\rangle| + \sum\nolimits_{\mathcal{G}_2} |\langle \mathcal{K}_\Theta \Delta_Q\, \varphi, \Delta_R\, \psi\rangle| + \left|\sum\nolimits_{\mathcal{G}_3} \langle \mathcal{K}_\Theta \Delta_Q\, \varphi, \Delta_R\, \psi\rangle\right|$$

and then proceed to show that each of the three right-hand terms of (†) can be estimated by constant multiples of the product of the L^2 norms of φ and ψ.

We can now explain the dropping of the estimate ($\star\star\star\star$) of the Third Reduction from the Final Good Function Estimate. In this paragraph the reader should continually take into account our notational change from (Q_1, Q_2) to (Q, R)! Note that the sum in ($\star\star\star\star$) is gotten from the sum in ($\star\star\star$) by dropping those terms corresponding to pairs (Q, R) for which $l(Q) = l(R)$. These pairs occur only in the two terms of (†) corresponding to $\mathcal{G}_1$ and $\mathcal{G}_2$. These two terms are sums of absolute values and so sums of nonnegative terms. Thus any upper estimates of these two terms of (†) will still be upper estimates if we drop some of their summands. In consequence, any upper estimate of the absolute value of the sum in ($\star\star\star$) gotten via (†) will also be an upper estimate of the absolute value of the sum in ($\star\star\star\star$).

To conclude, note that we have left the term corresponding to $\mathcal{G}_3$ in (†) as an absolute value of a sum and have not used the Triangle Inequality to convert it into a sum of absolute values (as we did with the $\mathcal{G}_1$ and $\mathcal{G}_2$ terms). Indeed, to have done so would have proven disastrous since the estimation of this term ultimately ends up depending on a crucial cancellation argument involving telescoping sums. Of course, some such thing was to be expected somewhere along the line since we are dealing with a singular integral here!

7.12 Interlude: The Schur Test

We state the Schur Test (L^2 version only) in a form most convenient for our needs. It is usually stated as a bound on the norm of an integral operator $\mathcal{K}$ arising from a kernel k. We want instead a bound on $|\langle \mathcal{K}\, \varphi, \psi\rangle|$ valid only for functions φ and ψ vanishing off certain sets ... thus the presence of X_1 and X_2 in our version. Lastly, given our future use of it, we state it solely in terms of k, omitting any reference to $\mathcal{K}$.

Proposition 7.24 (Schur Test) *Let μ be a positive measure on a set X. Suppose that $\lambda : X \mapsto [0, \infty]$, $k : X \times X \mapsto [0, \infty]$, $X_1, X_2 \subseteq X$, and $C_1, C_2 \in (0, \infty)$ together satisfy*

(1) $0 < \lambda(\zeta) < \infty$ *for all* $\zeta \in X_1$,

(2) $\displaystyle\int_{X_1} k(\zeta, z)\lambda(\zeta)\, d\mu(\zeta) \le C_1\lambda(z)$ *for all* $z \in X_2$, *and*

(3) $\displaystyle\int_{X_2} k(\zeta, z)\lambda(z)\, d\mu(z) \le C_2\lambda(\zeta)$ *for all* $\zeta \in X_1$.

Then for any $\varphi, \psi : X \mapsto [0, \infty]$ such that $\varphi = 0$ off X_1 and $\psi = 0$ off X_2 we have

$$\int_X \int_X k(\zeta, z)\varphi(\zeta)\psi(z)\, d\mu(\zeta)\, d\mu(z) \le \sqrt{C_1 C_2}\, \|\varphi\|_{L^2(\mu)}\, \|\psi\|_{L^2(\mu)}.$$

Proof Using the hypothesis on φ, (1), the Cauchy–Schwarz Inequality [RUD, 3.5], and (2) in succession, one sees that for any $z \in X_2$, one has

$$\begin{aligned}
(\star) \int_X k(\zeta, z)\varphi(\zeta)\, d\mu(\zeta) &= \int_{X_1} k(\zeta, z)\varphi(\zeta)\, d\mu(\zeta) \\
&= \int_{X_1} \sqrt{k(\zeta, z)\lambda(\zeta)} \sqrt{\frac{k(\zeta, z)}{\lambda(\zeta)}}\, \varphi(\zeta)\, d\mu(\zeta) \\
&\le \left[\int_{X_1} k(\zeta, z)\lambda(\zeta)\, d\mu(\zeta)\right]^{1/2} \left[\int_{X_1} \frac{k(\zeta, z)}{\lambda(\zeta)}\, \varphi^2(\zeta)\, d\mu(\zeta)\right]^{1/2} \\
&\le [C_1\lambda(z)]^{1/2} \left[\int_{X_1} \frac{k(\zeta, z)}{\lambda(\zeta)}\, \varphi^2(\zeta)\, d\mu(\zeta)\right]^{1/2}.
\end{aligned}$$

Thus using the hypothesis on ψ, the Cauchy–Schwarz Inequality [RUD, 3.5], ($\star$), Fubini's Theorem [RUD, 8.8], and (3) in succession, one obtains

$$\begin{aligned}
&\int_X \int_X k(\zeta, z)\varphi(\zeta)\psi(z)\, d\mu(\zeta)\, d\mu(z) \\
&= \int_{X_2} \left\{\int_X k(\zeta, z)\varphi(\zeta)\, d\mu(\zeta)\right\} \psi(z)\, d\mu(z) \\
&\le \left\{\int_{X_2} \left[\int_X k(\zeta, z)\varphi(\zeta)\, d\mu(\zeta)\right]^2 d\mu(z)\right\}^{1/2} \left\{\int_{X_2} \psi^2(z)\, d\mu(z)\right\}^{1/2} \\
&\le \left\{\int_{X_2} [C_1\lambda(z)] \left[\int_{X_1} \frac{k(\zeta, z)}{\lambda(\zeta)} \varphi^2(\zeta)\, d\mu(\zeta)\right] d\mu(z)\right\}^{1/2} \|\psi\|_{L^2(\mu)} \\
&= \left\{C_1 \int_{X_1} \frac{\varphi^2(\zeta)}{\lambda(\zeta)} \left[\int_{X_2} k(\zeta, z)\lambda(z)\, d\mu(z)\right] d\mu(\zeta)\right\}^{1/2} \|\psi\|_{L^2(\mu)}
\end{aligned}$$

$$\leq \left\{ C_1 \int_{X_1} \frac{\varphi^2(\zeta)}{\lambda(\zeta)} [C_2 \lambda(\zeta)] \, d\mu(\zeta) \right\}^{1/2} \|\psi\|_{L^2(\mu)} = \sqrt{C_1 C_2} \, \|\varphi\|_{L^2(\mu)} \, \|\psi\|_{L^2(\mu)}.$$

□

Setting λ equal to 1 and both X_1 and X_2 equal to X, we immediately obtain the following.

Corollary 7.25 *Let μ be a positive measure on a set X. Suppose that $k : X \times X \mapsto [0, \infty]$ and $C_1, C_2 \in (0, \infty)$ together satisfy*

(1) $\int_X k(\zeta, z) \, d\mu(\zeta) \leq C_1$ *for all $z \in X$ and*

(2) $\int_X k(\zeta, z) \, d\mu(z) \leq C_2$ *for all $\zeta \in X$.*

Then for any $\varphi, \psi : X \mapsto [0, \infty]$ we have

$$\int_X \int_X k(\zeta, z) \varphi(\zeta) \psi(z) \, d\mu(\zeta) \, d\mu(z) \leq \sqrt{C_1 C_2} \, \|\varphi\|_{L^2(\mu)} \, \|\psi\|_{L^2(\mu)}.$$

7.13 $\mathcal{G}_1$: The Crudely Handled Terms

Lemma 7.26 *Let Q be a dyadic square whose boundary is M'-negligible. Then for every $z \in \mathbb{C} \setminus \partial Q$,*

$$\int_{\mathbb{C} \setminus \partial Q} \min \left\{ \frac{1}{\text{dist}(\zeta, \partial Q)}, \frac{1}{\text{dist}(z, \partial Q)} \right\} \frac{1}{\sqrt{\text{dist}(\zeta, \partial Q)}} \, d\mu(\zeta) \leq \frac{12M'}{\sqrt{\text{dist}(z, \partial Q)}}.$$

Proof Set

$$A_n = \{\zeta \in \mathbb{C} \setminus \partial Q : \frac{1}{4^{n+1}} \text{dist}(z, \partial Q) < \text{dist}(\zeta, \partial Q) \leq \frac{1}{4^n} \text{dist}(z, \partial Q)\}$$

and

$$B_n = \{\zeta \in \mathbb{C} \setminus \partial Q : 4^n \text{dist}(z, \partial Q) < \text{dist}(\zeta, \partial Q) \leq 4^{n+1} \text{dist}(z, \partial Q)\}.$$

Then the left-hand side of our desired inequality is just

$$\sum_{n \geq 0} \int_{A_n} \frac{1}{\text{dist}(z, \partial Q)} \frac{1}{\sqrt{\text{dist}(\zeta, \partial Q)}} d\mu(\zeta) + \sum_{n \geq 0} \int_{B_n} \frac{1}{\text{dist}(\zeta, \partial Q)} \frac{1}{\sqrt{\text{dist}(\zeta, \partial Q)}} d\mu(\zeta).$$

Using the M'-negligibility of ∂Q and the geometric series formula, we can dominate the first term of the above by

$$\sum_{n\geq 0} \frac{1}{\operatorname{dist}(z,\partial Q)} \frac{2^{n+1}}{\sqrt{\operatorname{dist}(z,\partial Q)}} \cdot M' \frac{1}{4^n} \operatorname{dist}(z,\partial Q)$$
$$= \frac{2M'}{\sqrt{\operatorname{dist}(z,\partial Q)}} \sum_{n\geq 0} \frac{1}{2^n} = \frac{4M'}{\sqrt{\operatorname{dist}(z,\partial Q)}}$$

and the second term by

$$\sum_{n\geq 0} \frac{1}{4^n \operatorname{dist}(z,\partial Q)} \frac{1}{2^n \sqrt{\operatorname{dist}(z,\partial Q)}} \cdot M' 4^{n+1} \operatorname{dist}(z,\partial Q)$$
$$= \frac{4M'}{\sqrt{\operatorname{dist}(z,\partial Q)}} \sum_{n\geq 0} \frac{1}{2^n} = \frac{8M'}{\sqrt{\operatorname{dist}(z,\partial Q)}}.$$

The correctness of the desired estimate is now clear. □

Lemma 7.27 (Negligible Square) *Let Q be a dyadic square with M'-negligible boundary and let Θ be a Lip(1)-function for which* $\inf_{\mathbb{C}} \Theta > 0$. *Suppose that $\eta_1 \in L^2(\mu)$ vanishes off* cl Q *and $\eta_2 \in L^2(\mu)$ vanishes on* int Q. *Then*

$$|\langle \mathcal{K}_\Theta \eta_1, \eta_2 \rangle| \leq 12M' \, \|\eta_1\|_{L^2(\mu)} \, \|\eta_2\|_{L^2(\mu)}.$$

Proof Since ∂Q is M'-negligible, it follows that $\mu(\partial Q) = 0$ and so we may assume that η_1 vanishes off int Q and η_2 vanishes on cl Q. Note that

$$|\langle \mathcal{K}_\Theta \eta_1, \eta_2 \rangle| \leq \int_{\mathbb{C}} \int_{\mathbb{C}} |k_\Theta(\zeta, z)||\eta_1(\zeta)||\eta_2(z)| \, d\mu(\zeta) \, d\mu(z).$$

Thus to finish the proof it suffices to verify the hypotheses of the Schur Test (7.24) with $\lambda(\zeta) = 1/\sqrt{\operatorname{dist}(\zeta,\partial Q)}$, $k(\zeta, z) = |k_\Theta(\zeta, z)|$, $X_1 = \operatorname{int} Q$, $X_2 = \mathbb{C} \setminus \operatorname{cl} Q$, $C_1 = C_2 = 12M'$, $\varphi = |\eta_1|$, and $\psi = |\eta_2|$.

That hypothesis (1) holds is clear. To deal with hypothesis (2), consider $\zeta \in \operatorname{int} Q$ and $z \in \mathbb{C} \setminus \operatorname{cl} Q$. Let q be the intersection point of the segment $[\zeta, z]$ with ∂Q. Then by (c) of Proposition 6.10,

$$|k_\Theta(\zeta, z)| \leq \frac{1}{|\zeta - z|} \leq \min\left\{\frac{1}{|\zeta - q|}, \frac{1}{|q - z|}\right\} \leq \min\left\{\frac{1}{\operatorname{dist}(\zeta,\partial Q)}, \frac{1}{\operatorname{dist}(z,\partial Q)}\right\}$$

and so

$$\int_{\operatorname{int} Q} |k_\Theta(\zeta, z)| \frac{1}{\sqrt{\operatorname{dist}(\zeta,\partial Q)}} \, d\mu(\zeta)$$
$$\leq \int_{\mathbb{C}\setminus\partial Q} \min\left\{\frac{1}{\operatorname{dist}(\zeta,\partial Q)}, \frac{1}{\operatorname{dist}(z,\partial Q)}\right\} \frac{1}{\sqrt{\operatorname{dist}(\zeta,\partial Q)}} \, d\mu(\zeta).$$

The last lemma now gives us hypothesis (2). Hypothesis (3) is verified similarly. □

For the rest of this section, $\omega_1, \omega_2, \Theta, \varphi$, and ψ are as in the Final Good Function Estimate (7.23).

Lemma 7.28 *There exists a positive finite number U such that for all $(Q, R) \in \mathcal{G}_1$,*

$$|\langle \mathcal{K}_\Theta \Delta_Q \varphi, \Delta_R \psi \rangle| \le U \, \|\Delta_Q \varphi\|_{L^2(\mu)} \, \|\Delta_R \psi\|_{L^2(\mu)}.$$

Proof If $\Delta_Q \varphi = 0$, then our conclusion follows trivially, so by (4) of the Final Good Function Estimate (7.23) we may assume that Q is ω_2-good. Similarly we may also assume that R is ω_1-good. Let Q_j, $j = 1, 2, 3, 4$, denote the four children of Q labeled starting with the northeast one and proceeding counterclockwise for the sake of definiteness. Let R_k, $k = 1, 2, 3, 4$, be similarly related to R. A little fiddling around with the definitions of ω-good and $\mathcal{G}_1$ shows that ∂Q_j and ∂R_k are M'-negligible for all j and k.

Note that

$$\langle \mathcal{K}_\Theta \Delta_Q \varphi, \Delta_R \psi \rangle = \sum_{j,k=1}^{4} \langle \mathcal{K}_\Theta(\mathcal{X}_{Q_j} \Delta_Q \varphi), \mathcal{X}_{R_k} \Delta_R \psi \rangle.$$

and that for each j and k,

$$\begin{aligned} \langle \mathcal{K}_\Theta(\mathcal{X}_{Q_j} \Delta_Q \varphi), \mathcal{X}_{R_k} \Delta_R \psi \rangle &= \langle \mathcal{K}_\Theta(\mathcal{X}_{Q_j} \Delta_Q \varphi), \mathcal{X}_{R_k \setminus Q_j} \Delta_R \psi \rangle \leftarrow \text{(I)!} \\ &+ \langle \mathcal{K}_\Theta(\mathcal{X}_{Q_j \setminus R_k} \Delta_Q \varphi), \mathcal{X}_{Q_j \cap R_k} \Delta_R \psi \rangle \leftarrow \text{(II)!} \\ &+ \langle \mathcal{K}_\Theta(\mathcal{X}_{Q_j \cap R_k} \Delta_Q \varphi), \mathcal{X}_{Q_j \cap R_k} \Delta_R \psi \rangle. \leftarrow \text{(III)!} \end{aligned}$$

By the last lemma,

$$\begin{aligned} |\text{(I)}| &\le 12M' \, \|\mathcal{X}_{Q_j} \Delta_Q \varphi\|_{L^2(\mu)} \, \|\mathcal{X}_{R_k \setminus Q_j} \Delta_R \psi\|_{L^2(\mu)} \\ &\le 12M' \, \|\Delta_Q \varphi\|_{L^2(\mu)} \, \|\Delta_R \psi\|_{L^2(\mu)}. \end{aligned}$$

Using the antisymmetry of $\mathcal{K}_\Theta$ and then proceeding similarly we obtain the same bound on $|\text{(II)}|$. To deal with $|\text{(III)}|$ we must consider cases.

Case One (*Q_j is terminal*). Note that for $\zeta \in \operatorname{int} Q_j$ we have $\Theta(\zeta) \ge \delta \, \Phi_{\omega_1(\zeta)} \ge \delta \operatorname{dist}(\zeta, \partial Q_j)$. Thus by (d) of Proposition 6.10,

$$(\star) \; |k_\Theta(\zeta, z)| \le \frac{1}{\delta} \min\left\{\frac{1}{\Theta(\zeta)}, \frac{1}{\Theta(z)}\right\} \le \frac{1}{\delta} \min\left\{\frac{1}{\operatorname{dist}(\zeta, \partial Q_j)}, \frac{1}{\operatorname{dist}(z, \partial Q_j)}\right\}$$

for every $\zeta, z \in \operatorname{int} Q_j$. We now seek to apply the Schur Test (7.24) with $\lambda(\zeta) = 1/\sqrt{\operatorname{dist}(\zeta, \partial Q_j)}$, $k(\zeta, z) = |k_\Theta(\zeta, z)|$, $X_1 = X_2 = \operatorname{int}(Q_j \cap R_k)$, $C_1 = C_2 = 12M'/\delta$, $\varphi = |\mathcal{X}_{Q_j \cap R_k} \Delta_Q \varphi|$, and $\psi = |\mathcal{X}_{Q_j \cap R_k} \Delta_R \psi|$. Hypothesis

(1) of the Schur Test is clear while hypotheses (2) and (3) follow from ($\star$) and Lemma 7.26. It follows that

$$|(\mathrm{III})| \le \frac{12M'}{\delta} \, \|\mathcal{X}_{Q_j \cap R_k} \Delta_Q \,\varphi\|_{L^2(\mu)} \, \|\mathcal{X}_{Q_j \cap R_k} \Delta_R \,\psi\|_{L^2(\mu)}$$

$$\le \frac{12M'}{\delta} \, \|\Delta_Q \,\varphi\|_{L^2(\mu)} \, \|\Delta_R \,\psi\|_{L^2(\mu)}.$$

Case Two (*R_k is terminal*). Similar to Case One with the same bound obtained on $|(\mathrm{III})|$.

Case Three (*Q_j and R_k are transit*). In this case both $\mathcal{X}_{Q_j \cap R_k} \Delta_Q \,\varphi$ and $\mathcal{X}_{Q_j \cap R_k} \Delta_R \,\psi$ are constant multiples of $\tilde{h} = \mathcal{X}_{Q_j \cap R_k} h$, say $c_1 \tilde{h}$ and $c_2 \tilde{h}$ respectively. Thus $(\mathrm{III}) = \langle \mathcal{K}_\Theta c_1 \tilde{h}, c_2 \tilde{h} \rangle = c_1 c_2 \langle \mathcal{K}_\Theta \tilde{h}, \tilde{h} \rangle = 0$ by the antisymmetry of $\mathcal{K}_\Theta$ and the symmetry of our inner product $\langle \cdot\,, \cdot \rangle$.

Putting this all together, we see that the lemma holds with $U = 16(12M' + 12M' + 12M'/\delta) = 192M'(2 + 1/\delta)$. □

To state the next lemma we need some notation. Set $\mathrm{Dom} = \{Q : (Q, R) \in \mathcal{G}_1$ for some $R\}$ and given $Q \in \mathrm{Dom}$, set $\mathrm{Ran}(Q) = \{R : (Q, R) \in \mathcal{G}_1\}$. Also, set $\mathrm{Ran} = \{R : (Q, R) \in \mathcal{G}_1$ for some $Q\}$ and given $R \in \mathrm{Ran}$, set $\mathrm{Dom}(R) = \{Q : (Q, R) \in \mathcal{G}_1\}$.

Lemma 7.29 *There exists a positive finite number V such that $\#\mathrm{Ran}(Q) \le V$ for every $Q \in \mathrm{Dom}$ and there exists a positive finite number W such that $\#\mathrm{Dom}(R) \le W$ for every $R \in \mathrm{Ran}$.*

Proof If $Q \in \mathrm{Dom}$ and $R \in \mathrm{Ran}(Q)$, then $l(R) = 2^j \, l(Q)$ for $j = 0, 1, 2, \ldots, m$, so there are $m + 1$ choices for the size of R. Also $\mathrm{dist}(Q, R) < l(R)$, so $R \subseteq \mathrm{int}\{4 \cdot 2^m + 1\}Q$. Thus the number of squares $R \in \mathrm{Ran}(Q)$ of a given fixed size is at most

$$\frac{\mathcal{L}^2(\{4 \cdot 2^m + 1\}Q)}{\mathcal{L}^2(R)} = \frac{\{4 \cdot 2^m + 1\}^2 \, l(Q)^2}{l(R)^2} \le \{4 \cdot 2^m + 1\}^2$$

since $l(Q) \le l(R)$. It follows that $\#\mathrm{Ran}(Q) \le V = (m + 1)\{4 \cdot 2^m + 1\}^2$.

For $R \in \mathrm{Ran}$, a similar argument show that $\#\mathrm{Dom}(R) \le W = (m + 1)25(4^m)$. (The difference in V and W is because we now have $Q \subseteq \mathrm{int}\, 5R$ and $l(R) \le 2^m \, l(Q)$.) □

Proposition 7.30 (Estimate on the $\mathcal{G}_1$ Sum) *There exists a positive finite number $\widetilde{N}_1$ such that*

$$\sum\nolimits_{\mathcal{G}_1} |\langle \mathcal{K}_\Theta \Delta_Q \,\varphi, \Delta_R \,\psi \rangle| \le \widetilde{N}_1 \, \|\varphi\|_{L^2(\mu)} \, \|\psi\|_{L^2(\mu)}$$

for all ω_1, ω_2, Θ, φ, and ψ as in the Final Good Function Estimate (7.23).

Proof Denote the left-hand side of the inequality that we wish to prove by LHS! Then by Lemma 7.28 and the Cauchy–Schwarz Inequality [RUD, 3.5 – we inform the reader just this one time that the integral version of Cauchy–Schwartz easily implies the sequence version!],

$$\begin{aligned}
\text{LHS} &\le U \sum\nolimits_{(Q,R)\in\mathcal{G}_1} \|\Delta_Q\,\varphi\|_{L^2(\mu)}\, \|\Delta_R\,\psi\|_{L^2(\mu)} \\
&= U \sum\nolimits_{Q\in\text{Dom}} \|\Delta_Q\,\varphi\|_{L^2(\mu)} \Big[\sum\nolimits_{R\in\text{Ran}(Q)} \|\Delta_R\,\psi\|_{L^2(\mu)}\Big] \\
&\le U \Big\{\sum\nolimits_{Q\in\text{Dom}} \|\Delta_Q\,\varphi\|^2_{L^2(\mu)}\Big\}^{1/2} \Big\{\sum\nolimits_{Q\in\text{Dom}} \Big[\sum\nolimits_{R\in\text{Ran}(Q)} \|\Delta_R\,\psi\|_{L^2(\mu)}\Big]^2\Big\}^{1/2}.
\end{aligned}$$

Applying the Cauchy–Schwarz Inequality [RUD, 3.5] to the squared sum in the above and then using the first part of Lemma 7.29, we see that

$$\begin{aligned}
\Big[\sum\nolimits_{R\in\text{Ran}(Q)} \|\Delta_R\,\psi\|_{L^2(\mu)}\Big]^2 &\le \Big[\sum\nolimits_{R\in\text{Ran}(Q)} 1^2\Big] \Big[\sum\nolimits_{R\in\text{Ran}(Q)} \|\Delta_R\,\psi\|^2_{L^2(\mu)}\Big] \\
&= [\#\text{Ran}(Q)] \Big[\sum\nolimits_{R\in\text{Ran}(Q)} \|\Delta_R\,\psi\|^2_{L^2(\mu)}\Big] \\
&\le V \sum\nolimits_{R\in\text{Ran}(Q)} \|\Delta_R\,\psi\|^2_{L^2(\mu)}.
\end{aligned}$$

Putting the last two results together, we obtain

$$\begin{aligned}
\text{LHS} \le U\sqrt{V} &\Big\{\sum\nolimits_{Q\in\text{Dom}} \|\Delta_Q\,\varphi\|^2_{L^2(\mu)}\Big\}^{1/2} \\
&\Big\{\sum\nolimits_{Q\in\text{Dom}} \sum\nolimits_{R\in\text{Ran}(Q)} \|\Delta_R\,\psi\|^2_{L^2(\mu)}\Big\}^{1/2}.
\end{aligned}$$

Switching the order of the double sum in the above and then using the second part of Lemma 7.29, we see that

$$\begin{aligned}
\sum\nolimits_{Q\in\text{Dom}} \sum\nolimits_{R\in\text{Ran}(Q)} \|\Delta_R\,\psi\|^2_{L^2(\mu)} &= \sum\nolimits_{R\in\text{Ran}} \sum\nolimits_{Q\in\text{Dom}(R)} \|\Delta_R\,\psi\|^2_{L^2(\mu)} \\
&= \sum\nolimits_{R\in\text{Ran}} [\#\text{Dom}(R)] \|\Delta_R\,\psi\|^2_{L^2(\mu)} \\
&\le W \sum\nolimits_{R\in\text{Ran}} \|\Delta_R\,\psi\|^2_{L^2(\mu)}.
\end{aligned}$$

Putting the last two results together and invoking the Adapted Martingale Decomposition (7.15), we obtain

$$\text{LHS} \le U\sqrt{VW} \Big\{\sum\nolimits_{Q\in\text{Dom}} \|\Delta_Q\,\varphi\|^2_{L^2(\mu)}\Big\}^{1/2} \Big\{\sum\nolimits_{R\in\text{Ran}} \|\Delta_R\,\psi\|^2_{L^2(\mu)}\Big\}^{1/2}$$

$$\leq U\sqrt{VW}\ \{2\|\varphi\|^2_{L^2(\mu)}\}^{1/2}\ \{2\|\psi\|^2_{L^2(\mu)}\}^{1/2}.$$

Clearly $\widetilde{N}_1 = 2U\sqrt{VW}$ will do. □

We have called $\mathcal{G}_1$ "the crudely handled terms" not simply because we are estimating the sum of the absolute values of terms $\langle \mathcal{K}_\Theta \Delta_Q\,\varphi, \Delta_R\,\psi\rangle$ corresponding to (Q, R) from $\mathcal{G}_1$ as opposed to the absolute value of the sum of these terms, but rather because the absolute value of each term $\langle \mathcal{K}_\Theta \Delta_Q\,\varphi, \Delta_R\,\psi\rangle$ was estimated by $U\ \|\Delta_Q\,\varphi\|_{L^2(\mu)}\ \|\Delta_R\,\psi\|_{L^2(\mu)}$ with U *independent* of Q and R. This only succeeded in giving us an effective estimate when we summed over all (Q, R) from $\mathcal{G}_1$ because each Q involved in $\mathcal{G}_1$ interacted with at most a fixed constant number of Rs involved in $\mathcal{G}_1$ and vice versa. In the next section we turn to the $\mathcal{G}_2$ terms which cannot be handled so simply!

7.14 $\mathcal{G}_2$: The Distantly Interacting Terms

The pairs (Q, R) from $\mathcal{G}_2$ are of two types: those with Q not too small compared to R ... for which we know Q and R are "far apart" ... and those with Q very small compared to R ... for which we know Q and R are disjoint. For either type we will obtain an estimate of the form $|\langle \mathcal{K}_\Theta \Delta_Q\,\varphi, \Delta_R\,\psi\rangle| \leq 27\ T_{Q,R}\ \|\Delta_Q\,\varphi\|_{L^2(\mu)}\ \|\Delta_R\,\psi\|_{L^2(\mu)}$ with the numbers $T_{Q,R}$ being small enough to enable us to extract an appropriate estimate upon summing.

Given two dyadic squares Q and R, define the *long distance* $D(Q, R)$ between them by

$$D(Q, R) = l(Q) + \operatorname{dist}(Q, R) + l(R)$$

and set

$$T_{Q,R} = \frac{\sqrt{l(Q)}\sqrt{l(R)}}{D(Q, R)^2}\sqrt{\mu(Q)}\sqrt{\mu(R)}.$$

Lemma 7.31 (Distant Interaction) *Let Q and R be dyadic squares for which $l(Q) \leq l(R)$ and let Θ be a Lip(1)-function for which $\inf_{\mathbb{C}} \Theta > 0$. Suppose $\varphi_Q, \psi_R \in L^2(\mu)$ are such that $\varphi_Q = 0$ off Q and $\psi_R = 0$ off R with $\int_{\mathbb{C}} \varphi_Q\, d\mu = 0$ and* $\operatorname{dist}(Q, \operatorname{spt}(\psi_R)) \geq l(Q)^{1/4} l(R)^{3/4}$. *Then*

$$|\langle \mathcal{K}_\Theta \varphi_Q, \psi_R\rangle| \leq 27\ T_{Q,R}\ \|\varphi_Q\|_{L^2(\mu)}\ \|\psi_R\|_{L^2(\mu)}.$$

Note that we have not assumed that Q and R are "far apart" but rather that Q and $\operatorname{spt}(\psi_R)$ are "far apart." Indeed, in estimating some $\mathcal{G}_3$ terms later on we will in effect use this lemma through Proposition 7.34 on pairs (Q, R) for which Q and R intersect, yet Q and $\operatorname{spt}(\psi_R)$ are "far apart"!

Proof Denote the center of Q by ζ_Q. Given $z \in \operatorname{spt}(\psi_R)$, let q be the intersection point of the segment $[\zeta_Q, z]$ with ∂Q. Note that

$$\frac{1}{2}l(Q) \leq |\zeta_Q - q| \text{ and } l(Q) \leq l(Q)^{1/4}l(R)^{3/4} \leq \operatorname{dist}(Q, \operatorname{spt}(\psi_R)) \leq |q - z|.$$

Thus for any $\zeta \in Q$, we have

$$|\zeta_Q - \zeta| \leq \frac{1}{\sqrt{2}}l(Q) < \frac{3}{4}l(Q) = \frac{1}{2}\left\{\frac{1}{2}l(Q) + l(Q)\right\} \leq \frac{1}{2}\{|\zeta_Q - q| + |q - z|\} = \frac{1}{2}|\zeta_Q - z|.$$

Hence from (f) of Proposition 6.10,

$$|k_\Theta(\zeta, z) - k_\Theta(\zeta_Q, z)| \leq \frac{4|\zeta_Q - \zeta|}{|\zeta_Q - z|^2} \leq \frac{3\,l(Q)}{|\zeta_Q - z|^2} \leq \frac{3\,l(Q)}{\operatorname{dist}(Q, \operatorname{spt}(\psi_R))^2}.$$

It now follows from our hypothesis that φ_Q integrates to 0, the estimate just gotten, and the Cauchy–Schwarz Inequality [RUD, 3.5] that

$$\begin{aligned}
|\langle \mathcal{K}_\Theta \varphi_Q, \psi_R \rangle| &= \left| \int_{\mathbb{C}} \int_{\mathbb{C}} k_\Theta(\zeta, z) \varphi_Q(\zeta) \psi_R(z)\, d\mu(\zeta)\, d\mu(z) \right| \\
&= \left| \int_{\mathbb{C}} \int_{\mathbb{C}} \{k_\Theta(\zeta, z) - k_\Theta(\zeta_Q, z)\} \varphi_Q(\zeta) \psi_R(z)\, d\mu(\zeta)\, d\mu(z) \right| \\
&\leq \frac{3\,l(Q)}{\operatorname{dist}(Q, \operatorname{spt}(\psi_R))^2} \left\{ \int_{\mathbb{C}} |\varphi_Q|\, d\mu \right\} \left\{ \int_{\mathbb{C}} |\psi_R|\, d\mu \right\} \\
&\leq \frac{3\,l(Q)}{\operatorname{dist}(Q, \operatorname{spt}(\psi_R))^2} \left\{ \sqrt{\mu(Q)}\, \|\varphi_Q\|_{L^2(\mu)} \right\} \left\{ \sqrt{\mu(R)}\, \|\psi_R\|_{L^2(\mu)} \right\}.
\end{aligned}$$

Comparing this with what is to be proven, we see that it now suffices to show that

$$\frac{l(Q)}{\operatorname{dist}(Q, \operatorname{spt}(\psi_R))^2} \leq \frac{9\sqrt{l(Q)}\sqrt{l(R)}}{D(Q, R)^2}.$$

Case One ($l(R) \leq \operatorname{dist}(Q, \operatorname{spt}(\psi_R))$). Then $D(Q, R) \leq 3 \operatorname{dist}(Q, \operatorname{spt}(\psi_R))$, so

$$\frac{l(Q)}{\operatorname{dist}(Q, \operatorname{spt}(\psi_R))^2} \leq \frac{9\,l(Q)}{D(Q, R)^2} \leq \frac{9\sqrt{l(Q)}\sqrt{l(R)}}{D(Q, R)^2}.$$

Case Two ($l(R) > \operatorname{dist}(Q, \operatorname{spt}(\psi_R))$). Then $D(Q, R) \leq 3\,l(R)$, so

$$\frac{l(Q)}{\operatorname{dist}(Q, \operatorname{spt}(\psi_R))^2} \leq \frac{l(Q)}{l(Q)^{1/2}l(R)^{3/2}} = \frac{\sqrt{l(Q)}\sqrt{l(R)}}{l(R)^2} \leq \frac{9\sqrt{l(Q)}\sqrt{l(R)}}{D(Q, R)^2}.$$

□

An inspection of the last proof yields up the following result which will be needed toward the end of the chapter to prove Lemma 7.40.

Scholium 7.32 *Dropping the hypothesis $\psi_R = 0$ off R from the Distant Interaction Lemma (7.31), one has*

$$|\langle \mathcal{K}_\Theta \varphi_Q, \psi_R \rangle| \le 3\, l(Q) \sqrt{\mu(Q)}\, \|\varphi_Q\|_{L^2(\mu)} \left\{ \int_{\mathbb{C}} \frac{|\psi_R(z)|}{|\zeta_Q - z|^2}\, d\mu(z) \right\}$$

where ζ_Q denotes the center of Q.

For the rest of this section, $\sum_{(Q,R)}$ denotes a sum over all $(Q, R) \in \mathcal{D}^{\text{tran}}_{\omega_1} \times \mathcal{D}^{\text{tran}}_{\omega_2}$ such that $l(Q) \le l(R)$, i.e., a sum over all $(Q, R) \in \mathcal{G}$, $\sum_{Q}$ denotes a sum over all $Q \in \mathcal{D}^{\text{tran}}_{\omega_1}$, and $\sum_{R}$ denotes a sum over all $R \in \mathcal{D}^{\text{tran}}_{\omega_2}$.

Lemma 7.33 **($T_{Q,R}$)** *For all nonnegative "sequences" $\{a_Q\}_{Q \in \mathcal{D}^{\text{tran}}_{\omega_1}}$ and $\{b_R\}_{R \in \mathcal{D}^{\text{tran}}_{\omega_2}}$,*

$$\sum_{(Q,R)} T_{Q,R}\, a_Q\, b_R \le 180M \left\{ \sum_{Q} a_Q^2 \right\}^{1/2} \left\{ \sum_{R} b_R^2 \right\}^{1/2}.$$

Proof Given $j, k \ge 0$, $\sum_{(Q,R)}^{(j,k)}$ denotes a sum over all $(Q, R) \in \mathcal{D}^{\text{tran}}_{\omega_1} \times \mathcal{D}^{\text{tran}}_{\omega_2}$ such that $l(Q) = 2^{-j}\, l(R) = 2^{-(j+k)}$ (so $l(R)$ must be 2^{-k}), $\sum_{Q}^{j+k}$ denotes a sum over all $Q \in \mathcal{D}^{\text{tran}}_{\omega_1}$ such that $l(Q) = 2^{-(j+k)}$, and $\sum_{R}^{k}$ denotes a sum over all $R \in \mathcal{D}^{\text{tran}}_{\omega_2}$ such that $l(R) = 2^{-k}$.

Our proof will proceed in a backward manner!

Claim. To prove the lemma, it suffices to show

$$(\star) \quad \sum_{(Q,R)}^{(j,k)} T_{Q,R}\, a_Q\, b_R \le \frac{45M}{2^{j/2}} \left\{ \sum_{Q}^{j+k} a_Q^2 \right\}^{1/2} \left\{ \sum_{R}^{k} b_R^2 \right\}^{1/2}.$$

Verification. The left-hand side of $(\star)$ summed over all j and $k \ge 0$ becomes the left-hand side of what we wish to prove. Summing the right-hand side of $(\star)$ over all $k \ge 0$ and using the Cauchy–Schwarz Inequality [RUD, 3.5], we obtain

$$\frac{45M}{2^{j/2}}\sum_{k\geq 0}\left\{\sum_{Q}^{j+k} a_Q^2\right\}^{1/2}\left\{\sum_{R}^{k} b_R^2\right\}^{1/2} \leq \frac{45M}{2^{j/2}}\left\{\sum_{k\geq 0}\sum_{Q}^{j+k} a_Q^2\right\}^{1/2}\left\{\sum_{k\geq 0}\sum_{R}^{k} b_R^2\right\}^{1/2}$$

$$\leq \frac{45M}{2^{j/2}}\left\{\sum_{Q} a_Q^2\right\}^{1/2}\left\{\sum_{R} b_R^2\right\}^{1/2}.$$

Since $\sum_{j\geq 0} 1/2^{j/2} = 1/(1-1/\sqrt{2}) \leq 4$, we thus see that the right-hand side of ($\star$) summed over all j and $k \geq 0$ is less than or equal to the right-hand side of what we wish to prove.

Claim. ($\star$) can be reformulated as

$$(\star\star)\quad \int_{\mathbb{C}}\int_{\mathbb{C}} k_{(j,k)}(\zeta,z)\varphi_{j+k}(\zeta)\psi_k(z)\,d\mu(\zeta)\,d\mu(z) \leq \frac{45M}{2^{j/2}}\,\|\varphi_{j+k}\|_{L^2(\mu)}\,\|\psi_k\|_{L^2(\mu)}$$

where

$$k_{(j,k)}(\zeta,z) = \sum_{(Q,R)}^{(j,k)} \frac{T_{Q,R}}{\sqrt{\mu(Q)}\sqrt{\mu(R)}}\mathcal{X}_Q(\zeta)\mathcal{X}_R(z),$$

$$\varphi_{j+k}(\zeta) = \sum_{Q}^{j+k}\frac{a_Q}{\sqrt{\mu(Q)}}\mathcal{X}_Q(\zeta), \text{ and } \psi_k(z) = \sum_{R}^{k}\frac{b_R}{\sqrt{\mu(R)}}\mathcal{X}_R(z).$$

Verification. Since squares of the same size in a dyadic lattice are pairwise disjoint, we have $\varphi_{j+k}^2 = \sum_{Q}^{j+k}\frac{a_Q^2}{\mu(Q)}\mathcal{X}_Q$. Thus $\|\varphi_{j+k}\|_{L^2(\mu)} = \left\{\sum_{Q}^{j+k} a_Q^2\right\}^{1/2}$. Similarly, $\|\psi_k\|_{L^2(\mu)} = \left\{\sum_{R}^{k} b_R^2\right\}^{1/2}$. Lastly,

$$\int_{\mathbb{C}}\int_{\mathbb{C}} k_{(j,k)}(\zeta,z)\varphi_{j+k}(\zeta)\psi_k(z)\,d\mu(\zeta)\,d\mu(z)$$

$$= \sum_{(Q,R)}^{(j,k)}\frac{T_{Q,R}}{\sqrt{\mu(Q)}\sqrt{\mu(R)}}\left\{\int_Q \varphi_{j+k}(\zeta)\,d\mu(\zeta)\right\}\left\{\int_R \psi_k(z)\,d\mu(z)\right\}$$

$$= \sum_{(Q,R)}^{(j,k)}\frac{T_{Q,R}}{\sqrt{\mu(Q)}\sqrt{\mu(R)}}\left\{\frac{a_Q}{\sqrt{\mu(Q)}}\int_Q d\mu\right\}\left\{\frac{b_R}{\sqrt{\mu(R)}}\int_R d\mu\right\}$$

$$= \sum_{(Q,R)}^{(j,k)} T_{Q,R}\, a_Q\, b_R.$$

Claim. To prove $(\star\star)$, it suffices to show

$$(\star\star\star) \quad \int_{\mathbb{C}} \frac{r}{\{|\zeta - z| + r\}^2}\, d\mu(\zeta) \leq 5M$$

for all $z \in \mathbb{C}$ and $r > 0$.

Verification. Given $j, k \geq 0$, let $\zeta \in \mathbb{C}$ be such that it is contained in some square Q of $\mathcal{D}^{\text{tran}}_{\omega_1}$ with edge length $2^{-(j+k)}$ and let $z \in \mathbb{C}$ be such that it is contained in some square R of $\mathcal{D}^{\text{tran}}_{\omega_2}$ with edge length 2^{-k}. Then $|\zeta - z| + 2^{-k} = |\zeta - z| + l(R) \leq 3D(Q, R)$ and so

$$k_{(j,k)}(\zeta, z) = \frac{T_{Q,R}}{\sqrt{\mu(Q)}\sqrt{\mu(R)}} = \frac{\sqrt{l(Q)}\sqrt{l(R)}}{D(Q,R)^2} = \frac{2^{-j/2} \cdot 2^{-k}}{D(Q,R)^2} \leq \frac{9}{2^{j/2}} \cdot \frac{2^{-k}}{\{|\zeta - z| + 2^{-k}\}^2}.$$

Note that this estimate on $k_{(j,k)}(\zeta, z)$ is trivially true for all other $\zeta, z \in \mathbb{C}$. Thus we may use it and $(\star\star\star)$ with $r = 2^{-k}$, to obtain

$$(\dagger) \quad \int_{\mathbb{C}} k_{(j,k)}(\zeta, z)\, d\mu(\zeta) \leq \frac{9}{2^{j/2}} \int_{\mathbb{C}} \frac{2^{-k}}{\{|\zeta - z| + 2^{-k}\}^2}\, d\mu(\zeta) \leq \frac{45M}{2^{j/2}}$$

for all $z \in \mathbb{C}$. Since $(\star\star\star)$ is invariant upon switching ζ and z, we also have

$$(\ddagger) \quad \int_{\mathbb{C}} k_{(j,k)}(\zeta, z)\, d\mu(z) \leq \frac{9}{2^{j/2}} \int_{\mathbb{C}} \frac{2^{-k}}{\{|\zeta - z| + 2^{-k}\}^2}\, d\mu(z) \leq \frac{45M}{2^{j/2}}$$

for all $\zeta \in \mathbb{C}$. Because of $(\dagger)$ and $(\ddagger)$, the corollary to the Schur Test (7.25) now gives us $(\star\star)$.

Claim. $(\star\star\star)$ is indeed true!

Verification. Invoking the linear growth of μ, we quickly see that

$$\int_{B(z;r)} \frac{r}{\{|\zeta - z| + r\}^2}\, d\mu(\zeta) \leq \frac{r}{r^2} Mr = M.$$

Moreover, setting $A_k = B(z; 2^{k+1}r) \setminus B(z; 2^k r)$ and again invoking the linear growth of μ, we have

$$\begin{aligned}
\int_{\mathbb{C}\setminus B(z;r)} \frac{r}{\{|\zeta - z| + r\}^2}\, d\mu(\zeta) &\leq \sum_{k\geq 0} \int_{A_k} \frac{r}{|\zeta - z|^2}\, d\mu(\zeta) \\
&\leq \sum_{k\geq 0} \frac{r}{(2^k r)^2} M(2^{k+1} r) = 2M \sum_{k\geq 0} \frac{1}{2^k} = 4M.
\end{aligned}$$

Adding the last two displayed inequalities together, we obtain $(\star\star\star)$. □

For the rest of this section, ω_1, ω_2, Θ, φ, and ψ are as in the Final Good Function Estimate (7.23). The next proposition will end up being used thrice before the chapter's end!

Proposition 7.34 *Let $\mathcal{H} \subseteq \mathcal{G}$ be such that for each $(Q, R) \in \mathcal{H}$ there is an associated subset $A(Q, R)$ of $\mathbb{C}$. Suppose that* $\operatorname{dist}(Q, \operatorname{spt}(\mathcal{X}_{A(Q,R)}\Delta_R\,\psi)) \geq l(Q)^{1/4}l(R)^{3/4}$ *whenever $(Q, R) \in \mathcal{H}$ is such that Q is ω_2-good. Then*

$$\sum\nolimits_{\mathcal{H}} |\langle \mathcal{K}_\Theta \Delta_Q\,\varphi, \mathcal{X}_{A(Q,R)}\Delta_R\,\psi\rangle| \leq 9{,}720M\ \|\varphi\|_{L^2(\mu)}\ \|\psi\|_{L^2(\mu)}.$$

Proof We claim that for each $(Q, R) \in \mathcal{H}$,

$$(\star)\ |\langle \mathcal{K}_\Theta \Delta_Q\,\varphi, \mathcal{X}_{A(Q,R)}\Delta_R\,\psi\rangle| \leq 27\ T_{Q,R}\ \|\Delta_Q\,\varphi\|_{L^2(\mu)}\ \|\Delta_R\,\psi\|_{L^2(\mu)}.$$

If $\Delta_Q\,\varphi = 0$, then $(\star)$ is trivially true, so by (4) of the Final Good Function Estimate (7.23) we may assume that Q is ω_2-good. But then the hypothesis of the proposition allows us to apply the Distant Interaction Lemma (7.31) to $\varphi_Q = \Delta_Q\,\varphi$ and $\psi_R = \mathcal{X}_{A(Q,R)}\Delta_R\,\psi$ and invoke the trivial inequality $\|\mathcal{X}_{A(Q,R)}\Delta_R\,\psi\|_{L^2(\mu)} \leq \|\Delta_R\,\psi\|_{L^2(\mu)}$ to get $(\star)$.

Using $(\star)$, the $T_{Q,R}$ Lemma (7.33) applied to $a_Q = \|\Delta_Q\,\varphi\|_{L^2(\mu)}$ and $b_R = \|\Delta_R\,\psi\|_{L^2(\mu)}$, and the Adapted Martingale Decomposition (7.15), we see that

$$\begin{aligned}
&\sum\nolimits_{\mathcal{H}} |\langle \mathcal{K}_\Theta \Delta_Q\,\varphi, \mathcal{X}_{A(Q,R)}\Delta_R\,\psi\rangle| \\
&\quad\leq 27 \sum\nolimits_{\mathcal{H}} T_{Q,R}\ \|\Delta_Q\,\varphi\|_{L^2(\mu)}\ \|\Delta_R\,\psi\|_{L^2(\mu)} \\
&\quad\leq 27 \sum_{(Q,R)} T_{Q,R}\ \|\Delta_Q\,\varphi\|_{L^2(\mu)}\ \|\Delta_R\,\psi\|_{L^2(\mu)} \\
&\quad\leq 27(180M) \left\{\sum_Q \|\Delta_Q\,\varphi\|^2_{L^2(\mu)}\right\}^{1/2} \left\{\sum_R \|\Delta_R\,\psi\|^2_{L^2(\mu)}\right\}^{1/2} \\
&\quad\leq 27(180M)\{2\|\varphi\|^2_{L^2(\mu)}\}^{1/2}\{2\|\psi\|^2_{L^2(\mu)}\}^{1/2} \\
&\quad= 9{,}720M\ \|\varphi\|_{L^2(\mu)}\ \|\psi\|_{L^2(\mu)}.
\end{aligned}$$

□

Proposition 7.35 (Estimate on the $\mathcal{G}_2$ Sum) *There exists a positive finite number $\widetilde{N}_2$ such that*

$$\sum_{\mathcal{G}_2} |\langle \mathcal{K}_\Theta \Delta_Q \varphi, \Delta_R \psi \rangle| \le \widetilde{N}_2 \, \|\varphi\|_{L^2(\mu)} \, \|\psi\|_{L^2(\mu)}$$

for all ω_1, ω_2, Θ, φ, *and* ψ *as in the Final Good Function Estimate* (7.23).

Proof This is a simple application of the last proposition with $\mathcal{H} = \mathcal{G}_2$ and $A(Q, R) = \mathbb{C}$ always. Clearly $\widetilde{N}_2 = 9{,}720M$ will then work. Since $\operatorname{spt}(\mathcal{X}_{A(Q,R)} \Delta_R \psi) \subseteq R$, it suffices to check that $\operatorname{dist}(Q, R) \ge l(Q)^{1/4} l(R)^{3/4}$ whenever $(Q, R) \in \mathcal{G}_2$ is such that Q is ω_2-good.

Case One (($Q, R) \in \mathcal{G}_2$ *by the first clause of the definition*). Then $\operatorname{dist}(Q, R) \ge l(R)$. Since $l(R) \ge l(Q)$, we also have $l(R) \ge l(Q)^{1/4} l(R)^{3/4}$ and so we are done.

Case Two (($Q, R) \in \mathcal{G}_2$ by the second clause of the definition). Since $Q \cap R = \emptyset$, we have $\operatorname{dist}(Q, R) = \operatorname{dist}(Q, \partial R)$. Since $l(R) > 2^m\, l(Q)$ and Q is ω_2-good, we also have $\operatorname{dist}(Q, \partial R) \ge 16\, l(Q)^{1/4} l(R)^{3/4}$ and so we are again done. □

7.15 Splitting Up the $\mathcal{G}_3$ Terms

For any dyadic square R, let R_k, $k = 1, 2, 3, 4$, denote the four children of R labeled starting with the northeast one and proceeding counterclockwise for the sake of definiteness.

Considering any $(Q, R) \in \mathcal{G}_3$, we have $Q \cap R \ne \emptyset$ by definition and so $Q \cap R_k \ne \emptyset$ for some k. Define $R_Q = R_k$ where k is the smallest index such that $Q \cap R_k \ne \emptyset$.

Now suppose that in addition Q is ω_2-good. Since $(Q, R) \in \mathcal{G}_3$, we have $l(Q) < 2^{-m}\, l(R)$ by definition, and so $l(R_Q) \ge 2^m\, l(Q)$. Since Q is ω_2-good, this forces $\operatorname{dist}(Q, \partial R_Q) \ge 16\, l(Q)^{1/4} l(R_Q)^{3/4} > 0$. But Q is much smaller than R_Q, so Q must lie totally within the interior of R_Q and it is all the same whether we say $Q \cap R_Q \ne \emptyset$, $Q \subseteq R_Q$, or $Q \subseteq \operatorname{int} R_Q$. Thus when Q is ω_2-good, $(Q, R) \in \mathcal{G}_3$ if and only if $R \in \mathcal{D}_{\omega_2}^{\text{tran}}$, $l(Q) < 2^{-m}\, l(R)$, and Q is totally contained in one of the four children of R, namely R_Q. Note that in this situation we also have $\operatorname{dist}(Q, \partial R_Q) > 8\, l(Q)^{1/4} l(R)^{3/4}$.

We may now write $\mathcal{G}_3 = \mathcal{G}_3^{\text{term}} \cup \mathcal{G}_3^{\text{tran}}$ disjointly where

$$\mathcal{G}_3^{\text{term}} = \{(Q, R) \in \mathcal{G}_3 : R_Q \text{ is terminal}\} \text{ and } \mathcal{G}_3^{\text{tran}} = \{(Q, R) \in \mathcal{G}_3 : R_Q \text{ is transit}\}.$$

Thus to estimate $\left|\sum_{\mathcal{G}_3} \langle \mathcal{K}_\Theta \Delta_Q \varphi, \Delta_R \psi \rangle\right|$ it clearly suffices to separately estimate

$$\sum_{\mathcal{G}_3^{\text{term}}} |\langle \mathcal{K}_\Theta \Delta_Q \varphi, \Delta_R \psi \rangle| \text{ and } \left|\sum_{\mathcal{G}_3^{\text{tran}}} \langle \mathcal{K}_\Theta \Delta_Q \varphi, \Delta_R \psi \rangle\right|.$$

7.16 $\mathcal{G}_3^{\text{term}}$: The Suppressed Kernel Terms

To estimate the sum of the terms corresponding to $\mathcal{G}_3^{\text{term}}$ we need three lemmas, the first of which is a very special case of a well-known result from analysis and requires a preliminary definition and some notation.

Say that a dyadic square S is *nicely placed* with respect to another dyadic square R^* if $S \subseteq R^*$ and $\text{dist}(S, \partial R^*) = l(S)$. Recall that $2S$ denotes the dyadic square concentric with S whose edge length is double that of S. Note that when S is nicely placed with respect to R^*, we have $2S \subseteq R^*$ and $\text{dist}(2S, \partial R^*) = l(S)/2 = l(2S)/4$.

Lemma 7.36 (Whitney Decomposition for a Dyadic Square) *For any dyadic square R^* there exists a subcollection $W(R^*)$ of dyadic descendents of R^* with the following properties:*

(a) *$W(R^*)$ is a partition of* int R^*,
(b) *each $S \in W(R^*)$ is nicely placed with respect to R^*, and*
(c) *each point of $\mathbb{C}$ belongs to at most* 11 *of the doubled squares $2S$ corresponding to $S \in W(R^*)$.*

Proof We will work through the successive generations of R^* throwing into $W(R^*)$ all and only those squares of the generation under consideration that are nicely placed with respect to R^*.

None of the four children of R^* is nicely placed and so they contribute nothing to $W(R^*)$.

Of the 16 grandchildren of R^*, only the four central ones are nicely placed and so go into $W(R^*)$ giving us four pairwise disjoint dyadic descendents that cover all points of R^* whose distance to ∂R^* is greater than $l(R^*)/4$.

Of the 64 great-grandchildren of R^*, only the 20 which form a belt about the four previously selected squares are nicely placed and so go into $W(R^*)$ giving us 24 pairwise disjoint dyadic descendents that cover all points of R^* whose distance to ∂R^* is greater than $l(R^*)/8$.

Of the 256 great-great-grandchildren of R^*, only the 52 which form a belt about the 24 previously selected squares are nicely placed and so go into $W(R^*)$ giving us 76 pairwise disjoint dyadic descendents that cover all points of R^* whose distance to ∂R^* is greater than $l(R^*)/16$.

Proceeding through all the generations in this manner, we obtain $W(R^*)$. It is clear by our construction that (a) and (b) are satisfied (with regard to (a) it is important to note that our squares are half-closed as in the proof of Frostman's Lemma (2.9)).

Any point $z \notin \text{int } R^*$ is contained in none of the doubles of the squares of $W(R^*)$ since each of these doubles is contained in int R^*. So consider $z \in \text{int } R^*$. Let S be the square of $W(R^*)$ that contains z. Say that one square "touches" another if the closures of the two squares intersect. Clearly any square from $W(R^*)$ that touches S has a double that intersects S. Also, the squares of $W(R^*)$ that touch S, excluding

S itself, form a belt around S thick enough so that any square from $W(R^*)$ that does not touch S has a double that does not intersect S (here it is again important to note that our squares are half-closed as in the proof of Frostman's Lemma (2.9)). Thus the double of a square of $W(R^*)$ intersects S if and only if it touches S. It follows that the number of squares of $W(R^*)$ whose doubles contain z is at most the number of squares of $W(R^*)$ which touch S. Call a square of $W(R^*)$ "diagonal" if one of its diagonals is contained in one of the diagonals of R^*. Any diagonal square of $W(R^*)$ is touched by 11 squares of $W(R^*)$ (including itself). Any nondiagonal square of $W(R^*)$ is touched by nine squares of $W(R^*)$ (including itself). Thus S is touched by at most 11 squares of $W(R^*)$ and so (c) follows. □

Lemma 7.37 *Suppose that $\eta_1, \eta_2 \in L^2(\mu)$. Then for any $\mathcal{A} \subseteq \mathcal{D}_\omega^{\text{tran}}$,*

$$\sum_{Q\in\mathcal{A}} |\langle \Delta_Q\, \eta_1, \eta_2\rangle| \le \left\{2\sum_{Q\in\mathcal{A}} \|\Delta_Q\, \eta_1\|^2_{L^2(\mu)}\right\}^{1/2} \|\eta_2\|_{L^2(\mu)}.$$

Proof For each $Q \in \mathcal{A}$, choose a complex number c_Q with modulus one such that $c_Q\langle \Delta_Q\, \eta_1, \eta_2\rangle = |\langle \Delta_Q\, \eta_1, \eta_2\rangle|$. Then by the Cauchy–Schwarz Inequality [RUD, 3.5] and Proposition 7.17,

$$\begin{aligned}\sum_{Q\in\mathcal{A}} |\langle \Delta_Q\, \eta_1, \eta_2\rangle| &= \left\langle \sum_{Q\in\mathcal{A}} c_Q \Delta_Q\, \eta_1, \eta_2\right\rangle \\ &\le \left\|\sum_{Q\in\mathcal{A}} c_Q \Delta_Q\, \eta_1\right\|_{L^2(\mu)} \|\eta_2\|_{L^2(\mu)} \\ &\le \left\{2\sum_{Q\in\mathcal{A}} |c_Q|^2 \|\Delta_Q\, \eta_1\|^2_{L^2(\mu)}\right\}^{1/2} \|\eta_2\|_{L^2(\mu)} \\ &= \left\{2\sum_{Q\in\mathcal{A}} \|\Delta_Q\, \eta_1\|^2_{L^2(\mu)}\right\}^{1/2} \|\eta_2\|_{L^2(\mu)}.\end{aligned}$$

□

For the rest of this section, $\omega_1, \omega_2, \Theta, \varphi$, and ψ are as in the Final Good Function Estimate (7.23).

Lemma 7.38 (Nicely Placed Square) *Suppose that $R^* \in \mathcal{D}_{\omega_2}^{\text{term}}$ and S is a dyadic square nicely placed with respect to R^*. Then for any function $\psi \in L^2(\mu)$ such that $\psi = 0$ off $2S$,*

$$\|\mathcal{X}_{2S}\mathcal{K}_\Theta \psi\|_{L^2(\mu)} \le \frac{3M}{\delta}\|\psi\|_{L^2(\mu)}.$$

Proof For any $z \in 2S$, $\Theta(z) \ge \delta\, \Phi_{\omega_2}(z) \ge \delta\, \text{dist}(z, \partial R^*) \ge \delta\, \text{dist}(2S, \partial R^*) = \delta\, l(2S)/4$. Hence by (d) of Proposition 6.10, $|k_\Theta(\zeta, z)| \le 1/\Theta(z) \le 4/\{\delta\, l(2S)\}$ for every $\zeta \in \mathbb{C}$. Then by the Cauchy-Schwarz Inequality [RUD, 3.5],

$$|(\mathcal{K}_\Theta\psi)(z)| \leq \int_{2S} |k_\Theta(\zeta, z)||\psi(\zeta)|\, d\mu(\zeta) \leq \frac{4}{\delta\, l(2S)} \int_{2S} |\psi(\zeta)|\, d\mu(\zeta)$$
$$\leq \frac{4\sqrt{\mu(2S)}}{\delta\, l(2S)} \|\psi\|_{L^2(\mu)}.$$

This leads to

$$\|\mathcal{X}_{2S}\mathcal{K}_\Theta\psi\|_{L^2(\mu)} = \left\{\int_{2S} |(\mathcal{K}_\Theta\psi)(z)|^2\, d\mu(z)\right\}^{1/2} \leq \frac{4}{\delta} \cdot \frac{\mu(2S)}{l(2S)} \cdot \|\psi\|_{L^2(\mu)}.$$

Letting B denote the closed ball circumscribing $2S$ and invoking the linear growth of μ, we see that $\mu(2S) \leq \mu(B) \leq M\text{rad}(B) = M\{l(2S)/\sqrt{2}\} < \{3M/4\}l(2S)$ and so are done. □

Proposition 7.39 (Estimate on the $\mathcal{G}_3^{\text{term}}$ Sum) *There exists a positive finite number $\widetilde{N}_3^{\text{term}}$ such that*

$$\sum\nolimits_{\mathcal{G}_3^{\text{term}}} |\langle\mathcal{K}_\Theta\Delta_Q\,\varphi, \Delta_R\,\psi\rangle| \leq \widetilde{N}_3^{\text{term}}\, \|\varphi\|_{L^2(\mu)}\, \|\psi\|_{L^2(\mu)}$$

for all ω_1, ω_2, Θ, φ, and ψ as in the Final Good Function Estimate (7.23).

Proof Set $\mathcal{H} = \{(Q, R) \in \mathcal{G}_3^{\text{term}} : Q \text{ is } \omega_2\text{-good}\}$ and note that we need only estimate the sum of $|\langle\mathcal{K}_\Theta\Delta_Q\,\varphi, \Delta_R\,\psi\rangle|$ over all $(Q, R) \in \mathcal{H}$ by (4) of the Final Good Function Estimate (7.23). Recall from the last section that when $(Q, R) \in \mathcal{H}$, there exists a unique terminal child R_Q of R that contains Q. Moreover, we have $\text{dist}(Q, \partial R_Q) > 8\,l(Q)^{1/4}l(R)^{3/4}$. In this situation, denote the unique square of $W(R_Q)$ that contains ζ_Q, the center of Q, by $S(Q, R)$. Note that $8\,l(Q) \leq 8\,l(Q)^{1/4}l(R)^{3/4} < \text{dist}(Q, \partial R_Q) < \text{dist}(\zeta_Q, \partial R_Q) \leq 2l(S(Q, R))$. In consequence, $Q \subseteq 2S(Q, R)$ and, moreover,

$$\text{dist}(Q, \partial\{2S(Q, R)\}) \geq \frac{l(S(Q, R))}{2} - \frac{l(Q)}{2} > \frac{l(S(Q, R))}{4} > l(Q)^{1/4}l(R)^{3/4}.$$

Clearly

$$\sum\nolimits_{\mathcal{H}} |\langle\mathcal{K}_\Theta\Delta_Q\,\varphi, \Delta_R\,\psi\rangle| \leq \sum\nolimits_{\mathcal{H}} |\langle\mathcal{K}_\Theta\Delta_Q\,\varphi, \mathcal{X}_{R\setminus 2S(Q,R)}\Delta_R\,\psi\rangle| \leftarrow \text{(I)!}$$
$$+ \sum\nolimits_{\mathcal{H}} |\langle\mathcal{K}_\Theta\Delta_Q\,\varphi, \mathcal{X}_{2S(Q,R)}\Delta_R\,\psi\rangle|. \leftarrow \text{(II)!}$$

Since $\text{dist}(Q, \text{spt}(\mathcal{X}_{R\setminus 2S(Q,R)}\Delta_R\,\psi)) \geq \text{dist}(Q, \partial\{2S(Q, R)\}) > l(Q)^{1/4}l(R)^{3/4}$, we may apply Proposition 7.34 with $\mathcal{H}$ as defined above and $A(Q, R) = R \setminus 2S(Q, R)$ to deduce that

$$\text{(I)} \leq 9{,}720M\, \|\varphi\|_{L^2(\mu)}\, \|\psi\|_{L^2(\mu)}.$$

For the rest of this proof, we make the following notational conventions: first, $\sum_{R^*}$ denotes a sum over all $R^* \in \mathcal{D}_{\omega_2}^{\text{term}}$ (thus we are summing over all the terminal children R^* of all the squares $R \in \mathcal{D}_{\omega_2}^{\text{tran}}$); second, given such an R^*, $\sum_S$ denotes a sum over all $S \in W(R^*)$; and third, given such an R^* and such an S, $\sum_Q$ denotes a sum over all $Q \in \mathcal{D}_{\omega_1}^{\text{tran}}$ such that Q is ω_2-good, $l(Q) < 2^{-m}\{2\,l(R^*)\} = 2^{-m}\,l(R)$, and $\zeta_Q \in S$. The point of all this is the following:

$$(\dagger)\ \sum\nolimits_{\mathcal{H}} |\langle \mathcal{K}_\Theta \Delta_Q \varphi, \mathcal{X}_{2S(Q,R)} \Delta_R \psi \rangle| = \sum\nolimits_{R^*} \sum\nolimits_S \sum\nolimits_Q |\langle \mathcal{K}_\Theta \Delta_Q \varphi, \mathcal{X}_{2S} \Delta_R \psi \rangle|$$

with each term in the single sum on the left-hand side associated with a pair (Q, R) corresponding to the term in the triple sum on the right-hand side associated with $R^* = R_Q$, $S = S(Q, R)$, and $Q = Q$.

Given $R^* \in \mathcal{D}_{\omega_2}^{\text{term}}$ and $S \in W(R^*)$, using the antisymmetry of $\mathcal{K}_\Theta$, the containment of each Q in $2S$, and the last two lemmas, we obtain

$$(\ddagger)\ \sum\nolimits_Q |\langle \mathcal{K}_\Theta \Delta_Q \varphi, \mathcal{X}_{2S} \Delta_R \psi \rangle| = \sum\nolimits_Q |\langle \Delta_Q \varphi, \mathcal{X}_{2S} \mathcal{K}_\Theta (\mathcal{X}_{2S} \Delta_R \psi) \rangle|$$

$$\le \left\{ 2 \sum\nolimits_Q \|\Delta_Q \varphi\|_{L^2(\mu)}^2 \right\}^{1/2} \|\mathcal{X}_{2S} \mathcal{K}_\Theta (\mathcal{X}_{2S} \Delta_R \psi)\|_{L^2(\mu)}$$

$$\le \frac{3M}{\delta} \left\{ 2 \sum\nolimits_Q \|\Delta_Q \varphi\|_{L^2(\mu)}^2 \right\}^{1/2} \|\mathcal{X}_{2S}\, \Delta_R \psi\|_{L^2(\mu)}.$$

An immediate consequence of ($\dagger$), ($\ddagger$), and the Cauchy–Schwarz Inequality [RUD, 3.5] is

$$(\star)\quad \text{(II)} \le \frac{3M}{\delta} \sum\nolimits_{R^*} \sum\nolimits_S \left\{ 2 \sum\nolimits_Q \|\Delta_Q \varphi\|_{L^2(\mu)}^2 \right\}^{1/2} \left\{ \int_{2S} |\Delta_R \psi|^2 \, d\mu \right\}^{1/2}$$

$$\le \frac{3M}{\delta} \left\{ 2 \sum\nolimits_{R^*} \sum\nolimits_S \sum\nolimits_Q \|\Delta_Q \varphi\|_{L^2(\mu)}^2 \right\}^{1/2} \left\{ \sum\nolimits_{R^*} \sum\nolimits_S \int_{2S} |\Delta_R \psi|^2 \, d\mu \right\}^{1/2}.$$

Let (α) and (β) denote the first and second square roots occurring in the last line of ($\star$) respectively. Since the R^*s being summed over are pairwise disjoint and for each R^*, the Ss being summed over form a partition of int R^*, each $\|\Delta_Q \varphi\|_{L^2(\mu)}^2$ in (α) occurs only once and so

$$(\alpha) \le \left\{ 2 \sum_{Q \in \mathcal{D}_{\omega_1}^{\text{tran}}} \|\Delta_Q \varphi\|_{L^2(\mu)}^2 \right\}^{1/2} \le 2\, \|\varphi\|_{L^2(\mu)}$$

with the last inequality holding by the Adapted Martingale Decomposition (7.15).

By (c) of Lemma 7.36,

$$\sum\nolimits_S \int_{2S} |\Delta_R \psi|^2 \, d\mu \le 11 \int_{R^*} |\Delta_R \psi|^2 \, d\mu.$$

This, the pairwise disjointness of the R^*s being summed over, and another invocation of the Adapted Martingale Decomposition (7.15) allow us to deduce that

$$(\beta) \le \left\{ 11 \sum\nolimits_{R^*} \int_{R^*} |\Delta_R \psi|^2 \, d\mu \right\}^{1/2} \le \left\{ 11 \sum_{R \in \mathcal{D}_{\omega_2}^{\text{tran}}} \|\Delta_R \psi\|_{L^2(\mu)}^2 \right\}^{1/2} \le \sqrt{22} \, \|\psi\|_{L^2(\mu)}.$$

From $(\star)$ and our estimates on (α) and (β), it follows that

$$(\text{II}) \le \frac{6\sqrt{22}M}{\delta} \, \|\varphi\|_{L^2(\mu)} \, \|\psi\|_{L^2(\mu)} < \frac{30M}{\delta} \, \|\varphi\|_{L^2(\mu)} \, \|\psi\|_{L^2(\mu)}.$$

Clearly $\widetilde{N}_3^{\text{term}} = 9{,}720M + 30M/\delta = 30M(324 + 1/\delta)$ will do. □

7.17 $\mathcal{G}_3^{\text{tran}}$: The Telescoping Terms

For the whole of this section, ω_1, ω_2, Θ, φ, and ψ are as in the Final Good Function Estimate (7.23).

Before launching into three lemmas, we will first explain our basic approach to the estimation of the $\mathcal{G}_3^{\text{tran}}$ sum. Recall that when R_Q is transit we have defined $\Delta_R \psi$ to be $c_{Q,R} \, h$ on R_Q where

$$c_{Q,R} = \frac{\langle \psi \rangle_{R_Q}}{\langle h \rangle_{R_Q}} - \frac{\langle \psi \rangle_R}{\langle h \rangle_R}.$$

Now $\Delta_R \psi$ need not be this constant multiple of h on the rest of $\mathbb{C}$. We will make it so! Thus we will throw away $\Delta_R \psi|_{\mathbb{C}\setminus R_Q}$ and then add $c_{Q,R} \, h|_{\mathbb{C}\setminus R_Q}$ onto the remaining $\Delta_R \psi|_{R_Q} = c_{Q,R} \, h|_{R_Q}$ left behind in our $\mathcal{G}_3^{\text{tran}}$ sum to get $c_{Q,R} \, h$. Of course it must be shown that the sum of the terms removed and added are bounded by a constant times the L^2 norms of φ and ψ. Once this is done, we are left with a sum over appropriate pairs (Q, R) of terms of the form $\langle \mathcal{K}_\Theta \Delta_Q \varphi, c_{Q,R} \, h \rangle$. Using the antisymmetry of $\mathcal{K}_\Theta$ and then rewriting our sum over pairs as an iterated sum over appropriate Qs and then appropriate Rs, we obtain

$$-\sum\nolimits_Q \left\{ \sum\nolimits_R c_{Q,R} \right\} \langle \Delta_Q \varphi, \mathcal{K}_\Theta h \rangle.$$

Now the sum of the $c_{Q,R}$s over the appropriate Rs is a telescoping sum resulting in massive cancellation! When the smoke clears, we will put hypothesis (5) of Theorem 7.1 to crucial use on what remains to finish off the proof!

Lemma 7.40 *For* $(Q, R) \in \mathcal{G}_3^{\text{tran}}$,

$$|c_{Q,R}||\langle \mathcal{K}_\Theta \Delta_Q \varphi, \mathcal{X}_{\mathbb{C}\setminus R_Q} h\rangle| \le \frac{M}{4}\sqrt{\frac{l(Q)}{l(R)}}\sqrt{\frac{\mu(Q)}{\mu(R_Q)}}\, \|\Delta_Q \varphi\|_{L^2(\mu)}\, \|\Delta_R \psi\|_{L^2(\mu)}.$$

Proof If $\Delta_Q \varphi = 0$, then our conclusion follows trivially, so by (4) of the Final Good Function Estimate (7.23) we may assume that Q is ω_2-good. Recall from Section 7.15 that in this situation $\text{dist}(Q, \partial R_Q) > 8\, l(Q)^{1/4} l(R)^{3/4}$. Since R_Q is transit, by (a) of Lemma 7.10 and (c) of Lemma 7.6,

$$(\star)\ |c_{Q,R}| \le \frac{|c_{Q,R}||\langle h\rangle_{R_Q}|}{1-\delta} = \frac{|\langle \Delta_R \psi\rangle_{R_Q}|}{1-\delta} \le \frac{\sqrt{\langle |\Delta_R \psi|^2\rangle_{R_Q}}}{1-\delta} \le \frac{1}{1-\delta}\frac{\|\Delta_R \psi\|_{L^2(\mu)}}{\sqrt{\mu(R_Q)}}.$$

Since $\text{dist}(Q, \text{spt}(\mathcal{X}_{\mathbb{C}\setminus R_Q} h)) \ge \text{dist}(Q, \partial R_Q) > l(Q)^{1/4} l(R)^{3/4}$, we may apply Scholium 7.32 with $\varphi_Q = \Delta_Q \varphi$ and $\psi_R = \mathcal{X}_{\mathbb{C}\setminus R_Q} h$ to obtain

$$(\star\star)\ |\langle \mathcal{K}_\Theta \Delta_Q \varphi, \mathcal{X}_{\mathbb{C}\setminus R_Q} h\rangle| \le 3\, l(Q)\sqrt{\mu(Q)}\, \|\Delta_Q \varphi\|_{L^2(\mu)} \left\{\int_{\mathbb{C}\setminus R_Q} \frac{|h(z)|}{|\zeta_Q - z|^2}\, d\mu(z)\right\}.$$

Let $R^{(k)}$, $k = 0, 1, \ldots, N$, denote the sequence of squares from $\mathcal{D}_{\omega_2}^{\text{tran}}$ ascending from R_Q to R^0. Thus $R^{(0)} = R_Q$, $R^{(1)} = R$, each $R^{(k+1)}$ contains $R^{(k)}$ with $l(R^{(k+1)}) = 2l(R^{(k)})$, and $R^{(N)} = R^0$. Since Q is ω_2-good and $2^m\, l(Q) \le l(R_Q) \le l(R^{(k)})$, we have

$$(\dagger)\ \text{dist}(Q, R^{(k+1)} \setminus R^{(k)}) = \text{dist}(Q, \partial R^{(k)}) \ge 16\, l(Q)^{1/4} l(R^{(k)})^{3/4}$$
$$> 8\, l(Q)^{1/4} l(R^{(k+1)})^{3/4}.$$

Also, by (a) of Lemma 7.10 and the linear growth of μ,

$$(\dagger\dagger) \int_{R^{(k+1)}} |h| d\mu = \langle |h|\rangle_{R^{(k+1)}} \mu(R^{(k+1)}) < (1+\delta)\mu(R^{(k+1)}) \le (1+\delta) M l(R^{(k+1)})/\sqrt{2}.$$

Lastly,

$$(\dagger\dagger\dagger)\ l(R^{(k+1)}) = 2^k\, l(R).$$

Note that as far as μ is concerned $\mathbb{C}\setminus R_Q = R^{(N)} \setminus R_Q$ since $\text{spt}(\mu) \subseteq R^0 = R^{(N)}$. From this fact, $(\dagger)$, $(\dagger\dagger)$, $(\dagger\dagger\dagger)$, and the geometric series formula, we thus obtain

$$(\star\star\star)\quad \int_{\mathbb{C}\setminus R_Q} \frac{|h(z)|}{|\zeta_Q - z|^2}\, d\mu(z) = \sum_{k=0}^{N-1} \int_{R^{(k+1)}\setminus R^{(k)}} \frac{|h(z)|}{|\zeta_Q - z|^2}\, d\mu(z)$$
$$\leq \frac{(1+\delta)M}{64\sqrt{2}\sqrt{l(Q)}\sqrt{l(R)}} \sum_{k=0}^{N-1} \left(\frac{1}{\sqrt{2}}\right)^k$$
$$\leq \frac{(1+\delta)M}{64(\sqrt{2}-1)\sqrt{l(Q)}\sqrt{l(R)}}.$$

Putting together $(\star)$, $(\star\star)$, and $(\star\star\star)$, we see that $|c_{Q,R}||\langle \mathcal{K}_\Theta \Delta_Q \varphi, \mathcal{X}_{\mathbb{C}\setminus R_Q} h\rangle|$ is at most

$$\frac{1+\delta}{1-\delta} \cdot \frac{3}{64(\sqrt{2}-1)} \cdot M \sqrt{\frac{l(Q)}{l(R)}} \sqrt{\frac{\mu(Q)}{\mu(R_Q)}}\, \|\Delta_Q \varphi\|_{L^2(\mu)}\, \|\Delta_R \psi\|_{L^2(\mu)}.$$

Since $(1+\delta)/(1-\delta) < 2$ and $3/\{64(\sqrt{2}-1)\} < 1/8$, we are done. □

For the next result we return to the summation notations of Lemma 7.33.

Lemma 7.41 *($\widetilde{T}_{Q,R}$) Set* $\widetilde{T}_{Q,R} = \sqrt{\frac{l(Q)}{l(R)}}\sqrt{\frac{\mu(Q)}{\mu(R_1)}}$ *for those* $(Q, R) \in \mathcal{D}_{\omega_1}^{\text{tran}} \times \mathcal{D}_{\omega_2}^{\text{tran}}$ *for which* $Q \subseteq R_1$ *with* R_1 *transit and set* $\widetilde{T}_{Q,R} = 0$ *otherwise. Then*

$$\sum_{(Q,R)} \widetilde{T}_{Q,R}\, a_Q\, b_R \leq \frac{5}{2} \left\{\sum_Q a_Q^2\right\}^{1/2} \left\{\sum_R b_R^2\right\}^{1/2}$$

for all nonnegative "sequences" $\{a_Q\}_{Q\in\mathcal{D}_{\omega_1}^{\text{tran}}}$ *and* $\{b_R\}_{R\in\mathcal{D}_{\omega_2}^{\text{tran}}}$.

Proof Let $\sum_R^{R_1\,\text{tran}}$ denote a sum over all $R \in \mathcal{D}_{\omega_2}^{\text{tran}}$ for which R_1 is transit. Given $R \in \mathcal{D}_{\omega_2}^{\text{tran}}$ with R_1 transit and $k \geq 1$, let $\sum_Q^{k,R}$ denote a sum over all $Q \in \mathcal{D}_{\omega_1}^{\text{tran}}$ for which $Q \subseteq R_1$ and $l(Q) = 2^{-k}\, l(R)$. Finally, given $k \geq 1$, let $\sum_{(Q,R)}^{k}$ denote a sum over all $(Q, R) \in \mathcal{D}_{\omega_1}^{\text{tran}} \times \mathcal{D}_{\omega_2}^{\text{tran}}$ for which $Q \subseteq R_1$ with R_1 transit and $l(Q) = 2^{-k}\, l(R)$. Clearly, $(\star)$ $\sum_{(Q,R)}^{k} \widetilde{T}_{Q,R}\, a_Q\, b_R = \sum_R^{R_1\,\text{tran}} \sum_Q^{k,R} \widetilde{T}_{Q,R}\, a_Q\, b_R$.

By the Cauchy–Schwarz Inequality [RUD, 3.5],

$$\sum_{Q}^{k,R}\sqrt{\frac{\mu(Q)}{\mu(R_1)}}\,a_Q \le \left\{\sum_{Q}^{k,R}\frac{\mu(Q)}{\mu(R_1)}\right\}^{1/2}\left\{\sum_{Q}^{k,R}a_Q^2\right\}^{1/2} \le \left\{\sum_{Q}^{k,R}a_Q^2\right\}^{1/2}$$

with the last inequality true since all the Qs occurring in our sums are pairwise disjoint and contained in R_1. It now follows that

$$\sum_{Q}^{k,R}\widetilde{T}_{Q,R}\,a_Q\,b_R \le 2^{-k/2}\left\{\sum_{Q}^{k,R}a_Q^2\right\}^{1/2} b_R$$

since for all the Qs occurring in our sums we have $\sqrt{l(Q)/l(R)} = 2^{-k/2}$. Summing the last inequality over all $R \in \mathcal{D}_{\omega_2}^{\text{tran}}$ for which R_1 is transit, using $(\star)$ and then the Cauchy–Schwarz Inequality [RUD, 3.5] once more, we obtain

$$\begin{aligned}\sum_{(Q,R)}^{k}\widetilde{T}_{Q,R}\,a_Q\,b_R &\le 2^{-k/2}\sum_{R}^{R_1\,\text{tran}}\left\{\sum_{Q}^{k,R}a_Q^2\right\}^{1/2} b_R\\ &\le 2^{-k/2}\left\{\sum_{R}^{R_1\,\text{tran}}\sum_{Q}^{k,R}a_Q^2\right\}^{1/2}\left\{\sum_{R}^{R_1\,\text{tran}}b_R^2\right\}^{1/2}\\ &\le 2^{-k/2}\left\{\sum_{Q}a_Q^2\right\}^{1/2}\left\{\sum_{R}b_R^2\right\}^{1/2}.\end{aligned}$$

(In passing from the middle line of the displayed array above to the last, the reader should be sure to understand why any Q occurring in the double sum of the middle line occurs there only once!)

Summing the last inequality over all $k \ge 1$ and noting that $\sum_{k\ge 1}2^{-k/2} = 1/(\sqrt{2}-1) < 5/2$, we are done. □

Lemma 7.42 *Suppose that $\eta_1, \eta_2 \in L^2(\mu)$. Then for any $\mathcal{B} \subseteq \{Q \in \mathcal{D}_{\omega}^{\text{tran}} : \|\Delta_Q\,\eta_1\|_{L^2(\mu)} > 0\}$,*

$$\sum_{Q\in\mathcal{B}}\frac{|\langle\Delta_Q\,\eta_1,\eta_2\rangle|^2}{\|\Delta_Q\,\eta_1\|_{L^2(\mu)}^2} \le 2\,\|\eta_2\|_{L^2(\mu)}^2.$$

Proof By the Cauchy–Schwarz Inequality [RUD, 3.5] and Proposition 7.17,

$$(\star)\quad \left|\sum_{Q\in\mathcal{B}}c_Q\,\langle\Delta_Q\,\eta_1,\eta_2\rangle\right|^2 = \left|\left\langle\sum_{Q\in\mathcal{B}}c_Q\,\Delta_Q\,\eta_1,\eta_2\right\rangle\right|^2$$

$$\leq \left\| \sum\nolimits_{Q \in \mathcal{B}} c_Q \, \Delta_Q \, \eta_1 \right\|_{L^2(\mu)}^2 \, \|\eta_2\|_{L^2(\mu)}^2$$

$$\leq \left\{ 2 \sum\nolimits_{Q \in \mathcal{B}} |c_Q|^2 \|\Delta_Q \, \eta_1\|_{L^2(\mu)}^2 \right\} \, \|\eta_2\|_{L^2(\mu)}^2 .$$

Substituting $c_Q = \dfrac{\overline{\langle \Delta_Q \, \eta_1, \eta_2 \rangle}}{\|\Delta_Q \, \eta_1\|_{L^2(\mu)}^2}$ into ($\star$), we obtain

$$\left\{ \sum\nolimits_{Q \in \mathcal{B}} \frac{|\langle \Delta_Q \, \eta_1, \eta_2 \rangle|^2}{\|\Delta_Q \, \eta_1\|_{L^2(\mu)}^2} \right\}^2 \leq \left\{ 2 \sum\nolimits_{Q \in \mathcal{B}} \frac{|\langle \Delta_Q \, \eta_1, \eta_2 \rangle|^2}{\|\Delta_Q \, \eta_1\|_{L^2(\mu)}^2} \right\} \|\eta_2\|_{L^2(\mu)}^2 .$$

Dividing both sides by the sum, we are done. □

Proposition 7.43 (Estimate on the $\mathcal{G}_3^{\text{tran}}$ Sum) *There exists a positive finite number $\widetilde{N}_3^{\text{tran}}$ such that*

$$\left| \sum\nolimits_{\mathcal{G}_3^{\text{tran}}} \langle \mathcal{K}_\Theta \Delta_Q \, \varphi, \Delta_R \, \psi \rangle \right| \leq \widetilde{N}_3^{\text{tran}} \, \|\varphi\|_{L^2(\mu)} \, \|\psi\|_{L^2(\mu)}$$

for all ω_1, ω_2, Θ, φ, and ψ as in the Final Good Function Estimate (7.23).

Proof Putting into effect the strategy outlined at the beginning of this section, by simple algebra we obtain

$$\left| \sum\nolimits_{\mathcal{G}_3^{\text{tran}}} \langle \mathcal{K}_\Theta \Delta_Q \, \varphi, \Delta_R \, \psi \rangle \right| \leq \sum\nolimits_{\mathcal{G}_3^{\text{tran}}} |\langle \mathcal{K}_\Theta \Delta_Q \, \varphi, \mathcal{X}_{\mathbb{C} \setminus R_Q} \Delta_R \, \psi \rangle| \leftarrow \text{(I)!}$$

$$+ \sum\nolimits_{\mathcal{G}_3^{\text{tran}}} |c_{Q,R}| |\langle \mathcal{K}_\Theta \Delta_Q \, \varphi, \mathcal{X}_{\mathbb{C} \setminus R_Q} h \rangle| \leftarrow \text{(II)!}$$

$$+ \left| \sum\nolimits_{\mathcal{G}_3^{\text{tran}}} c_{Q,R} \, \langle \Delta_Q \, \varphi, \mathcal{K}_\Theta h \rangle \right| . \leftarrow \text{(III)!}$$

Estimation of (I). Since $\text{dist}(Q, \text{spt}(\mathcal{X}_{\mathbb{C} \setminus R_Q} \Delta_R \, \psi)) \geq \text{dist}(Q, \partial R_Q) > l(Q)^{1/4} l(R)^{3/4}$ whenever Q is ω_2-good, we may apply Proposition 7.34 with $\mathcal{H} = \mathcal{G}_3^{\text{tran}}$ and $A(Q, R) = \mathbb{C} \setminus R_Q$ to deduce that

$$\text{(I)} \leq 9{,}720 M \, \|\varphi\|_{L^2(\mu)} \, \|\psi\|_{L^2(\mu)} .$$

Estimation of (II). We may apply Lemma 7.40 to each term of (II). We may then apply Lemma 7.41 with $a_Q = \|\Delta_Q \, \varphi\|_{L^2(\mu)}$ and $b_Q = \|\Delta_R \, \psi\|_{L^2(\mu)}$ to those terms of (II) corresponding pairs (Q, R) for which $R_Q = R_1$. Although Lemma 7.41 was stated with reference to R_1, the northeast child of R, it clearly still holds when R_1 is replaced by R_2, R_3 or R_4, the northwest, southwest, or southeast child of R respectively. Apply these variants of Lemma 7.41 to those terms of (II) corresponding pairs (Q, R) for which $R_Q = R_2$, $R_Q = R_3$, and $R_Q = R_4$ in turn and then add the resulting four estimates together to see that

$$(\mathrm{II}) \leq 4 \cdot \frac{M}{4} \cdot \frac{5}{2} \left\{ \sum_Q \|\Delta_Q \varphi\|^2_{L^2(\mu)} \right\}^{1/2} \left\{ \sum_R \|\Delta_R \psi\|^2_{L^2(\mu)} \right\}^{1/2} \leq 5M \, \|\varphi\|_{L^2(\mu)} \, \|\psi\|_{L^2(\mu)}$$

with the last inequality being a consequence of the Adapted Martingale Decomposition (7.15).

Estimation of (III). Let Dom $= \{Q \in \mathcal{D}^{\mathrm{tran}}_{\omega_1} : (Q, R) \in \mathcal{G}^{\mathrm{tran}}_3$ for some $R \in \mathcal{D}^{\mathrm{tran}}_{\omega_2}$ and $\Delta_Q \varphi \neq 0\}$, and given such a Q, let $\mathrm{Ran}(Q) = \{R \in \mathcal{D}^{\mathrm{tran}}_{\omega_2} : (Q, R) \in \mathcal{G}^{\mathrm{tran}}_3\}$. Then, since the inner product occurring in (III) does not involve R and vanishes whenever $\Delta_Q \varphi$ does, we have

$$(\mathrm{III}) = \left| \sum\nolimits_{Q \in \mathrm{Dom}} \left\{ \sum\nolimits_{R \in \mathrm{Ran}(Q)} c_{Q,R} \right\} \langle \Delta_Q \varphi, \mathcal{K}_\Theta h \rangle \right|.$$

Now, given a $Q \in$ Dom, what are the appropriate Rs over which we are summing, i.e., what exactly is $\mathrm{Ran}(Q)$? Note that Q must be ω_2-good by (4) of the Final Good Function Estimate (7.23). For such a Q it follows from Section 7.15 that $R \in \mathrm{Ran}(Q)$ if and only if $R \in \mathcal{D}^{\mathrm{tran}}_{\omega_2}$, $l(Q) < 2^{-m} l(R)$, and $Q \subseteq R_Q$ with R_Q transit. Letting $R(Q)$ denote the smallest square from $\mathcal{D}^{\mathrm{tran}}_{\omega_2}$ that contains Q and has edge length greater than *or equal to* $2^m \, l(Q)$, we see that the Rs from $\mathrm{Ran}(Q)$ form a chain in $\mathcal{D}^{\mathrm{tran}}_{\omega_2}$ starting just above $R(Q)$ and proceeding upward through $\mathcal{D}^{\mathrm{tran}}_{\omega_2}$, doubling in size at each step, to end at $R^0_{\omega_2}$, the largest square of $\mathcal{D}^{\mathrm{tran}}_{\omega_2}$. Since $c_{Q,R} = \langle \psi \rangle_{R_Q} / \langle h \rangle_{R_Q} - \langle \psi \rangle_R / \langle h \rangle_R$, the sum over all appropriate Rs of the $c_{Q,R}$s is a telescoping series and so

$$\sum\nolimits_{R \in \mathrm{Ran}(Q)} c_{Q,R} = \frac{\langle \psi \rangle_{R(Q)}}{\langle h \rangle_{R(Q)}} - \frac{\langle \psi \rangle_{R^0_{\omega_2}}}{\langle h \rangle_{R^0_{\omega_2}}} = \frac{\langle \psi \rangle_{R(Q)}}{\langle h \rangle_{R(Q)}}$$

with the last equality holding by (3) of the Final Good Function Reduction (7.23). It follows that

$$(\mathrm{III}) = \left| \sum\nolimits_{Q \in \mathrm{Dom}} \frac{\langle \psi \rangle_{R(Q)}}{\langle h \rangle_{R(Q)}} \langle \Delta_Q \varphi, \mathcal{K}_\Theta h \rangle \right| \leq \sum\nolimits_{Q \in \mathrm{Dom}} \left| \frac{\langle \psi \rangle_{R(Q)}}{\langle h \rangle_{R(Q)}} \langle \Delta_Q \varphi, \mathcal{K}_\Theta h \rangle \right|.$$

Note how after the orgy of cancellation involving the telescoping sums we have simply crashed the absolute value crudely through the remaining sum over the Qs. Because of this we need not be too picky anymore about exactly which Qs we are summing over. We are free now to sum up "too many" Qs since the terms now being summed are all nonnegative! Set $\mathcal{A} = \{Q \in \mathcal{D}^{\mathrm{tran}}_{\omega_1} : l(Q) < 2^{-m}$ and $\Delta_Q \varphi \neq 0\}$. Clearly Dom $\subseteq \mathcal{A}$. Any $Q \in \mathcal{A}$, being transit, must intersect $\mathrm{spt}(\mu) \subseteq B(0; 1/8)$ (see hypothesis (1) of Theorem 7.1) and have $l(Q) < 2^{-m} \leq 1/8$ (see the proof of Lemma 7.18). Thus such a Q is always contained in $[-1/4, 1/4) \times [-1/4, 1/4)$.

But for any $\omega_2 \in \Omega$, $[-1/4, 1/4) \times [-1/4, 1/4)$ is always contained in $R_{\omega_2}^0$ (see Section 7.2). It follows that $R(Q)$ still makes perfect sense for any $Q \in \mathcal{A}$ since $R_{\omega_2}^0$ is always a square from $\mathcal{D}_{\omega_2}^{\text{tran}}$ that contains Q and has edge length greater than or equal to $2^m\, l(Q)$. Thus by (a) of Lemma 7.10 and the Cauchy–Schwarz Inequality [RUD, 3.5],

$$(\star)\quad \text{(III)} \leq \frac{1}{1-\delta} \sum_{Q\in\mathcal{A}} \|\Delta_Q\,\varphi\|_{L^2(\mu)} \cdot \frac{|\langle \Delta_Q\,\varphi, \mathcal{K}_\Theta h\rangle|}{\|\Delta_Q\,\varphi\|_{L^2(\mu)}}\, |\langle\psi\rangle_{R(Q)}|$$

$$\leq \frac{1}{1-\delta} \left\{\sum_{Q\in\mathcal{A}} \|\Delta_Q\,\varphi\|_{L^2(\mu)}^2\right\}^{1/2} \left\{\sum_{Q\in\mathcal{A}} \frac{|\langle \Delta_Q\,\varphi, \mathcal{K}_\Theta h\rangle|^2}{\|\Delta_Q\,\varphi\|_{L^2(\mu)}^2}\, |\langle\psi\rangle_{R(Q)}|^2\right\}^{1/2}.$$

Let (α) and (β) denote the insides of the first and second square roots occurring in the last line of $(\star)$.

By the Adapted Martingale Decomposition (7.15), $(\alpha) \leq 2\,\|\varphi\|_{L^2(\mu)}^2$.

To handle (β) we shall invoke the Dyadic Carleson Imbedding Inequality (7.8) on the full dyadic lattice $\mathcal{F}$ corresponding to ω_2. Thus $\mathcal{F}$ consists of the square $\omega_2 + [-1/2, 1/2) \times [-1/2, 1/2)$ along with all its dyadic descendents. The Dyadic Carleson Imbedding Inequality (7.8) was stated in terms of squares Q and R from $\mathcal{F}$, but we are presently using Q to denote squares from $\mathcal{D}_{\omega_1}^{\text{tran}}$, a dyadic lattice corresponding to ω_1. Because of this we shall replace the Qs and Rs of the Dyadic Carleson Imbedding Inequality (7.8) with Rs and Ss respectively. Apologies to the reader!

Given $R \in \mathcal{F}$, set $\mathcal{A}(R) = \{Q \in \mathcal{A} : R(Q) = R\}$ and

$$a_R = \sum_{Q\in\mathcal{A}(R)} \frac{|\langle \Delta_Q\,\varphi, \mathcal{K}_\Theta h\rangle|^2}{\|\Delta_Q\,\varphi\|_{L^2(\mu)}^2}.$$

We here adopt the convention that a sum over the empty set is 0, so $a_R = 0$ whenever $\mathcal{A}(R) = \emptyset$. In particular, $a_R = 0$ for all $R \in \mathcal{F} \setminus \mathcal{D}_{\omega_2}^{\text{tran}}$. Thus we have

$$(\beta) = \sum_{R\in\mathcal{F}} a_R\, |\langle\psi\rangle_R|^2.$$

We recall a bit of notation from the Dyadic Carleson Imbedding Inequality (7.8): $\mathcal{F}(R) = \{S \in \mathcal{F} : S \subseteq R\}$ for $R \in \mathcal{F}$. In addition, we introduce a new bit of notation: $\mathcal{B}(R) = \{Q \in \mathcal{A} : Q \subseteq R\}$ for $R \in \mathcal{F}$. Note two things: first, $Q \in \mathcal{B}(R)$ whenever $Q \in \mathcal{A}(S)$ and $S \in \mathcal{F}(R)$, and second, $\mathcal{A}(S) \cap \mathcal{A}(S') = \emptyset$ whenever $S \neq S'$. Next note that hypothesis (5) of Theorem 7.1 implies that $\|\mathcal{K}_\Theta h\|_{L^\infty(\mu)} \leq \widetilde{M}$ (see the first paragraph of the proof of the First Reduction in Section 7.11 for the details). From these observations and the last lemma, we see that for any $R \in \mathcal{F}$,

$$\sum_{S\in\mathcal{F}(R)} a_S = \sum_{S\in\mathcal{F}(R)} \sum_{Q\in\mathcal{A}(S)} \frac{|\langle \Delta_Q\,\varphi, \mathcal{K}_\Theta h\rangle|^2}{\|\Delta_Q\,\varphi\|_{L^2(\mu)}^2}$$

$$\leq \sum_{Q \in \mathcal{B}(R)} \frac{|\langle \Delta_Q \varphi, \mathcal{X}_R \mathcal{K}_\Theta h\rangle|^2}{\|\Delta_Q \varphi\|^2_{L^2(\mu)}}$$

$$\leq 2 \, \|\mathcal{X}_R \mathcal{K}_\Theta h\|^2_{L^2(\mu)}$$

$$= 2 \int_R |\mathcal{K}_\Theta h|^2 d\mu \;\; \leq \;\; 2\widetilde{M}^2 \mu(R).$$

Having verified the hypothesis of the Dyadic Carleson Imbedding Inequality (7.8) with $A = 2\widetilde{M}^2$, we may apply it to conclude that $(\beta) \leq 8\widetilde{M}^2 \, \|\psi\|^2_{L^2(\mu)}$.

From $(\star)$, our estimates on (α) and (β), and the fact that $1/(1-\delta) \leq 2$, we conclude that

$$\text{(III)} \leq 2\,\{2\,\|\varphi\|^2_{L^2(\mu)}\}^{1/2}\{8\widetilde{M}^2 \,\|\psi\|^2_{L^2(\mu)}\}^{1/2} = 8\widetilde{M} \, \|\varphi\|_{L^2(\mu)} \, \|\psi\|_{L^2(\mu)}.$$

Clearly $\widetilde{N}_3^{\text{tran}} = 9{,}720M + 5M + 8\widetilde{M} = 9{,}725M + 8\widetilde{M}$ will do. □

Putting Propositions 7.30, 7.35, 7.39, and 7.43 together, we see that the Final Reduced Good Function Estimate (7.23) holds with $\widetilde{N} = \widetilde{N}_1 + \widetilde{N}_2 + \widetilde{N}_3^{\text{term}} + \widetilde{N}_3^{\text{tran}}$. We have thus proved Theorem 7.1.

Chapter 8
The Curvature Theorem of David and Léger

8.1 Restatement of the Result and an Initial Reduction

The goal of this very long chapter is to prove Theorem 6.16, the second of the two difficult results needed to complete the resolution of Vitushkin's Conjecture. Our treatment here is from [LÉG]. The theorem is actually stronger than we have stated in two ways. First, it is a result dealing with $\mathbb{R}^n$ and not just $\mathbb{R}^2 = \mathbb{C}$. We however are interested in the result only for $\mathbb{C}$ and so will stay there and feel free to avail ourselves of any conveniences that two dimensions gives us (e.g., the nonexistence of skew lines). The second way that it is stronger, which we do wish to incorporate into our restatement, requires a comment. By Corollary 4.23, any Lipschitz graph is the graph of a rectifiable curve. Of course the converse fails. So the stronger version of Theorem 6.16 that we state now, the proof of which is the whole point of this chapter, is the following.

Theorem 8.1 (David and Léger) *Suppose E is an $\mathcal{H}^1$-measurable subset of $\mathbb{C}$ such that $\mathcal{H}^1(E) < \infty$ and $c^2(\mathcal{H}^1_E) < \infty$. Then there exists a countable family of Lipschitz graphs whose union Γ satisfies*

$$\mathcal{H}^1(E \setminus \Gamma) = 0.$$

Theorem 8.1 immediately leads to strengthened versions of Reduction 6.3 and Theorems 6.4 and 6.5 in which the graphs of rectifiable curves are replaced by Lipschitz graphs.

In this section we will show how Theorem 8.1 follows from a reduction whose lengthy proof will occupy the rest of the chapter. Before stating this reduction we need the following.

Lemma 8.2 *Suppose M' and κ' are positive finite numbers. Then, given any compact subset F of $\mathbb{C}$ such that $\mathcal{H}^1_F$ is nontrivial, $\mathcal{H}^1_F$ has linear growth with bound M', and $c^2(\mathcal{H}^1_F) < \infty$, there exists a compact subset K of F such that $\mathcal{H}^1(K) > |K|/4$ and $c^2(\mathcal{H}^1_K) < \kappa'|K|$. Of course it is then also the case that $\mathcal{H}^1_K$ is nontrivial and has linear growth with bound M'.*

J.J. Dudziak, *Vitushkin's Conjecture for Removable Sets*, Universitext,
DOI 10.1007/978-1-4419-6709-1_8,

Proof Given $r > 0$, set $F(r) = \{(\zeta, \eta, \xi) \in F \times F \times F : |\zeta - \eta|, |\zeta - \xi|$, and $|\eta - \xi| < r\}$. Since $c^2(\mathcal{H}^1_F) < \infty$, by Lebesgue's Dominated Convergence Theorem [RUD, 1.34] we may choose an $r_0 > 0$ so small that

$$(\star) \quad \iiint_{F(r_0)} c^2(\zeta, \eta, \xi)\, d\mathcal{H}^1(\zeta)\, d\mathcal{H}^1(\eta)\, d\mathcal{H}^1(\xi) < \frac{\kappa'}{8M'} \mathcal{H}^1(F).$$

Denote the collection of all closed balls B centered on F for which $|B| < r_0$ and $\mathcal{H}^1(F \cap B) > |B|/4$ by $\mathcal{V}$. By Lemma 6.12, $\mathcal{V}$ is a Vitali class of closed subsets for a subset G of F such that $\mathcal{H}^1(F \setminus G) = 0$. Thus by (b) of Vitali's Covering Lemma (5.14), there exists a countable pairwise disjoint subcollection $\{B_n\}$ of $\mathcal{V}$ such that

$$\mathcal{H}^1(G) < \sum_n |B_n| + \frac{1}{4}\mathcal{H}^1(G).$$

Clearly,

$$(\star\star) \quad \sum_n |B_n| > \frac{3}{4}\mathcal{H}^1(G) = \frac{3}{4}\mathcal{H}^1(F).$$

The collection $\{(F \cap B_n) \times (F \cap B_n) \times (F \cap B_n)\}$ is a pairwise disjoint collection of subsets of $F(r_0)$. So from $(\star)$ we may conclude that

$$\sum_n c^2\left(\mathcal{H}^1_{F \cap B_n}\right) < \frac{\kappa'}{8M'}\mathcal{H}^1(F).$$

Set $\mathcal{G} = \{n : c^2(\mathcal{H}^1_{F \cap B_n}) < (\kappa'/4M')|B_n|\}$ and $\mathcal{B} = \{n : c^2(\mathcal{H}^1_{F \cap B_n}) \geq (\kappa'/4M')|B_n|\}$. From the last displayed inequality it follows that

$$\sum_{n \in \mathcal{B}} |B_n| \leq \frac{4M'}{\kappa'} \sum_{n \in \mathcal{B}} c^2\left(\mathcal{H}^1_{F \cap B_n}\right) \leq \frac{4M'}{\kappa'} \sum_n c^2\left(\mathcal{H}^1_{F \cap B_n}\right) < \frac{1}{2}\mathcal{H}^1(F).$$

But then from $(\star\star)$ it follows that $\mathcal{G}$ must be nonempty. So choose and fix an element $n \in \mathcal{G}$. For this n we have

$$\mathcal{H}^1(F \cap B_n) > \frac{1}{4}|B_n| \geq \frac{1}{4}|F \cap B_n|$$

and

$$c^2(\mathcal{H}^1_{F \cap B_n}) < \frac{\kappa'}{4M'}|B_n| < \frac{\kappa'}{M'}\mathcal{H}^1(F \cap B_n) \leq \kappa'|F \cap B_n|$$

with the last inequality following from the triviality that any set is contained in a disc of radius the diameter of the set and the linear growth of the measure $\mathcal{H}^1_F$. Clearly $K = F \cap B_n$ works. □

Reduction 8.3 *For any positive finite number M, there exists a positive finite number κ such that the following holds:*

If μ is any positive Borel measure on $\mathbb{C}$ for which

(1) $|\mathrm{spt}(\mu)| < 1$,

(2) $\mu(\mathbb{C}) > 1$,

(3) *μ has linear growth with bound M, and*

(4) $c^2(\mu) < \kappa$,

then there exists a Lipschitz graph Γ such that

$$\mu(\Gamma) > \frac{99}{100}\mu(\mathbb{C}).$$

In the reduction's conclusion, the percentage of the mass of μ that the Lipschitz graph must at least cover can be anything short of 100%. We have arbitrarily chosen 99% and, to be on the safe side, our proof will actually produce a Lipschitz graph covering at least 99.99% of the mass of the μ! The use to which this reduction will now be put simply requires that the percentage be anything exceeding 0%! Finally, it is worth noting that μ is automatically regular by [RUD, 2.18].

Proof of David and Léger's Theorem (8.1) *from the Reduction* (8.3) Set

$$\alpha = \sup \left\{ \mathcal{H}^1(E \cap \Gamma) : \Gamma \text{ is a countable union of Lipschitz graphs} \right\}.$$

Since a countable union of sets each of which is a countable union of Lipschitz graphs is again a countable union of Lipschitz graphs, the supremum defining α is easily seen to be a maximum. So let Γ be a countable union of Lipschitz graphs which attains this maximum. Thus $\mathcal{H}^1(E \cap \Gamma) = \alpha$. To finish it suffices to prove that $\mathcal{H}^1(E \setminus \Gamma) = 0$. This we do by contradiction.

So assume $\mathcal{H}^1(E \setminus \Gamma) > 0$. By Proposition 5.21 and Corollary 6.2, there exists a compact subset F of $E \setminus \Gamma$ and a positive finite number M' such that $\mathcal{H}^1_F$ is nontrivial and has linear growth with bound M'. Let κ be the number produced by Reduction 8.3 corresponding to $M = 8M'$. Setting $\kappa' = \kappa/256$, Lemma 8.2 produces a compact subset K of F such that $\mathcal{H}^1(K) > |K|/4$ and $c^2(\mathcal{H}^1_K) < \kappa'|K|$. Of course $\mathcal{H}^1_K$ is nontrivial and has linear growth with bound M'.

Let $\varphi : \mathbb{C} \mapsto \mathbb{C}$ be the similarity transformation defined by $z \mapsto 2|K|z$. Define a positive Borel measure μ on $\mathbb{C}$ by setting $\mu = (4/|K|)\mathcal{H}^1_K \circ \varphi$, i.e., set

$$\mu(G) = \frac{4}{|K|}\mathcal{H}^1_K(\varphi(G))$$

for any Borel subset G of $\mathbb{C}$. The reader should have no trouble checking that μ satisfies hypotheses (1), (2), and (3) of the reduction. To check hypothesis (4), first note that by the usual approximation argument that starts with characteristic functions and finishes with increasing limits of nonnegative simple functions, we have

$$\int f\, d(\mathcal{H}_K^1 \circ \varphi) = \int (f \circ \varphi^{-1})\, d\mathcal{H}_K^1$$

for any nonnegative Borel function f. Second note that we obviously have

$$c^2(\varphi^{-1}(\zeta), \varphi^{-1}(\eta), \varphi^{-1}(\xi)) = c^2\left(\frac{\zeta}{2|K|}, \frac{\eta}{2|K|}, \frac{\xi}{2|K|}\right) = 4|K|^2 c^2(\zeta, \eta, \xi)$$

since the curvature of a point triple is the reciprocal of the radius of the circumcircle of the triangle determined by the point triple. Thus we have

$$\begin{aligned}
c^2(\mu) &= \iiint c^2(\zeta, \eta, \xi)\, d\mu(\zeta)\, d\mu(\eta)\, d\mu(\xi) \\
&= \frac{64}{|K|^3} \iiint c^2(\varphi^{-1}(\zeta), \varphi^{-1}(\eta), \varphi^{-1}(\xi))\, d\mathcal{H}_K^1(\zeta)\, d\mathcal{H}_K^1(\eta)\, d\mathcal{H}_K^1(\xi) \\
&= \frac{256}{|K|} \iiint c^2(\zeta, \eta, \xi)\, d\mathcal{H}_K^1(\zeta)\, d\mathcal{H}_K^1(\eta)\, d\mathcal{H}_K^1(\xi) \\
&= \frac{256}{|K|}\, c^2(\mathcal{H}_K^1) < 256\kappa' \;=\; \kappa,
\end{aligned}$$

i.e., μ satisfies hypothesis (4).

The reduction now supplies us with a Lipschitz graph Γ^* such that $\mu(\Gamma^*) > 0$, i.e., $\mathcal{H}^1(K \cap \varphi(\Gamma^*)) > 0$. But $\varphi(\Gamma^*)$ is in turn a Lipschitz graph, so $\Gamma \cup \varphi(\Gamma^*)$ is a countable union of Lipschitz graphs. Since $K \subseteq E \setminus \Gamma$, it follows that

$$\begin{aligned}
\mathcal{H}^1(E \cap \{\Gamma \cup \varphi(\Gamma^*)\}) &= \mathcal{H}^1(E \cap \Gamma) + \mathcal{H}^1(\{E \setminus \Gamma\} \cap \varphi(\Gamma^*)) \\
&\geq \mathcal{H}^1(E \cap \Gamma) + \mathcal{H}^1(K \cap \varphi(\Gamma^*)) > \alpha.
\end{aligned}$$

This contradicts the way α was defined and so finishes the proof. □

8.2 Two Lemmas Concerning High-Density Balls

Stipulation 8.4 *For the rest to this chapter, we shall assume that M is a positive finite number and that μ is a positive Borel measure on $\mathbb{C}$ satisfying* (1), (2), *and* (3) *of Reduction* 8.3. *Moreover, we will denote the support of μ by K. Thus $K = \operatorname{spt}(\mu)$.*

We note in passing that our number M is not just positive but actually is greater than 1. Also, our measure μ is not just positive but also compactly supported and of finite total mass. That this is so follows easily from (1), (2), and (3) of the reduction and is left to the reader to verify. In the next four sections we state and prove a number of geometric/measure-theoretic facts concerning our measure μ that will be of basic use to us throughout. After that we have a section which, given our M, produces the positive number κ asserted to exist in the reduction. The remainder of the chapter then assumes that μ also satisfies (4) of the reduction and is devoted to constructing a Lipschitz graph which threads its way through at least 99% of the mass of μ.

For any nontrivial closed ball $B = B(z; r)$ of $\mathbb{C}$, the *density* of μ in B is defined to be

$$\delta(B) = \delta(z; r) = \frac{\mu(B(z; r))}{r}.$$

We have control from above over this density since (3) of the reduction says that $\delta(B) \leq M$ always. What makes the reduction so difficult to prove is the lack of any control from below over this density. This leads us to introduce a density threshold parameter $\delta > 0$, which will be specified precisely later and then denoted δ_0. The first half of our proof will use balls B for which $\delta(B) \geq \delta_0$ to construct the desired Lipschitz graph and show that it contains a subset K_0 of K. The second half of our proof establishes and then exploits the fact that $\delta(B) \geq \delta_0$ does not fail "too often" to show that $K \setminus K_0$ has less than 1% of the mass of μ.

This section sets forth two basic geometric lemmas concerning high density balls that are consequences of Besicovitch's Covering Lemma (5.26). Recall that throughout this book we have used $\mathrm{rad}(B)$ to denote the radius of a closed ball B. From this point onward, it will be convenient to similarly denote the center of B by $\mathrm{cen}(B)$.

Lemma 8.5 *Given* $0 < \delta < 1$, *set*

$$c_1(M, \delta) = \frac{1}{10^4} \cdot \frac{\delta}{M} \text{ and } c_2(M, \delta) = \frac{c_1(M, \delta)^2}{1{,}000} \cdot \delta = \frac{1}{10^{11}} \cdot \frac{\delta^3}{M^2}.$$

Then for any nontrivial closed ball B for which $\delta(B) \geq \delta$, there exist two nontrivial closed balls B_1 and B_2 centered on $K \cap B$ such that

(a) $\mathrm{rad}(B_1)$ *and* $\mathrm{rad}(B_2) = c_1(M, \delta)\,\mathrm{rad}(B)$,

(b) $\mu(B_1 \cap B)$ *and* $\mu(B_2 \cap B) > c_2(M, \delta)\,\mathrm{rad}(B)$, *and*

(c) $\mathrm{dist}(B_1, B_2) > 10\,c_1(M, \delta)\,\mathrm{rad}(B)$.

Moreover, if there exists even a single nontrivial closed ball B for which $\delta(B) \geq \delta$, then $c_1(M, \delta) \leq 10^{-4}$ and $c_2(M, \delta) \leq 10^{-11}$.

The small "constants" $c_1(M, \delta)$ and $c_2(M, \delta)$ just introduced will be more concisely referred to as c_1 and c_2 in what follows when the values of M and δ are clear from context or have been made clear by stipulation. Of course c_1 and c_2 are not really constants at all due to their dependence on the parameters M and δ. We will be introducing more parameters in what follows and will be defining more "constants," small and large, depending on this enlarged repertoire of parameters as the chapter proceeds. When the "constant" is large, a "C" will be used in place of a "c." When first introduced, such "constants" will have their parameter dependence explicitly indicated; afterward this dependence will usually be suppressed when the parameters are clear from context or have been made clear by stipulation. Eventually all these parameters will be fixed absolutely or in terms of M, the one parameter given to us by nature and so not open to our free choice.

Proof We verify the last assertion of the lemma first. For the ball B stipulated to exist we have $\delta \leq \delta(B) \leq M$. Thus $\delta/M \leq 1$ and so the assertion concerning c_1 and c_2 follows from the way these constants have been defined and the fact that $\delta < 1$.

We now turn to proving the main assertion of the lemma. Applying (a) of Besicovitch's Covering Lemma (5.26) to the collection of closed balls centered on $K \cap B$ and of radius $c_1 \operatorname{rad}(B)$, we obtain a countable subfamily $\mathcal{F}$ of such balls which covers $K \cap B$ and has overlap at most 125.

Set $\mathcal{G} = \{\widetilde{B} \in \mathcal{F} : \mu(\widetilde{B} \cap B) > c_2 \operatorname{rad}(B)\}$. To finish it suffices to show there exist $B_1, B_2 \in \mathcal{G}$ satisfying (c). This we do by supposing not and getting a contradiction.

Claim. $\displaystyle\sum_{\widetilde{B} \in \mathcal{G}} \mu(\widetilde{B}) \leq 2{,}500\, M\, c_1 \operatorname{rad}(B).$

Without loss of generality, $\mathcal{G} \neq \emptyset$. So choose and fix a $B' \in \mathcal{G}$. Then for any $\widetilde{B} \in \mathcal{G}$ we have $\widetilde{B} \subseteq 20B'$ with room to spare since we are assuming that (c) fails for $\widetilde{B}$ and B'. Thus $\sum_{\widetilde{B} \in \mathcal{G}} \mathcal{X}_{\widetilde{B}} \leq 125\, \mathcal{X}_{20B'}$. Integrating with respect to μ, one obtains the Claim:

$$\sum_{\widetilde{B} \in \mathcal{G}} \mu(\widetilde{B}) \leq 125\, \mu(20B') \leq 125\, M \operatorname{rad}(20B') = 2{,}500\, M\, c_1 \operatorname{rad}(B).$$

Claim. $\displaystyle\sum_{\widetilde{B} \in \mathcal{F} \setminus \mathcal{G}} \mu(\widetilde{B} \cap B) \leq \frac{500\, c_2}{c_1^2} \operatorname{rad}(B).$

Since $c_1 \leq 10^{-4}$, for any $\widetilde{B} \in \mathcal{F}$ we have $\widetilde{B} \subseteq 2B$ with room to spare and so $\sum_{\widetilde{B} \in \mathcal{F}} \mathcal{X}_{\widetilde{B}} \leq 125\, \mathcal{X}_{2B}$. Integrating with respect to area, one sees that $(\#\mathcal{F}) \cdot \pi \{c_1 \operatorname{rad}(B)\}^2 \leq 125 \cdot \pi \operatorname{rad}(2B)^2 = 125 \cdot \pi \{2 \operatorname{rad}(B)\}^2$, i.e., $\#\mathcal{F} \leq 500/c_1^2$. The Claim now follows easily:

$$\sum_{\widetilde{B} \in \mathcal{F} \setminus \mathcal{G}} \mu(\widetilde{B} \cap B) \leq (\#\mathcal{F}) \cdot c_2 \operatorname{rad}(B) \leq \frac{500 c_2}{c_1^2} \operatorname{rad}(B).$$

From our density assumption, the fact that $\mathcal{F}$ covers $K \cap B$, and the two Claims, we obtain

$$\delta\,\mathrm{rad}(B) \le \mu(B) \le \sum_{\widetilde{B}\in\mathcal{F}} \mu(\widetilde{B}\cap B) \le 2{,}500\,M\,c_1\,\mathrm{rad}(B) + \frac{500\,c_2}{c_1^2}\,\mathrm{rad}(B),$$

i.e., $\delta \le 2{,}500\,M\,c_1 + 500\,c_2/c_1^2$. But c_1 was chosen to make $2{,}500\,M\,c_1 = \delta/4$ and c_2 was chosen to make $500\,c_2/c_1^2 = \delta/2$, so our desired contradiction is at hand. □

The next lemma just involves applying the last lemma three times over and doing the necessary bookkeeping.

Lemma 8.6 *Given* $0 < \delta < 1$, *set*

$$c_3(M,\delta) = c_1(M,\delta)\;c_1\!\left(M, \frac{c_1(M,\delta)}{1{,}000}\cdot\delta\right) = \frac{1}{10^{15}}\cdot\frac{\delta^3}{M^3}$$

and

$$c_4(M,\delta) = c_1(M,\delta)\;c_2\!\left(M, \frac{c_1(M,\delta)}{1{,}000}\cdot\delta\right) = \frac{1}{10^{36}}\cdot\frac{\delta^7}{M^6}.$$

Then for any nontrivial closed ball B for which $\delta(B) \ge \delta$, there exist four nontrivial closed balls B_1, B_2, B_3, and B_4 centered on $K \cap B$ such that

(a) $\mathrm{rad}(B_j) = c_3(M,\delta)\,\mathrm{rad}(B)$ *for each* j,

(b) $\mu(B_j \cap B) > c_4(M,\delta)\,\mathrm{rad}(B)$ *for each* j, *and*

(c) $\mathrm{dist}(B_j, B_k) > 10\,c_3(M,\delta)\,\mathrm{rad}(B)$ *for each pair* (j,k) *with* $j \ne k$.

Moreover, if there exists even a single nontrivial closed ball B for which $\delta(B) \ge \delta$, then $c_3(M,\delta) \le 10^{-15}$ and $c_4(M,\delta) \le 10^{-36}$.

Proof Apply the last lemma to μ and B to get two balls which we choose to label B_1' and B_2' instead of B_1 and B_2. Let c_1 and c_2 denote $c_1(M,\delta)$ and $c_2(M,\delta)$, respectively.

Next apply the last lemma to $\mu' = \mu_B$ and $B' = B_1'$ to get two balls which we choose to label B_1 and B_2. Note that B_1 and B_2 are centered on $\mathrm{spt}(\mu') \cap B' \subseteq \mathrm{spt}(\mu') \subseteq \mathrm{spt}(\mu) \cap B = K \cap B$. But what are M', δ', c_1', and c_2' in this situation? Clearly we can take $M' = M$. However,

$$\frac{\mu'(B')}{\mathrm{rad}(B')} = \frac{\mu(B_1'\cap B)}{\mathrm{rad}(B_1')} > \frac{c_2\,\mathrm{rad}(B)}{c_1\,\mathrm{rad}(B)} = \frac{c_2}{c_1},$$

so we can take $\delta' = c_2/c_1 = (c_1/1{,}000)\cdot\delta$. It follows that

$$c_1' = c_1(M',\delta') = c_1\!\left(M, \frac{c_1}{1{,}000}\cdot\delta\right) \text{ and } c_2' = c_2(M',\delta') = c_2\!\left(M, \frac{c_1}{1{,}000}\cdot\delta\right).$$

In the statement of this lemma, we have defined c_3 and c_4 to be c_1c_1' and c_1c_2', respectively. Hence for $j = 1$ and 2, $\mathrm{rad}(B_j) = c_1' \,\mathrm{rad}(B') = c_1'c_1 \,\mathrm{rad}(B) = c_3 \,\mathrm{rad}(B)$ and $\mu(B_j \cap B) = \mu'(B_j) \geq \mu'(B_j \cap B') > c_2' \,\mathrm{rad}(B') = c_2'c_1 \,\mathrm{rad}(B) = c_4 \,\mathrm{rad}(B)$. Also, $\mathrm{dist}(B_1, B_2) > 10\, c_1' \,\mathrm{rad}(B') = 10\, c_1'c_1 \,\mathrm{rad}(B) = 10\, c_3 \,\mathrm{rad}(B)$.

Finally, apply the last lemma to $\mu' = \mu_B$ and $B' = B_2'$ to get two balls which we choose to label B_3 and B_4. Arguing with B_3 and B_4 as we just did with B_1 and B_2, all that is left to prove is that $\mathrm{dist}(B_j, B_k) > 10\, c_3 \,\mathrm{rad}(B)$ for $j = 1, 2$ and $k = 3, 4$. But clearly in this situation

$$\mathrm{dist}(B_j, B_k) \geq \mathrm{dist}(B_1', B_2') - \mathrm{rad}(B_j) - \mathrm{rad}(B_k) > 10\, c_1 \,\mathrm{rad}(B) - 2\, c_3 \,\mathrm{rad}(B),$$

so it suffices to verify that $10\, c_1 - 2\, c_3 \geq 10\, c_3$, i.e., $c_3 \leq (5/6)c_1$, i.e., $c_1'c_1 \leq (5/6)c_1$, i.e., $c_1' \leq 5/6$.

This last inequality and the last assertion of this lemma are verified just as the last assertion of the previous lemma was verified. □

8.3 The Beta Numbers of Peter Jones

Given a closed ball B, recall that kB denotes the closed ball concentric with B and with radius k times that of B. We need a measure of how well the support of μ can be approximated by straight lines at a given point $z \in \mathbb{C}$ and on a given scale determined by $r > 0$. This measure is provided by the L^p *beta numbers* of Peter Jones: Given a closed ball $B = B(z; r)$ of $\mathbb{C}$ and $1 \leq p < \infty$, set

$$\beta_p(B) = \beta_p(z; r) = \inf\{\beta_p^L(B) : L \text{ is a line in } \mathbb{C}\}$$

where

$$\beta_p^L(B) = \beta_p^L(z; r) = \left\{ \frac{1}{\mathrm{rad}(B)} \int_{kB} \left\{ \frac{\mathrm{dist}(\zeta, L)}{\mathrm{rad}(B)} \right\}^p d\mu(\zeta) \right\}^{1/p}.$$

Note the dependence of these quantities on a parameter $k \geq 1$ as well as on μ. This dependence will usually be notationally suppressed.

Jones first introduced, not these beta numbers, but rather a dyadic supremum norm version of these numbers, denoted $\beta_\infty(Q)$, in [JON2]. They were defined with reference to a bounded subset E rather than a measure μ and Jones used them to obtain a sufficient condition for E to lie within some rectifiable curve. The L^p beta numbers defined above were introduced later and used by Guy David and Stephen Semmes in their theory of uniformly rectifiable sets in [DS1] and [DS2]. The result of most interest for us is a result of Hervé Pajot from [PAJ1] and states: Given $1 \leq p < \infty$, if E is a compact Ahlfors–David regular subset of $\mathbb{C}$ such that

$$\int_0^{|E|} \int_{\mathbb{C}} \beta_p^2(z; r)\, d\mathcal{H}_E^1(z) \,\frac{dr}{r} < \infty,$$

then E is contained in a rectifiable curve. Two comments should be made here: one minor, the other major. First, the minor comment: the L^p beta numbers in the integral above are with respect to the measure $\mathcal{H}^1_E$. Second, the major comment: for us this result falls short because of the assumption of Ahlfors–David regularity. We have the upper control on density that is assumed by Ahlfors–David regularity but not the lower control (if the reader has forgotten the definition, see the end of Section 6.5). David and Léger's proof overcomes this obstacle and at the same time ensures that a similar integral to the one above converges by bounding it above by a multiple of the curvature of our measure squared (see Proposition 8.10 below).

For what we wish to do only the L^1 version of the beta numbers will be needed (except for a brief appearance of the L^2 version in the next section). The number $\beta_1^L(B)$ is supposed to be a scale-invariant measure of the average distance within kB of the support of μ from L. For this really to be the case we should have had $1/\mu(kB)$ instead of $1/\text{rad}(B)$ in front of the integral in the definition of $\beta_1^L(B)$. When $\mu(kB)$ and $\text{rad}(B)$ are comparable this objection becomes moot. By (3) of Reduction 8.3, we always have $\mu(kB) \leq Mk\,\text{rad}(B)$. Thus we need only worry about comparability from below. Since $\mu(kB) \geq \mu(B) \geq \delta\,\text{rad}(B)$ whenever $\delta(B) \geq \delta$, this can be secured by insisting that B be a high density ball as determined by the density threshold δ. Thus in what follows we are only interested in beta numbers for high density balls.

The point of the next proposition is that given two points not too far apart and with small beta numbers in a high density situation, any two lines almost realizing these beta numbers will be "close" to one another.

Proposition 8.7 *Given* $0 < \delta < 1$ *and* $k \geq 2$, *set*

$$c_5(M,\delta) = \frac{c_1(M,\delta)\,c_2(M,\delta)}{10^6} = \frac{1}{10^{21}} \cdot \frac{\delta^4}{M^3} \text{ and}$$

$$C_1(M,\delta) = \frac{1}{10\,c_5(M,\delta)} = 10^{20} \cdot \frac{M^3}{\delta^4}.$$

Suppose $z, w \in \mathbb{C}$ *and* $r > 0$ *satisfy* $\delta(z;r) \geq \delta$ *and* $|z - w| \leq (k/2)r$. *Let* L_1 *and* L_2 *be lines in* $\mathbb{C}$ *such that*

$$\beta_1^{L_1}(z;r) \leq 5{,}000\,\varepsilon \text{ and } \beta_1^{L_2}(w;r) \leq 5{,}000\,\varepsilon$$

where $0 < \varepsilon \leq c_5(M,\delta)$. *Then*

(a) $\text{dist}(z, L_1)$ *and* $\text{dist}(z, L_2) \leq 2r$,

(b) $\text{dist}(\zeta, L_2) \leq C_1(M,\delta)\,\varepsilon\,\{|\zeta - z| + r\}$ *for all* $\zeta \in L_1$, *and*

(c) $0 \leq \angle(L_1, L_2) \leq \pi/40$ *with* $\sin(\angle(L_1, L_2)) \leq \angle(L_1, L_2) \leq \tan(\angle(L_1, L_2)) \leq C_1(M,\delta)\,\varepsilon$.

Moreover, if there exists even a single nontrivial closed ball B for which $\delta(B) \geq \delta$, then $c_5(M, \delta) \leq 10^{-21}$ and $C_1(M, \delta) \geq 10^{20}$.

With regard to (c), our convention is that the angle between two distinct intersecting lines is to be acute or at most right whereas the angle between two parallel or identical lines is to be 0. (The reader may be wondering why the bound of $5{,}000\,\varepsilon$ on the beta numbers in the enunciation of this proposition. If so, see the proof of Proposition 8.33 where $4{,}800\,\varepsilon$ is needed!)

Proof Apply Lemma 8.5 to $B = B(z; r)$ to get closed balls B_1 and B_2 as stated. Since $k \geq 2$, $B_1 \cap B$ and $B_2 \cap B$ are contained in $B(z; kr)$ and $B(w; kr)$. Set

$$G_j = \{\zeta \in B_j \cap B : \operatorname{dist}(\zeta, L_1) + \operatorname{dist}(\zeta, L_2) \leq (2 \times 10^4/c_2)\,\varepsilon r\}$$

for $j = 1, 2$. Then

$$\begin{aligned} \mu(B_j \cap B \setminus G_j) &\leq \int_{B_j \cap B \setminus G_j} \frac{\operatorname{dist}(\zeta, L_1) + \operatorname{dist}(\zeta, L_2)}{(2 \times 10^4/c_2)\,\varepsilon r}\, d\mu(\zeta) \\ &\leq \frac{c_2 r}{(2 \times 10^4)\,\varepsilon} \left\{\beta_1^{L_1}(z; r) + \beta_1^{L_2}(z; r)\right\} \leq \frac{c_2}{2} r \end{aligned}$$

and so $\mu(G_j) = \mu(B_j \cap B) - \mu(B_j \cap B \setminus G_j) > c_2 r - (c_2/2) r > 0$ for $j = 1, 2$. Thus we may pick $z_1 \in G_1$ and $z_2 \in G_2$.

We already have (a): By our choice of c_5, $(2 \times 10^4/c_2)\,\varepsilon \leq (2 \times 10^4/c_2)\, c_5 = c_1/50 \leq 1$. Thus the mere existence of the point $z_1 \in G_1$ (or $z_2 \in G_2$ for that matter) implies that for $j = 1, 2$,

$$\operatorname{dist}(z, L_j) \leq |z - z_1| + \operatorname{dist}(z_1, L_j) \leq r + \frac{2 \times 10^4}{c_2} \cdot \varepsilon r \leq 2r.$$

Denote the orthogonal projections of z_1 and z_2 onto L_1 by z_{11} and z_{21}, respectively.

Claim. $|z_{11} - z_{21}| > c_1 r$.

Fix a line $L_1^{\perp}$ perpendicular to L_1 and denote the orthogonal projections of z_1 and z_2 onto $L_1^{\perp}$ by $z_{11}^{\perp}$ and $z_{21}^{\perp}$ respectively. We proceed by contradiction. Since $|z_1 - z_2| \geq \operatorname{dist}(B_1, B_2) > 10 c_1 r$, the falsity of the Claim would imply that $|z_{11}^{\perp} - z_{21}^{\perp}| > \sqrt{99} c_1 r$ by the Pythagorean Theorem (Euclid I.47). But then we would have $\operatorname{dist}(B_1^{\perp}, B_2^{\perp}) > (\sqrt{99} - 4) c_1 r$ where $B_1^{\perp}$ and $B_2^{\perp}$ denote the orthogonal projections of B_1 and B_2 onto $L_1^{\perp}$ respectively. It would then follow that $\operatorname{dist}(B_j, L_1) > (\sqrt{99} - 4) c_1 r/2 > 2 c_1 r$ for either $j = 1$ or 2. For this j, on the one hand, one would have that

$$\frac{1}{r} \int_{B_j \cap B} \frac{\operatorname{dist}(\zeta, L_1)}{r}\, d\mu(\zeta) \geq \frac{2c_1}{r} \mu(B_j \cap B) > 2 c_1 c_2.$$

On the other hand, by our choice of c_5 one certainly has that

$$\frac{1}{r}\int_{B_j\cap B}\frac{\operatorname{dist}(\zeta,L_1)}{r}\,d\mu(\zeta)\le\beta_1^{L_1}(z;r)\le 5{,}000\,c_5=c_1c_2/200.$$

This contradiction establishes the Claim.

Given $\zeta\in L_1$, since z_{11} and z_{21} are points of L_1 that are *distinct* (by the Claim), we may write $\zeta=tz_{11}+(1-t)z_{21}$ for some $t\in\mathbb{R}$. Recall that the distance of a point (x_0,y_0) to a line $ax+by=c$ is given by $|ax_0+by_0-c|/\sqrt{a^2+b^2}$. From this it follows that

$$\begin{aligned}\operatorname{dist}(\zeta,L_2)&\le|t|\operatorname{dist}(z_{11},L_2)+|1-t|\operatorname{dist}(z_{21},L_2)\\&\le|t|\{\operatorname{dist}(z_{11},L_2)+\operatorname{dist}(z_{21},L_2)\}+\operatorname{dist}(z_{21},L_2).\end{aligned}$$

Now $\operatorname{dist}(z_{11},L_2)\le|z_{11}-z_1|+\operatorname{dist}(z_1,L_2)=\operatorname{dist}(z_1,L_1)+\operatorname{dist}(z_1,L_2)\le(2\times10^4/c_2)\,\varepsilon r$. Similarly, $\operatorname{dist}(z_{21},L_2)\le(2\times10^4/c_2)\,\varepsilon r$. Thus

$$(\star)\ \operatorname{dist}(\zeta,L_2)\le\frac{2\times10^4}{c_2}\varepsilon r\{2|t|+1\}.$$

We now make three observations. First, since $|\zeta-z_{21}|=|t||z_{11}-z_{21}|$, the Claim implies that $|t|<|\zeta-z_{21}|/(c_1r)$. Second, $|z-z_2|\le r$ since $z_2\in B$. Third, $|z_2-z_{21}|=\operatorname{dist}(z_2,L_1)\le(2\times10^4/c_2)\,\varepsilon r\le(2\times10^4/c_2)\,c_5r\le c_1r/2$. Putting these observations together, we obtain

$$\begin{aligned}(\star\star)\ 2|t|+1&<\frac{2}{c_1r}\{|\zeta-z|+|z-z_2|+|z_2-z_{21}|\}+1\\&\le\frac{2\{|\zeta-z|+(1+c_1)r\}}{c_1r}\\&\le\frac{(5/2)\{|\zeta-z|+r\}}{c_1r}.\end{aligned}$$

It easily follows from $(\star)$ and $(\star\star)$ that

$$\operatorname{dist}(\zeta,L_2)\le\frac{5\times10^4}{c_1c_2}\,\varepsilon\,\{|\zeta-z|+r\}=\frac{1}{20c_5}\,\varepsilon\,\{|\zeta-z|+r\}=\frac{C_1}{2}\,\varepsilon\,\{|\zeta-z|+r\}.$$

Thus (b) has been established with some wiggle room to spare. This wiggle room will now be used to establish (c)!

In proving (c), we may as well assume that L_1 and L_2 are distinct intersecting lines and so have a unique intersection point ζ_0. Then for any $\zeta\in L_1$ with $|\zeta-z|>|\zeta_0-z|$, by what has just been proven we have

$$\sin(\angle(L_1,L_2))=\frac{\operatorname{dist}(\zeta,L_2)}{|\zeta-\zeta_0|}\le\frac{C_1}{2}\,\varepsilon\,\frac{|\zeta-z|+r}{|\zeta-z|-|\zeta_0-z|}.$$

Upon letting $|\zeta - z| \to \infty$ while keeping $\zeta \in L_1$, we see that

$$\sin(\angle(L_1, L_2)) \le \frac{C_1}{2}\,\varepsilon.$$

From the simple inequality $\sin x \ge (2/\pi)x$ valid for $0 \le x \le \pi/2$, we then obtain

$$\angle(L_1, L_2) \le \frac{\pi}{2}\sin(\angle(L_1, L_2)) \le \frac{\pi}{4}\,C_1\,\varepsilon.$$

From this and the way we have specified C_1 with respect to c_5,

$$\angle(L_1, L_2) \le \frac{\pi}{4}\,C_1\,c_5 = \frac{\pi}{40}.$$

But now the simple inequality $\tan x \le (4/\pi)x$ valid for $0 \le x \le \pi/4$ yields

$$\tan(\angle(L_1, L_2) \le C_1\,\varepsilon.$$

Since $\sin x \le x \le \tan x$ for all $0 \le x < \pi/2$, we have established (c).

The very last assertion of the proposition is true for the usual reasons. □

8.4 Domination of Beta Numbers by Local Curvature

In this section we prove a difficult result that is the first of two steps which will allow us to get enough balls with small beta numbers from a measure with small enough global curvature. The stepping stone will be through a notion that localizes curvature.

Given a closed ball $B = B(z; r)$, set

$$c(B) = c(z; r) = \left\{\iiint_{\Delta(z;r)} c^2(\zeta, \eta, \xi)\, d\mu(\zeta)\, d\mu(\eta)\, d\mu(\xi)\right\}^{1/2}$$

where

$$\Delta(B) = \Delta(z; r) = \{(\zeta, \eta, \xi) \in k'B \times k'B \times k'B : |\zeta - \eta|, |\zeta - \xi|,\ \text{and } |\eta - \xi| \ge \mathrm{rad}(B)/k'\}.$$

Note the dependence of this quantity, the *local curvature* of μ in B, on a new parameter $k' \ge 1$ as well as on μ. This dependence will usually be notationally suppressed.

Proposition 8.8 *Given* $0 < \delta < 1$ *and* $k \ge 1$, *set*

$$C_2(M, \delta, k) = Mk\left[\frac{(k+1)^4}{4}\left\{\frac{4}{c_4(M,\delta)}\right\}^2 + 81\left\{\frac{4}{c_4(M,\delta)}\right\}^2 + \frac{9(k+1)^4 M}{50\, c_3(M,\delta)}\left\{\frac{4}{c_4(M,\delta)}\right\}^3\right].$$

Suppose that $k' \geq \max\{k, 1/c_3(M,\delta)\}$. Then for any closed ball B for which $\delta(B) \geq \delta$, one has

$$\beta_1^2(B) \leq Mk\,\beta_2^2(B) \leq C_2(M,\delta,k)\,\frac{c^2(B)}{\mathrm{rad}(B)}.$$

Proof The first inequality is an easy consequence of the Cauchy–Schwarz Inequality [RUD, 3.5] and (3) of Reduction 8.3. Thus we may turn our attention to the second inequality.

Without loss of generality, $c^2(B) < \infty$. Let α be finite and greater than $c^2(B)$ (so $\alpha \neq 0$). Apply Lemma 8.6 to B to get closed balls B_1, B_2, and B_3 as stated (we will not need B_4).

Given $\zeta \in \mathbb{C}$, set

$$\Delta_1(\zeta) = \{(\eta,\xi) : (\zeta,\eta,\xi) \in \Delta(B)\}$$

and let G_1 denote the set of all points $\zeta \in B_1 \cap B$ such that

$$f_1(\zeta) = \iint_{\Delta_1(\zeta)} c^2(\zeta,\eta,\xi)\,d\mu(\eta)\,d\mu(\xi) < \frac{4}{c_4}\frac{\alpha}{\mathrm{rad}(B)}.$$

Then

$$\frac{4}{c_4}\frac{\alpha}{\mathrm{rad}(B)}\mu(B_1 \cap B \setminus G_1) \leq \int_{B_1\cap B\setminus G_1} f_1(\zeta)\,d\mu(\zeta) \leq c^2(B) < \alpha,$$

i.e., $\mu((B_1 \cap B \setminus G_1) < (c_4/4)\,\mathrm{rad}(B)$. Since $\mu(B_1 \cap B) > c_4\,\mathrm{rad}(B)$, it follows that $\mu(G_1) > 0$. We may thus choose and fix a point $z_1 \in G_1$.

Given $\eta \in \mathbb{C}$, set

$$\Delta_{12}(z_1,\eta) = \{\xi : (z_1,\eta,\xi) \in \Delta(B)\}$$

and let G_{12} denote the set of all points $\eta \in B_2 \cap B$ such that

$$f_{12}(z_1,\eta) = \iint_{\Delta_{12}(z_1,\eta)} c^2(z_1,\eta,\xi)\,d\mu(\xi) < \left\{\frac{4}{c_4}\right\}^2 \frac{\alpha}{\mathrm{rad}^2(B)}.$$

Then

$$\left\{\frac{4}{c_4}\right\}^2 \frac{\alpha}{\mathrm{rad}^2(B)}\mu(B_2 \cap B \setminus G_{12}) \leq \int_{B_2\cap B\setminus G_{12}} f_{12}(z_1,\eta)\,d\mu(\eta) \leq f_1(z_1) < \frac{4}{c_4}\frac{\alpha}{\mathrm{rad}(B)},$$

i.e., $\mu((B_2 \cap B \setminus G_{12}) < (c_4/4)\,\mathrm{rad}(B)$. Since $\mu(B_2 \cap B) > c_4\,\mathrm{rad}(B)$, it follows that $\mu(G_{12}) > 0$. We may thus choose and fix a point $z_2 \in G_{12}$.

So we now have points $z_1 \in B_1 \cap B$ and $z_2 \in B_2 \cap B$ such that

$$\int_{\Delta_{12}(z_1,z_2)} c^2(z_1, z_2, \xi)\, d\mu(\xi) < \left\{\frac{4}{c_4}\right\}^2 \frac{\alpha}{\mathrm{rad}^2(B)}.$$

Let L be the line through z_1 and z_2. This line is well defined, i.e., $z_1 \neq z_2$, since B_1 and B_2 are disjoint. Because k' is greater than or equal to k and $1/c_3$, we have $kB \setminus \{2B_1 \cup 2B_2\} \subseteq \Delta_{12}(z_1, z_2)$. From Proposition 4.1 it follows that for any $\xi \in kB$,

$$\mathrm{dist}(\xi, L) = \frac{c(z_1, z_2, \xi)}{2} \cdot |\xi - z_1| \cdot |\xi - z_2| \leq \frac{c(z_1, z_2, \xi)}{2} \cdot \{(k+1)\,\mathrm{rad}(B)\}^2.$$

Thus

$$(\star) \int_{kB\setminus\{2B_1\cup 2B_2\}} \left\{\frac{\mathrm{dist}(\xi, L)}{\mathrm{rad}(B)}\right\}^2 d\mu(\xi) \leq$$
$$\frac{(k+1)^4}{4}\mathrm{rad}^2(B) \int_{\Delta_{12}(z_1,z_2)} c^2(z_1, z_2, \xi)\, d\mu(\xi) < \frac{(k+1)^4}{4}\left\{\frac{4}{c_4}\right\}^2 \alpha.$$

To finish we need similar estimates for the corresponding integrals over $2B_1$ and $2B_2$.

Given $\xi \in \mathbb{C}$, set

$$\Delta_{13}(z_1, \xi) = \{\eta : (z_1, \eta, \xi) \in \Delta(B)\}$$

and let G_{13} denote the set of all points $\xi \in B_3 \cap B$ such that

$$f_{13}(z_1, \xi) = \int_{\Delta_{13}(z_1,\xi)} c^2(z_1, \eta, \xi)\, d\mu(\eta) < \left\{\frac{4}{c_4}\right\}^2 \frac{\alpha}{\mathrm{rad}^2(B)}.$$

Arguing as we did with G_{12}, we see that $\mu(B_3 \cap B \setminus G_{13}) < (c_4/4)\,\mathrm{rad}(B)$.

Given $\xi \in \mathbb{C}$, set

$$\Delta_{23}(z_2, \xi) = \{\zeta : (\zeta, z_2, \xi) \in \Delta(B)\}$$

and let G_{23} denote the set of all points $\xi \in B_3 \cap B$ such that

$$f_{23}(z_2, \xi) = \int_{\Delta_{23}(z_2,\xi)} c^2(\zeta, z_2, \xi)\, d\mu(\zeta) < \left\{\frac{4}{c_4}\right\}^2 \frac{\alpha}{\mathrm{rad}^2(B)}.$$

Arguing as we did with G_{12}, we see that $\mu(B_3 \cap B \setminus G_{23}) < (c_4/4)\,\mathrm{rad}(B)$.

Finally, let G_3 denote the set of points $\xi \in B_3 \cap B$ such that

$$\left\{\frac{\operatorname{dist}(\xi, L)}{\operatorname{rad}(B)}\right\}^2 < \frac{(k+1)^4}{4}\left\{\frac{4}{c_4}\right\}^3 \frac{\alpha}{\operatorname{rad}(B)}.$$

Noting that $B_3 \cap B \subseteq kB \setminus \{2B_1 \cup 2B_2\}$, we see that $(\star)$ implies

$$\begin{aligned}\frac{(k+1)^4}{4}\left\{\frac{4}{c_4}\right\}^3 \frac{\alpha}{\operatorname{rad}(B)}\mu(B_3 \cap B \setminus G_3) &\leq \int_{B_3 \cap B \setminus G_3}\left\{\frac{\operatorname{dist}(\xi, L)}{\operatorname{rad}(B)}\right\}^2 d\mu(\xi)\\ &\leq \int_{kB\setminus\{2B_1\cup 2B_2\}}\left\{\frac{\operatorname{dist}(\xi, L)}{\operatorname{rad}(B)}\right\}^2 d\mu(\xi)\\ &< \frac{(k+1)^4}{4}\left\{\frac{4}{c_4}\right\}^2 \alpha,\end{aligned}$$

i.e., $\mu(B_3 \cap B \setminus G_3) < (c_4/4)\operatorname{rad}(B)$.

Since $\mu(B_3 \cap B) > c_4 \operatorname{rad}(B)$, it follows from the last three paragraphs that $\mu(G_{13} \cap G_{23} \cap G_3) > 0$. We may thus choose and fix a point $z_3 \in G_{13} \cap G_{23} \cap G_3$. So we have a point $z_3 \in B_3 \cap B$ such that

$$\int_{\Delta_{13}(z_1,z_3)} c^2(z_1, \eta, z_3)\, d\mu(\eta) < \left\{\frac{4}{c_4}\right\}^2 \frac{\alpha}{\operatorname{rad}^2(B)},$$

$$\int_{\Delta_{23}(z_2,z_3)} c^2(\zeta, z_2, z_3)\, d\mu(\zeta) < \left\{\frac{4}{c_4}\right\}^2 \frac{\alpha}{\operatorname{rad}^2(B)}, \quad \text{and}$$

$$\left\{\frac{\operatorname{dist}(z_3, L)}{\operatorname{rad}(B)}\right\}^2 < \frac{(k+1)^4}{4}\left\{\frac{4}{c_4}\right\}^3 \frac{\alpha}{\operatorname{rad}(B)}.$$

Let L' be the line through z_2 and z_3. This line is well defined, i.e., $z_2 \neq z_3$, since B_2 and B_3 are disjoint. Because k' is greater than or equal to $1/c_3$, we have $2B_1 \subseteq \Delta_{23}(z_2, z_3)$. From Proposition 4.1 it follows that for any $\zeta \in 2B_1$,

$$\operatorname{dist}(\zeta, L') = \frac{c(\zeta, z_2, z_3)}{2} \cdot |\zeta - z_2| \cdot |\zeta - z_3| \leq \frac{c(\zeta, z_2, z_3)}{2} \cdot \{3\operatorname{rad}(B)\}^2.$$

Thus

$$(\dagger)\quad \begin{aligned}\int_{2B_1}\left\{\frac{\operatorname{dist}(\zeta, L')}{\operatorname{rad}(B)}\right\}^2 d\mu(\zeta) &\leq \frac{81}{4}\operatorname{rad}^2(B)\int_{\Delta_{23}(z_2,z_3)} c^2(\zeta, z_2, z_3)\, d\mu(\zeta)\\ &< \frac{81}{4}\left\{\frac{4}{c_4}\right\}^2 \alpha.\end{aligned}$$

Given $\zeta \in 2B_1$, let ζ' be the foot of the perpendicular from ζ to L'. By Similar Triangles,

$$\mathrm{dist}(\zeta', L) = \frac{|\zeta' - z_2|}{|z_3 - z_2|}\,\mathrm{dist}(z_3, L) \leq \frac{|\zeta - z_2|}{|z_3 - z_2|}\,\mathrm{dist}(z_3, L) \leq \frac{3\,\mathrm{rad}(B)}{10\,c_3\,\mathrm{rad}(B)}\mathrm{dist}(z_3, L).$$

Thus

$$\left\{\frac{\mathrm{dist}(\zeta', L)}{\mathrm{rad}(B)}\right\}^2 \leq \frac{9}{100\,c_3^2}\left\{\frac{\mathrm{dist}(z_3, L)}{\mathrm{rad}(B)}\right\}^2 < \frac{9}{100\,c_3^2}\frac{(k+1)^4}{4}\left\{\frac{4}{c_4}\right\}^3\frac{\alpha}{\mathrm{rad}(B)}.$$

From this and the estimate $\mu(2B_1) \leq M\,\mathrm{rad}(2B_1) = M(2\,c_3\,\mathrm{rad}(B))$, a consequence of (3) of Reduction 8.3, we obtain

$$(\ddagger)\quad \int_{2B_1}\left\{\frac{\mathrm{dist}(\zeta', L)}{\mathrm{rad}(B)}\right\}^2 d\mu(\zeta) < \frac{9(k+1)^4 M}{200\,c_3}\left\{\frac{4}{c_4}\right\}^3 \alpha.$$

Now note that $\{\mathrm{dist}(\zeta, L)\}^2 \leq 2|\zeta - \zeta'|^2 + 2\{\mathrm{dist}(\zeta', L)\}^2 = 2\{\mathrm{dist}(\zeta, L')\}^2 + 2\{\mathrm{dist}(\zeta', L)\}^2$. From this inequality, (†), and (‡), we obtain

$$\int_{2B_1}\left\{\frac{\mathrm{dist}(\zeta, L)}{\mathrm{rad}(B)}\right\}^2 d\mu(\zeta) < \left[\frac{81}{2}\left\{\frac{4}{c_4}\right\}^2 + \frac{9(k+1)^4 M}{100\,c_3}\left\{\frac{4}{c_4}\right\}^3\right]\alpha.$$

This last estimate, a corresponding and similarly obtained estimate for the integral over $2B_2$, and ($\star$) imply that

$$\beta_2^2(B) \leq \{\beta_2^L(B)\}^2 = \frac{1}{\mathrm{rad}(B)}\int_{kB}\left\{\frac{\mathrm{dist}(\zeta, L)}{\mathrm{rad}(B)}\right\}^2 d\mu(\zeta) < \frac{C_2}{Mk}\frac{\alpha}{\mathrm{rad}(B)}.$$

Letting $\alpha \downarrow c^2(B)$, we are done. □

8.5 Domination of Local Curvature by Global Curvature

The next result is the second step from beta numbers to global curvature.

Proposition 8.9 *For* $k' \geq 1$, $\displaystyle\int_0^\infty \int_{\mathbb{C}} c^2(z; r)\, d\mu(z)\, \frac{dr}{r^2} \leq Mk' \ln(2k'^2)\, c^2(\mu)$.

Proof Given $(\zeta, \eta, \xi) \in \mathbb{C}^3$, set $\widetilde{\Delta}(\zeta, \eta, \xi) = \{(z, r) \in \mathbb{C} \times (0, \infty) : (\zeta, \eta, \xi) \in \Delta(z; r)\}$. Then, by Fubini's Theorem [RUD, 8.8], the left-hand side of the inequality that we wish to establish is just

$$\iiint_{\mathbb{C}^3}\left\{\iint_{\widetilde{\Delta}(\zeta,\eta,\xi)} d\mu(z)\,\frac{dr}{r^2}\right\} c^2(\zeta, \eta, \xi)\, d\mu(\zeta)\, d\mu(\eta)\, d\mu(\xi).$$

Thus to finish it suffices to show that the double integral inside the curly brackets above is always at most $Mk' \ln(2k'^2)$.

Given $(z, r) \in \widetilde{\Delta}(\zeta, \eta, \xi)$, note the following:

1. $|\zeta - \eta| \geq r/k'$, i.e., $r \leq k'|\zeta - \eta|$,
2. $|\zeta - \eta| \leq |\zeta - z| + |z - \eta| \leq 2k'r$, i.e., $r \geq |\zeta - \eta|/(2k')$, and
3. $|\zeta - z| \leq k'r$, i.e., $z \in B(\zeta; k'r)$.

Moreover,

4. $\displaystyle\int_{B(\zeta;k'r)} d\mu(z) = \mu(B(\zeta; k'r)) \leq Mk'r$ by (3) of Reduction 8.3.

Hence

$$\iint_{\widetilde{\Delta}(\zeta,\eta,\xi)} d\mu(z)\, \frac{dr}{r^2} \leq \int_{|\zeta-\eta|/(2k')}^{k'|\zeta-\eta|} \int_{B(\zeta;k'r)} d\mu(z)\, \frac{dr}{r^2}$$

$$\leq Mk' \int_{|\zeta-\eta|/(2k')}^{k'|\zeta-\eta|} \frac{dr}{r} = Mk' \ln(2k'^2).$$

□

Putting the last two propositions together we could get a result whose punch line would read as follows:

$$\int_0^\infty \int_{\mathbb{C}} \beta_1^2(z; r)\, \mathcal{X}_{\{\delta(z;r)\geq\delta\}}\, d\mu(z)\, \frac{dr}{r} \leq C_3\, c^2(\mu)$$

for an appropriate constant C_3. This gives us control over the β_1s corresponding to discs with high density. We will need something a little more general however – control over the β_1s corresponding to discs that may not have a high density, but are "close" to a disc of the same radius with high density. Thus a regional version of density must be introduced.

Given a closed ball $B = B(z; r)$, set

$$\tilde{\delta}(B) = \tilde{\delta}(z; r) = \sup\{\delta(B') : \text{cen}(B') \in k^* B \text{ and } \text{rad}(B') = \text{rad}(B)\}.$$

Note the dependence of this quantity, the *regional density* of μ about B, on a new parameter $k^* \geq 0$ as well as on μ. This dependence will usually be notationally suppressed.

The next proposition is the whole point of this and the previous section.

Proposition 8.10 *Given* $0 < \delta < 1$, $k \geq 1$, *and* $k^* \geq 0$, *set* $k' = \max\{k + k^*, 1/c_3(M, \delta)\}$ *and*

$$C_3(M, \delta, k, k^*) = M\{k' + k^*\} \ln(2\{k' + k^*\}^2)\, C_2(M, \delta, k + k^*).$$

Then

$$\int_0^\infty \int_{\mathbb{C}} \beta_1^2(z;r)\,\mathcal{X}_{\{\tilde{\delta}(z;r)\geq\delta\}}\,d\mu(z)\,\frac{dr}{r} \leq C_3(M,\delta,k,k^*)\,c^2(\mu).$$

Proof Local curvature, although not occurring in the enunciation of the proposition, will occur in its proof. What shall we take the local curvature's parameter k' to be? To use Proposition 8.8 as we wish below, it turns out that $k' = \max\{k+k^*, 1/c_3\}$ will suffice. Thus our choice of k' in the proposition's enunciation.

Consider any z and r for which $\tilde{\delta}(z;r) \geq \delta$. Then almost (but not quite) by definition, we obtain a point $w \in B(z;k^*r)$ such that $\delta(w;r) \geq \delta$ (one needs to show that the supremum defining $\tilde{\delta}(z;r)$ is actually a maximum – this is left to the reader and involves a use of the Lower Continuity Property of Measures [RUD, 1.19(e)]). At this point it is convenient to subscript the beta numbers, local curvatures, and auxiliary sets Δ involved in the definition of local curvature with the parameters upon which they depend and not suppress them as usual. Note that $B(z;kr) \subseteq B(w;(k+k^*)r)$ and $\Delta_{k'}(w;r) \subseteq \Delta_{k'+k^*}(z;r)$. Then, by Proposition 8.8,

$$\begin{aligned}\beta_{1,k}^2(z;r) \leq \beta_{1,k+k^*}^2(w;r) &\leq C_2(M,\delta,k+k^*)\,\frac{c_{k'}^2(w;r)}{r}\\ &\leq C_2(M,\delta,k+k^*)\,\frac{c_{k'+k^*}^2(z;r)}{r}.\end{aligned}$$

Proposition 8.9 now finishes off the proof. □

8.6 Selection of Parameters for the Construction

The construction of the Lipschitz graph involves eight parameters: M, δ, k, k^*, α, ε, κ, and θ. The last parameter θ occurs in a technical context toward the very end of our construction and is thus rather awkward to explain just now. Accordingly we will simply fix its value in this section and offer no explanation here of what it is. The other seven parameters, whose values will also be fixed in this section, can be more easily explained. Indeed, the reader has already encountered the first four of the seven. Nevertheless, for the convenience of the reader, we now reintroduce these four familiar parameters ... and then newly introduce the three remaining strangers:

- M is a possibly quite large positive upper bound on the density of our measure holding for all balls without exception,
- δ is a small positive lower density threshold determining which balls are to be counted as having high density,
- k is a large positive number occurring in the definition of the beta numbers,
- k^* is a large positive number occurring in the definition of the regional densities,
- α is a small positive number whose double will be an upper bound on the angles that certain lines associated with certain "good" balls will make with a baseline L_0 to be selected in the next section,

- ε is a small positive number whose triple will be an upper bound on the beta numbers of these "good" balls, and
- κ is the small positive upper bound on the global curvature of the measure occurring in the enunciation of Reduction 8.3. Indeed, to prove Reduction 8.3 in a constructive fashion, one must produce a candidate for κ and then show that the purported candidate actually works!

Of these parameters, M is the one that is given to us fixed and so unalterable. The other seven parameters are open to our free choice and must be very carefully chosen. This we now do and as their values are fixed we subscript each parameter with a zero.

In order to do this, however, we must first specify more "constants." Recall that in Lemma 8.5, Lemma 8.6, and Proposition 8.7 we defined five small quantities $c_i(M,\delta)$, $i = 1,2,3,4,5$, and one large one $C_1(M,\delta)$. Moreover, in Proposition 8.8 and Proposition 8.10 we defined two large quantities, $C_2(M,\delta,k)$ and $C_3(M,\delta,k,k^*)$. We now need to define five more large quantities of which the first three depend on the familiar numbers M and δ while the last two depend on a smooth function φ compactly supported on $\mathbb{R}$:

$$C_4(M,\delta) = 6\left\{2 + \frac{6}{\delta} + C_1(M,\delta)\left[2 + \frac{3}{\delta}\right]\right\},$$

$$C_5(M,\delta) = \max\left\{\frac{1}{10\delta}\left[\frac{4}{\delta} + \frac{10^{13}M^2\,C_1(M,\delta)}{c_2(M,\delta)^2}\right]^2, 2[1{,}201\,C_1(M,\delta)]^2\right\},$$

$$C_6(M,\delta) = 10^{20}\,C_1(M,\delta)^2 + 11{,}522\,C_5(M,\delta),$$

$$C_7(\varphi) = \frac{\{\|\varphi\|_\infty^2 + \|\varphi''\|_\infty^2\}\,\|\varphi\|_\infty^2}{4\pi}, \text{ and}$$

$$C_8(\varphi) = 16\sqrt{71\{1 + 2\,\|\varphi\|_1\|\varphi\|_\infty\}} + \frac{71}{2}\|\varphi'\|_1\|\varphi\|_\infty.$$

These quantities will make their first appearance, in the order we have listed them above, in Lemma 8.19, Proposition 8.53, Proposition 8.59, Proposition 8.71, and Proposition 8.74 to follow.

First, fix the value of the parameter δ by setting $\delta_0 = 1/(54 \times 10^9)$. The full force of this choice will be needed later to prove Proposition 8.50. At most other places it will suffice simply to know that $0 < \delta_0 < 1$.

Second, fix the value of the parameter α by setting $\alpha_0 = 1/(32 \times 71)$. The full force of this choice will be needed later to prove Proposition 8.52. At most other places it will suffice simply to know that $0 < \alpha_0 \leq \pi/8$.

Third, fix the value of the parameter k by setting $k_0 = 880\,C_4(M,\delta_0)/\alpha_0$. The full force of this choice will be needed later to prove Lemma 8.38. At most other places it will suffice simply to know that $k_0 = 1{,}999{,}360\, C_4(M, \delta_0)$ due to our choice of α_0.

Fourth, fix the value of the parameter k^* by setting $k_0^* = 40$. The full force of this choice will be needed later to prove Proposition 8.53.

Fifth, fix the value of the parameter θ by setting $\theta_0 = \min\{1/2, c_1(M, \delta_0)/[12\, C_8(\varphi)]\}$ where φ is a fixed function as in Lemma 8.60 to follow. The full force of this choice will be needed later to prove Lemmas 8.74 and 8.79. It bears emphasizing that Lemma 8.60 in no way depends on the construction we are now embarking on – for if it did, there would be some danger of circularity! So the reader should convince him- or herself, now or sometime later, that we could easily have stated and proved this lemma right here. Indeed, here is the purely logical place to put the lemma; we have instead elected to put it later for essentially aesthetic reasons of narrative flow.

Our last two parameters must be more elaborately chosen to satisfy a number of inequalities! A parenthesized item after an inequality indicates where the full force of the inequality is needed.

Sixth, fix the value of the parameter ε by choosing $\varepsilon_0 > 0$ but so small that it satisfies the following seven conditions:

(1) $\varepsilon_0 \le c_5(M, \delta_0)$ (Lemma 8.19 and any other place where Proposition 8.7 is used),

(2) $[1 + 3\alpha_0]\, C_4(M, \delta_0)\sqrt{\varepsilon_0} \le [3 - (8/\pi)]\, \alpha_0$ (Corollary 8.24),

(3) $6{,}000{,}000\, C_1(M, \delta_0)\, \varepsilon_0 \le \alpha_0$ (Lemma 8.37),

(4) $(3 \times 10^{30})\, M \dfrac{C_4(M, \delta_0)^5}{c_2(M, \delta_0)} \sqrt{\varepsilon_0} \le 10^{-5}$ (Proposition 8.49),

(5) $\{1 + \ln(1/\theta_0)\}\, C_1(M, \delta_0)\, \varepsilon_0 \le \alpha_0/6$ (Lemma 8.76),

(6) $2\sqrt{\varepsilon_0} \le C_8(\varphi)\, \theta_0\, \alpha_0$ (Lemma 8.79), and

(7) $8{,}004{,}000\, M \dfrac{C_6(M, \delta_0)\, C_7(\varphi)}{\theta_0^4\, \alpha_0^2}\, \varepsilon_0^2 \le 10^{-5}$ (Proposition 8.82).

Seventh and last, fix the value of the parameter κ by choosing $\kappa_0 > 0$ but so small that it satisfies the following four conditions:

(1) $C_2(M, \delta_0, k_0)\, \kappa_0 \le \varepsilon_0^2$ (Construction 8.12),

(2) $C_3(M, \delta_0/4, k_0, k_0^*)\, \kappa_0 \le \{\varepsilon_0/64\}^2\, (\ln 2)\, (2 \times 10^{-5})$ (Proposition 8.43),

(3) $10^{11} M \dfrac{C_4(M, \delta_0)^4}{c_2(M, \delta_0)^3}\, \kappa_0 \le 10^{-5}$ (Proposition 8.45), and

(4) $C_3(M, 20\, c_2(M, \delta_0), k_0, k_0^*)\, \kappa_0 \le \varepsilon_0^2$ (Proposition 8.59).

Our choices for δ_0, k_0, k_0^*, and ε_0 let us apply Lemmas 8.5 and 8.6 and Propositions 8.7, 8.8, and 8.10 as needed. Also, note that by Proposition 8.7, Condition

(1) for the selection of ε_0 implies that $\varepsilon_0 \le 10^{-21}$, a fact that will be of use to us a number of times in what follows.

Stipulation 8.11 *For the rest to this chapter, we shall assume that the measure* μ *of Stipulation* 8.4 *also satisfies* (4) *of Reduction* 8.3 *with the number* κ *there having the value* κ_0 *just chosen. Thus we are assuming that* $c^2(\mu) < \kappa_0$.

In what lies ahead we will be defining and constructing a profusion of auxiliary objects. For the sake of ease of future reference, these definitions and constructions will henceforth be enumerated.

8.7 Construction of a Baseline L_0

Construction 8.12 (A Baseline L_0) *Arbitrarily fix a point* $z_0 \in K$. *By* (1) *and* (2) *of Reduction* 8.3 *and our choice of* δ_0, $\delta(z_0; 1) = \mu(\mathbb{C}) > 1 \ge \delta_0$. *Because of this and our choice of* k_0, *we may apply Proposition* 8.8 (*with* $k' = \max\{k_0, 1/c_3(M, \delta_0)\}$) *to get*

$$\beta_1^2(z_0; 1) \le C_2(M, \delta_0, k_0)\, c^2(z_0; 1) \le C_2(M, \delta_0, k_0)\, c^2(\mu).$$

But (4) *of Reduction* 8.3 *and Condition* (1) *for the selection of* κ_0 *imply that* $C_2(M, \delta_0, k_0)\, c^2(\mu) < \varepsilon_0^2$. *Hence* $\beta_1(z_0; 1) < \varepsilon_0$. *The definition of* β_1 *now allows us to choose and fix a line* L_0 *in* $\mathbb{C}$ *such that* $\beta_1^{L_0}(z_0; 1) < \varepsilon_0$.

We now engage in some conveniences available to us since we have elected to work only in $\mathbb{C} = \mathbb{R}^2$ *and not more generally in* $\mathbb{R}^n$. *Via a rotation and vertical translation of* $\mathbb{C}$ *we may assume that* L_0 *is the real axis of* $\mathbb{C}$. *As a general convention for the rest of this chapter, given* $a > 0$, *set* $U_a = (-a, a) \subseteq \mathbb{R} = L_0$. *Let* π *denote the orthogonal projection of* $\mathbb{C}$ *onto* L_0. *Then, by* (1) *of Reduction* 8.3, *via a horizontal translation of* $\mathbb{C}$ *we may also assume that* $\pi(K) \subseteq U_{1/2}$. *The imaginary axis of* $\mathbb{C}$ *will now serve as a line perpendicular to* L_0 *and be denoted* $L_0^\perp$. *Let* $\pi^\perp$ *denote the orthogonal projection of* $\mathbb{C}$ *onto* $L_0^\perp$. *We have thus conveniently arranged things so that* $\pi(z) = \operatorname{Re} z$ *and* $\pi^\perp(z) = i \operatorname{Im} z$ *for all* $z \in \mathbb{C}$.

The line L_0 is the baseline referred to in the last section in connection with the parameter α_0; it will serve as the x-axis for the Lipschitz graph that will eventually be constructed.

8.8 Definition of a Stopping-Time Region S_0

The presence of the 3 and 2 in (ii) and (iii) of the following definition is to avoid the presence of fractions much later in Definition 8.41.

Definition 8.13 (The Total Region) *Let* S_{total} *denote the set of all ordered pairs* $(z, r) \in K \times (0, 3)$ *which satisfy the following three conditions*:

(i) $\delta(z; r) \ge \delta_0$,

(ii) $\beta_1(z; r) < 3\varepsilon_0$, *and*
(iii) $\exists$ *a line* L *such that* $\beta_1^L(z; r) < 3\varepsilon_0$ *and* $\angle(L, L_0) < 2\alpha_0$.

Proposition 8.14 *For any* $(z, r) \in K \times [1, 3)$, $\delta(z; r) > \delta_0$ *and* $\beta_1^{L_0}(z; r) < \varepsilon_0$. *Thus* $K \times [1, 3) \subseteq S_{\text{total}}$.

Proof For $(z, r) \in K \times [1, 3)$, by (1) and (2) of Reduction 8.3 and our choice of δ_0, $\delta(z; r) > \mu(\mathbb{C})/3 > 1/3 \geq \delta_0$. Also, by our choice of k_0, (1) of Reduction 8.3, and Construction 8.12,

$$\beta_1^{L_0}(z; r) = \frac{1}{r} \int_{B(z;k_0 r)} \frac{\text{dist}(\zeta, L_0)}{r} d\mu(\zeta) \leq \int_{\mathbb{C}} \text{dist}(\zeta, L_0) d\mu(\zeta) = \beta_1^{L_0}(z_0; 1) < \varepsilon_0.$$

Since $\angle(L_0, L_0) = 0$, the final assertion of the proposition is now clear. □

What would be highly desirable for what we wish to do is for S_{total} to have a property called *coherence* in [DS1] and [DS2], i.e., for it to be the case that $(z, r') \in S_{\text{total}}$ whenever $(z, r) \in S_{\text{total}}$ and $r' \in [r, 3)$. This is unfortunately not necessarily the case. Accordingly, we will find and drop down to a subset S_0 of S_{total} which is coherent. In [DS1] and [DS2] such a collection is called a *stopping-time region* – hence the title of this section. The subset S_0 will be defined in terms of an important auxiliary function h to which we now turn.

Definition 8.15 (The First Auxiliary Function) Given $z \in K$, set

$$h(z) = 2 \sup\{s \in (0, 1) : \exists\, w \in K \text{ such that } z \in B(w; s/4) \text{ and } (w, s) \notin S_{\text{total}}\}.$$

A few remarks are in order. First, here we adopt the convention that $\sup \emptyset = 0$. Thus $0 \leq h(z) \leq 2$ always. Second, in the definition we are restricting s to $(0, 1)$. By the last proposition we could have restricted s to $(0, 3)$ instead with no change at all in the resulting value of $h(z)$. Third, given a closed ball $B = B(w; s)$, think of z as being "deep within" B when $z \in B(w; s/4)$ and think of B as "bad" when $(w, s) \notin S_{\text{total}}$. Then $h(z)$ is a measure of how big the bad balls that z lies deep within can get. So points where h vanishes are best for what we wish to do.

We are now ready for the following.

Definition 8.16 (The Stopping-Time Region) $S_0 = \{(z, r) \in S_{\text{total}} : r \geq h(z)\}$.

Proposition 8.17 *If* $z \in K$, $r \in (0, 3)$, *and* $r \geq h(z)$, *then* $(z, r) \in S_0$.

Proof Proceeding by contradiction, suppose $(z, r) \notin S_0$. Then we must have $(z, r) \notin S_{\text{total}}$. By the last proposition, $r \in (0, 1)$. Thus we may set $w = z$ and $s = r$ in the Definition of h (8.15) to conclude that $h(z) \geq 2r$. But $r \geq h(z)$ by assumption, so $r \geq 2r$. This is absurd since $r > 0$. □

That S_0 is coherent and so a stopping-time region is just (c) of the following.

Corollary 8.18 (a) *If* $h(z) = 0$ *where* $z \in K$, *then* $(z, r) \in S_0$ *for all* $r \in (0, 3)$.
(b) *If* $h(z) > 0$ *where* $z \in K$, *then* $(z, r) \in S_0$ *for all* $r \in [h(z), 3)$.
In particular, $(z, h(z)) \in S_0$ *when* $z \in K$ *and* $h(z) > 0$.
(c) *If* $(z, r) \in S_0$ *where* $z \in K$, *then* $(z, r') \in S_0$ *for all* $r' \in [r, 3)$.

The sets S_{total} and S_0 will frequently be thought of as sets of closed balls in $\mathbb{C}$. Thus when it is written that $B \in S_0$ where B is a closed ball, what is really meant is that $B = B(z; r)$ and $(z, r) \in S_0$. Clearly the balls of S_0 are good for constructing the sought-after Lipschitz graph. The next lemma makes this observation concrete and precise. We call it a lemma since its sole purpose in life is to be superseded by Proposition 8.22 in the next section!

Lemma 8.19 *Suppose* $z, w \in K$, $r > 0$, *and* $N \geq 1$. *Then*

$$|\pi^{\perp}(z) - \pi^{\perp}(w)| \leq (8/\pi)\,\alpha_0\,|\pi(z) - \pi(w)| + (C_4(M, \delta_0)/3)\{\sqrt{\varepsilon_0} + 1/N\}|z - w|$$

whenever $B(z; r) \in S_0$, $B(w; r) \in S_0$, *and* $|z - w| \geq Nr$.

Recall that the quantity $C_4(M, \delta)$ was defined back in Section 8.6.

Proof For now and forevermore, note that Proposition 8.7 may be applied with δ, k, and ε set equal to δ_0, k_0, and ε_0, respectively, because of our choices of δ_0 and k_0 and Condition (1) for the selection of ε_0.

Setting $B_1 = B(z; |z - w|)$, we have $B_1 \in S_0$ by (c) of Corollary 8.18 since $N \geq 1$. Then the Definitions of S_{total} and S_0 (8.13 and 8.16) imply that $\delta(B_1) \geq \delta_0$ and that there exists a line L_1 such that $\beta_1^{L_1}(B_1) < 3\varepsilon_0$. Now set $B_1' = B(z; \sqrt{\varepsilon_0}\,|z - w| + r)$. Using (c) of Corollary 8.18 and the Definitions of S_{total} and S_0 (8.13 and 8.16) again, we see that $\delta(B_1') \geq \delta_0$ and so $\mu(B_1') \geq \delta_0\sqrt{\varepsilon_0}\,|z - w|$. Let z' be a point of B_1' closest to L_1. Then, by our choice of k_0,

$$\operatorname{dist}(z', L_1)\,\delta_0\sqrt{\varepsilon_0}\,|z - w| \leq \int_{B_1'} \operatorname{dist}(\zeta, L_1)\,d\mu(\zeta) \leq \beta_1^{L_1}(B_1)|z - w|^2 \leq 3\varepsilon_0|z - w|^2,$$

i.e., $\operatorname{dist}(z', L_1) \leq (3/\delta_0)\sqrt{\varepsilon_0}\,|z - w|$.

Setting $B_2 = B(w; |z - w|)$ and $B_2' = B(w; \sqrt{\varepsilon_0}\,|z - w| + r)$, we may proceed similarly to get a line L_2 such that $\beta_1^{L_2}(B_2) < 3\varepsilon_0$, $\angle(L_2, L_0) < 2\alpha_0$, and a point $w' \in B_2'$ such that $\operatorname{dist}(w', L_2) \leq (3/\delta_0)\sqrt{\varepsilon_0}\,|z - w|$. (We could also have arranged for $\angle(L_1, L_0) < 2\alpha_0$ above but that turns out not to be needed below.) Clearly

$$|z - z'| \text{ and } |w - w'| \leq \sqrt{\varepsilon_0}\,|z - w| + r \leq \{\sqrt{\varepsilon_0} + 1/N\}|z - w|.$$

Let z_1' be the foot of the perpendicular from z' to L_1 and w_2' be the foot of the perpendicular from w' to L_2. Clearly

$$|z' - z_1'| \text{ and } |w' - w_2'| \leq (3/\delta_0)\sqrt{\varepsilon_0}\,|z - w| \leq (3/\delta_0)\{\sqrt{\varepsilon_0} + 1/N\}|z - w|.$$

Moreover, letting z'_{12} be the foot of the perpendicular from z'_1 to L_2, applying (b) of Proposition 8.7 to the lines L_1 and L_2 obtained from the balls B_1 and B_2, and using the last two displayed inequalities, we see that

$$\begin{aligned}|z'_1 - z'_{12}| &= \mathrm{dist}(z'_1, L_2)\\ &\le C_1\,\varepsilon_0\,\{|z'_1 - z| + |z - w|\}\\ &\le C_1\{|z'_1 - z| + \{\sqrt{\varepsilon_0} + 1/N\}|z - w|\}\\ &\le C_1\{2 + (3/\delta_0)\}\{\sqrt{\varepsilon_0} + 1/N\}|z - w|.\end{aligned}$$

Finally, the last three displayed inequalities imply

$$(\star)\ |z - z'_{12}| + |w - w'_2| \le (C_4/6)\{\sqrt{\varepsilon_0} + 1/N\}|z - w|.$$

Note that both z'_{12} and w'_2 lie on L_2 which makes an angle of at most $2\alpha_0$ with L_0. Also, recall the easy inequality $0 \le \tan x \le (4/\pi)x$ valid for $0 \le x \le \pi/4$. Thus, by our choice of α_0,

$$(\star\star)\ |\pi^\perp(z'_{12}) - \pi^\perp(w'_2)| \le (\tan 2\alpha_0)|\pi(z'_{12}) - \pi(w'_2)| \le (8/\pi)\,\alpha_0\,|\pi(z'_{12}) - \pi(w'_2)|.$$

From both $(\star)$ and $(\star\star)$ we obtain

$$\begin{aligned}(\dagger)\ &|\pi^\perp(z) - \pi^\perp(w)|\\ &\le |\pi^\perp(z) - \pi^\perp(z'_{12})| + |\pi^\perp(z'_{12}) - \pi^\perp(w'_2)| + |\pi^\perp(w'_2) - \pi^\perp(w)|\\ &\le |z - z'_{12}| + |\pi^\perp(z'_{12}) - \pi^\perp(w'_2)| + |w'_2 - w|\\ &\le (8/\pi)\,\alpha_0\,|\pi(z'_{12}) - \pi(w'_2)| + (C_4/6)\{\sqrt{\varepsilon_0} + 1/N\}|z - w|,\end{aligned}$$

whereas from $(\star)$ alone we obtain

$$\begin{aligned}(\ddagger)\ |\pi(z'_{12}) - \pi(w'_2)| &\le |\pi(z'_{12}) - \pi(z)| + |\pi(z) - \pi(w)| + |\pi(w) - \pi(w'_2)|\\ &\le |z'_{12} - z| + |\pi(z) - \pi(w)| + |w - w'_2|\\ &\le |\pi(z) - \pi(w)| + (C_4/6)\{\sqrt{\varepsilon_0} + 1/N\}|z - w|.\end{aligned}$$

The inequality of interest to us now follows easily from $(\dagger)$, and $(\ddagger)$:

$$\begin{aligned}|\pi^\perp(z) - \pi^\perp(w)| &\le (8/\pi)\,\alpha_0\,|\pi(z) - \pi(w)|\\ &\quad+\{(8/\pi)\,\alpha_0 + 1\}\,(C_4/6)\{\sqrt{\varepsilon_0} + 1/N\}|z - w|\\ &\le (8/\pi)\,\alpha_0\,|\pi(z) - \pi(w)| + (C_4/3)\{\sqrt{\varepsilon_0} + 1/N\}|z - w|.\end{aligned}$$

□

8.9 Definition of a Lipschitz Set K_0 over L_0

In this section we will define a closed subset K_0 of K and then show it to be a Lipschitz set over the base line L_0. Eventually we will construct a Lipschitz graph over L_0 which treads its way through K_0. Our proof of Reduction 8.3 will then be completed by showing that K_0 carries more than 99% of the mass of μ.

A glance at (a) of Corollary 8.18 and Lemma 8.19 suggests that we take K_0 to be $\{z \in K : h(z) = 0\}$. It turns out that, although this choice of K_0 is indeed a Lipschitz set through which we can thread a Lipschitz graph, we would then have difficulty showing that K_0 carries more than 99% of the mass of μ. The difficulty would stem from a lack of effective control on our part over where h is positive. So we now turn to a function d, essentially a smoothed out version of h, which is dominated by h and will take its place in defining K_0.

Definition 8.20 (The Second Auxiliary Function) *Given $z \in \mathbb{C}$, set*

$$d(z) = \inf\{|z - \tilde{z}| + r : B(\tilde{z}; r) \in S_0\}.$$

The elementary, but very useful, facts of interest to us concerning d are summarized in the following.

Proposition 8.21 (a) *The function d is Lipschitz on $\mathbb{C}$ with bound* 1.
(b) *For all $z \in K$, $0 \le d(z) \le h(z)$.*
(c) *For all $z \in \mathbb{C}$, $\operatorname{dist}(z, K) \le d(z) \le \operatorname{dist}(z, K) + 2$.*

Proof (a) Note that for each $B(\tilde{z}; r) \in S_0$, the function $z \in \mathbb{C} \mapsto |z - \tilde{z}| + r \in [0, \infty)$ is clearly Lipschitz on $\mathbb{C}$ with bound 1. Since the infimum of any number of nonnegative Lipschitz functions with bound 1 on a metric space is again a Lipschitz function with bound 1 on the metric space, we are done.

(b) Given $r \in (h(z), 3)$, $B(z; r) \in S_0$ by Proposition 8.17. Thus, by definition, $d(z) \le |z - z| + r = r$. Letting $r \downarrow h(z)$, we are done.

(c) If $B(\tilde{z}; r) \in S_0$, then $\tilde{z} \in K$ and so $\operatorname{dist}(z, K) \le |z - \tilde{z}| < |z - \tilde{z}| + r$. Upon infing over all $B(\tilde{z}; r) \in S_0$, we see that $\operatorname{dist}(z, K) \le d(z)$.

Let $\tilde{z} \in K$ be such that $|z - \tilde{z}| = \operatorname{dist}(z, K)$. Recall that $h(\tilde{z}) \le 2$ by the Definition of h (8.15). Thus $B(\tilde{z}; 2) \in S_0$ by Proposition 8.17. Hence $d(z) \le |z - \tilde{z}| + 2 = \operatorname{dist}(z, K) + 2$. □

A deeper fact about the function d, essentially an improvement of Lemma 8.19, follows.

Proposition 8.22 *Suppose $z, w \in K$ and $N \ge 1$. Then*

$$|\pi^{\perp}(z) - \pi^{\perp}(w)| \le (8/\pi)\,\alpha_0\,|\pi(z) - \pi(w)| + C_4(M, \delta_0)\{\sqrt{\varepsilon_0} + 2/N\}|z - w|$$

whenever $d(z) < |z - w|/(N + 2)$ and $d(w) < |z - w|/(N + 2)$.

Proof Set $r = |z - w|/(N + 2)$, so $d(z) < r$ and $d(w) < r$. By the Definition of d (8.20), there exist $B(\tilde{z}; s) \in S_0$ and $B(\tilde{w}; t) \in S_0$ such that $|z - \tilde{z}| + s < r$ and $|w - \tilde{w}| + t < r$. From (c) of Corollary 8.18 it follows that $B(\tilde{z}; r) \in S_0$ and $B(\tilde{w}; r) \in S_0$. Clearly $|\tilde{z} - \tilde{w}| \ge |z - w| - \{|z - \tilde{z}| + |w - \tilde{w}|\} \ge (N+2)r - 2r = Nr$. Lemma 8.19 applied to $\tilde{z}$, $\tilde{w}$, r, and N now yields

$$|\pi^{\perp}(\tilde{z}) - \pi^{\perp}(\tilde{w})| \leq (8/\pi)\,\alpha_0\,|\pi(\tilde{z}) - \pi(\tilde{w})| + (C_4/3)\{\sqrt{\varepsilon_0} + 1/N\}|\tilde{z} - \tilde{w}|.$$

Moreover, $|z - \tilde{z}| + |w - \tilde{w}| \leq 2r \leq 2|z - w|/N$, and so

$$\begin{aligned}|\pi^{\perp}(z) - \pi^{\perp}(w)| &\leq |\pi^{\perp}(z) - \pi^{\perp}(\tilde{z})| + |\pi^{\perp}(\tilde{z}) - \pi^{\perp}(\tilde{w})| + |\pi^{\perp}(\tilde{w}) - \pi^{\perp}(w)|\\ &\leq |z - \tilde{z}| + |\pi^{\perp}(\tilde{z}) - \pi^{\perp}(\tilde{w})| + |\tilde{w} - w|\\ &\leq |\pi^{\perp}(\tilde{z}) - \pi^{\perp}(\tilde{w})| + 2|z - w|/N.\end{aligned}$$

Similarly

$$|\pi(\tilde{z}) - \pi(\tilde{w})| < |\pi(z) - \pi(w)| + 2|z - w|/N$$

and

$$|\tilde{z} - \tilde{w}| \leq |z - w| + 2|z - w|/N \leq 3|z - w|.$$

From the last four displayed inequalities one gets

$$\begin{aligned}|\pi^{\perp}(z) - \pi^{\perp}(w)| \leq{}& (8/\pi)\,\alpha_0\,|\pi(z) - \pi(w)|\\ &+ [C_4\{\sqrt{\varepsilon_0} + 1/N\} + \{(8/\pi)\,\alpha_0 + 1\}(2/N)]|z - w|.\end{aligned}$$

A look at the definition of C_4 in Section 8.6 makes it clear that $C_4 \geq 12$. Thus, by our choice of α_0, we see that $\{(8/\pi)\,\alpha_0 + 1\}(2/N) \leq C_4/N$ and so are done. □

Of course the conclusion of this proposition is nontrivial only when $C_4\{\sqrt{\varepsilon_0} + 2/N\} \ll 1$ or even just < 1. We will now arrange for these two events to happen in the next two corollaries! To state the first corollary we need to first define our Lipschitz set.

Definition 8.23 (The Lipschitz Set) $K_0 = \{z \in K : d(z) = 0\}$.

By (a) of Proposition 8.21, K_0 is closed and so compact. That it is a Lipschitz set with bound $3\alpha_0$ is the content of the following.

Corollary 8.24 *For any $z, w \in K_0$,*

$$|\pi^{\perp}(z) - \pi^{\perp}(w)| \leq 3\alpha_0|\pi(z) - \pi(w)|.$$

Proof Without loss of generality, $z \neq w$. Since $d(z) = d(w) = 0$, we may apply Proposition 8.22 to z and w with N arbitrarily large to obtain

$$\begin{aligned}|\pi^{\perp}(z) - \pi^{\perp}(w)| &\leq (8/\pi)\,\alpha_0\,|\pi(z) - \pi(w)| + C_4\sqrt{\varepsilon_0}\,|z - w|\\ &\leq [(8/\pi)\,\alpha_0 + C_4\sqrt{\varepsilon_0}\,]|\pi(z) - \pi(w)|\\ &\quad + C_4\sqrt{\varepsilon_0}\,|\pi^{\perp}(z) - \pi^{\perp}(w)|.\end{aligned}$$

Condition (2) for the selection of ε_0 and our choice of α_0 imply that $C_4\sqrt{\varepsilon_0} < 1$. Thus the last displayed inequality may be rearranged to yield

$$|\pi^{\perp}(z) - \pi^{\perp}(w)| \leq \frac{(8/\pi)\,\alpha_0 + C_4\sqrt{\varepsilon_0}}{1 - C_4\sqrt{\varepsilon_0}}\,|\pi(z) - \pi(w)|.$$

Since Condition (2) for the selection of ε_0 also implies that

$$\frac{(8/\pi)\,\alpha_0 + C_4\sqrt{\varepsilon_0}}{1 - C_4\sqrt{\varepsilon_0}} \leq 3\alpha_0,$$

we are done. □

Construction 8.25 (A Function ℓ on $\pi(K_0)$) *From Corollary* 8.24 *we see that* $\pi : K_0 \mapsto L_0$ *is injective and so we can define a function* $\ell : \pi(K_0) \subseteq L_0 = \mathbb{R} \mapsto \mathbb{R} = L_0^{\perp}/i$ *by* $\pi(z) \mapsto \pi^{\perp}(z)/i$ *for* $z \in K_0$. *(Recall that in Constuction 8.12 we conveniently arranged for* L_0 *and* $L_0^{\perp}$ *to be the real and imaginary axes of* $\mathbb{C}$, *respectively.)*

The corollary also implies that ℓ is Lipschitz on $\pi(K_0)$ with bound $3\alpha_0$. Moreover, K_0 is contained in $\{x + i\ \ell(x) : x \in \pi(K_0)\}$, i.e., the graph of ℓ over K_0. Of course for K_0 to be really contained in a genuine Lipschitz graph, we would need to have ℓ extended in a Lipschitz manner to some subinterval of $L_0 = \mathbb{R}$ containing $\pi(K_0)$. It may occur to reader that we could proceed in the simple-minded fashion of the proof of Lemma 4.24 where we encountered another Lipschitz set. This would indeed result in a genuine Lipschitz graph, but we would then have trouble completing the proof of Reduction 8.3 by showing that K_0 has substantial μ measure. To accomplish this it turns out that we need to extend ℓ in a more devious manner; indeed, in a manner that results in "most" of K lying "near" the graph of ℓ. The next few sections are preparatory to this more devious manner of extending ℓ.

But before turning to that, we close this section with a second corollary to the last proposition. This one, unlike the first corollary, is valid for all points of K, not just those from K_0.

Corollary 8.26 *Suppose* $z, w \in K$ *and* $r > 0$ *are such that* $d(z) \leq r$, $d(w) \leq r$, *and* $|\pi(z) - \pi(w)| \leq r$. *Then*

$$|z - w| \leq 9\,C_4(M, \delta_0)\,r.$$

Proof Condition (2) for the selection of ε_0 and our choice of α_0 imply that $C_4\sqrt{\varepsilon_0} < 1/4$. So for $N = 8\,C_4$ we have $C_4\{\sqrt{\varepsilon_0} + 2/N\} \leq 1/2$. A look at the definition of C_4 in section 8.6 makes it clear that $C_4 \geq 12$.

Thus, on the one hand, if $r < |z - w|/(N + 2)$, we may apply Proposition 8.22 and use our choice of α_0 to obtain

$$|\pi^{\perp}(z) - \pi^{\perp}(w)| \leq (8/\pi)\,\alpha_0\,|\pi(z) - \pi(w)| + C_4\{\sqrt{\varepsilon_0} + 2/N\}|z - w|$$

$$\leq |\pi(z) - \pi(w)| + (1/2)\{|\pi(z) - \pi(w)| + |\pi^{\perp}(z) - \pi^{\perp}(w)|\}.$$

This may be easily manipulated to yield $|\pi^{\perp}(z) - \pi^{\perp}(w)| \leq 3|\pi(z) - \pi(w)|$, from which we have $|z - w| \leq 4|\pi(z) - \pi(w)| \leq 4r \leq 9\,C_4\,r$.

On the other hand, if $r \geq |z - w|/(N+2)$, then $|z - w| \leq (N+2)\,r \leq 9\,C_4\,r$. □

8.10 Construction of Adapted Dyadic Intervals $\{I_n\}$

The intervals we seek are defined through an auxiliary function D, involving d, which will enable us to associate a nicely placed ball of S_0 to each point of L_0.

Definition 8.27 (The Third Auxiliary Function) Given $p \in L_0$, set

$$D(p) = \inf\{d(z) : z \in \pi^{-1}(p)\}.$$

The elementary, but very useful, facts of interest to us concerning D are summarized in the following.

Proposition 8.28 (a) *For* $p \in L_0$, $D(p) = \min\{d(z) : z \in \pi^{-1}(p)\}$.
(b) *For* $p \in L_0$, $D(p) = \inf\{|p - \pi(\tilde{z})| + r : B(\tilde{z}; r) \in S_0\}$.
(c) *The function D is Lipschitz on L_0 with bound* 1.
(d) $L_0 \setminus \pi(K_0) = \{p \in L_0 : D(p) > 0\}$.

Proof (a) Choose $z_n \in \pi^{-1}(p)$ such that $d(z_n) \downarrow D(p)$. By (c) of Proposition 8.21, the sequence $\{z_n\}$ is bounded. Thus we may drop to a subsequence and assume that $\{z_n\}$ converges to a point $z \in \mathbb{C}$. By (a) of Proposition 8.21, $d(z_n) \to d(z)$. Clearly $z \in \pi^{-1}(p)$ and $d(z) = D(p)$, i.e., the infimum defining D is actually a minimum.

(b) For brevity's sake, denote $\inf\{|p - \pi(\tilde{z})| + r : B(\tilde{z}; r) \in S_0\}$ by $\widetilde{D}(p)$. Given $B(\tilde{z}; r) \in S_0$, let z be the point of $\mathbb{C}$ such that $\pi(z) = p$ and $\pi^{\perp}(z) = \pi^{\perp}(\tilde{z})$. Then $D(p) \leq d(z) \leq |z - \tilde{z}| + r = |p - \pi(\tilde{z})| + r$. Upon infing over all $B(\tilde{z}; r) \in S_0$, we see that $D(p) \leq \widetilde{D}(p)$.

For the reverse direction, given $\eta > 0$, from the Definitions of d and D (8.20 and 8.27) we obtain a point $z \in \pi^{-1}(p)$ and a ball $B(\tilde{z}; r) \in S_0$ such that $|z - \tilde{z}| + r < D(p) + \eta$. But $\widetilde{D}(p) \leq |p - \pi(\tilde{z})| + r = |\pi(z) - \pi(\tilde{z})| + r \leq |z - \tilde{z}| + r$, so $\widetilde{D}(p) < D(p) + \eta$. Upon letting $\eta \downarrow 0$, we obtain $\widetilde{D}(p) \leq D(p)$.

(c) Note that for each $B(\tilde{z}; r) \in S_0$, the function $p \in L_0 \mapsto |p - \pi(\tilde{z})| + r \in [0, \infty)$ is clearly Lipschitz on L_0 with bound 1. Since the infimum of any number of nonnegative Lipschitz functions with bound 1 on a metric space is again a Lipschitz function with bound 1 on the metric space, we are done by (b) just shown.

(d) We show $\pi(K_0) = \{p \in L_0 : D(p) = 0\}$ instead. So suppose $p \in \pi(K_0)$. Then $p = \pi(z)$, i.e., $z \in \pi^{-1}(p)$, for some $z \in K_0$, i.e., for some $z \in K$ such that $d(z) = 0$. Thus $0 \leq D(p) \leq d(z) = 0$.

Conversely, suppose $p \in L_0$ is such that $D(p) = 0$. By (a) already proven, $d(z) = 0$ for some $z \in \pi^{-1}(p)$. By (c) of Proposition 8.21, $d(z) = 0$ forces z to be an element of K and thus K_0 also. Hence $p = \pi(z) \in \pi(K_0)$. □

Recall that in Construction 8.12 we made the convention that $U_a = (-a, a)$ for $a > 0$.

Construction 8.29 (Adapted Dyadic Intervals $\{I_n\}$) *In Construction* 8.12 *we arranged for* $\pi(K) \subseteq U_{1/2}$. *Thus* $\pi(K_0) \subseteq U_5 \subseteq L_0$. *Consider the collection of dyadic intervals on* L_0 *of length at most* 1 *generated by the interval* U_5, *i.e., consider the collection of all intervals of the form*

$$(-5, -5 + 1/2^n) \text{ and } [-5 + m/2^n, -5 + (m+1)/2^n)$$

where $n = 0, 1, 2, \ldots$ *and* $m = 1, 2, 3, \ldots, 10 \cdot 2^n - 1$. *For* $p \in U_5 \setminus \pi(K_0)$, *let* J_p *denote the largest of these dyadic intervals containing* p *for which*

$$(\star)\ |J_p| \le \frac{1}{8} \inf\{D(q) : q \in J_p\}.$$

The interval J_p *exists by* (c) *and* (d) *of Proposition* 8.28. *From* (c) *of Proposition* 8.21 *it follows that* $\inf\{D(q) : q \in J_p\} = \inf\{d(z) : z \in \pi^{-1}(J_p)\} \le \inf\{\mathrm{dist}(z, K) + 2 : z \in \pi^{-1}(J_p)\} < 8$. *But then* $|J_p| < 1$ *by* $(\star)$ *and so* J_p *has length at most* $1/2$. *Thus* J_p *has a dyadic parent which we denote by* J_p'. *Clearly* J_p' *satisfies*

$$(\star\star)\ |J_p'| > \frac{1}{8} \inf\{D(q') : q' \in J_p'\}.$$

Now the collection $\{J_p : p \in U_5 \setminus \pi(K_0)\}$ *is actually countable and so may be relabeled* $\{I_n : n = 0, 1, 2, \ldots\}$. *These intervals* I_n *are our adapted dyadic intervals.*

As an exercise using the lower bound on $d(z)$ in (c) of Proposition 8.21, the reader may show that the open interval $(-5, -5+1/2)$ is an adapted dyadic interval. By (a) of the next proposition, all other adapted dyadic intervals are half-open. As a matter of convenience we shall frequently act as if all the adapted dyadic intervals are half-open. No harm will come from this practice.

The elementary, but very useful, facts of interest to us concerning the adapted dyadic intervals are summarized in the following.

Proposition 8.30 (a) *The collection* $\{I_n\}$ *is a partition of* $U_5 \setminus \pi(K_0)$.
(b) *For each* n, $4I_n \cap \pi(K_0) = \emptyset$.
(c) *For each* n *and each* $q \in 4I_n$, $6|I_n| \le D(q) < 20|I_n|$.
(d) *For each* m *and* n, *if* $4I_m \cap 4I_n \ne \emptyset$, *then either* $|I_m| = |I_n|/2$, $|I_m| = |I_n|$, *or* $|I_m| = 2|I_n|$.
(e) *For each* n, *there are at most* 25 *intervals* I_m *such that* $4I_m \cap 4I_n \ne \emptyset$.

Proof (a) If $I_n \cap I_m \ne \emptyset$, then, since these are dyadic intervals, we must have either $I_n \subseteq I_m$ or $I_m \subseteq I_n$. Without loss of generality, suppose $I_n \subseteq I_m$. By construction, $I_n = J_p$ for some $p \in U_5 \setminus \pi(K_0)$. But now both I_n and I_m qualify as the largest interval J_p containing p and satisfying $(\star)$ of Construction 8.29. Thus we must have $I_n = I_m$ and pairwise disjointness is shown.

Clearly $U_5 \setminus \pi(K_0) \subseteq \bigcup_n I_n \subseteq U_5$ by construction. Thus to finish it suffices to show that $I_n \cap \pi(K_0) = \emptyset$ for each n. This clearly follows from (b) to whose proof we now turn.

(b) By construction, $I_n = J_p$ for some $p \in U_5 \setminus \pi(K_0)$. By (d) of Proposition 8.28, $D(p) > 0$. Then for any $q \in 4I_n = 4J_p$, by (c) of Proposition 8.28 and ($\star$) of Construction 8.29 we have

$$|D(p) - D(q)| \leq |p - q| \leq 4|J_p| \leq \frac{1}{2}D(p).$$

This forces $D(q) > 0$. Another application of (d) of Proposition 8.28 yields $q \notin \pi(K_0)$.

(c) By construction, $I_n = J_p$ for some $p \in U_5 \setminus \pi(K_0)$. Let $\tilde{q}$ be the center of J_p. Then by ($\star$) of Construction 8.29 and (c) of Proposition 8.28 we have $8|J_p| \leq D(\tilde{q}) = D(\tilde{q}) - D(q) + D(q) \leq |\tilde{q} - q| + D(q) \leq 2|J_p| + D(q)$, i.e., $D(q) \geq 6|J_p| = 6|I_n|$.

For the other inequality, use ($\star\star$) of Construction 8.29 to find a point $q' \in J'_p$ such that $D(q') < 8|J'_p| = 16|J_p|$. Then (c) of Proposition 8.28 and this inequality yield $D(q) = D(q) - D(q') + D(q') < |q - q'| + 16|J_p| < 20|J_p| = 20|I_n|$.

(d) Let $q \in 4I_m \cap 4I_n$. Then by (c) just proven, $6|I_m| \leq D(q) < 20|I_n|$, i.e., $|I_m|/|I_n| < 10/3$. By symmetry, $|I_m|/|I_n| > 3/10$. But since we are dealing with dyadic intervals here, $|I_m|/|I_n|$ is an integral power, positive, zero, or negative, of 2. It follows that we have only three possibilities: either $|I_m|/|I_n| = 1/2$, 1, or 2.

(e) Assume $4I_m \cap 4I_n \neq \emptyset$. By translating and rescaling we may assume that $I_n = [0, 1)$. From (d) just proven, either $|I_m| = 1/2$, $|I_m| = 1$, or $|I_m| = 2$.

Consider the first case. Here $I_m = [j/2, (j+1)/2)$ for some integer j. It is left to the reader to verify that if $[c_A - r_A, c_A + r_A) \cap [c_B - r_B, c_B + r_B) \neq \emptyset$, then $|c_A - c_B| < r_A + r_B$. Thus we must have $|(j/2 + 1/4) - 1/2| < 4/4 + 4/2$. This solves to $-11/2 < j < 13/2$, i.e., $j = -5, -4, \ldots, 5, 6$. Thus there are at most 12 intervals I_m of length 1/2 with $4I_m$ intersecting $4I_n$.

One can similarly check that there are at most seven intervals I_m of length 1 with $4I_m$ intersecting $4I_n$ and that there are at most six intervals I_m of length 2 with $4I_m$ intersecting $4I_n$. Since $12 + 7 + 6 = 25$, we are done. □

8.11 Assigning a Good Linear Function ℓ_n to Each I_n

Construction 8.31 (A Good Ball B_n Assigned to Each I_n) *Consider an adapted dyadic interval I_n and denote its center by p_n. By (c) of Proposition 8.30, $D(p_n) < 20|I_n|$. So applying (b) of Proposition 8.28, we obtain a ball $B(z_n; \tilde{r}_n) \in S_0$ such that $|p_n - \pi(z_n)| + \tilde{r}_n < 20|I_n|$. Set $r_n = \max\{\tilde{r}_n, |I_n|\}$. Then we have the following:*

(i) $B_n = B(z_n; r_n) \in S_0$,
(ii) $|I_n| \leq r_n < 20|I_n|$, *and*
(iii) $|\pi(z_n) - p_n| < 20|I_n|$.

Note that (i) is an immediate consequence of coherence, i.e., (c) of Corollary 8.18; whereas (ii) and (iii) are clear by construction.

Construction 8.32 (A Good Linear Function ℓ_n Assigned to Each I_n) *From* (i) *of Construction* 8.31 *and the Definitions of* S_{total} *and* S_0 (8.13 *and* 8.16), *we see that for each ball* B_n *we have* $\delta(B_n) \geq \delta_0$ *and we may choose and fix a line* L_n *such that* $\beta_1^{L_n}(B_n) < 3\varepsilon_0$ *and* $\angle(L_n, L_0) < 2\alpha_0$. *Let* $\ell_n : L_0 = \mathbb{R} \mapsto \mathbb{R} = L_0^{\perp}/i$ *be the linear function whose graph is* L_n. *By our choice of* α_0 *and the easy inequality* $0 \leq \tan x \leq (4/\pi)x$ *valid for* $0 \leq x \leq \pi/4$, *we have* $0 \leq \tan 2\alpha_0 \leq (4/\pi)2\alpha_0 < 3\alpha_0$. *Thus each* ℓ_n *is a Lipschitz function on* L_0 *with bound* $3\alpha_0$.

The following estimates will prove useful.

Proposition 8.33 *If* $4I_m \cap 4I_n \neq \emptyset$, *then*

(a) $|z_m - z_n| \leq 200\, C_4(M, \delta_0)\{|I_m| + |I_n|\} \leq 600\, C_4(M, \delta_0) \min\{|I_m|, |I_n|\}$,

(b) $|\ell_m' - \ell_n'| \leq 2\, C_1(M, \delta_0)\, \varepsilon_0$, *and*

(c) *for any* $p \in 4I_m \cup 4I_n$,

$$\begin{aligned} |\ell_m(p) - \ell_n(p)| &\leq 400\, C_1(M, \delta_0)\, \varepsilon_0\, \{|I_m| + |I_n|\} \\ &\leq 1{,}200\, C_1(M, \delta_0)\, \varepsilon_0\, \min\{|I_m|, |I_n|\}. \end{aligned}$$

Proof In proving (a) and (c) below we will show only the first inequalities since the second inequalities will then follow by (d) of Proposition 8.30.

By the Definition of d (8.20) and (i) and (ii) of Construction 8.31, $d(z_m) \leq r_m < 20|I_m|$. Similarly, $d(z_n) < 20|I_n|$. Moreover, by (iii) of Construction 8.31 and the assumption that $4I_m \cap 4I_n \neq \emptyset$,

$$\begin{aligned} |\pi(z_m) - \pi(z_n)| &\leq |\pi(z_m) - p_m| + |p_m - p_n| + |p_n - \pi(z_n)| \\ &\leq 20|I_m| + 2\{|I_m| + |I_n|\} + 20|I_n| \\ &= 22\{|I_m| + |I_n|\}. \end{aligned}$$

Thus we may apply Corollary 8.26 with $z = z_m$, $w = z_n$, and $r = 22\{|I_m| + |I_n|\}$ to obtain (a):

$$|z_m - z_n| \leq 9\, C_4 \cdot 22\{|I_m| + |I_n|\} \leq 200\, C_4\{|I_m| + |I_n|\}.$$

Without loss of generality, suppose $r_m \leq r_n$. To prove (b) and (c) we need to apply Proposition 8.7 with z, w, r, L_1, and L_2 of that proposition being z_m, z_n, r_m, L_m, and L_n respectively. Clearly $\delta(z_m, r_m) \geq \delta_0$ and $\beta_1^{L_m}(z_m; r_m) \leq 3\varepsilon_0 \leq 5{,}000\, \varepsilon_0$ by Construction 8.32. From (a) just proven, (ii) of Construction 8.31, and our choice of k_0, it follows that $|z_m - z_n| \leq (k_0/2)r_m$. Since $r_n < 20|I_n| \leq 40|I_m| \leq 40r_m$ (by (ii) of Construction 8.31 and (d) of Proposition 8.30), we also have

$$\beta_1^{L_n}(z_n; r_m) \leq \left\{\frac{r_n}{r_m}\right\}^2 \beta_1^{L_n}(z_n; r_n) \leq 40^2(3\varepsilon_0) = 4{,}800\, \varepsilon_0 \leq 5{,}000\, \varepsilon_0.$$

Thus Proposition 8.7 applies here as desired (and this situation is what prompted us to state that proposition in terms of $5{,}000\,\varepsilon$ instead of just ε).

Turning to (b), its proof is a chain of in/equalites which we exhibit now and explain immediately thereafter:

$$\begin{aligned}
|\ell'_m - \ell'_n| &= |\tan(\angle(L_m, L_0)) - \tan(\angle(L_n, L_0))| \\
&= (\sec^2 \vartheta^*)\,|\angle(L_m, L_0) - \angle(L_n, L_0)| \\
&\leq 2\,|\angle(L_m, L_0) - \angle(L_n, L_0)| \\
&= 2\,|\angle(L_m, L_n)| \\
&\leq 2\,C_1\,\varepsilon_0.
\end{aligned}$$

Up till now we have thought of the angle ϑ between two distinct intersecting lines L_1 and L_2 as being invariably positive and non-obtuse by fiat. For the course of proving (b) we think of our angles as being signed: our signed $\angle(L_1, L_2)$ is positive and still equal to ϑ if an anti-clockwise non-obtuse rotation of the complex plane about the intersection point of L_1 and L_2 carries L_2 onto L_1 but now negative and so equal to $-\vartheta$ otherwise. Of course, the angle between two parallel or identical lines continues to be zero. With these stipulations, the first equality above should now be obvious.

The second equality above is simply the Mean Value Theorem. Thus the number ϑ^* appearing there is between $\angle(L_m, L_0)$ and $\angle(L_n, L_0)$. Since each of these two angles has absolute value at most $2\alpha_0$, the same is true of ϑ^*. It follows that all three of these angles has absolute value at most $\pi/4$ by our choice of α_0. The first inequality above now follows.

Now it is not always true that $\angle(L_m, L_0) = \angle(L_m, L_n) + \angle(L_n, L_0)$ for arbitrary lines L_m, L_n, and L_0. Considering the case where $\angle(L_m, L_n) = \angle(L_n, L_0) = \pi/3$, we see what can go wrong: the composition of two non-obtuse rotations need not be non-obtuse! This is not a problem here since, as noted in the last paragraph, each of $\angle(L_m, L_0)$ and $\angle(L_n, L_0)$ has absolute value at most $\pi/4$. The third equality above now follows.

The second inequality above follows from (c) of Proposition 8.7 and so (b) of this proposition is established.

Finally, turning to (c), let ζ_m and ζ_n be the points of L_m and L_n, respectively, such that $\pi(\zeta_m) = \pi(\zeta_n) = p$. Let ζ_{mn} be the foot of the perpendicular from ζ_m to L_n. Using elementary geometry, one can see that the angle at ζ_m in the right triangle with vertices ζ_m, ζ_{mn}, and ζ_n is congruent to the angle between L_n and L_0 (there are a number of cases to consider depending on which of the collinear points ζ_m, ζ_n, and p is between the other two). Thus

$$|\ell_m(p) - \ell_n(p)| = |\zeta_m - \zeta_n| = \sec(\angle(L_n, L_0))\,|\zeta_m - \zeta_{mn}| \leq 2\,\mathrm{dist}(\zeta_m, L_n)$$

since $0 \leq \angle(L_n, L_0) < \pi/3$ by our choice of α_0. From (b) of Proposition 8.7 one obtains

$$\mathrm{dist}(\zeta_m, L_n) \leq C_1\, \varepsilon_0\, \{|\zeta_m - z_m| + r_m\}.$$

Letting z_{mm} be the foot of the perpendicular from z_m to L_m, from (a) of Proposition 8.7 one obtains

$$|\pi(z_m) - \pi(z_{mm})| \leq |z_m - z_{mm}| = \mathrm{dist}(z_m, L_m) \leq 2r_m.$$

Clearly then

$$|\zeta_m - z_m| \leq |\zeta_m - z_{mm}| + |z_{mm} - z_m| \leq |\zeta_m - z_{mm}| + 2r_m.$$

Since $0 \leq \angle(L_m, L_0) < \pi/3$ by our choice of α_0,

$$\begin{aligned}
|\zeta_m - z_{mm}| &= \sec(\angle(L_m, L_0))\, |p - \pi(z_{mm})| \\
&\leq 2|p - \pi(z_{mm})| \\
&\leq 2\{|p - p_m| + |p_m - \pi(z_m)| + |\pi(z_m) - \pi(z_{mm})|\}.
\end{aligned}$$

Since $p \in 4I_m \cup 4I_n$ and $4I_m \cap 4I_n \neq \emptyset$, we have

$$|p - p_m| \leq \max\{2|I_m|, 4|I_n| + 2|I_m|\} \leq 4\{|I_m| + |I_n|\}.$$

By (iii) of Construction 8.31,

$$|p_m - \pi(z_m)| < 20\{|I_m| + |I_n|\}.$$

Finally, by (ii) of Construction 8.31,

$$r_m < 20\{|I_m| + |I_n|\}.$$

From the last eight displayed equations it follows that $|\ell_m(p) - \ell_n(p)| \leq 376\, C_1\, \varepsilon_0\, \{|I_m| + |I_n|\}$ and so (c) has been proven with some room to spare. □

8.12 Construction of a Function ℓ Whose Graph Γ Contains K_0

For any open subset U of $\mathbb{R}$, let $C^\infty(U)$ denote the set of functions that are defined and smooth, i.e., defined and infinitely differentiable, on U. Consider $f : \mathbb{R} \mapsto [0, 1)$ defined by

$$f(x) = \begin{cases} e^{-1/x} & \text{when } x > 0 \\ 0 & \text{when } x \leq 0 \end{cases}.$$

This function is clearly infinitely differentiable on $\mathbb{R} \setminus \{0\}$. In fact, it is also infinitely differentiable at $x = 0$ with $f^{(n)}(0) = 0$ for all $n \geq 0$! This follows by a simple induction from the following two facts: (a) $f(x)/x^n \to 0$ as $x \downarrow 0$ for all $n \geq 0$ (use the change of variables $u = 1/x$ and then apply L'Hôpital's Rule n times) and (b) on the interval $(0, \infty)$, any derivative of f is of the form f times a polynomial in $1/x$ (another simple induction). Thus we have produced a function $f \in C^\infty(\mathbb{R})$ with $f = 0$ on $(-\infty, 0]$ and $f > 0$ on $(0, \infty)$.

Next consider $g : \mathbb{R} \mapsto [0, e^{-1}]$ defined by

$$g(x) = \begin{cases} e^{-1/(1-x^2)} & \text{when} \quad |x| < 1 \\ 0 & \text{when} \quad |x| \geq 1 \end{cases}.$$

Since $g(x) = f(2(1-x)) \cdot f(2(1+x))$, we have produced a function $g \in C^\infty(\mathbb{R})$ with $g = 0$ on $(-\infty, -1] \cup [1, \infty)$ and $g > 0$ on $(-1, 1)$. Clearly $\|g\|_\infty = g(0) = e^{-1}$. Moreover,

$$\int_{-\infty}^{\infty} g(x)\, dx \geq \int_{-1/2}^{1/2} g(x)\, dx \geq e^{-4/3}[1/2 - (-1/2)] = e^{-4/3}$$

and, via the change of variables $u = 1/(1-x^2)$,

$$\|g'\|_\infty = \left\| \frac{-2x}{(1-x^2)^2} e^{-1/(1-x^2)} \right\|_{(-1,1)} \leq 2 \left\| \frac{e^{-1/(1-x^2)}}{(1-x^2)^2} \right\|_{(-1,1)} = 2\|u^2 e^{-u}\|_{(1,\infty)} = 8e^{-2}.$$

Finally consider $h : \mathbb{R} \mapsto [0, 1]$ defined by

$$h(x) = \int_{-\infty}^{x} g(x)\, dx \Big/ \int_{-\infty}^{\infty} g(x)\, dx.$$

We have produced a function $h \in C^\infty(\mathbb{R})$ with $h = 0$ on $(-\infty, -1]$, h strictly increasing on $(-1, 1)$, and $h = 1$ on $[1, \infty)$. Clearly $\|h\|_\infty = 1$. Moreover,

$$\|h'\|_\infty = \|g\|_\infty \Big/ \int_{-\infty}^{\infty} g(x)\, dx \leq e^{-1}/e^{-4/3} = e^{1/3}$$

and

$$\|h''\|_\infty = \|g'\|_\infty \Big/ \int_{-\infty}^{\infty} g(x)\, dx \leq 8e^{-2}/e^{-4/3} = 8e^{-2/3}.$$

With this function h in hand, we can now easily prove the following result.

Lemma 8.34 *Suppose $l > 0$ and $a \leq b$. Then there exists a function $\widetilde{\varphi} \in C^\infty(\mathbb{R})$ with $\widetilde{\varphi} = 0$ on $(-\infty, a-l]$, $\widetilde{\varphi}$ strictly increasing on $[a-l, a]$, $\widetilde{\varphi} = 1$ on $[a, b]$, $\widetilde{\varphi}$ strictly decreasing on $[b, b+l]$, and $\widetilde{\varphi} = 0$ on $[b+l, \infty)$. Moreover, and just as importantly, one can also arrange that*

$$\|\widetilde{\varphi}'\|_\infty < 3/l \text{ and } \|\widetilde{\varphi}''\|_\infty < 20/l^2.$$

Proof Define $\widetilde{\varphi} : \mathbb{R} \mapsto [0, 1]$ as follows:

$$\widetilde{\varphi}(x) = h\left(1 + 2\left\{\frac{x-a}{l}\right\}\right) \cdot h\left(1 - 2\left\{\frac{x-b}{l}\right\}\right).$$

All assertions except the norm estimates are immediately clear. To see these estimates, note that $\widetilde{\varphi}' = \widetilde{\varphi}'' = 0$ off $(a-l, a) \cup (b, b+l)$. Moreover, for all $x \in (a-l, a)$,

$$\widetilde{\varphi}(x) = h\left(1 + 2\left\{\frac{x-a}{l}\right\}\right).$$

Hence we have $|\widetilde{\varphi}'| = |h'| \cdot (2/l) \leq e^{1/3} \cdot (2/l) < 3/l$ and $|\widetilde{\varphi}''| = |h''| \cdot (2/l)^2 \leq 8e^{-2/3} \cdot (2/l)^2 < 20/l^2$ on the interval $(a-l, a)$. The interval $(b, b+l)$ is handled similarly and so we are done. □

From this last lemma and Proposition 8.30 we will now prove a proposition which yields a partition of unity nicely suited to our purposes. In what follows we will refer to a collection of functions as being of *multiplicity* $N < \infty$ on an open set. By this we mean that each point of the set has a neighborhood on which at most N of the functions of the collection are not identically zero. Clearly the sum of a collection of functions smooth on an open set and with finite multiplicity on the set is again smooth on the set. Moreover, the derivative of order n of such a sum is simply the sum of the derivatives of order n.

Recall that we arranged for $\pi(K) \subseteq U_{1/2}$ in Construction 8.12). Thus $\pi(K_0) \subseteq U_5 \subseteq \mathbb{R}$.

Proposition 8.35 *There exists a sequence of functions* $\{\varphi_n\} \subseteq C^\infty(U_5 \setminus \pi(K_0))$ *such that*

(a) $0 \leq \varphi_n \leq 1$ *on* $U_5 \setminus \pi(K_0)$ *with* $\varphi_n = 0$ *on* $\{U_5 \setminus \pi(K_0)\} \setminus \mathrm{int}(4I_n)$ *for each* n,
(b) $|\varphi_n'| \leq 200/|I_n|$ *on* $U_5 \setminus \pi(K_0)$ *for each* n,
(c) $|\varphi_n''| \leq 50{,}000/|I_n|^2$ *on* $U_5 \setminus \pi(K_0)$ *for each* n,
(d) $\sum_n \varphi_n = 1$ *on* $U_5 \setminus \pi(K_0)$, *and*
(e) $\{\varphi_n\}$ *has multiplicity* 25 *on* $U_5 \setminus \pi(K_0)$.

Proof For each n, let $\widetilde{\varphi}_n \in C^\infty(\mathbb{R})$ be the function produced by the last lemma when l is set equal to $|I_n|$, a is set equal to the left endpoint of $2I_n$, and b is set equal to the right endpoint of $2I_n$. Then set

$$\varphi_n = \frac{\widetilde{\varphi}_n}{\sum_m \widetilde{\varphi}_m}$$

for each n. By (a) of Proposition 8.30, $\sum_m \widetilde{\varphi}_m \geq 1$ on $U_5 \setminus \pi(K_0)$ and so the φ_ns are well defined there. It also follows from (a) of Proposition 8.30 that the collection $\{\text{int}(4I_n)\}$ is an open cover of $U_5 \setminus \pi(K_0)$. Given n, set $\mathcal{A}_n = \{m : \widetilde{\varphi}_m \not\equiv 0 \text{ on } \text{int}(4I_n)\}$. Then $\#\mathcal{A}_n \leq 25$ by (e) of Proposition 8.30. Thus $\{\widetilde{\varphi}_m\}$ has multiplicity 25 on $U_5 \setminus \pi(K_0)$. It is now clear that each φ_n is smooth on $U_5 \setminus \pi(K_0)$ and that (a), (d), and (e) of the proposition hold.

It remains to verify the estimates (b) and (c). From (d) of Proposition 8.30 and the last lemma, we see that on $\text{int}(4I_n)$,

$$\left|\left\{\sum_m \widetilde{\varphi}_m\right\}'\right| = \left|\sum_{m\in\mathcal{A}_n} \widetilde{\varphi}'_m\right| \leq \sum_{m\in\mathcal{A}_n} |\widetilde{\varphi}'_m| \leq \sum_{m\in\mathcal{A}_n} \frac{3}{|I_m|} \leq \#\mathcal{A}_n \cdot \frac{6}{|I_n|} \leq \frac{150}{|I_n|}.$$

Estimate (b) of the proposition now follows immediately, with room to spare, since on $\{U_5 \setminus \pi(K_0)\} \cap \text{int}(4I_n)$ we now have

$$\begin{aligned}|\varphi'_n| &= \left|\widetilde{\varphi}'_n \cdot \left\{\sum_m \widetilde{\varphi}_m\right\}^{-1} - \widetilde{\varphi}_n \cdot \left\{\sum_m \widetilde{\varphi}_m\right\}^{-2} \cdot \left\{\sum_m \widetilde{\varphi}_m\right\}'\right| \\ &\leq \frac{3}{|I_n|} \cdot \{1\}^{-1} + 1 \cdot \{1\}^{-2} \cdot \frac{150}{|I_n|} = \frac{153}{|I_n|}.\end{aligned}$$

The verification of estimate (c) is similar, but grubbier, and left to the reader. As a help to and check for the reader, we note that being careful and stingy one actually gets $|\varphi''_n| \leq 47{,}920/|I_n|^2$ on $U_5 \setminus \pi(K_0)$. So again we have some room to spare. □

Construction 8.36 (An Extension of ℓ to U_5) *Recall that in Construction 8.25 we defined a Lipschitz function ℓ on the set $\pi(K_0) \subseteq U_{1/2}$ whose graph contains K_0. We now extend ℓ to a function defined on all of U_5:*

Given $p \in U_5 \setminus \pi(K_0)$, set

$$\ell(p) = \sum_n \ell_n(p)\varphi_n(p)$$

where the functions ℓ_n and φ_n are as in Construction 8.32 and Proposition 8.35 respectively. Let Γ be the graph of ℓ. Thus $\Gamma = \{p + i\ell(p) : p \in U_5\}$.

Of course it is not at all obvious that Γ is a Lipschitz graph, i.e., that ℓ is a Lipschitz function on U_5. The next section is devoted to establishing this fact.

8.13 Verification That ℓ is Lipschitz

Lemma 8.37 *For any $p, q \in 4I_m \cap U_5$,*

$$|\ell(p) - \ell(q)| \leq 4\alpha_0|p - q|.$$

Proof By (b) of Proposition 8.30, $4I_m \cap U_5 \subseteq U_5 \setminus \pi(K_0)$. Thus

$$\begin{aligned}
|\ell(p) - \ell(q)| &= \left|\sum_n \ell_n(p)\varphi_n(p) - \sum_n \ell_n(q)\varphi_n(q)\right| \\
&\le \sum_n |\ell_n(p) - \ell_n(q)|\varphi_n(p) \leftarrow \text{(I)!} \\
&+ \sum_n |\ell_n(q) - \ell_m(q)||\varphi_n(p) - \varphi_n(q)| \leftarrow \text{(II)!} \\
&+ |\ell_m(q)| \left|\sum_n \{\varphi_n(p) - \varphi_n(q)\}\right|. \leftarrow \text{(III)!}
\end{aligned}$$

Since each ℓ_n is a Lipschitz on L_0 with bound $3\alpha_0$ (see Construction 8.32), by (d) of Proposition 8.35 we have

$$\text{(I)} \le \sum_n 3\alpha_0|p - q|\varphi_n(p) = 3\alpha_0|p - q|.$$

Next, note that if $\varphi_n(p) - \varphi_n(q) \neq 0$ in (II), then $4I_n \cap 4I_m \neq \emptyset$ because of (a) of Proposition 8.35. By (e) of Proposition 8.30 there are thus at most 25 such n. Moreover, by (c) of Proposition 8.33, the Mean Value Theorem, and (b) of Proposition 8.35, each such n has its corresponding summand in (II) bounded by

$$1{,}200\, C_1\, \varepsilon_0\, |I_n| \cdot \frac{200}{|I_n|}|p - q| = 240{,}000\, C_1\, \varepsilon_0\, |p - q|.$$

Thus

$$\text{(II)} \le 25 \times 240{,}000\, C_1\, \varepsilon_0\, |p - q| = 6{,}000{,}000\, C_1\, \varepsilon_0\, |p - q| \le \alpha_0|p - q|$$

with the last inequality following from Condition (3) for the selection of ε_0.

Finally, (III) $=$ 0 by (d) of Proposition 8.35. The desired conclusion is now clear. □

Lemma 8.38 *For any $p \in 4I_m \cap U_5$ and any $q \in \pi(K_0)$,*

$$|\ell(p) - \ell(q)| \le 71\alpha_0|p - q|.$$

Proof The reader should reread Constructions 8.31, 8.32, and 8.36 since many facts therein will be used in this proof without mention. By (c) of Proposition 8.30 and (d) and (c) of Proposition 8.28, $6|I_m| \le D(p) = |D(p) - D(q)| \le |p - q|$. Thus $|I_m| \le |p - q|/6$, a fact that will be freely used in the rest of the proof without mention.

Recall that $\delta(B_m) \ge \delta_0$, i.e., $\mu(B_m) \ge \delta_0\, r_m$. Let z'_m be a point of B_m closest to L_m. Then

$$\operatorname{dist}(z'_m, L_m)\,\delta_0\,r_m \le \int_{B_m} \operatorname{dist}(\zeta, L_m)\,d\mu(\zeta) \le \beta_1^{L_m}(B_m)\,r_m^2 \le 3\varepsilon_0\,r_m^2,$$

i.e., $\operatorname{dist}(z'_m, L_m) \le (3/\delta_0)\varepsilon_0\,r_m$. Note that $|\pi(z'_m) - p_m| \le |\pi(z'_m) - \pi(z_m)| + |\pi(z_m) - p_m| < r_m + 20|I_m| < 40|I_m|$, a fact that will be freely used in the rest of the proof without mention.

Clearly,

$$\begin{aligned} |\ell(p) - \ell(q)| \le\ & |\ell(p) - \ell(p_m)| \leftarrow \text{(I)!} \\ &+ |\ell(p_m) - \ell_m(p_m)| \leftarrow \text{(II)!} \\ &+ |\ell_m(p_m) - \ell_m(\pi(z'_m))| \leftarrow \text{(III)!} \\ &+ |\ell_m(\pi(z'_m)) - \pi^{\perp}(z'_m)/i| \leftarrow \text{(IV)!} \\ &+ |\pi^{\perp}(z'_m)/i - \ell(q)|. \leftarrow \text{(V)!} \end{aligned}$$

By the last lemma,

$$\text{(I)} \le 4\alpha_0|p - p_m| \le 8\alpha_0|I_m| \le (4/3)\alpha_0|p - q|.$$

From (d) and (a) of Proposition 8.35, (c) of Proposition 8.33, and Condition (3) for the selection of ε_0, we see that

$$\begin{aligned} \text{(II)} &\le \sum_n |\ell_n(p_m) - \ell_m(p_m)|\varphi_n(p_m) \;\le\; 1{,}200\,C_1\,\varepsilon_0\,|I_m| \\ &\le 200\,C_1\,\varepsilon_0\,|p - q| \;\le\; (1/3)\alpha_0|p - q|. \end{aligned}$$

Since each ℓ_m is Lipschitz on L_0 with bound $3\alpha_0$,

$$\text{(III)} \le 3\alpha_0|p_m - \pi(z'_m)| \le 3\alpha_0 \times 40|I_m| \le 20\alpha_0|p - q|.$$

Let $\tilde{z}_m$ be the point of L_m such that $\pi(\tilde{z}_m) = \pi(z'_m)$ and let z'_{mm} be the foot of the perpendicular from z'_m to L_m . Using elementary geometry, one can see that the angle at z'_m in the right triangle with vertices z'_m, $\tilde{z}_m$, and z'_{mm} is congruent to the angle between L_m and L_0 and that $|\ell_m(\pi(z'_m)) - \pi^{\perp}(z'_m)/i| = |\tilde{z}_m - z'_m|$ (there are a number of cases to consider depending on which of the collinear points z'_m, $\tilde{z}_m$, and $\pi(z'_m)$ is between the other two). Of course, $\operatorname{dist}(z'_m, L_m) = |z'_m - z'_{mm}|$. Thus, since $0 \le \angle(L_m, L_0) < \pi/3$ by our choice of α_0, we have

$$\begin{aligned} \text{(IV)} &= \sec(\angle(L_m, L_0))\operatorname{dist}(z'_m, L_m) \;\le\; 2\,(3/\delta_0)\,\varepsilon_0\,r_m \\ &\le (120/\delta_0)\,\varepsilon_0\,|I_m| \;\le\; (20/\delta_0)\,\varepsilon_0\,|p - q| \;\le\; (1/3)\alpha_0|p - q| \end{aligned}$$

with the last inequality true by Condition (3) for the selection of ε_0 since the specification of C_1 in Proposition 8.7 makes it clear that $C_1 \ge 10^{20}/\delta_0$.

To estimate (V), let w be the point of K_0 for which $\pi(w) = q$ and $\pi^{\perp}(w) = i\,\ell(q)$. Then

$$(\dagger)\ (\mathrm{V}) = |\pi^{\perp}(z'_m)/i - \ell(q)| = |\pi^{\perp}(z'_m) - \pi^{\perp}(w)|$$

and

$$\begin{aligned}(\ddagger)\ |\pi(z'_m) - \pi(w)| &= |\pi(z'_m) - q| \\ &\leq |\pi(z'_m) - p_m| + |p_m - p| + |p - q| \\ &< 40|I_m| + 2|I_m| + |p - q| \ \leq\ 8|p - q|.\end{aligned}$$

Case One ($|z'_m - w| > (N+2)2r_m$ *where* $N = 44\,C_4/\alpha_0$). Recall that Archimedes showed that $3\frac{10}{71} < \pi < 3\frac{10}{70} = 3\frac{1}{7}$. Thus by condition (2) for the selection of ε_0 and our choice of N, we have

$$C_4\{\sqrt{\varepsilon_0} + 2/N\} \leq [3 - (8/\pi)]\alpha_0 + 2C_4/N \leq [3 - 56/22 + 1/22]\alpha_0 = (1/2)\alpha_0.$$

Since $z'_m \in B(z_m; r_m) \in S_0$, by the Definition of d (8.20) we have $d(z'_m) \leq |z'_m - z_m| + r_m \leq 2r_m < |z'_m - w|/(N+2)$. Clearly $d(w) = 0 < |z'_m - w|/(N+2)$. We may thus apply Proposition 8.22, to obtain

$$\begin{aligned}|\pi^{\perp}(z'_m) - \pi^{\perp}(w)| &\leq (8/\pi)\alpha_0|\pi(z'_m) - \pi(w)| + (1/2)\alpha_0|z'_m - w|. \\ &\leq (568/223 + 1/2)\alpha_0|\pi(z'_m) - \pi(w)| \\ &\quad +(1/2)|\pi^{\perp}(z'_m) - \pi^{\perp}(w)|.\end{aligned}$$

Thus

$$|\pi^{\perp}(z'_m) - \pi^{\perp}(w)| \leq (1{,}136/223 + 1)\alpha_0|\pi(z'_m) - \pi(w)|.$$

From ($\dagger$), the last displayed inequality, and ($\ddagger$), we conclude that

$$(\mathrm{V}) \leq 8(1{,}136/223 + 1)\alpha_0|p - q| \leq 49\alpha_0|p - q|.$$

Case Two ($|z'_m - w| \leq (N+2)2r_m$ *where* $N = 44\,C_4/\alpha_0$). Unfortunately neither Lemma 8.19 nor Proposition 8.22 help us here and we seem to be forced into repeating the proof of Lemma 8.19 with small changes. The author apologizes to the reader for this but sees no less painful way to proceed!

Since $d(w) = 0$, there exists a ball $B(w'; r') \in S_0$ for which $|w - w'| + r' < \sqrt{\varepsilon_0}\, r_m$ (see Definition 8.20). By (c) of Corollary 8.18, $B(w'; \sqrt{\varepsilon_0}\, r_m)$ and $B(w'; r_m) \in S_0$. The Definitions of S_{total} and S_0 (8.13 and 8.16) now let us choose a line L_* such that $\beta_1^{L_*}(w'; r_m) < 3\varepsilon_0$ and $\angle(L_*, L_0) < 2\alpha_0$. They also tell us that $\delta(w'; \sqrt{\varepsilon_0}\, r_m) \geq \delta_0$, i.e., $\mu(B(w'; \sqrt{\varepsilon_0}\, r_m)) \geq \delta_0 \sqrt{\varepsilon_0}\, r_m$. Thus

$$\begin{aligned}&\{\mathrm{dist}(w', L_*) - \sqrt{\varepsilon_0}\, r_m\}\,\delta_0\sqrt{\varepsilon_0}\, r_m \leq \\ &\qquad \int_{B(w';\sqrt{\varepsilon_0}\, r_m)} \mathrm{dist}(\zeta, L_*)\, d\mu(\zeta) \leq \beta_1^{L}(w'; r_m)\, r_m^2 \leq 3\varepsilon_0 r_m^2,\end{aligned}$$

i.e., $\operatorname{dist}(w', L_*) \le \{1 + (3/\delta_0)\}\sqrt{\varepsilon_0}\, r_m$.

Let w'_* be the foot of the perpendicular from w' to L_*. Clearly

$$|z'_m - z'_{mm}| \le (3/\delta_0)\sqrt{\varepsilon_0}\, r_m \text{ and } |w - w'_*| \le \{2 + (3/\delta_0)\}\sqrt{\varepsilon_0}\, r_m.$$

By the case hypothesis and our choice of k_0, $|z_m - w'| \le |z_m - z'_m| + |z'_m - w| + |w - w'| \le r_m + (N+2)2r_m + r_m \le 10Nr_m \le (k_0/2)r_m$. This allows us to apply (b) of Proposition 8.7 to the two lines L_m and L_* obtained from the balls $B(z_m; r_m)$ and $B(w'; r_m)$. Letting z'_{mm*} denote the foot of the perpendicular from z'_{mm} to L_*, we thus see that

$$\begin{aligned}
|z'_{mm} - z'_{mm*}| &= \operatorname{dist}(z'_{mm}, L_*) \\
&\le C_1\,\varepsilon_0\,\{|z'_{mm} - z_m| + r_m\} \\
&\le C_1\,\varepsilon_0\,\{|z'_{mm} - z'_m| + |z'_m - z_m| + r_m\} \\
&\le C_1\,\varepsilon_0\,\{(3/\delta_0)\, r_m + r_m + r_m\} \\
&\le C_1\,\{2 + (3/\delta_0)\}\sqrt{\varepsilon_0} r_m.
\end{aligned}$$

The last two displayed inequalities imply

$$(\star)\ |z'_m - z'_{mm*}| + |w - w'_*| \le (C_4/6)\sqrt{\varepsilon_0} r_m.$$

Note that both z'_{mm*} and w'_* lie on L_* which makes an angle of at most $2\alpha_0$ with L_0. Also, recall that $0 \le \tan 2\alpha_0 \le (4/\pi)2\alpha_0 \le 3\alpha_0$ since $2\alpha_0 \le \pi/4$ by our choice of α_0. Thus

$$(\star\star)\ |\pi^\perp(z'_{mm*}) - \pi^\perp(w'_*)| \le 3\alpha_0|\pi(z'_{mm*}) - \pi(w'_*)|.$$

From $(\star)$ and $(\star\star)$ we obtain

$$\begin{aligned}
|\pi^\perp(z'_m) - \pi^\perp(w)| &\le |\pi^\perp(z'_m) - \pi^\perp(z'_{mm*})| + |\pi^\perp(z'_{mm*}) - \pi^\perp(w'_*)| \\
&\quad + |\pi^\perp(w'_*) - \pi^\perp(w)| \\
&\le |z'_m - z'_{mm*}| + |\pi^\perp(z'_{mm*}) - \pi^\perp(w'_*)| + |w'_* - w| \\
&\le 3\alpha_0|\pi(z'_{mm*}) - \pi(w'_*)| + (C_4/6)\sqrt{\varepsilon_0} r_m,
\end{aligned}$$

whereas from $(\ddagger)$ and $(\star)$ we obtain

$$\begin{aligned}
|\pi(z'_{mm*}) - \pi(w'_*)| &\le |\pi(z'_{mm*}) - \pi(z'_m)| + |\pi(z'_m) - \pi(w)| + |\pi(w) - \pi(w'_*)| \\
&\le |z'_{mm*} - z'_m| + |\pi(z'_m) - \pi(w)| + |w - w'_*| \\
&\le 8|p - q| + (C_4/6)\sqrt{\varepsilon_0} r_m.
\end{aligned}$$

If we invoke $(\dagger)$, combine the last two displayed inequalities, appeal to Condition (2) for the selection of ε_0, and use the fact that $r_m \le 20|I_m| \le 20|p - q|/6$, we again obtain

$$\begin{aligned}(\mathrm{V}) &\leq 24\alpha_0|p-q| + \{1+3\alpha_0\}(C_4/6)\sqrt{\varepsilon_0}\cdot r_m \\ &\leq 24\alpha_0|p-q| + (3\alpha_0/6)\cdot 20|p-q|/6 \;\leq\; 49\alpha_0|p-q|.\end{aligned}$$

Adding up our estimates of the various pieces, the desired conclusion is now clear. □

Proposition 8.39 (The Function ℓ is Lipschitz) *For any $p, q \in U_5$,*

$$|\ell(p) - \ell(q)| \leq 71\alpha_0|p-q|.$$

Proof Note that ℓ is smooth on the open set $U_5 \setminus \pi(K_0)$ since it is the sum of a collection of functions smooth on this set and with finite multiplicity on this set (by (e) of Proposition 8.35). But then $|\ell'| \leq 4\alpha_0$ everywhere there by Lemma 8.37 and (a) of Proposition 8.30. The Mean Value Theorem may now be used to conclude that $|\ell(p) - \ell(q)| \leq 4\alpha_0|p-q|$ whenever p and q come from the same component of $U_5 \setminus \pi(K_0)$.

If, however, p and q come from different components of $U_5 \setminus \pi(K_0)$, then there is a point $u \in \pi(K_0)$ that lies between them. Then Lemma 8.38 applied twice yields

$$|\ell(p)-\ell(q)| \leq |\ell(p)-\ell(u)| + |\ell(u)-\ell(q)| \leq 71\alpha_0|p-u| + 71\alpha_0|u-q| = 71\alpha_0|p-q|.$$

The remaining cases where one or both of p and q come from $\pi(K_0)$ are covered by Lemma 8.38 and Corollary 8.24, respectively. □

We conclude this section with an estimate that will be of use to us later.

Lemma 8.40 *If $p \in 4I_m \cap U_5$, then*

$$|\ell''(p)| \leq 10^{10} C_1(M, \delta_0)\, \varepsilon_0/|I_m|.$$

Proof Recall that $4I_m \cap U_5 \subseteq U_5 \setminus \pi(K_0)$ by (b) of Proposition 8.30. From the linearity of the ℓ_ns and the definition of ℓ in Construction 8.36 as a sum with finite multiplicity on $U_5 \setminus \pi(K_0)$ (the last true by (e) of Proposition 8.35), we have that

$$\ell'' = \sum_n 2\ell_n' \varphi_n' + \sum_n \ell_n \varphi_n''$$

on $U_5 \setminus \pi(K_0)$. Moreover, $\sum_n \varphi_n' = \sum_n \varphi_n'' = 0$ on $U_5 \setminus \pi(K_0)$ by (d) and (e) of Proposition 8.35. Thus

$$|\ell''(p)| \leq \sum_n 2|\ell_n'(p) - \ell_m'(p)||\varphi_n'(p)| + \sum_n |\ell_n(p) - \ell_m(p)||\varphi_n''(p)|.$$

Each sum in this inequality has locally at most 25 terms by (e) of Proposition 8.35. Estimates on the four absolute values in the sums are provided by (b) and (c) of Propositions 8.33 and 8.35 (with an appeal to (a) of Proposition 8.35 to justify using Proposition 8.33). Finally, the $1/|I_n|$s that result can be replaced by $2/|I_m|$s by (d) of Proposition 8.30. Putting this all to use, we actually conclude that $|\ell''(p)| \leq 3{,}000{,}040{,}000\, C_1\, \varepsilon_0/|I_m|$ and so have some room to spare. □

8.14 A Partition of $K\backslash K_0$ into Three Sets: K_1, K_2, and K_3

In this section $K\backslash K_0$ will be partitioned into three subsets K_1, K_2, and K_3. As we have succeeded in threading a Lipschitz graph through K_0, the proof of Reduction 8.3 will then be completed by showing that the μ-measure of each of these three subsets is very small.

Definition 8.41 (The Sets K_1, K_2, and K_3) *Our three sets are defined as follows*:

$$K_1 = \{z \in K\backslash K_0 : \exists\, w \in K \textit{ and } s \in [h(z)/4, 3h(z)/4] \textit{ such that } z \in B(w; s/2) \textit{ and } \delta(w; s) < 2\delta_0\},$$

$$K_2 = \{z \in K \setminus \{K_0 \cup K_1\} : \exists\, w \in K \textit{ and } s \in [h(z)/4, 3h(z)/4] \textit{ such that } z \in B(w; s/2) \textit{ and } \beta_1(w; s) \geq \varepsilon_0\}, \textit{ and}$$

$$K_3 = \{z \in K\backslash\{K_0 \cup K_1 \cup K_2\} : \exists\, w \in K \textit{ and } s \in [h(z)/4, 3h(z)/4] \textit{ such that } z \in B(w; s/2) \textit{ and } \forall \textit{ line } L,\ \angle(L, L_0) \geq \alpha_0 \textit{ whenever } \beta_1^L(w; s) < \varepsilon_0\}.$$

Proposition 8.42 (K_1, K_2, and K_3 Partition $K\backslash K_0$) $K\backslash K_0 = K_1 \cup K_2 \cup K_3$ *disjointly.*

Proof Given $z \in K \setminus K_0$, it suffices to show that $z \in K_1 \cup K_2 \cup K_3$. From the Definition K_0 (8.23) and (b) of Proposition 8.21, we see that $h(z) > 0$. Thus, applying the Definition of h (8.15), we see that there exist sequences $\{s_n\}$ and $\{w_n\}$ from $(0, 1)$ and K respectively such that $s_n \uparrow h(z)/2$ with $z \in B(w_n; s_n/4)$ and $B(w_n, s_n) \notin S_{\text{total}}$ for each n. Then, from the definition of S_{total} (8.13), it follows that for each n either

(i) $\delta(w_n; s_n) < \delta_0$,
(ii) $\delta(w_n; s_n) \geq \delta_0$ and $\beta_1(w_n; s_n) \geq 3\varepsilon_0$, or
(iii) $\delta(w_n; s_n) \geq \delta_0$, $\beta_1(w_n; s_n) < 3\varepsilon_0$, and $\forall$ line L, $\angle(L, L_0) \geq 2\alpha_0$ whenever $\beta_1^L(w_n; s_n) < 3\varepsilon_0$.

Dropping to a subsequence, we may assume by the Pigeonhole Principle that either (i) holds for all n, (ii) holds for all n, or (iii) holds for all n. Dropping to a subsequence once more, we may also assume by the compactness of K that $w_n \to w$ for some $w \in K$. Note that $z \in B(w; h(z)/8) \subseteq B(w; s/2)$ whenever $s \geq h(z)/4$.

Case One ((i) *holds for all* n). We show that $z \in K_1$. To this end, choose and fix an n such that $|w_n - w| \leq h(z)/8$ and $s_n \geq 3h(z)/8$. Then we have $B(w; h(z)/4) \subseteq B(w_n; s_n)$ and so, by (i),

$$\delta(w; h(z)/4) \leq \frac{s_n}{h(z)/4} \cdot \delta(w_n; s_n) < \frac{h(z)/2}{h(z)/4} \cdot \delta_0 = 2\delta_0.$$

Thus for $s = h(z)/4$ we have $s \in [h(z)/4, 3h(z)/4]$, $z \in B(w; s/2)$, and $\delta(w; s) < 2\,\delta_0$, i.e., $z \in K_1$.

Case Two ((ii) *holds for all n and* $z \notin K_1$). We show that $z \in K_2$. To this end, note that for all n sufficiently large we have $|w_n - w| \le k_0\, h(z)/4$ and so $B(w; k_0\, 3h(z)/4) \supseteq B(w_n; k_0\, s_n)$. Hence, by (ii),

$$\beta_1(w; 3h(z)/4) \ge \left\{\frac{s_n}{3h(z)/4}\right\}^2 \beta_1(w_n; s_n) \ge \left\{\frac{s_n}{3h(z)/4}\right\}^2 3\varepsilon_0$$

for these n. Letting $n \to \infty$, we obtain

$$\beta_1(w; 3h(z)/4) \ge \left\{\frac{h(z)/2}{3h(z)/4}\right\}^2 3\varepsilon_0 = \frac{4}{3}\varepsilon_0 > \varepsilon_0.$$

Thus for $s = 3h(z)/4$ we have $s \in [h(z)/4, 3h(z)/4]$, $z \in B(w; s/2)$, and $\beta_1(w; s) \ge \varepsilon_0$, i.e., $z \in K_2$.

Case Three ((iii) *holds for all n and* $z \notin K_1 \cup K_2$). We show that $z \in K_3$. To this end, choose and fix an n such that $|w_n - w| \le k_0\, h(z)/8$ and $s_n \ge h(z)/4$. Since $z \notin K_1$, $\delta(w; s_n) \ge 2\delta_0$. By (iii), there exists a line L_n satisfying $\beta_1^{L_n}(w_n; s_n) < 3\varepsilon_0$ and ($\star$) $\angle(L_n, L_0) \ge 2\alpha_0$.

Consider any line L such that $\beta_1^L(w; s_n) < \varepsilon_0$. Applying (c) of Proposition 8.7 with z, w, and r replaced by w, w_n, and s_n, respectively, we deduce ($\star\star$) $\angle(L, L_n) \le C_1\, \varepsilon_0 \le \alpha_0$ by Condition (3) for the selection of ε_0. From ($\star$) and ($\star\star$) it now follows that $\angle(L, L_0) \ge \alpha_0$.

Thus for $s = s_n$ we have $s \in [h(z)/4, 3h(z)/4]$, $z \in B(w; s/2)$, and for each line L, $\angle(L, L_0) \ge \alpha_0$ whenever $\beta_1^L(w; s) < \varepsilon_0$, i.e., $z \in K_3$. □

8.15 The Smallness of K_2

Of the three sets that we must show to have small measure, K_2 is the easiest to handle. This section, which deals with K_2, is short and sweet since the hard work has already been done in sections 8.4 and 8.5!

Proposition 8.43 (The Smallness of K_2) $\mu(K_2) \le 2 \times 10^{-5}$.

Proof Suppose that $z \in K_2$ and $r \in [h(z), 2h(z)]$. From the Definition of K_2 (8.41), we obtain $w \in K$ and $s \in [h(z)/4, 3h(z)/4]$ such that $z \in B(w; s/2)$ and $\beta_1(w; s) \ge \varepsilon_0$. Note that $B(z; k_0 r) \supseteq B(w; k_0 s)$ by our choice of k_0. One also has $s/r \ge 1/8$. Thus

$$\beta_1(z; r) \ge \left\{\frac{s}{r}\right\}^2 \beta_1(w; s) \ge \frac{\varepsilon_0}{64}.$$

Next note that $\delta(w; s) \ge 2\delta_0$ since $z \notin K_1$ and that $w \in B(z; k_0^* r)$ by our choice of k_0^*. Thus

$$\tilde{\delta}(z;r) \geq \delta(w;r) \geq \frac{s}{r}\delta(w;s) \geq \frac{\delta_0}{4}.$$

It follows that

$$(\star)\ \int_{\mathbb{C}}\int_0^\infty \beta_1^2(z;r)\,\mathcal{X}_{\{\tilde{\delta}(z;r)\geq\delta_0/4\}}\,\frac{dr}{r}\,d\mu(z) \geq \int_{K_2}\int_{h(z)}^{2h(z)} \left\{\frac{\varepsilon_0}{64}\right\}^2 \frac{dr}{r}\,d\mu(z)$$
$$= \left\{\frac{\varepsilon_0}{64}\right\}^2 (\ln 2)\,\mu(K_2).$$

However, by Fubini's Theorem [RUD, 8.8] and Proposition 8.10, we also have

$$(\star\star)\ \int_{\mathbb{C}}\int_0^\infty \beta_1^2(z;r)\,\mathcal{X}_{\{\tilde{\delta}(z;r)\geq\delta_0/4\}}\,\frac{dr}{r}\,d\mu(z) \leq C_3(M,\delta_0/4,k_0,k_0^*)\,c^2(\mu).$$

Combining $(\star)$ and $(\star\star)$, the desired conclusion now follows by invoking (4) of Reduction 8.3 and Condition (2) for the selection of κ_0. □

The next three sections will be devoted to showing that K_1 has small measure. The most difficult set to deal with, K_3, is left to last and requires six sections to polish off.

8.16 The Smallness of a Horrible Set *H*

The author had thought of using "*B*" to denote the "bad" set of this section but then thought better of it since "*B*" already does duty denoting a typical closed ball. Instead he settled on using "*H*" to denote the "horrible" set of this section! Without further ado ...

Definition 8.44 (The Horrible Set)

$$H = \{z \in K \setminus K_0 : \pi(z) \in \pi(K_0) \text{ or } z \notin 1{,}980C_4(M,\delta_0)\,B_m \text{ whenever } \pi(z) \in 4I_m\}.$$

Proposition 8.45 (The Smallness of *H*) $\mu(H) \leq 10^{-5}$. *More strongly, we even have*

$$\frac{250\,M}{c_2(M,\delta_0)}\,\mu(H) \leq 10^{-5}.$$

The stronger conclusion will be needed to prove the stronger conclusion of Proposition 8.49 to follow – which in turn will be needed later to prove Lemma 8.78.

Proof On the one hand, consider a point $z \in H$ for which $\pi(z) \in \pi(K_0)$. Let $\tilde{z} \in K_0$ be such that $\pi(\tilde{z}) = \pi(z)$. Note that $\tilde{z} \neq z$, so $r_0 = |\tilde{z} - z|/\{1{,}980C_4 + 1\} > 0$. Thus, since $d(\tilde{z}) = 0$ by the Definition of K_0 (8.23), we can find a ball $B(z_0;r) \in S_0$

such that $|\tilde{z} - z_0| + r < r_0$ by the Definition of d (8.20). By (c) of Corollary 8.18, $B_0 = B(z_0; r_0) \in S_0$. Moreover, $|\pi(z) - \pi(z_0)| = |\pi(\tilde{z}) - \pi(z_0)| \leq |\tilde{z} - z_0| \leq 22r_0$ and $|z - z_0| \geq |\tilde{z} - z| - |\tilde{z} - z_0| > \{1{,}980C_4 + 1\} r_0 - r_0 = 1{,}980C_4 r_0$.

On the other hand, consider a point $z \in H$ for which $\pi(z) \notin \pi(K_0)$. Since $\pi(K) \subseteq U_{1/2}$, we must have $\pi(z) \in 4I_m$ for some $m \geq 1$ by (a) of Proposition 8.30. Fixing such an m, we must then have $z \notin 1{,}980C_4\, B_m$. Note too that by (iii) and (ii) of Construction 8.31, $|\pi(z) - \pi(z_m)| \leq |\pi(z) - p_m| + |p_m - \pi(z_m)| \leq 2|I_m| + 20|I_m| \leq 22r_m$.

Thus in either case we have produced a ball $B_m = B(z_m; r_m) \in S_0$ ($m = 0$ or $m \geq 1$) such that $|\pi(z) - \pi(z_m)| \leq 22r_m$ and $|z - z_m| > 1{,}980C_4\, r_m$.

Since $|z - z_m| > 9C_4 \cdot 220r_m$, we must have at least one of $d(z)$, $d(z_m)$ or $|\pi(z) - \pi(z_m)|$ greater than $220r_m$ by Corollary 8.26. But $d(z_m) \leq r_m$ by the Definition of d (8.20) and (i) of Construction 8.31. Also, $|\pi(z) - \pi(z_m)| \leq 22r_m$. Thus we must have $d(z) > 220r_m$.

We draw a number of consequences from the last paragraph's conclusion. First,

$$(\star)\ |\pi(z) - \pi(z_m)| < d(z)/10.$$

Second, $d(z) \leq |z - z_m| + r_m < |z - z_m| + d(z)/220$, so we certainly have

$$(\star\star)\ |z - z_m| > 19d(z)/20.$$

Third, $d(z)$, $d(z_m)$, and $|\pi(z) - \pi(z_m)|$ are all less than or equal to $d(z)$, so applying Corollary 8.26 again we have

$$(\star\star\star)\ |z - z_m| \leq 9C_4\, d(z).$$

Fourth and last, since $B(z_m; r_m) \in S_0$ and $r_m \leq d(z)/20$, it follows that $B = B(z_m; d(z)/20) \in S_0$ by (c) of Corollary 8.18. Thus, from the Definitions of S_{total} and S_0 (8.13 and 8.16), we may select a line L such that $\beta_1^L(B) < 3\varepsilon_0$ and $\angle(L, L_0) < 2\alpha_0$. We may also get balls B_1 and B_2 by applying Lemma 8.5 to B. (Do not confuse these two balls with the balls B_1 and B_2 of Construction 8.31!)

For $j = 1, 2$, set

$$G_j = \{\eta \in B_j \cap B : \text{dist}(\eta, L) < (6\varepsilon_0/c_2) \cdot d(z)/20\}.$$

Note that

$$\frac{6\varepsilon_0/c_2}{d(z)/20} \cdot \mu((B_j \cap B) \setminus G_j) \leq \frac{1}{d(z)/20} \int_{(B_j \cap B) \setminus G_j} \frac{\text{dist}(\eta, L)}{d(z)/20}\, d\mu(\eta) \leq \beta_1^L(B) < 3\varepsilon_0,$$

i.e., $\mu((B_j \cap B) \setminus G_j) < (c_2/2) \cdot d(z)/20$. Since $\mu(B_j \cap B) > c_2 \cdot d(z)/20$, we obtain

$$(\dagger)\ \mu(G_j) > (c_2/2) \cdot d(z)/20 \text{ for } j = 1, 2.$$

Now consider any two points $\eta \in G_1$ and $\xi \in G_2$. Let L_* be the line through η and ξ and z_* be the foot of the perpendicular from z to L_*. Consider the following three angles with vertex z: $\angle(\eta, z, \pi(z))$, $\angle(\pi(z), z, z_*)$, and $\angle(\eta, z, z_*)$. Denote their measures by ϑ_1, ϑ_2, and ϑ, respectively. Since $|\eta - z_m| \leq d(z)/20$, from $(\star)$ and $(\star\star)$ we deduce that

$$\sin\vartheta_1 = \frac{|\pi(\eta) - \pi(z)|}{|\eta - z|} \leq \frac{|\pi(\eta) - \pi(z_m)| + |\pi(z_m) - \pi(z)|}{|z - z_m| - |\eta - z_m|} < \frac{d(z)/20 + d(z)/10}{19d(z)/20 - d(z)/20} = \frac{1}{6}.$$

Next note that by (c) of Lemma 8.5,

$$\sin\angle(L_*, L) = \frac{|\mathrm{dist}(\eta, L) - \mathrm{dist}(\xi, L)|}{|\eta - \xi|} < \frac{2\cdot(6\varepsilon_0/c_2)\cdot d(z)/20}{10c_1\cdot d(z)/20} = \frac{6\varepsilon_0}{5c_1c_2}.$$

By the specification of c_5 in terms of c_1 and c_2 in Proposition 8.7 and Condition (1) for the selection of ε_0, the rightmost item of this last inequality is less than or equal to $6/(5\times 10^6) \leq 1/6$. Use of the simple inequality $\sin x \geq (2/\pi)x$ for $0 \leq x \leq \pi/2$ now lets us conclude that $\angle(L_*, L) \leq \pi/12$. But then, since $\angle(L, L_0) < \pi/4$ by our choice of α_0, it follows that $\angle(L_*, L_0) < \pi/3$. From elementary geometry one sees that $\vartheta_2 = \angle(L_*, L_0)$ or $\pi - \angle(L_*, L_0)$. Thus

$$\sin\vartheta_2 < \sqrt{3}/2.$$

Since ϑ is one of $\vartheta_1 + \vartheta_2$, $\vartheta_1 - \vartheta_2$, or $\vartheta_2 - \vartheta_1$, the Cosine Addition/Subtraction Formulas and our estimates on $\sin\vartheta_1$ and $\sin\vartheta_2$ imply that

$$\cos\vartheta \geq \cos\vartheta_1\cos\vartheta_2 - \sin\vartheta_1\sin\vartheta_2 > \frac{\sqrt{35} - \sqrt{3}}{12} > \frac{1}{9}$$

with the last inequality leaving a lot of room to spare.

As a consequence of the last paragraph, $(\star\star)$, and the fact that $|\eta - z_m| \leq d(z)/20$, we have

$$\begin{aligned}\mathrm{dist}(z, L_*) &= |z - z_*| = |z - \eta|\cos\vartheta\\ &> \frac{|z - z_m| - |\eta - z_m|}{9} > \frac{19d(z)/20 - d(z)/20}{9} = \frac{d(z)}{10}.\end{aligned}$$

Since both $|\eta - z_m|$ and $|\xi - z_m| \leq d(z)/20 \leq C_4\,d(z)$, it follows from $(\star\star\star)$ that

$$|z - \eta| \text{ and } |z - \xi| \leq 10C_4\,d(z).$$

Proposition 4.1 and the last two displayed inequalities now yield

$$(\ddagger)\ c(z,\eta,\xi) = \frac{2\,\mathrm{dist}(z,L_*)}{|z-\eta||z-\xi|} \geq \frac{1}{500\,C_4^2}\cdot\frac{1}{d(z)} \text{ for all } \eta \in G_1 \text{ and } \xi \in G_2.$$

From (†) and (‡), one sees that for any $z \in H$,

$$\iint c^2(z,\eta,\xi)\,d\mu(\eta)\,d\mu(\xi) \geq \int_{G_1}\int_{G_2} \frac{1}{(25\times 10^4)\,C_4^4}\cdot\frac{1}{d(z)^2}\,d\mu(\eta)\,d\mu(\xi)$$
$$\geq \frac{c_2^2}{(4\times 10^8)\,C_4^4}.$$

Integrating both sides of this inequality with respect to μ over all points $z \in H$ and performing a bit of algebra, we obtain

$$\frac{250\,M}{c_2}\,\mu(H) \leq \frac{10^{11} M\,C_4^4}{c_2^3}\,c^2(\mu).$$

Invoking (4) of Reduction 8.3 and Condition (3) for the selection of κ_0, we are done. □

8.17 Most of K Lies in the Vicinity of Γ

Given $p \in L_0$ and $r \geq 0$, from here on in we will adopt the following notational convention: $I(p;r)$ denotes the closed subinterval of L_0 with center p and radius r.

Lemma 8.46 *For any $z \in K$ and any $r \geq d(z)/10$,*

$$\int_{B(z;r)\setminus H} |\zeta - \{\pi(\zeta) + i\,\ell(\pi(\zeta))\}|\,d\mu(\zeta) \leq (24\times 10^8)\,\varepsilon_0\,r^2.$$

Proof Without loss of generality, $r > 0$. Since K is the support of μ and the integrand vanishes on K_0, we need only integrate over $[B(z;r)\cap K]\setminus[H\cup K_0]$. For any $\zeta \in [B(z;r)\cap K]\setminus[H\cup K_0]$, there exists an index m such that $\pi(\zeta) \in 4I_m$ and $\zeta \in 1{,}980C_4\,B_m$ by (a) of Proposition 8.30 and the Definition of H (8.44). Setting $\mathcal{M} = \{m : \pi([B(z;r)\cap K]\setminus[H\cup K_0])\cap 4I_m \neq \emptyset\}$ and $\mathcal{N}(m) = \{n : 4I_n\cap 4I_m \neq \emptyset\}$, we see, with the help of (a) and (d)of Proposition 8.35, that

$$\int_{B(z;r)\setminus H} |\zeta - \{\pi(\zeta) + i\,\ell(\pi(\zeta))\}|\,d\mu(\zeta)$$
$$\leq \sum_{m\in\mathcal{M}} \int_{\pi^{-1}(4I_m)\cap 1980C_4 B_m} |\zeta - \{\pi(\zeta) + i\,\ell(\pi(\zeta))\}|\,d\mu(\zeta)$$

$$= \sum_{m\in\mathcal{M}} \int_{\pi^{-1}(4I_m)\cap 1980C_4 B_m} \left| \sum_{n\in\mathcal{N}(m)} [\zeta - \{\pi(\zeta) + i\,\ell_n(\pi(\zeta))\}]\,\varphi_n(\pi(\zeta)) \right| d\mu(\zeta)$$

$$\le \sum_{m\in\mathcal{M}} \sum_{n\in\mathcal{N}(m)} \int_{1980C_4 B_m} |\zeta - \{\pi(\zeta) + i\,\ell_n(\pi(\zeta))\}|\, d\mu(\zeta).$$

Fix $m \in \mathcal{M}$ and $n \in \mathcal{N}(m)$. We take note of a number of facts. First, $1{,}980C_4\,B_m \subseteq k_0 B_n$ since $1{,}980C_4\,r_m + |z_m - z_n| \le 1{,}980C_4 \cdot 20 \cdot 2|I_n| + 600C_4 \cdot |I_n| \le 79{,}800C_4\,r_n \le k_0\,r_n$ by (ii) from Construction 8.31, (d) of Proposition 8.30, (a) of Proposition 8.33, and our choice of k_0. Second, $|\zeta - \{\pi(\zeta) + i\,\ell_n(\pi(\zeta))\}| \le 2\,\mathrm{dist}(\zeta, L_n)$ since in the triangle with vertices $\pi(\zeta) + i\,\ell_n(\pi(\zeta))$, ζ, and the foot of the perpendicular from ζ to L_n, the angle at ζ is congruent to the angle between L_n and L_0, and so at most $\pi/3$ by Construction 8.32 and our choice of α_0. Third, $\beta_1^{L_n}(B_n) < 3\varepsilon_0$ by Construction 8.32. Fourth, $r_n \le 40\,r_m$ by (ii) from Construction 8.31 and (d) of Proposition 8.30. Fifth and finally, $\#\mathcal{N}(m) \le 25$ by (e) of Proposition 8.30. Thus

$$(\star) \int_{B(z;r)\setminus H} |\zeta - \{\pi(\zeta) + i\,\ell(\pi(\zeta))\}|\, d\mu(\zeta) \le 2 \sum_{m\in\mathcal{M}} \sum_{n\in\mathcal{N}(m)} \int_{k_0 B_n} \mathrm{dist}(\zeta, L_n)\, d\mu(\zeta)$$

$$= 2 \sum_{m\in\mathcal{M}} \sum_{n\in\mathcal{N}(m)} \beta_1^{L_n}(B_n)\, r_n^2$$

$$\le 6\varepsilon_0 \sum_{m\in\mathcal{M}} \sum_{n\in\mathcal{N}(m)} r_n^2$$

$$\le 240{,}000\,\varepsilon_0 \sum_{m\in\mathcal{M}} r_m^2.$$

Given $m \in \mathcal{M}$, there exists a point $\zeta \in [B(z;r) \cap K] \setminus [H \cup K_0]$ such that $\pi(\zeta) \in 4I_m$. Hence

$$r_m \le 20|I_m| \le \frac{20}{6} D(\pi(\zeta)) \le \frac{10}{3} d(\zeta) \le \frac{10}{3}\{d(z) + |\zeta - z|\} \le \frac{10}{3}\{10r + r\} \le 40r$$

by (ii) from Construction 8.31, (c) of Proposition 8.30, the Definition of D (8.27), (a) of Proposition 8.21, and the hypothesis of the lemma. Dropping the initial r_m from the this displayed chain of inequalities, we also see that each $|I_m| \le 2r$. Since each $4I_m$ intersects $\pi(B(z;r)) = I(\pi(z);r)$, it now follows that $\{I_m : m \in \mathcal{M}\}$ is a pairwise disjoint collection of half-open intervals whose union is contained in $I(\pi(z); 6r)$. Thus

$$(\star\star) \sum_{m\in\mathcal{M}} r_m^2 \le 40r \sum_{m\in\mathcal{M}} r_m \le 40r \sum_{m\in\mathcal{M}} 20|I_m| \le 800r \cdot 12r \le 10{,}000\,r^2.$$

From $(\star)$ and $(\star\star)$ we conclude that

$$\int_{B(z;r)\setminus H} |\zeta-\{\pi(\zeta)+i\,\ell(\pi(\zeta))\}|d\mu(\zeta) \le 240{,}000\,\varepsilon_0\cdot 10{,}000\,r^2 = (24\times 10^8)\,\varepsilon_0\,r^2,$$

and so are done. □

Lemma 8.47 *For any $z \in K \setminus H$,*

$$\frac{d(z)}{10^4\,C_4(M,\delta_0)} \le D(\pi(z)) \le d(z).$$

Proof The second inequality is trivially true by the Definition of D (8.27).

If $z \in K_0$, then $d(z) = 0$ by the Definition of K_0 (8.23) and the first inequality obviously holds.

If $z \notin K_0$, then, using the Definition of H (8.44), $z \notin H$ forces $\pi(z) \notin \pi(K_0)$. Since it is also the case that $\pi(K) \subseteq U_{1/2}$ by Construction 8.12, we must have $\pi(z) \in 4I_m$ for a number of ms by (a) of Proposition 8.30. But then, using the Definition of H (8.44) again, $z \notin H$ now forces $z \in 1{,}980C_4\,B_m$ for at least one these ms. Applying (c) of Proposition 8.30 to such an m and then invoking (ii) of Construction 8.31, we see that $D(\pi(z)) \ge 6|I_m| \ge (3/10)\,r_m$. Clearly, using the Definition of d (8.20), we see that $d(z) \le |z - z_m| + r_m \le 1{,}980C_4\,r_m + r_m \le 3{,}000C_4\,r_m$. The desired conclusion now follows. □

The following is a set of "good" points, i.e., a set each of whose points lies in the "vicinity" of our Lipschitz graph:

Definition 8.48 (The Good Set) $G = \{z \in K \setminus H : |z - \{\pi(z) + i\,\ell(\pi(z))\}| \le \sqrt{\varepsilon_0}\,d(z)\}$.

That "most" of K lies in the vicinity of our Lipschitz graph translates into saying that $K \setminus G$ has small μ-measure. Establishing this is the main point of the present section. It is accomplished in the following.

Proposition 8.49 (The Smallness of $K \setminus G$) $\mu(K \setminus G) \le 2 \times 10^{-5}$. *More strongly, we even have*

$$\frac{250\,M}{c_2(M,\delta_0)}\,\mu(K \setminus G) \le 2 \times 10^{-5}.$$

The stronger conclusion will be needed later to prove Lemma 8.78.

Proof Since $K \setminus G \subseteq \{K \setminus [G \cup H]\} \cup H$, by the stronger conclusion of Proposition 8.45 it suffices to show that

$$\frac{250\,M}{c_2(M,\delta_0)}\,\mu(K \setminus [G \cup H]) \le 10^{-5}.$$

Without loss of generality, $K\setminus[G\cup H] \ne \emptyset$. Since $K_0 \subseteq G$ by Construction 8.25 and the Definitions of H and G (8.44 and 8.48), it follows from the Definition of K_0 (8.23) that $d > 0$ everywhere on $K\setminus[G\cup H]$. In consequence we may apply part (a) of Besicovitch's Covering Lemma (5.26) to the collection $\{B(z; d(z)/10) : z \in K \setminus [G \cup H]\}$ to extract a subcollection $\mathcal{B}$ which still covers $K \setminus [G \cup H]$ but, in addition, now has a controlled overlap of at most 125. By (a) of Proposition 8.21, given any $\zeta \in B \in \mathcal{B}$, $d(\zeta) \geq d(\mathrm{cen}(B)) - |\zeta - \mathrm{cen}(B)| \geq (9/10)d(\mathrm{cen}(B)) = 9\,\mathrm{rad}(B)$. Thus

$$
\begin{aligned}
(\star)\ \sqrt{\varepsilon_0}\,\mu(K \setminus [G \cup H]) &\leq \int_{K\setminus[G\cup H]} \frac{|\zeta - \{\pi(\zeta) + i\,\ell(\pi(\zeta))\}|}{d(\zeta)}\, d\mu(\zeta) \\
&\leq \sum_{B\in\mathcal{B}} \int_{B\setminus H} \frac{|\zeta - \{\pi(\zeta) + i\,\ell(\pi(\zeta))\}|}{d(\zeta)}\, d\mu(\zeta) \\
&\leq \frac{1}{9} \sum_{B\in\mathcal{B}} \frac{1}{\mathrm{rad}(B)} \int_{B\setminus H} |\zeta - \{\pi(\zeta) + i\,\ell(\pi(\zeta))\}|\, d\mu(\zeta) \\
&\leq (3 \times 10^8)\, \varepsilon_0 \sum_{B\in\mathcal{B}} \mathrm{rad}(B)
\end{aligned}
$$

with the last inequality being an application of Lemma 8.46.

We now turn to estimating the sum of the radii of the balls from $\mathcal{B}$. For that we need the following.

Claim. Given $B \in \mathcal{B}$, set $\mathcal{B}(B) = \{B' \in \mathcal{B} : \mathrm{rad}(B') \leq 2\,\mathrm{rad}(B)$ and $\pi(cB') \cap \pi(cB) \neq \emptyset\}$ where $c = 1/\{10^4\, C_4\}$. Then

$$
\#\mathcal{B}(B) \leq (2 \times 10^{15})\, C_4^4.
$$

To see this, suppose that $B = B(z; d(z)/10)$ and $B' = B(w; d(w)/10)$ for the sake of convenience. Then $B' \in \mathcal{B}(B)$ implies that $d(w) \leq 2\,d(z)$ and $|\pi(z) - \pi(w)| \leq c\{d(z)/10 + d(w)/10\} \leq (c/2)d(z)$. We may thus apply Corollary 8.26 with $r = 2\,d(z)$ to conclude that $|z - w| \leq 18C_4\, d(z)$. This implies that $B' \subseteq 200C_4\, B$ for all $B' \in \mathcal{B}(B)$ and so

$$
(\dagger)\ \sum_{B'\in\mathcal{B}(B)} \mathcal{X}_{B'} \leq 125\, \mathcal{X}_{200C_4\, B}.
$$

Next note that by Lemma 8.47 and (c) of Proposition 8.28, $c\,d(z) \leq D(\pi(z)) \leq D(\pi(w)) + |\pi(z) - \pi(w)| \leq d(w) + (c/2)d(z)$. Thus $d(w) \geq (c/2)d(z)$ and so

$$
(\ddagger)\ \mathrm{rad}(B') \geq (c/2)\mathrm{rad}(B) \text{ for all } B' \in \mathcal{B}(B).
$$

Integrating both sides of $(\dagger)$ with respect to area and utilizing $(\ddagger)$ on the left side of the resulting inequality, we obtain

$$
\#\mathcal{B}(B) \cdot \pi\,(c/2)^2\,\mathrm{rad}^2(B) \leq \sum_{B'\in\mathcal{B}(B)} \pi\,\mathrm{rad}^2(B') \leq 125\,\pi(200C_4)^2\mathrm{rad}^2(B).
$$

The Claim's conclusion is now clear.

Select a ball $B_1 \in \mathcal{B}$ with $\text{rad}(B_1) \geq (1/2)\sup\{\text{rad}(B') : B' \in \mathcal{B}\}$ and set $\mathcal{B}_1 = \{B' \in \mathcal{B} : \pi(cB') \cap \pi(cB_1) \neq \emptyset\}$. Note that $\mathcal{B}_1 = \mathcal{B}(B_1)$. If $\mathcal{B} \setminus \mathcal{B}_1 \neq \emptyset$, select a ball $B_2 \in \mathcal{B} \setminus \mathcal{B}_1$ with $\text{rad}(B_2) \geq (1/2)\sup\{\text{rad}(B') : B' \in \mathcal{B} \setminus \mathcal{B}_1\}$ and set $\mathcal{B}_2 = \{B' \in \mathcal{B} \setminus \mathcal{B}_1 : \pi(cB') \cap \pi(cB_2) \neq \emptyset\}$. Note that $\mathcal{B}_2 \subseteq \mathcal{B}(B_2)$ and that $\pi(cB_2) \cap \pi(cB_1) = \emptyset$. If $\mathcal{B} \setminus (\mathcal{B}_1 \cup \mathcal{B}_2) \neq \emptyset$, select a ball $B_3 \in \mathcal{B} \setminus (\mathcal{B}_1 \cup \mathcal{B}_2)$ with $\text{rad}(B_3) \geq (1/2)\sup\{\text{rad}(B') : B' \in \mathcal{B} \setminus (\mathcal{B}_1 \cup \mathcal{B}_2)\}$ and set $\mathcal{B}_3 = \{B' \in \mathcal{B} \setminus (\mathcal{B}_1 \cup \mathcal{B}_2) : \pi(cB') \cap \pi(cB_3) \neq \emptyset\}$. Note that $\mathcal{B}_3 \subseteq \mathcal{B}(B_3)$ and that $\pi(cB_3) \cap \{\pi(cB_1) \cup \pi(cB_2)\} = \emptyset$. Continuing in this manner for as long as possible, we obtain a sequence $\{B_n\}$ of balls from $\mathcal{B}$ and a corresponding sequence $\{\mathcal{B}_n\}$ of subfamilies of $\mathcal{B}$. These sequences may be finite or infinite. Since the finite case is simpler, let us assume the two sequences are infinite. By construction the subintervals $\pi(cB_n)$ of L_0 are pairwise disjoint. Their centers are contained in $\pi(K) \subseteq U_{1/2}$ by Construction 8.12 and their radii are at most $c/5 \leq 1/2$ by (c) of Proposition 8.21. So clearly $\sum_n 2c\,\text{rad}(B_n) \leq 2$, i.e., $\sum_n 2\,\text{rad}(B_n) \leq 2/c = (2 \times 10^4)\,C_4$. This implies that $\text{rad}(B_n) \to 0$ as $n \to \infty$, which in turn implies that $\mathcal{B} = \bigcup_n \mathcal{B}_n$ (if one had some $B \in \mathcal{B} \setminus \bigcup_n \mathcal{B}_n$, then one would have $\text{rad}(B_n) \geq \text{rad}(B)/2$ for all n). Finally, each $\mathcal{B}_n \subseteq \mathcal{B}(B_n)$. From these facts and the Claim we obtain

$$(\star\star)\quad \sum_{B\in\mathcal{B}} \text{rad}(B) = \sum_n \sum_{B\in\mathcal{B}_n} \text{rad}(B) \leq (2\times 10^{15})\,C_4^4 \sum_n 2\,\text{rad}(B_n) \leq (4\times 10^{19})\,C_4^5.$$

From $(\star)$ and $(\star\star)$ one sees that

$$\frac{250\,M}{c_2}\,\mu(K \setminus [G \cup H]) \leq \frac{(3 \times 10^{30})\,M\,C_4^5}{c_2}\,\sqrt{\varepsilon_0}.$$

Condition (4) for the selection of ε_0 now finishes off the proof. □

8.18 The Smallness of K_1

Proposition 8.50 (The Smallness of K_1) $\mu(K_1) \leq 3 \times 10^{-5}$.

Proof Since $K_1 \subseteq \{K_1 \cap G\} \cup \{K \setminus G\}$, by Proposition 8.49 it suffices to show that $\mu(K_1 \cap G) \leq 10^{-5}$.

Without loss of generality, $K_1 \cap G \neq \emptyset$. Since $K_0 \cap K_1 = \emptyset$ by the Definition of K_1 (8.41), it follows from the Definition of K_0 (8.23) and (b) of Proposition 8.21 that $h > 0$ everywhere on $K_1 \cap G$. In consequence we may apply part (a) of Besicovitch's Covering Lemma (5.26) to the collection $\{B(z; h(z)/8) : z \in K_1 \cap G\}$ to extract a subcollection $\mathcal{B}$ which still covers $K_1 \cap G$ but, in addition, now has a controlled overlap of at most 125.

First Claim. For any $B \in \mathcal{B}$, $\mu(B) < 12\delta_0\,\text{rad}(B)$.

To see this, note that $B = B(z; h(z)/8)$ where $z \in K_1$. Thus by the Definition of K_1 (8.41) there exist $w \in K$ and $s \in [h(z)/4, 3h(z)/4]$ such that $z \in B(w; s/2)$ and $\delta(w; s) < 2\delta_0$. Then $B \subseteq B(z; s/2) \subseteq B(w; s)$ and so

$$\mu(B) \leq \mu(B(w; s)) < 2\delta_0 s \leq 2\delta_0 3h(z)/4 = 12\delta_0 \,\text{rad}(B).$$

Second Claim. For any $B \in \mathcal{B}$, $\text{rad}(B) \leq \mathcal{H}^1(\Gamma \cap B)$.

To see this, note that $B = B(z; h(z)/8)$ where $z \in G$. Setting $w = \pi(z) + i\,\ell(\pi(z))$, we have $B(w; h(z)/16) \subseteq B$ by the Definition of G (8.48), (b) of Proposition 8.21, and the fact that $\sqrt{\varepsilon_0} \leq 10^{-21/2} \leq 1/16$ by Condition (1) for the selection of ε_0. Since $\pi(w) \in \pi(K) \subseteq U_{1/2}$ by Construction 8.12 and $h(z)/16 \leq 1/8$ by the Definition of h (8.15), the interval $I(\pi(w); h(z)/16)$ is contained in U_5, the interval of definition of ℓ. Let p_1 be the largest number in $[\pi(w) - h(z)/16, \pi(w)]$ such that $|\{p_1 + i\,\ell(p_1)\} - w| = h(z)/16$. That p_1 exists and is well defined is a consequence of the Intermediate Value Theorem and the continuity of ℓ. Clearly $\Gamma_1 = \{p + i\,\ell(p) : p \in [p_1, \pi(w)]\}$ is a continuum of diameter at least $h(z)/16$. Another application of the Intermediate Value Theorem and the continuity of ℓ shows that Γ_1 is contained in $\Gamma \cap B(w; h(z)/16) \subseteq \Gamma \cap B$. We may similarly get a point p_2 in $[\pi(w), \pi(w) + h(z)/16]$ such that $\Gamma_2 = \{p + i\,\ell(p) : p \in [\pi(w), p_2]\}$ is a continuum of diameter at least $h(z)/16$ contained in $\Gamma \cap B$. Since Γ_1 and Γ_2 intersect at the single point w, which has $\mathcal{H}^1$-measure zero, we may invoke Proposition 5.5 and Lemma 5.6 to conclude that $\mathcal{H}^1(\Gamma \cap B) \geq \mathcal{H}^1(\Gamma_1 \cup \Gamma_2) = \mathcal{H}^1(\Gamma_1) + \mathcal{H}^1(\Gamma_2) \geq h(z)/16 + h(z)/16 = \text{rad}(B)$.

Third Claim. $\mathcal{H}^1(\Gamma) \leq 360$.

We prove this by going back to the definition of Hausdorff measure. Given a Lipschitz graph Γ over an interval I with bound M and an arbitrary $\delta > 0$, choose $n \geq 1$ so that $\sqrt{1 + M^2}\,|I|/n < \delta$. We can clearly cover Γ with closed rectangles R_1, R_2, ..., R_n each of dimensions $|I|/n \times M|I|/n$ and so of diameter $\sqrt{1 + M^2}\,|I|/n$. Thus

$$\mathcal{H}^1_\delta(\Gamma) \leq \sum_{m=1}^{n} |R_m| = n \cdot \sqrt{1 + M^2}\,|I|/n = \sqrt{1 + M^2}\,|I|.$$

Letting $\delta \downarrow 0$, we have $\mathcal{H}^1(\Gamma) \leq \sqrt{1 + M^2}\,|I|$. In our present situation we may take I to be U_5 and M to be $71\alpha_0$ by Proposition 8.39. Thus by our choice of α_0,

$$\begin{aligned}\mathcal{H}^1(\Gamma) &\leq \sqrt{1 + M^2}\,|I| \leq \sqrt{1 + \{71\pi/8\}^2} \cdot 10 \leq 5\sqrt{4 + \{71\pi/4\}^2} \\ &\leq 5\sqrt{4 + 71^2} \leq 360.\end{aligned}$$

From the properties of our Besicovitch cover $\mathcal{B}$ and these three Claims we see that

$$\mu(K_1 \cap G) \leq \sum_{B \in \mathcal{B}} \mu(B) \quad < \quad \sum_{B \in \mathcal{B}} 12\delta_0 \,\text{rad}(B)$$

$$\leq 12\delta_0 \sum_{B \in \mathcal{B}} \mathcal{H}^1(\Gamma \cap B)$$
$$\leq 12\delta_0\{125\,\mathcal{H}^1(\Gamma)\} \;\leq\; 540{,}000\,\delta_0.$$

Our choice of δ_0 now finishes off the proof. □

8.19 Gamma Functions Associated with ℓ

To prove that K_3 has small μ-measure we need to introduce and estimate two functions, $\gamma(p; r)$ and $\tilde{\gamma}(p; r)$, associated with our Lipschitz graph. In this section we shall define both functions and establish that they are comparable. In the next section we shall derive a point estimate for the second of these functions; while in the section thereafter this point estimate will be upped, via comparability, to a global estimate for the first of these functions which is similar to the estimate obtained in Proposition 8.10 for $\beta_1(z; r)$. Since what goes without saying frequently goes much better when actually said, we note that neither the $\gamma(p; r)$ nor the $\tilde{\gamma}(p; r)$ about to be defined is the analytic capacity of any set!

Recall that in Construction 8.12 we made the convention that $U_a = (-a, a)$ for $a > 0$.

Definition 8.51 (The Gamma Functions) *Given $p \in U_4 \subseteq \mathbb{R} = L_0$ and $r \in (0, 1)$, note that $I(p; r) \subseteq U_5$, the domain of definition of ℓ. Thus, for such a p and such an r, we may set*

$$\gamma(p; r) = \inf\{\gamma^l(p; r) : l \text{ is a linear function from } L_0 = \mathbb{R} \text{ to } \mathbb{R} = L_0^{\perp}/i\}$$

where

$$\gamma^l(p; r) = \frac{1}{r}\int_{I(p;r)} \frac{|\ell(t) - l(t)|}{r}\,dt$$

and

$$\tilde{\gamma}(p; r) = \inf\{\tilde{\gamma}^L(p; r) : L \text{ is a line}\}$$

where

$$\tilde{\gamma}^L(p; r) = \frac{1}{r}\int_{I(p;r)} \frac{\mathrm{dist}(t + i\,\ell(t), L)}{r}\,dt.$$

The comparability of these two functions that is the goal of this section follows.

Proposition 8.52 *For any $p \in U_4$ and any $r \in (0, 1)$, $\tilde{\gamma}(p; r) \leq \gamma(p; r) \leq \sqrt{2}\,\tilde{\gamma}(p; r)$.*

Proof Given any linear function $l : L_0 = \mathbb{R} \mapsto \mathbb{R} = L_0^\perp/i$, let L be the graph of l, so $L = \{t + i\,l(t) : t \in \mathbb{R}\}$. For convenience we will use ϑ to denote the angle formed by L and L_0. Then, by elementary geometry, in the triangle with vertices $t + i\,\ell(t), t + i\,l(t)$, and the foot of the perpendicular from $t + i\,\ell(t)$ to L, the angle at $t + i\,\ell(t)$ is also ϑ. It follows that $|\ell(t) - l(t)| = (\sec\vartheta)\,\mathrm{dist}(t + i\,\ell(t), L)$ and so we have

$$(\star)\ \gamma^l(p;r) = (\sec\vartheta)\,\tilde{\gamma}^L(p;r).$$

Since $\sec\vartheta \geq 1$ always, $(\star)$ immediately implies the first inequality of the proposition. The second inequality of the proposition will also follow from $(\star)$, but not so immediately.

What does follow from $(\star)$ immediately is

$$(\star\star)\ \gamma(p;r) \leq \sqrt{2}\,\inf\{\tilde{\gamma}^L(p;r) : L \text{ is a line with } 0 \leq \vartheta \leq \pi/4\}.$$

So now consider the case where L has $\pi/4 < \vartheta < \pi/2$. By Proposition 8.39,

$$|\ell(t) - \ell(p)| \leq 71\alpha_0\,r$$

for all $t \in I(p;r)$. Since $\tan\vartheta > 1$, $|l(t) - \ell(p)| \leq 71\alpha_0\,r$ only holds for t in a subinterval of $\mathbb{R}$ with length at most $142\alpha_0\,r$. Moreover, $142\alpha_0\,r \leq r$ by our choice of α_0. Thus $I(p;r)$ contains a subinterval I of length $r/2$ such that either $l(t) \geq \ell(p) + 71\alpha_0\,r$ for all $t \in I$ or $l(t) \leq \ell(p) - 71\alpha_0\,r$ for all $t \in I$. Without loss of generality, suppose the former to be the case. Letting q denote the center of I, we then have

$$|l(t) - \ell(p)| = l(t) - \ell(p) = (\pm\tan\vartheta)(t-q) + l(q) - \ell(p) \geq (\pm\tan\vartheta)(t-q) + 71\alpha_0\,r$$

for all $t \in I$. In the above we use the plus sign if l is positively sloped and the minus sign otherwise. From the last two displayed inequalities we now conclude that

$$|l(t) - \ell(t)| \geq |l(t) - \ell(p)| - |\ell(t) - \ell(p)| > (\pm\tan\theta)(t - q)$$

for all $t \in I$. Denoting the right and left halves of I by I^+ and I^-, respectively, and noting that $\sin\vartheta \geq 1/\sqrt{2}$, we then conclude that

$$\begin{aligned}\tilde{\gamma}^L(p;r) = (\cos\vartheta)\,\gamma^l(p;r) &\geq \frac{\cos\vartheta}{r}\int_{I^\pm}\frac{(\pm\tan\vartheta)(t-q)}{r}\,dt \\ &= \frac{\sin\vartheta}{r^2}\cdot\frac{(r/4)^2}{2} \geq \frac{1}{32\sqrt{2}}.\end{aligned}$$

It is left to the reader to show that this inequality also holds in the case where L has $\vartheta = \pi/2$, i.e., where L is a vertical line.

Consideration of the linear function with slope 0 and y-intercept $\ell(p)$ and a use of Proposition 8.39 yield

$$\gamma(p;r) \leq \frac{1}{r}\int_{I(p;r)} \frac{|\ell(t)-\ell(p)|}{r}\,dt \leq \frac{1}{r}\int_{p-r}^{p+r} \frac{71\alpha_0|t-p|}{r}\,dt = 71\alpha_0.$$

From the last two displayed inequalities and our choice of α_0 we obtain

$$(\star\star\star)\ \gamma(p;r) \leq \sqrt{2}\,\inf\{\tilde{\gamma}^L(p;r) : L \text{ is a line with } \pi/4 < \vartheta \leq \pi/2\}.$$

The second inequality of the proposition now follows from $(\star\star)$ and $(\star\star\star)$. □

8.20 A Point Estimate on One of the Gamma Functions

The point estimate on $\tilde{\gamma}(p;r)$ that is the goal of this section follows.

Proposition 8.53 *Suppose $p \in U_4$ and $r \in (0,1)$ are such that $D(p) < 20r$. Set*

$$\mathcal{N}(p;r) = \{n : I_n \cap I(p;r) \neq \emptyset\}.$$

Then

$$\tilde{\gamma}^2(p;r) \leq C_5(M,\delta_0)\left[\int_{\pi^{-1}(I(p;40r))} \frac{\beta_1^2(\zeta;r)}{r}\,\mathcal{X}_{\{\tilde{\delta}(\zeta;r)\geq 20\,c_2(M,\delta_0)\}}\,d\mu(\zeta) + \left\{\varepsilon_0 \sum_{n\in\mathcal{N}(p;r)} \left(\frac{|I_n|}{r}\right)^2\right\}^2\right].$$

Recall that the quantity $C_5(M,\delta)$ was defined back in Section 8.6.

The proof of this proposition will follow upon the heels of a series of lemmas.

Given $p \in U_4$ and $r \in (0,1)$ such that $D(p) < 20r$, there exists a ball $B(\tilde{z};\tilde{r}) \in S_0$ such that $|p-\pi(\tilde{z})|+\tilde{r} < 20r$ by (b) of Proposition 8.28. *A standing assumption for the rest of this section is that p, r, $\tilde{z}$, and $\tilde{r}$ are as in the last sentence!*

Lemma 8.54 *Suppose $\zeta \in B(\tilde{z};20r)\cap K$ and L_* is any line. Then*

$$\frac{1}{r}\int_{I(p;r)\cap\pi(K_0)} \frac{\operatorname{dist}(t+i\,\ell(t),L_*)}{r}\,dt \leq \frac{4}{\delta_0}\,\beta_1^{L_*}(\zeta;r).$$

Proof Recall that in Section 2.1 we showed that $\mathcal{H}^1$ and $\mathcal{L}^1$ coincide for linear sets. From this and Proposition 2.2 (note that orthogonal projection onto a line decreases distances), we see that for any Borel subset E of $\mathbb{R}$,

$$\mathcal{L}^1(E\cap\pi(K_0)) = \mathcal{H}^1(E\cap\pi(K_0)) = \mathcal{H}^1(\pi(\pi^{-1}(E)\cap K_0)) \leq \mathcal{H}^1(\pi^{-1}(E)\cap K_0).$$

To continue we must next relate $\mathcal{H}^1$ on K_0 to μ on K_0. To do this we need the following.

Claim. For any $z \in K_0$ and $0 < r < 1$, $\mu(B(z;r)) \geq (\delta_0/4)|B(z;r)|$.

Since $d(z) = 0$ by the Definition of K_0 (8.23), there exists a ball $B(w;s) \in S_0$ such that $|z - w| + s < r/2$ by the Definition of d (8.20). Applying (c) of Corollary 8.18, we see that $B(w;r/2) \in S_0$ and so $\delta(B(w;r/2)) \geq \delta_0$ by the Definitions of S_{total} and S_0 (8.13 and 8.16). Hence $\mu(B(z;r)) \geq \mu(B(w;r/2)) \geq \delta_0(r/2) = (\delta_0/4)|B(z;r)|$, establishing the Claim.

Given $0 < \delta < 2$, by (a) of Besicovitch's Covering Lemma (5.26) we may easily obtain a δ-cover $\{B_n\}$ of K_0 consisting of closed balls centered on K_0 and having a controlled overlap of 125. (Do not confuse these balls with the balls B_n of Construction 8.31!) Applying the Claim to each B_n, it is now clear that

$$\mathcal{H}^1_\delta(K_0) \leq \sum_n |B_n| \leq (4/\delta_0) \sum_n \mu(B_n) \leq (4/\delta_0) 125\, \mu(\mathbb{C}).$$

Letting $\delta \downarrow 0$, we see that $\mathcal{H}^1(K_0) < \infty$. This allows us to apply Lemma 5.14 and Proposition 5.21 in the next paragraph.

Given $0 < \varepsilon < 1$ and F a compact subset of K_0, by (b) of Vitali's Covering Lemma (5.14) we may easily obtain a countable pairwise disjoint collection $\{B_n\}$ of closed balls centered on F with radius less than ε such that $\mathcal{H}^1(F) < \sum_n |B_n| + \varepsilon$. (Do not confuse these balls with the balls B_n of Construction 8.31 ... or the balls B_n of the last paragraph!) Applying the Claim to each B_n, it is now clear that

$$\begin{aligned} \mathcal{H}^1(F) < \sum_n |B_n| + \varepsilon &\leq (4/\delta_0) \sum_n \mu(B_n) + \varepsilon \\ &\leq (4/\delta_0)\mu(\{z \in \mathbb{C} : \operatorname{dist}(z, F) < \varepsilon\}) + \varepsilon. \end{aligned}$$

Letting $\varepsilon \downarrow 0$, we obtain $\mathcal{H}^1(F) \leq (4/\delta_0)\mu(F)$ from the Lower Continuity Property of Measures [RUD, 1.19(e)]. Proposition 5.21 now allows us to easily extend this inequality to any Borel subset F of K_0.

Setting $F = \pi^{-1}(E) \cap K_0$ in the inequality established in the last paragraph and combining the result with the inequality established in the first paragraph, we have

$$\mathcal{L}^1(E \cap \pi(K_0)) \leq (4/\delta_0)\mu(\pi^{-1}(E) \cap K_0)$$

for any Borel subset E of $\mathbb{R}$.

We have thus verified the formula

$$\int_{\pi(K_0)} \varphi \, d\mathcal{L}^1 \leq \frac{4}{\delta_0} \int_{K_0} \varphi \circ \pi \, d\mu$$

for φ the characteristic function of any Borel subset E of $\mathbb{R}$. By the usual approximation argument involving increasing limits of nonnegative simple functions, we conclude that this formula holds for all nonnegative Borel functions. Setting

$\varphi(t) = \mathrm{dist}(t+i\,\ell(t), L_*)\cdot \mathcal{X}_{I(p;r)}(t)$ in this formula, noting that $\pi(\xi)+i\,\ell(\pi(\xi)) = \xi$ for $\xi \in K_0$, and then dividing by r^2, one obtains

$$(\star)\ \frac{1}{r}\int_{I(p;r)\cap\pi(K_0)} \frac{\mathrm{dist}(t+i\,\ell(t), L_*)}{r}\,dt \le \frac{4}{\delta_0}\left\{\frac{1}{r}\int_{\pi^{-1}(I(p;r))\cap K_0} \frac{\mathrm{dist}(\xi, L_*)}{r}\,d\mu(\xi)\right\}.$$

For any $\xi \in \pi^{-1}(I(p;r)) \cap K_0$, we have $|\pi(\xi) - \pi(\tilde{z})| \le |\pi(\xi) - p| + |p - \pi(\tilde{z})| < r + 20r = 21r$, $d(\xi) = 0$ by the Definition of K_0 (8.23), and $d(\tilde{z}) \le \tilde{r} < 20r$ by the Definition of d (8.20). Corollary 8.26 thus applies and tells us that $|\xi - \tilde{z}| \le 9C_4 \cdot 21r$. Since $|\tilde{z} - \zeta| \le 20r$, our choice of k_0 now yields

$$(\star\star)\ \pi^{-1}(I(p;r)) \cap K_0 \subseteq B(\zeta; k_0 r).$$

The proposition now follows from $(\star)$, $(\star\star)$, and the definition of $\beta_1^{L_*}(\zeta; r)$. □

Lemma 8.55 *Suppose L_* is any line. Then for any n,*

$$\frac{1}{r}\int_{I_n} \frac{\mathrm{dist}(t+i\,\ell(t), L_*)}{r}\,dt \le$$

$$1{,}201\, C_1\, \varepsilon_0 \left(\frac{|I_n|}{r}\right)^2 + \frac{C_1}{c_2^2}\,\frac{|I_n|}{r^2}\left\{\frac{1}{|I_n|}\int_{B_n} \mathrm{dist}(\xi, L_*)^{1/3}\,d\mu(\xi)\right\}^3$$

where $C_1 = C_1(M, \delta_0)$ and $c_2 = c_2(M, \delta_0)$.

Proof Given $t \in I_n$, let w_n be the foot of the perpendicular from $w = t + i\,\ell(t)$ to L_n. Clearly

$$\mathrm{dist}(t+i\,\ell(t), L_*) \le |t + i\,\ell(t) - w_n| + \mathrm{dist}(w_n, L_*).$$

Thus the lemma will follow easily from the First and Third of the Claims to follow.

First Claim. $|t + i\,\ell(t) - w_n| \le 1{,}200\, C_1\, \varepsilon_0\, |I_n|$.

Appealing to the fact that $t + i\,\ell_n(t) \in L_n$, the Construction of ℓ on $U_5 \setminus K_0$ (8.36), (a) and (d) of Proposition 8.35, and (c) of Proposition 8.33, we see that

$$|t + i\,\ell(t) - w_n| \le |\ell(t) - \ell_n(t)| \le \sum_m |\ell_m(t) - \ell_n(t)|\,\varphi_m(t) \le 1{,}200\, C_1\, \varepsilon_0\, |I_n|,$$

thus establishing the First Claim.

Second Claim. $|w_n - z_n| \le 130\,|I_n|$.

From the First Claim, Condition (3) for the selection of ε_0, and our choice of α_0, we see that

$$|t - \pi(w_n)| \leq |t + i\,\ell(t) - w_n| \leq 1{,}200\, C_1\, \varepsilon_0\, |I_n| \leq |I_n|/2.$$

Let z_{nn} be the foot of the perpendicular from z_n to L_n. By Construction 8.32 we may apply (a) of Proposition 8.7 and then invoke (ii) of Construction 8.31 to obtain

$$|\pi(z_n) - \pi(z_{nn})| \leq |z_n - z_{nn}| = \operatorname{dist}(z_n, L_n) \leq 2r_n \leq 40|I_n|.$$

The last two displayed inequalities and (iii) of Construction 8.31 now imply that

$$\begin{aligned} |\pi(w_n) - \pi(z_{nn})| &\leq |\pi(w_n) - t| + |t - p_n| + |p_n - \pi(z_n)| + |\pi(z_n) - \pi(z_{nn})| \\ &\leq |I_n|/2 + |I_n|/2 + 20|I_n| + 40|I_n| \;=\; 61|I_n|. \end{aligned}$$

By Construction 8.32 and our choice of α_0, $\angle(L_n, L_0) < \pi/4$. Since both w_n and z_{nn} lie on L_n, it then follows that

$$|w_n - z_{nn}| \leq \sqrt{2}\,|\pi(w_n) - \pi(z_{nn})| \leq 90|I_n|.$$

Finally,

$$|w_n - z_n| \leq |w_n - z_{nn}| + |z_{nn} - z_n| \leq 90|I_n| + 40|I_n| = 130|I_n|,$$

thus establishing the Second Claim.

Third Claim. $\operatorname{dist}(w_n, L_*) \leq C_1\, \varepsilon_0\, |I_n| + \dfrac{C_1}{c_2^2} \left\{ \dfrac{1}{|I_n|} \displaystyle\int_{B_n} \operatorname{dist}(\xi, L_*)^{1/3}\, d\mu(\xi) \right\}^3.$

Let B_1 and B_2 denote the two balls gotten by applying Lemma 8.5 to the ball $B = B_n \ldots$ which we may do by Construction 8.32. (Do not confuse these two balls with the balls B_1 and B_2 of Construction 8.31!) Set

$$G_j = \{\xi \in B_j \cap B_n : \operatorname{dist}(\xi, L_n) < (120/c_2)\, \varepsilon_0\, |I_n|\}$$

for $j = 1, 2$. Then

$$\begin{aligned} \mu([B_j \cap B_n] \setminus G_j) &\leq \int_{[B_j \cap B_n] \setminus G_j} \frac{\operatorname{dist}(\xi, L_n)}{(120/c_2)\, \varepsilon_0\, |I_n|}\, d\mu(\xi) \\ &\leq \frac{r_n^2}{(120/c_2)\, \varepsilon_0\, |I_n|}\, \beta_1^{L_n}(B_n) < \frac{c_2}{2} r_n \end{aligned}$$

since $\beta_1^{L_n}(B_n) < 3\varepsilon_0$ and $r_n < 20|I_n|$ by Construction 8.32 and (ii) of Construction 8.31 respectively. It then follows from (b) of Lemma 8.5 and (ii) of Construction 8.31 that $\mu(G_j) = \mu(B_j \cap B_n) - \mu([B_j \cap B_n] \setminus G_j) > c_2 r_n - (c_2/2) r_n = (c_2/2) r_n \geq (c_2/2)|I_n|$ for $j = 1, 2$.

Now consider any pair of points $\xi_1 \in G_1$ and $\xi_2 \in G_2$. Let ξ_{jn} be the foot of the perpendicular from ξ_j to L_n for $j = 1, 2$. Clearly $|\xi_1 - \xi_2| \geq 10 c_1 r_n \geq 10 c_1 |I_n|$ by (c) of Lemma 8.5 and (ii) of Construction 8.31. Also

$$
\begin{aligned}
|\xi_1 - \xi_2| &\le |\xi_1 - \xi_{1n}| + |\xi_{1n} - \xi_{2n}| + |\xi_{2n} - \xi_2| \\
&\le |\xi_{1n} - \xi_{2n}| + (240/c_2)\,\varepsilon_0\,|I_n| \\
&\le |\xi_{1n} - \xi_{2n}| + c_1|I_n|
\end{aligned}
$$

with the very last inequality true by Condition (1) for the selection of ε_0 since $c_5 = c_1c_2/10^6$ by Proposition 8.7. Thus $|\xi_{1n} - \xi_{2n}| \ge 9c_1|I_n|$.

Since w_n, ξ_{1n}, and ξ_{2n} are all points of L_n, with the last two points being distinct, we may write $w_n = s\,\xi_{1n}+(1-s)\,\xi_{2n}$ for some $s \in \mathbb{R}$. Then $w_n - \xi_{2n} = s\,(\xi_{1n} - \xi_{2n})$. Moreover, by the Second Claim and (ii) of Construction 8.31,

$$
\begin{aligned}
|w_n - \xi_{2n}| &\le |w_n - z_n| + |z_n - \xi_2| + |\xi_2 - \xi_{2n}| \\
&\le 130\,|I_n| + r_n + (120/c_2)\,\varepsilon_0\,|I_n| \\
&\le 150\,|I_n| + (120/c_2)\,\varepsilon_0\,|I_n| \quad\le\quad 153\,|I_n|
\end{aligned}
$$

with the very last inequality true by Condition (1) for the selection of ε_0 since $c_5 = c_1c_2/10^6 \le c_2/10^{10}$ by Proposition 8.7 and Lemma 8.5. It now follows that $|s| = |w_n - \xi_{2n}|/|\xi_{1n} - \xi_{2n}| \le 17/c_1$ and so $1 + |s| \le 20/c_1$.

Letting ξ_{jn*} be the foot of the perpendicular from ξ_{jn} to L_* for $j = 1, 2$, we obtain

$$
\begin{aligned}
(\star)\ \operatorname{dist}(w_n, L_*) &\le |w_n - [s\,\xi_{1n*} + (1-s)\,\xi_{2n*}]| \\
&\le |s|\,|\xi_{1n} - \xi_{1n*}| + |1-s|\,|\xi_{2n} - \xi_{2n*}| \\
&\le \{1 + |s|\}\,\{|\xi_{1n} - \xi_1| + |\xi_1 - \xi_{1n*}| \ + |\xi_{2n} - \xi_2| + |\xi_2 - \xi_{2n*}|\} \\
&\le \{20/c_1\}\,\{(120/c_2)\,\varepsilon_0\,|I_n| \\
&\qquad + \operatorname{dist}(\xi_1, L_*) + (120/c_2)\,\varepsilon_0\,|I_n| + \operatorname{dist}(\xi_2, L_*)\} \\
&= (4{,}800/c_1c_2)\,\varepsilon_0\,|I_n| + (20/c_1)\operatorname{dist}(\xi_1, L_*) \\
&\qquad + (20/c_1)\operatorname{dist}(\xi_2, L_*) \\
&\le C_1\,\varepsilon_0\,|I_n| + (C_1c_2/16)\operatorname{dist}(\xi_1, L_*) + (C_1c_2/16)\operatorname{dist}(\xi_2, L_*)
\end{aligned}
$$

with the very last inequality true since $C_1 = 10^5/(c_1c_2)$ by Proposition 8.7.

Execute the following operations in order on the inequality ($\star$): first, shift the first and third terms on the right-hand side to the left; second, take cube roots of both sides; third, integrate both sides with respect to μ over all points $\xi_1 \in G_1$; fourth, utilize the inequality $(c_2/2)|I_n| \le \mu(G_1)$ on left-hand side; fifth, replace G_1 by B_n and ξ_1 by ξ on the right-hand side; sixth, divide both sides by $(c_2/2)|I_n|$; seventh and finally take cubes of both sides. Your result should be the following:

$$(\star\star)\ \operatorname{dist}(w_n, L_*) - C_1\,\varepsilon_0\,|I_n| - \frac{C_1 c_2}{16}\operatorname{dist}(\xi_2, L_*) \le \frac{C_1}{2c_2^2}\left\{\frac{1}{|I_n|}\int_{B_n}\operatorname{dist}(\xi, L_*)^{1/3}\,d\mu(\xi)\right\}^3.$$

Execute the following operations in order on the inequality $(\star\star)$: first, shift the third term on the left-hand side to the right and the original single term on the right to the left; second, take cube roots of both sides; third, integrate both sides with respect to μ over all points $\xi_2 \in G_2$; fourth, utilize the inequality $(c_2/2)|I_n| \le \mu(G_2)$ on left-hand side; fifth, replace G_2 by B_n and ξ_2 by ξ on the right-hand side; sixth, divide both sides by $(c_2/2)|I_n|$; seventh and finally, take cubes of both sides. Your result should be the following:

$$(\star\star\star)\ \operatorname{dist}(w_n, L_*) - C_1\,\varepsilon_0\,|I_n| - \frac{C_1}{2c_2^2}\left\{\frac{1}{|I_n|}\int_{B_n}\operatorname{dist}(\xi, L_*)^{1/3}\,d\mu(\xi)\right\}^3 \le \frac{C_1}{2c_2^2}\left\{\frac{1}{|I_n|}\int_{B_n}\operatorname{dist}(\xi, L_*)^{1/3}\,d\mu(\xi)\right\}^3.$$

With a little rearrangement, the inequality $(\star\star\star)$ establishes the Third Claim and hence the lemma. □

Lemma 8.56 *Suppose $\zeta \in B(\tilde{z}; 20r) \cap K$ and L_* is any line. Then*

$$\sum_{n\in\mathcal{N}(p;r)} \frac{|I_n|}{r^2}\left\{\frac{1}{|I_n|}\int_{B_n}\operatorname{dist}(\xi, L_*)^{1/3}\,d\mu(\xi)\right\}^3 \le 10^{13} M^2\,\beta_1^{L_*}(\zeta; r).$$

Proof Given $n \in \mathcal{N}(p; r)$, set $\mathcal{N}(n) = \{m \in \mathcal{N}(p; r) : |B_m| \le |B_n| \text{ and } B_m \cap B_n \ne \emptyset\}$ and define

$$N_n(\xi) = \sum_{m\in\mathcal{N}(n)} \mathcal{X}_{B_m}(\xi) = \#\{m \in \mathcal{N}(p; r) : \xi \in B_m,\ |B_m| \le |B_n|, \text{ and } B_m \cap B_n \ne \emptyset\}.$$

First Claim. For any $n \in \mathcal{N}(p; r)$, $\int_{B_n} N_n(\xi)\,d\mu(\xi) \le 18{,}800\,M\,|I_n|$.

If $m \in \mathcal{N}(n)$, then by (ii) of Construction 8.31, $|I_m| \le r_m \le r_n \le 20|I_n|$ and $|z_m - z_n| \le r_m + r_n \le 2r_n \le 40|I_n|$. Hence, by (iii) of Construction 8.31,

$$\begin{aligned}|p_m - p_n| &\le |p_m - \pi(z_m)| + |\pi(z_m) - \pi(z_n)| + |\pi(z_n) - p_n|\\ &\le 20|I_m| + |z_m - z_n| + 20|I_n|\\ &\le 20(20|I_n|) + 40|I_n| + 20|I_n| \quad = \quad 460|I_n|\end{aligned}$$

and so $I_m \subseteq 940 I_n$ for such m. Since the intervals I_m are disjoint by (a) of Proposition 8.30, we deduce that

$$\sum_{m\in\mathcal{N}(n)} |I_m| \le 940|I_n|.$$

From (3) of Reduction 8.3, (ii) of Construction 8.31, and the last displayed inequality it now follows that

$$\begin{aligned}\int_{B_n} N_n(\xi)\, d\mu(\xi) &\le \sum_{m\in\mathcal{N}(n)} \mu(B_m) \le \sum_{m\in\mathcal{N}(n)} M r_m \\ &\le 20M \sum_{m\in\mathcal{N}(n)} |I_m| \le 18{,}800\, M\, |I_n|,\end{aligned}$$

thus establishing the First Claim.

Second Claim. $\sum_{n\in\mathcal{N}(p;r)} \mathcal{X}_{B_n}(\xi) N_n^{-2}(\xi) \le 6{,}440$ for μ-a.e. $\xi \in \bigcup\{B_n : n \in \mathcal{N}(p;r)\}$.

Fix a point $\xi \in \bigcup\{B_n : n \in \mathcal{N}(p;r)\}$ and a finite positive integer k for which the collection $\mathcal{N}[\xi,k] = \{n \in \mathcal{N}(p;r) : \xi \in B_n$ and $N_n(\xi) = k\}$ is nonempty. Consider any two $n, \tilde{n} \in \mathcal{N}[\xi,k]$. Without loss of generality, $|B_n| \le |B_{\tilde{n}}|$. Clearly then

$$(\star)\ \{m \in \mathcal{N}(p;r) : \xi \in B_m \text{ and } |B_m| \le |B_n|\} \subseteq \{m \in \mathcal{N}(p;r) : \xi \in B_m \text{ and } |B_m| \le |B_{\tilde{n}}|\}.$$

The left-hand set in $(\star)$ can also be described as $\{m \in \mathcal{N}(p;r) : \xi \in B_m,\ |B_m| \le |B_n|$, and $B_m \cap B_n \ne \emptyset\}$ (one automatically has $B_m \cap B_n \ne \emptyset$ for any m such that $\xi \in B_m$ since we also have $\xi \in B_n$). Thus the cardinality of the left-hand set in $(\star)$ is just $N_n(\xi) = k$. Similarly, the cardinality of the right-hand set in $(\star)$ is just $N_{\tilde{n}}(\xi) = k$. Now k is *finite*! This forces the two sets to actually be *identical*. Since $\tilde{n}$ is an element of the right-hand set, it is thus also in the left-hand set. In particular, $|B_{\tilde{n}}| \le |B_n|$. We conclude that $|B_n| = |B_{\tilde{n}}|$, i.e., $r_n = r_{\tilde{n}}$, for any two $n, \tilde{n} \in \mathcal{N}[\xi,k]$.

Now fix $\tilde{n} \in \mathcal{N}[\xi,k]$ with $|I_{\tilde{n}}|$ maximal and consider any $n \in \mathcal{N}[\xi,k]$. Since B_n and $B_{\tilde{n}}$ intersect, both containing ξ, we have $|z_n - z_{\tilde{n}}| \le r_n + r_{\tilde{n}} = 2r_{\tilde{n}} \le 40|I_{\tilde{n}}|$ by (ii) of Construction 8.31. Hence, by (iii) of Construction 8.31,

$$\begin{aligned}|p_n - p_{\tilde{n}}| &\le |p_n - \pi(z_n)| + |\pi(z_n) - \pi(z_{\tilde{n}})| + |\pi(z_{\tilde{n}}) - p_{\tilde{n}}| \\ &\le 20|I_n| + |z_n - z_{\tilde{n}}| + 20|I_{\tilde{n}}| \\ &\le 20|I_{\tilde{n}}| + 40|I_{\tilde{n}}| + 20|I_{\tilde{n}}| \quad = \quad 80|I_{\tilde{n}}|\end{aligned}$$

and so $I_n \subseteq 161 I_{\tilde{n}}$ for such n. Since the intervals I_n are disjoint by (a) of Proposition 8.30, we deduce that

$$\sum_{n \in \mathcal{N}[\xi,k]} |I_n| \leq 161|I_{\tilde{n}}|.$$

However by (ii) of Construction 8.31, one also has $|I_n| \geq r_n/20 = r_{\tilde{n}}/20 \geq |I_{\tilde{n}}|/20$ for each of these n. Thus

$$\sum_{n \in \mathcal{N}[\xi,k]} |I_n| \geq \{\#\mathcal{N}[\xi,k]\} \cdot |I_{\tilde{n}}|/20.$$

From the last two displayed inequalities we conclude that $\#\mathcal{N}[\xi,k] \leq 3{,}220$.

Note that $N_n \geq 1$ on B_n for each $n \in \mathcal{N}(p;r)$ (since $n \in \mathcal{N}(n)$ always) and so there is no division by zero in each term of the sum that we wish to estimate in this Claim. Also, $N_n < \infty$ μ-a.e. on B_n for each $n \in \mathcal{N}(p;r)$ by the First Claim. Thus for the purpose of establishing this Claim we need only consider a point $\xi \in \bigcup\{B_n : n \in \mathcal{N}(p;r)\}$ for which $N_n(\xi) < \infty$ for all $n \in \mathcal{N}(p;r)$. But for such a point,

$$\sum_{n \in \mathcal{N}(p;r)} \mathcal{X}_{B_n}(\xi) N_n^{-2}(\xi) = \sum_{k \geq 1} \frac{1}{k^2} \cdot \#\mathcal{N}[\xi,k] \leq 3{,}220 \sum_{k \geq 1} \frac{1}{k^2} \leq 6{,}440,$$

thus establishing the Second Claim.

Using Hölder's Inequality [RUD, 3.5] with $p = 3/2$ and the First Claim, we have

$$\begin{aligned}
&\int_{B_n} \operatorname{dist}(\xi, L_*)^{1/3}\, d\mu(\xi) \\
&\quad = \int_{B_n} N_n(\xi)^{2/3} \cdot \operatorname{dist}(\xi, L_*)^{1/3}\, N_n(\xi)^{-2/3}\, d\mu(\xi) \\
&\quad \leq \left\{ \int_{B_n} N_n(\xi)\, d\mu(\xi) \right\}^{2/3} \left\{ \int_{B_n} \operatorname{dist}(\xi, L_*)\, N_n(\xi)^{-2}\, d\mu(\xi) \right\}^{1/3} \\
&\quad \leq \{18{,}800\, M\, |I_n|\}^{2/3} \left\{ \int_{B_n} \operatorname{dist}(\xi, L_*)\, N_n(\xi)^{-2}\, d\mu(\xi) \right\}^{1/3}.
\end{aligned}$$

From this and the Second Claim it now follows that

$$\begin{aligned}
&\sum_{n \in \mathcal{N}(p;r)} \frac{|I_n|}{r^2} \left\{ \frac{1}{|I_n|} \int_{B_n} \operatorname{dist}(\xi, L_*)^{1/3}\, d\mu(\xi) \right\}^3 \\
&\quad \leq \{18{,}800\, M\}^2 \sum_{n \in \mathcal{N}(p;r)} \frac{1}{r} \int_{B_n} \frac{\operatorname{dist}(\xi, L_*)}{r}\, N_n(\xi)^{-2}\, d\mu(\xi) \\
&\quad = \{18{,}800\, M\}^2 \cdot \frac{1}{r} \int \frac{\operatorname{dist}(\xi, L_*)}{r} \sum_{n \in \mathcal{N}(p;r)} \mathcal{X}_{B_n}(\xi)\, N_n^{-2}(\xi)\, d\mu(\xi)
\end{aligned}$$

$$\leq (6{,}440 \times 18{,}800^2)\, M^2 \cdot \frac{1}{r} \int_{\bigcup\{B_n : n \in \mathcal{N}(p;r)\}} \frac{\operatorname{dist}(\xi, L_*)}{r}\, d\mu(\xi)$$
$$\leq 10^{13} M^2 \cdot \frac{1}{r} \int_{\bigcup\{B_n : n \in \mathcal{N}(p;r)\}} \frac{\operatorname{dist}(\xi, L_*)}{r}\, d\mu(\xi).$$

To finish we need only show that $\bigcup\{B_n : n \in \mathcal{N}(p;r)\} \cap K \subseteq B(\zeta; k_0 r)$ and invoke the definition of $\beta_1^{L_*}(\zeta; r)$.

So suppose $\xi \in B_n \cap K$ where $n \in \mathcal{N}(p;r)$. Since $n \in \mathcal{N}(p;r)$, there exists a point $q \in I_n \cap I(p;r)$. By (b) of Proposition 8.28, $D(q) \leq |q - \pi(\tilde{z})| + \tilde{r} \leq |q - p| + |p - \pi(\tilde{z})| + \tilde{r} \leq r + 20r = 21r$. Then, by Construction 8.29, $|I_n| \leq (1/8)\inf\{D(q) : q \in I_n\} \leq 21r/8 < 3r$. Of course by (ii) of Construction 8.31 we then also have $r_n \leq 20|I_n| < 60r$. Hence, invoking (iii) of Construction 8.31,

$$\begin{aligned}
|\pi(\xi) - \pi(\zeta)| &\leq |\pi(\xi) - \pi(z_n)| + |\pi(z_n) - p_n| + |p_n - q| \\
&\quad + |q - p| + |p - \pi(\tilde{z})| + |\pi(\tilde{z}) - \pi(\zeta)| \\
&\leq |\xi - z_n| + 20|I_n| + |I_n| + r + 20r + |\tilde{z} - \zeta| \\
&\leq r_n + 21|I_n| + 21r + 20r \quad < \quad 164r.
\end{aligned}$$

Note too that by the Definition of d (8.20), $d(\xi) \leq |\xi - z_n| + r_n \leq 2r_n < 120r$ and $d(\zeta) \leq |\zeta - \tilde{z}| + \tilde{r} \leq 20r + 20r = 40r$. Corollary 8.26 thus applies and tells us that $|\xi - \zeta| \leq 9C_4 \cdot 164r$. Our choice of k_0 now implies that $\xi \in B(\zeta; k_0 r)$ and so we are done. □

Lemma 8.57 *For any $\zeta \in K$ and any $r > 0$, there exists a line L_* such that $\beta_1^{L_*}(\zeta; r) = \beta_1(\zeta; r)$.*

Proof Choose lines L_n such that $\beta_1^{L_n}(\zeta; r) \downarrow \beta_1(\zeta; r)$. If a line L_n does not intersect $B(\zeta; k_0 r)$, then a smaller beta number results if one replaces L_n by the line between L_n and ζ that is parallel to L_n and tangent to $B(\zeta; k_0 r)$. We may thus in addition assume that each L_n intersects $B(\zeta; k_0 r)$. Associate to each line L_n a pair (z_n, ϑ_n) where z_n is any point in the intersection of L_n with $B(\zeta; k_0 r)$ and $\vartheta_n \in [0, \pi)$ is the counterclockwise angle L_n makes with the real axis, i.e., the angle through which the real axis must be rotated counterclockwise to end up parallel with L_n. By compactness, we may drop to a subsequence and assume that $z_n \to z$ and $\vartheta_n \to \vartheta$ for some $z \in B(\zeta; k_0 r)$ and $\vartheta \in [0, \pi]$. Let L_* be the line through z whose counterclockwise angle with the real axis is ϑ. Thinking in terms of vectors in $\mathbb{R}^2$ now and letting $\times$ denote the cross product, we see that for any point η, $\operatorname{dist}(\eta, L_n) = \|(\eta - z_n) \times (\cos\vartheta_n, \sin\vartheta_n)\| \to \|(\eta - z) \times (\cos\vartheta, \sin\vartheta)\| = \operatorname{dist}(\eta, L_*)$. Applying Fatou's Lemma [RUD, 1.28], we now have $\beta_1^{L_*}(\zeta; r) \leq \liminf_{n\to\infty} \beta_1^{L_n}(\zeta; r) = \beta_1(\zeta; r)$. Since we clearly have $\beta_1(\zeta; r) \leq \beta_1^{L_*}(\zeta; r)$, we are done. □

Proof of Proposition 8.53 Since $B(\tilde{z}; \tilde{r}) \in S_0$ and $\tilde{r} < 20r$, we have $B(\tilde{z}; 20r) \in S_0$ by (c) of Corollary 8.18 if, on the one hand, $20r < 3$. Thus, by the Definitions of S_{total} and S_0 (8.13 and 8.16),

$$(\star)\ \mu(B(\tilde{z}; 20r)) \geq \delta_0 \cdot 20r = 20\delta_0\, r.$$

If, on the other hand, $20r \geq 3$, then $\mu(B(\tilde{z}; 20r)) = \mu(K) > 1$ by (1) and (2) of Reduction 8.3 and $20\delta_0 r \leq 20\delta_0 \leq 1$ by our selection of δ_0. Thus ($\star$) holds for any $r \in (0, 1)$.

Consider any $\zeta \in B(\tilde{z}; 20r) \cap K$ and use Lemma 8.57 to obtain a line L_* such that $\beta_1^{L_*}(\zeta; r) = \beta_1(\zeta; r)$. Invoking (a) of Proposition 8.30 and Lemmas 8.54, 8.55, and 8.56, we have

$$
\begin{aligned}
(\star\star)\ \tilde{\gamma}(p; r) &\leq \frac{1}{r} \int_{I(p;r)} \frac{\operatorname{dist}(t + i\,\ell(t), L_*)}{r}\, dt \\
&= \frac{1}{r} \int_{I(p;r)\cap\pi(K_0)} \frac{\operatorname{dist}(t + i\,\ell(t), L_*)}{r}\, dt \\
&\quad + \sum_{n\in\mathcal{N}(p;r)} \frac{1}{r} \int_{I(p;r)\cap I_n} \frac{\operatorname{dist}(t + i\,\ell(t), L_*)}{r}\, dt \\
&\leq \frac{4}{\delta_0}\beta_1^{L_*}(\zeta; r) + 1{,}201\, C_1\, \varepsilon_0 \sum_{n\in\mathcal{N}(p;r)} \left(\frac{|I_n|}{r}\right)^2 + \frac{C_1}{c_2^2} \cdot 10^{13} M^2 \beta_1^{L_*}(\zeta; r) \\
&= \left[\frac{4}{\delta_0} + \frac{10^{13} M^2 C_1}{c_2^2}\right] \beta_1(\zeta; r) + 1{,}201\, C_1\, \varepsilon_0 \sum_{n\in\mathcal{N}(p;r)} \left(\frac{|I_n|}{r}\right)^2.
\end{aligned}
$$

Execute the following operations in order on the inequality ($\star\star$): first, square both sides; second, utilize the trivial inequality $(a + b)^2 \leq 2a^2 + 2b^2$ on the right-hand side; third, integrate both sides with respect to μ over all points $\zeta \in B(\tilde{z}; 20r) \cap K$; fourth, divide both sides by $\mu(B(\tilde{z}; 20r))$; fifth and finally, use ($\star$) to eliminate the single remaining occurrence of $\mu(B(\tilde{z}; 20r))$ (it will be in a denominator on the right-hand side). Your result should be that $\tilde{\gamma}^2(p; r)$ is bounded above by

$$
\begin{aligned}
&\frac{1}{10\delta_0}\left[\frac{4}{\delta_0} + \frac{10^{13} M^2 C_1}{c_2^2}\right]^2 \int_{B(\tilde{z};20r)} \frac{\beta_1^2(\zeta; r)}{r}\, d\mu(\zeta) \\
&\qquad + 2(1{,}201\, C_1)^2 \left\{\varepsilon_0 \sum_{n\in\mathcal{N}(p;r)} \left(\frac{|I_n|}{r}\right)^2\right\}^2.
\end{aligned}
$$

Given our specification of $C_5(M, \delta_0)$, to finish we need only show that $\zeta \in \pi^{-1}(I(p; 40r))$ and $\tilde{\delta}(\zeta; r) \geq 20\, c_2$ whenever $\zeta \in B(\tilde{z}; 20r)$. The first of these is easy: $|\pi(\zeta) - p| \leq |\pi(\zeta) - \pi(\tilde{z})| + |\pi(\tilde{z}) - p| \leq |\zeta - \tilde{z}| + |\pi(\tilde{z}) - p| \leq 20r + 20r = 40r$. To prove the second note that by ($\star$) we may apply Lemma 8.5 to $B = B(\tilde{z}; 20r)$. The result is a ball $B(w; s)$ with $w \in B(\tilde{z}; 20r)$, $s = c_1\, 20r$, and $\mu(B(w; s)) > c_2\, 20r$. (The lemma actually gives us more than this, but this is all we need.) Note that $s \leq r$ since $c_1 \leq 10^{-4}$. Thus $\delta(w; r) \geq \mu(B(w; s))/r \geq 20\, c_2$.

Finally, $|w-\zeta| \le |w-\tilde{z}| + |\tilde{z}-\zeta| \le 20r + 20r = 40r$, so $w \in B(\zeta; k_0^* r)$ by our choice of k_0^*. It follows that $\tilde{\delta}(\zeta; r) \ge \delta(w; r) \ge 20\, c_2$ and so we are done. □

8.21 A Global Estimate on the Other Gamma Function

Lemma 8.58 $\displaystyle\sum_n \int_0^{|I_n|} \int_{I_n \cap U_4} \gamma^2(p; r)\, dp\, \frac{dr}{r} \le 10^{20}\, C_1(M, \delta_0)^2\, \varepsilon_0^2.$

Proof For $p \in I_n \cap U_4$ and $r \in (0, |I_n|]$, consider the linear function $l(t) = \ell(p) + \ell'(p)(t-p)$. Then any $t \in I(p; r)$ is also in $3I_n \cap U_5$, so by Taylor's Theorem and Lemma 8.40,

$$|\ell(t) - l(t)| \le \frac{1}{2} \cdot \sup\{|\ell''(q)| : q \in 3I_n \cap U_5\} \cdot (t-p)^2 \le \frac{10^{10}}{2}\, C_1\, \frac{\varepsilon_0}{|I_n|}\, (t-p)^2.$$

Thus

$$\begin{aligned} \gamma(p; r) &\le \frac{1}{r} \int_{I(p;r)} \frac{|\ell(t) - l(t)|}{r}\, dt \le \frac{10^{10}}{2}\, C_1\, \frac{\varepsilon_0}{|I_n|} \int_{p-r}^{p+r} \left(\frac{t-p}{r}\right)^2 dt \\ &= \frac{10^{10}}{3}\, C_1\, \frac{\varepsilon_0}{|I_n|}\, r \end{aligned}$$

and so

$$\begin{aligned} \sum_n \int_0^{|I_n|} \int_{I_n \cap U_4} \gamma^2(p; r)\, dp\, \frac{dr}{r} &\le \frac{10^{20}}{9}\, C_1^2\, \varepsilon_0^2 \sum_n \frac{1}{|I_n|^2} \int_0^{|I_n|} \int_{I_n} r\, dp\, dr \\ &= \frac{10^{20}}{18}\, C_1^2\, \varepsilon_0^2 \sum_n |I_n|. \end{aligned}$$

Since $\sum_n |I_n| \le |U_5| = 10$, we are done. □

The whole point of this section and the last two is the following.

Proposition 8.59 $\displaystyle\int_0^1 \int_{U_4} \gamma^2(p; r)\, dp\, \frac{dr}{r} \le C_6(M, \delta_0)\, \varepsilon_0^2.$

Recall that the quantity $C_6(M, \delta)$ was defined back in Section 8.6.

Proof Because of the last lemma we only need to appropriately estimate

$$\sum_n \int_{|I_n|}^1 \int_{I_n \cap U_4} \gamma^2(p; r)\, dp\, \frac{dr}{r} + \int_0^1 \int_{\pi(K_0)} \gamma^2(p; r)\, dp\, \frac{dr}{r}.$$

On the one hand, for $(p, r) \in \{I_n \cap U_4\} \times [|I_n|, 1)$, we have $D(p) < 20|I_n| \le 20r$ by (c) of Lemma 8.30. On the other hand, for $(p, r) \in \pi(K_0) \times (0, 1)$, we have

$D(p) = 0 < 20r$ by the Definitions of K_0 and D (8.23 and 8.27). Thus each pair (p, r) being integrated over in any of the integrals in the above displayed expression satisfies $D(p) < 20r$ and so we only need to appropriately estimate

$$\int_0^1 \int_{\{p \in U_4 : D(p) < 20r\}} \gamma^2(p; r)\, dp\, \frac{dr}{r}.$$

By Propositions 8.52 and 8.53, this quantity is less than or equal to $2\, C_5[(\mathrm{I}) + (\mathrm{II})]$ where

$$(\mathrm{I}) = \int_0^1 \int_{\{p \in U_4 : D(p) < 20r\}} \int_{\pi^{-1}(I(p;40r))} \frac{\beta_1^2(\zeta; r)}{r}\, \mathcal{X}_{\{\tilde{\delta}(\zeta;r) \geq 20\, c_2\}}\, d\mu(\zeta)\, dp\, \frac{dr}{r}$$

and

$$(\mathrm{II}) = \varepsilon_0^2 \int_0^1 \int_{\{p \in U_4 : D(p) < 20r\}} \left\{ \sum_{n \in \mathcal{N}(p;r)} \left(\frac{|I_n|}{r} \right)^2 \right\}^2 dp\, \frac{dr}{r}.$$

Estimating (I) is short and sweet. By Fubini's Theorem [RUD, 8.8], Proposition 8.10, (4) of Reduction 8.3, and Condition (4) for the selection of κ_0,

$$\begin{aligned}
(\mathrm{I}) &\leq \int_0^1 \int_{\mathbb{C}} \frac{\beta_1^2(\zeta; r)}{r}\, \mathcal{X}_{\{\tilde{\delta}(\zeta;r) \geq 20\, c_2\}} \left[\int_{\pi(\zeta)-40r}^{\pi(\zeta)+40r} dp \right] d\mu(\zeta)\, \frac{dr}{r} \\
&= 80 \int_0^1 \int_{\mathbb{C}} \beta_1^2(\zeta; r)\, \mathcal{X}_{\{\tilde{\delta}(\zeta;r) \geq 20\, c_2\}}\, d\mu(\zeta)\, \frac{dr}{r} \\
&\leq 80\, C_3(M, 20\, c_2(M, \delta_0), k_0, k_0^*)\, c^2(\mu) \\
&< 80\, C_3(M, 20\, c_2(M, \delta_0), k_0, k_0^*)\, \kappa_0 \;\leq\; \varepsilon_0^2.
\end{aligned}$$

Estimating (II) will take a bit more ink. First, note that for $p \in U_4$, $r \in (D(p)/20, 1)$, and $n \in \mathcal{N}(p; r)$, we have $I_n \cap I(p; r) \neq \emptyset$, so there exists a point $q \in I_n$ such that $|q - p| \leq r$. Thus, by Construction 8.29 and (c) of Proposition 8.28, we have

$$(\star)\ |I_n| \leq \frac{1}{8} D(q) \leq \frac{1}{8}\{D(p) + |q - p|\} \leq \frac{1}{8}\{20r + r\} < 3r.$$

Second, note that as a consequence of I_n intersecting $I(p; r)$ and $(\star)$, $I_n \subseteq I(p; 4r)$ for each of our n. Since the intervals I_n are disjoint by (a) of Proposition 8.30, we deduce that $\sum_{n \in \mathcal{N}(p;r)} |I_n| \leq 8r$. From this and $(\star)$, we obtain

$$(\star\star)\quad \sum_{n\in\mathcal{N}(p;r)} \left(\frac{|I_n|}{r}\right)^2 \le \frac{3r}{r^2} \sum_{n\in\mathcal{N}(p;r)} |I_n| \le 24.$$

Third and finally, $I_n \subseteq I(p; 4r)$ implies

$$(\star\star\star)\ p \in I(p_n; 4r).$$

Applying $(\star\star)$ and then $(\star)$ and $(\star\star\star)$ to our expression for (II), we see that

$$\begin{aligned}
\text{(II)} &\le 24\,\varepsilon_0^2 \int_0^1 \int_{L_0} \sum_{n\in\mathcal{N}(p;r)} \left(\frac{|I_n|}{r}\right)^2 \mathcal{X}_{\{r>|I_n|/3\}}\, \mathcal{X}_{\{p\in I(p_n;4r)\}}\, dp\, \frac{dr}{r} \\
&\le 24\,\varepsilon_0^2 \sum_n |I_n|^2 \int_{|I_n|/3}^1 \int_{p_n-4r}^{p_n+4r} dp\, \frac{dr}{r^3} \\
&\le 24\,\varepsilon_0^2 \sum_n |I_n|^2 \cdot \frac{24}{|I_n|} \;=\; 576\,\varepsilon_0^2 \sum_n |I_n| \;\le\; 5{,}760\,\varepsilon_0^2
\end{aligned}$$

with the very last inequality holding since $\sum_n |I_n| \le |U_5| = 10$. Collecting all of our estimates together and consulting our specification of $C_6(M, \delta_0)$, we are done. □

8.22 Interlude: Calderón's Formula

This section will make use of the Fourier transform and so we revert to the notational conventions established in section 4.3. In addition, given an open subset U of $\mathbb{R}$, we use $C_c^\infty(U)$, $L_c^\infty(U)$, and $L_c^1(U)$ to denote the sets of compactly supported functions from $C^\infty(U)$, $L^\infty(U)$, and $L^1(U)$, respectively.

A last tool needed to handle K_3 is a formula due to Alberto Calderón. The author has found the original sources difficult to penetrate. What unlocked the result for him were pages 382–383 of [FRAZ]. Calderón's formula holds for a number of function spaces. The version we need, Proposition 8.65 below, holds for the space of compactly supported Lipschitz functions on the real line. The formula involves an auxiliary smooth function whose Fourier transform is well behaved. This function is constructed in the following.

Lemma 8.60 *There exists a real-valued, even, smooth function φ on $\mathbb{R}$ with mean value zero and support contained in $(-1, 1)$ for which $\widehat{\varphi}$ is a real-valued, even function satisfying*

$$\int_0^\infty \{\widehat{\varphi}(\xi)\}^2\, \frac{d\xi}{\xi} = 1.$$

Moreover, for any such function φ one has

$$\int_0^\infty \xi\,\{\widehat{\varphi}(\xi)\}^2\,d\xi \le \frac{\|\varphi\|_\infty^2 + \|\varphi''\|_\infty^2}{2}.$$

Proof Let $\widetilde{\varphi}$ be gotten via Lemma 8.34 with $a = -1/10$, $b = 1/10$, and $l = 1/10$. Set

$$\varphi(x) = \{\widetilde{\varphi}(x - 1/4) + \widetilde{\varphi}(-x - 1/4)\} - \{\widetilde{\varphi}(x - 3/4) + \widetilde{\varphi}(-x - 3/4)\}.$$

Clearly φ is a nonzero, real-valued, even function in $C^\infty(\mathbb{R})$ with mean value zero and support contained in $(-1, 1)$. So to prove all but the last assertion of the lemma it suffices to show that $\widehat{\varphi}$ is a real-valued, even function satisfying

$$(\star)\; 0 < \int_0^\infty \{\widehat{\varphi}(\xi)\}^2\,\frac{d\xi}{\xi} < \infty.$$

Indeed, one then simply normalizes φ by dividing it by the square root of the integral in $(\star)$.

Using De Moivre's formula, the evenness of φ, and the containment of the support of φ in $(-1, 1)$, one sees that for any $\xi \in \mathbb{R}$,

$$(\star\star)\; \widehat{\varphi}(\xi) = \int_{-\infty}^{\infty} e^{-i\xi t}\varphi(t)\,\frac{dt}{\sqrt{2\pi}} = \int_{-\infty}^{\infty} \{\cos(\xi t) - i\sin(\xi t)\}\,\varphi(t)\,\frac{dt}{\sqrt{2\pi}}$$
$$= \int_{-1}^{1} \cos(\xi t)\,\varphi(t)\,\frac{dt}{\sqrt{2\pi}}.$$

From this it easily follows that $\widehat{\varphi}$ is real-valued and even.

If the integral in $(\star)$ were 0, then $\widehat{\varphi}(\xi)$ would have to vanish for almost every $\xi > 0$. But $\widehat{\varphi}$ is even, so we would then have $\widehat{\varphi} = 0$ almost everywhere. Plancherel's Theorem (4.10) would then force φ to be the zero function, which it is not. Thus the integral in $(\star)$ is positive.

For any $t \in \mathbb{R}$, define

$$\Phi(t) = \int_{-1}^{t} \varphi(x)\,dx.$$

Using integration by parts once on $(\star\star)$, one sees that for any $\xi > 0$,

$$\widehat{\varphi}(\xi) = \frac{1}{\sqrt{2\pi}}\left[\cos(\xi t)\,\Phi(t) + \xi\int \sin(\xi t)\,\Phi(t)\,dt\right]_{-1}^{1} = \xi\int_{-1}^{1} \sin(\xi t)\,\Phi(t)\,\frac{dt}{\sqrt{2\pi}}$$

with $\Phi(1) = 0$ since φ has mean value zero. But then $|\widehat{\varphi}(\xi)| \le \xi\,2\,\|\Phi\|_\infty/\sqrt{2\pi} \le 2\,\xi$ and so

$$\int_0^1 \{\widehat{\varphi}(\xi)\}^2 \frac{d\xi}{\xi} \leq 2.$$

Finally, by another appeal to Plancherel's Theorem (4.10),

$$\int_1^\infty \{\widehat{\varphi}(\xi)\}^2 \frac{d\xi}{\xi} \leq \int_1^\infty \{\widehat{\varphi}(\xi)\}^2 \, d\xi \leq \|\widehat{\varphi}\|_2^2 = \|\varphi\|_2^2 \leq 2$$

and so the integral in ($\star$) is at most 4 and thus finite.

We now turn to the last assertion of the lemma. Using integration by parts twice on ($\star\star$), one sees that for any $\xi > 0$,

$$\begin{aligned}\widehat{\varphi}(\xi) &= \frac{1}{\sqrt{2\pi}} \left[\frac{\sin(\xi t)\, \varphi(t)}{\xi} + \frac{\cos(\xi t)\, \varphi'(t)}{\xi^2} - \int \frac{\cos(\xi t)\, \varphi''(t)}{\xi^2}\, dt \right]_{-1}^{1} \\ &= -\int_{-1}^{1} \frac{\cos(\xi t)\, \varphi''(t)}{\xi^2} \frac{dt}{\sqrt{2\pi}}.\end{aligned}$$

But then $|\widehat{\varphi}(\xi)| \leq 2\, \|\varphi''\|_\infty / (\xi^2 \sqrt{2\pi}) \leq \|\varphi''\|_\infty / \xi^2$ and so

$$\int_1^\infty \xi\, \{\widehat{\varphi}(\xi)\}^2 \, d\xi \leq \frac{\|\varphi''\|_\infty^2}{2}.$$

Lastly, since $|\widehat{\varphi}(\xi)| \leq 2\, \|\varphi\|_\infty / \sqrt{2\pi} \leq \|\varphi\|_\infty$, we have

$$\int_0^1 \xi\, \{\widehat{\varphi}(\xi)\}^2 \, d\xi \leq \frac{\|\varphi\|_\infty^2}{2}$$

and thus are done. □

Note how in this proof we established the finiteness of the integral in ($\star$) above for the pre-normalized φ by getting an effective upper bound on it, namely 4. However, we did not establish the positiveness of the integral in ($\star$) for the pre-normalized φ by getting an effective lower bound on it, but rather via a Reductio ad Absurdum argument. This leads to the lack of an effective upper bound on various quantities involving the normalized φ such as $\|\varphi\|_\infty$, $\|\varphi''\|_\infty$, $\|\varphi\|_1$, and $\|\varphi'\|_1$. It is because of this that certain conditions on our parameters have been stated, not with reference to absolute constants, but with reference to these quantities. This is innocuous but the author wonders if this state of affairs can be avoided without an unreasonable expenditure of labor.

At this point the author wishes to urge the reader who has not already done so to convince him- or herself that Lemma 8.60 and Lemma 8.34, which was used to prove Lemma 8.60, could both have been stated and proved right in Section 8.6 where the first was explicitly and the second was implicitly called upon in fixing the value of the mysterious eighth parameter θ!

Stipulation 8.61 *For the rest of this chapter, φ will denote a fixed function as in the last lemma. Moreover, given $r > 0$, we will let $\varphi_r : \mathbb{R} \mapsto \mathbb{R}$ denote the function with action $x \mapsto (1/r)\varphi(x/r)$. Note that $\varphi_r \in C_c^\infty(\mathbb{R}) \subseteq L_c^\infty(\mathbb{R}) \cap L_c^1(\mathbb{R})$ with $\|\varphi_r\|_\infty = \|\varphi\|_\infty/r$ and $\|\varphi_r\|_1 = \|\varphi\|_1$. It is at this point convenient to also record the fact that $\widehat{\varphi_r}(\xi) = \widehat{\varphi}(\xi r)$.*

Recall that the usual convolution of two functions f and g at a point x is defined by

$$(f * g)(x) = \int_{-\infty}^{\infty} f(x-y)g(y)\,\frac{dy}{\sqrt{2\pi}}.$$

When does this definition make sense? One situation, the usual one but not of interest to us here, is when f and g are elements of $L^1(\mathbb{R})$. Then the integral defining $f * g$ converges absolutely for almost every x in $\mathbb{R}$ and the resulting function $f * g$ is an element of $L^1(\mathbb{R})$ with $\|f * g\|_1 \leq \|f\|_1\|g\|_1/\sqrt{2\pi}$ [RUD, 8.14]. Another situation, the one of real interest to us here, is when f is an element of $L^\infty(\mathbb{R})$ and g is an element of $L^1(\mathbb{R})$. Then the integral defining $f * g$ converges absolutely for every x in $\mathbb{R}$ and the resulting function $f * g$ is an element of $L^\infty(\mathbb{R})$ with $\|f * g\|_\infty \leq \|f\|_\infty\|g\|_1/\sqrt{2\pi}$. Once sure that a convolution is well defined, either almost everywhere or everywhere, many properties of it such as commutativity and associativity follow immediately from its definition and will be used without comment. One worth mentioning now, to be invoked in the next paragraph, is that if f and g vanish almost everywhere off the sets X and Y respectively, then $f * g$ vanishes everywhere off the set $X + Y = \{x + y : x \in X \text{ and } y \in Y\}$.

Let us now restrict f to be an element of $L_c^\infty(\mathbb{R})$. For any $N > 0$, let the function $\Phi_N f : \mathbb{R} \mapsto \mathbb{R}$ be defined by

$$(\Phi_N f)(x) = \int_{1/N}^{N} (f * \varphi_r * \varphi_r)(x)\,\frac{dr}{r}.$$

Note that $(\Phi_N f)(x)$ has actually just been defined as a triple integral since each of the two convolutions above hides an integral. This triple integral converges triply absolutely since from the previous paragraph it easily follows that

$$\int_{1/N}^{N} (|f| * |\varphi_r| * |\varphi_r|)(x)\,\frac{dr}{r} \leq \frac{\ln N}{\pi}\|f\|_\infty\|\varphi\|_1^2 < \infty$$

for every point $x \in \mathbb{R}$. It also easily follows from the previous paragraph that if $[a, b]$ is an interval off which f vanishes almost everywhere, then

$$\int_{1/N}^{N} (|f| * |\varphi_r| * |\varphi_r|)(x)\,\frac{dr}{r} = 0$$

for every point $x \in \mathbb{R} \setminus [a - 2N, b + 2N]$. Thus $\int_{1/N}^{N} (|f| * |\varphi_r| * |\varphi_r|)(x)\, dr/r$ converges for every $x \in \mathbb{R}$ and the resulting function of x is an element of $L_c^\infty(\mathbb{R})$. Of course it then follows that $\Phi_N f$ is a well-defined element of $L_c^\infty(\mathbb{R})$. These comments justify the applications of Fubini's Theorem [RUD, 8.8] which will be made in the remainder of this section.

Lemma 8.62 *For any* $f \in L_c^\infty(\mathbb{R})$,

$$\lim_{N \to \infty} \|\Phi_N f - f\|_2 = 0.$$

Proof Since f and $\Phi_N f \in L_c^\infty(\mathbb{R}) \subseteq L^1(\mathbb{R}) \cap L^2(\mathbb{R})$, it suffices, by Plancherel's Theorem (4.10), to show that $\|(\Phi_N f)\hat{} - \hat{f}\|_2^2 \to 0$ as $N \to \infty$. To this end we compute the Fourier transform of $\Phi_N f$:

$$\begin{aligned}(\Phi_N f)\hat{}(\xi) &= \int_{1/N}^{N} (f * \varphi_r * \varphi_r)\hat{}(\xi)\, \frac{dr}{r} = \int_{1/N}^{N} \hat{f}(\xi)\{\widehat{\varphi_r}(\xi)\}^2\, \frac{dr}{r} \\ &= \hat{f}(\xi) \int_{1/N}^{N} \{\widehat{\varphi}(\xi r)\}^2\, \frac{dr}{r}.\end{aligned}$$

Using the definitions of $\Phi_N f$ and the Fourier transform, one sees that the first equality is just an application of Fubini's Theorem [RUD, 8.8]. The second equality is a use of the well-known (and easily verified) fact that the Fourier transform of a convolution is the product of the transforms. Finally, the third equality involves the last fact noted in Stipulation 8.61.

Pressing forward, via the evenness of $\widehat{\varphi}$ and the change of variables $s = |\xi| r$, one sees that

$$(\Phi_N f)\hat{}(\xi) = \hat{f}(\xi) \int_{|\xi|/N}^{|\xi| N} \{\widehat{\varphi}(s)\}^2\, \frac{ds}{s},$$

for all $\xi \neq 0$ (although not needed, the formula is also true and easily verified for $\xi = 0$).

The upshot of this all is

$$(\star)\ \|(\Phi_N f)\hat{} - \hat{f}\|_2^2 = \int_{-\infty}^{\infty} |\hat{f}(\xi)|^2 \left| \int_{|\xi|/N}^{|\xi| N} \{\widehat{\varphi}(s)\}^2\, \frac{ds}{s} - 1 \right|^2 d\xi.$$

By Lemma 8.60, the integrand of the $d\xi$-integral of $(\star)$ is bounded by $|\hat{f}|^2 \in L^1(\mathbb{R})$. Moreover, this same $d\xi$-integrand converges to 0 as $N \to \infty$ for all $\xi \neq 0$ by an application of the Monotone Convergence Theorem [RUD, 1.26] to the ds-integral of $(\star)$. We may thus invoke Lebesgue's Dominated Convergence Theorem [RUD, 1.34] as $N \to \infty$ in $(\star)$ to obtain our desired conclusion: $\|(\Phi_N f)\hat{} - \hat{f}\|_2^2 \to 0$. □

Now let f be a compactly supported Lipschitz function on $\mathbb{R}$ with bound M. One can easily verify that such an f is an element of $L_c^\infty(\mathbb{R})$ and that for any $g \in L_c^1(\mathbb{R})$, $f * g$ is a compactly supported Lipschitz function on $\mathbb{R}$ with bound $M\|g\|_1/\sqrt{2\pi}$. From this it follows that $\Phi_N f$ is a compactly supported Lipschitz function on $\mathbb{R}$ with bound $\{\ln N/\pi\}M\|\varphi\|_1^2$. By Lemma 4.15, $\Phi_N f$ has a derivative almost everywhere from which $\Phi_N f$ can be recovered by integration.

Lemma 8.63 *Suppose f be is a Lipschitz function on $\mathbb{R}$ with bound M that vanishes outside the interval $[a, b]$. Then*

$$\|(\Phi_N f)'\|_2 \leq M\sqrt{b-a}.$$

The point of this lemma is that the estimate gotten on the L^2 norm of the derivative does not depend on N!

Proof Since $(\Phi_N f)' \in L_c^\infty(\mathbb{R}) \subseteq L^1(\mathbb{R}) \cap L^2(\mathbb{R})$, it suffices, by Plancherel's Theorem (4.10), to show that $\|\{(\Phi_N f)'\}\hat{}\|_2^2 \leq M^2(b-a)$. To this end we invoke Proposition 4.9 and the computation of $(\Phi_N f)\hat{}(\xi)$ from the last proof to see that

$$\begin{aligned}\{(\Phi_N f)'\}\hat{}(\xi) = i\xi\, (\Phi_N f)\hat{}(\xi) &= i\xi\, \hat{f}(\xi) \int_{|\xi|/N}^{|\xi|N} \{\widehat{\varphi}(s)\}^2 \frac{ds}{s} \\ &= \widehat{f'}(\xi) \int_{|\xi|/N}^{|\xi|N} \{\widehat{\varphi}(s)\}^2 \frac{ds}{s}.\end{aligned}$$

Hence, by Lemma 8.60, Plancherel's Theorem (4.10), and Lemma 4.15,

$$\begin{aligned}\|\{(\Phi_N f)'\}\hat{}\|_2^2 &= \int_{-\infty}^{\infty} |\widehat{f'}(\xi)|^2 \left\{\int_{|\xi|/N}^{|\xi|N} \{\widehat{\varphi}(s)\}^2 \frac{ds}{s}\right\}^2 d\xi \\ &\leq \int_{-\infty}^{\infty} |\widehat{f'}(\xi)|^2 \left\{\int_{0}^{\infty} \{\widehat{\varphi}(s)\}^2 \frac{ds}{s}\right\}^2 d\xi \\ &= \int_{-\infty}^{\infty} |\widehat{f'}(\xi)|^2 d\xi \;=\; \int_{-\infty}^{\infty} |f'(x)|^2 dx \;\leq\; M^2(b-a).\end{aligned}$$

□

We are really interested in

$$(\Phi_\infty f)(x) = \int_0^\infty (f * \varphi_r * \varphi_r)(x)\, \frac{dr}{r},$$

but the absolute convergence of this integral is not immediately apparent – both 0 and ∞ seem to present problems. For this reason we have dealt with $\Phi_N f$ with $N < \infty$ up till now. (Another technical advantage of $\Phi_N f$ with $N < \infty$ as opposed to $\Phi_\infty f$ is that the former, unlike the latter, is easily seen to have compact support and so be an element of $L_c^\infty(\mathbb{R}) \subseteq L^1(\mathbb{R}) \cap L^2(\mathbb{R})$.) The next lemma disposes of these absolute convergence problems at 0 and ∞ and is phrased in more generality than needed just now with a view to use in the following section.

Lemma 8.64 *Suppose $a > 0$ and $x \in \mathbb{R}$. Then*

(a) $\displaystyle\int_a^\infty (|f| * |\varphi_r| * |\varphi_r|)(x)\, \frac{dr}{r} \le \frac{\|f\|_1 \|\varphi\|_\infty \|\varphi\|_1}{2\pi a}$ *for any $f \in L^1(\mathbb{R})$ and*

(b) $\displaystyle\int_0^a (\{\mathcal{X}_E\, |f * \varphi_r|\} * |\varphi_r|)(x)\, \frac{dr}{r} \le \frac{M \|\varphi\|_1^2\, a}{2\pi}$ *for any Borel subset E of $\mathbb{R}$ and any Lipschitz function f on $\mathbb{R}$ with bound M.*

Thus, for any Borel subset E of $\mathbb{R}$, any compactly supported Lipschitz function f on $\mathbb{R}$, and any point x of $\mathbb{R}$, we have

$$\int_0^\infty (\{\mathcal{X}_E\, |f * \varphi_r|\} * |\varphi_r|)(x)\, \frac{dr}{r} < \infty.$$

Proof (a) Clearly

$$(|f| * |\varphi_r| * |\varphi_r|)(x) \le \frac{\||f| * |\varphi_r|\|_\infty \|\varphi_r\|_1}{\sqrt{2\pi}} \le \frac{\|f\|_1 \|\varphi_r\|_\infty \|\varphi_r\|_1}{2\pi}$$
$$= \frac{\|f\|_1 \|\varphi\|_\infty \|\varphi\|_1}{2\pi} \cdot \frac{1}{r}$$

for all $r > 0$. Thus

$$\int_a^\infty (|f| * |\varphi_r| * |\varphi_r|)(x)\, \frac{dr}{r} \le \frac{\|f\|_1 \|\varphi\|_\infty \|\varphi\|_1}{2\pi} \int_a^\infty \frac{dr}{r^2} = \frac{\|f\|_1 \|\varphi\|_\infty \|\varphi\|_1}{2\pi a}.$$

(b) Since φ_r has mean value zero, f is Lipschitz on $\mathbb{R}$ with bound M, and $\operatorname{spt}(\varphi_r) \subseteq [-r, r]$, we have

$$\begin{aligned}
|(f * \varphi_r)(y)| &= \left| \int_{-\infty}^\infty f(y - z)\varphi_r(z)\, \frac{dz}{\sqrt{2\pi}} - f(y) \int_{-\infty}^\infty \varphi_r(z)\, \frac{dz}{\sqrt{2\pi}} \right| \\
&\le \int_{-\infty}^\infty |f(y - z) - f(y)||\varphi_r(z)|\, \frac{dz}{\sqrt{2\pi}} \\
&\le \int_{-\infty}^\infty M|z||\varphi_r(z)|\, \frac{dz}{\sqrt{2\pi}} \\
&\le \frac{Mr\|\varphi_r\|_1}{\sqrt{2\pi}} = \frac{Mr\|\varphi\|_1}{\sqrt{2\pi}}.
\end{aligned}$$

for all $r > 0$ and $y \in \mathbb{R}$. Thus

$$\int_0^a (\{\mathcal{X}_E \, |f * \varphi_r|\} * |\varphi_r|)(x) \, \frac{dr}{r} \leq \int_0^a \frac{Mr\|\varphi\|_1}{\sqrt{2\pi}} \, \frac{\|\varphi_r\|_1}{\sqrt{2\pi}} \, \frac{dr}{r} = \frac{M\|\varphi\|_1^2 \, a}{2\pi}.$$

□

With all the necessary pieces assembled, we may proceed to the goal of this section.

Proposition 8.65 (Compactly Supported Lipschitz Version of Calderón's Formula) *For any compactly supported Lipschitz function f on $\mathbb{R}$ and any point x of $\mathbb{R}$,*

$$f(x) = \int_0^\infty (f * \varphi_r * \varphi_r)(x) \, \frac{dr}{r}$$

with the integral converging absolutely in the sense that

$$\int_0^\infty (|f * \varphi_r| * |\varphi_r|)(x) \, \frac{dr}{r} < \infty.$$

Proof Let f have Lipschitz constant M and vanish outside the interval $[a, b]$. Recall that we denote the integral occurring in the enunciation of the theorem by $(\Phi_\infty f)(x)$ – see the paragraph preceding the last lemma. The absolute convergence of the integral defining $(\Phi_\infty f)(x)$ follows from this last lemma. Thus, by Lebesgue's Dominated Convergence Theorem [RUD, 1.34],

$$(\star) \quad \lim_{N \to \infty} (\Phi_N f)(x) = (\Phi_\infty f)(x)$$

for all points $x \in \mathbb{R}$. Recall that L^2 convergence implies pointwise almost everywhere convergence of a subsequence [RUD, 3.12]. Thus Lemma 8.62 implies there exists a subsequence $\{N_m\}$ and a subset E of $\mathbb{R}$ with $\mathcal{L}^1(\mathbb{R} \setminus E) = 0$ such that

$$(\star\star) \quad \lim_{m \to \infty} (\Phi_{N_m} f)(y) = f(y)$$

for all points $y \in E$.

Since E is dense in $\mathbb{R}$, given $x \in \mathbb{R}$ and $\varepsilon > 0$, we may choose a point $y \in E$ such that

$$M|x - y| < \varepsilon/4 \text{ and } M\sqrt{b-a}\sqrt{|y-x|} < \varepsilon/4.$$

Having fixed y, by $(\star\star)$ and $(\star)$ we may now choose m so large that

$$|f(y) - (\Phi_{N_m} f)(y)| < \varepsilon/4 \text{ and } |(\Phi_{N_m} f)(x) - (\Phi_\infty f)(x)| < \varepsilon/4.$$

Note that, on the one hand, since f is Lipschitz with constant M, we have

$$|f(x) - f(y)| \leq M|x - y|$$

while, on the other hand, by Lemma 4.15, Cauchy–Schwarz [RUD, 3.5], and Lemma 8.63, we have

$$\begin{aligned}|(\Phi_{N_m} f)(y) - (\Phi_{N_m} f)(x)| &= \left| \int_x^y (\Phi_{N_m} f)'(u)\, du \right| \leq \|(\Phi_{N_m} f)'\|_2 \sqrt{|y - x|} \\ &\leq M\sqrt{b - a}\sqrt{|y - x|}.\end{aligned}$$

Thus, given the way we have chosen first y and then m, we clearly have

$$\begin{aligned}|f(x) - (\Phi_\infty f)(x)| &\leq |f(x) - f(y)| + |f(y) - (\Phi_{N_m} f)(y)| \\ &\quad + |(\Phi_{N_m} f)(y) - (\Phi_{N_m} f)(x)| \\ &\quad + |(\Phi_{N_m} f)(x) - (\Phi_\infty f)(x)| \ < \ \varepsilon\end{aligned}$$

and so are done. □

8.23 A Decomposition of ℓ

Construction 8.66 (A Decomposition of ℓ) *Recall that our function ℓ is defined on the open interval U_5 (8.36) and is Lipschitz there with bound $71\alpha_0$ (8.39). A routine Cauchy sequence argument shows that we may define ℓ at the endpoints ± 5 of U_5 in such a way that the extended ℓ is Lipschitz on the closure of U_5 with bound $71\alpha_0$. Define ℓ on $[5, \infty)$ by having it at first increase or decrease linearly with slope $\pm 71\alpha_0$ until it becomes 0 after which point it is to constantly remain 0. Define ℓ similarly on $(-\infty, -5]$. For the rest of this chapter we consider ℓ as so extended. It is thus a function defined, compactly supported, and Lipschitz on all of $\mathbb{R} = L_0$ with bound $71\alpha_0$.*

For any $p \in L_0$, set

$$\ell_1(p) = \int_1^\infty (\ell * \varphi_r * \varphi_r)(p)\, \frac{dr}{r} + \int_0^1 (\{\mathcal{X}_{L_0 \setminus U_4} (\ell * \varphi_r)\} * \varphi_r)(p)\, \frac{dr}{r}$$

and

$$\ell_2(p) = \int_0^1 (\{\mathcal{X}_{U_4} (\ell * \varphi_r)\} * \varphi_r)(p)\, \frac{dr}{r}.$$

(Do not confuse these two functions with the functions ℓ_1 and ℓ_2 of Construction 8.32!) By Lemma 8.64, the three integrals involved in the definition of ℓ_1 and ℓ_2 converge absolutely and so our functions are well defined.

Moreover by Calderón's Formula (8.65), *for all points* $p \in L_0$ *we have*

$$\ell(p) = \ell_1(p) + \ell_2(p).$$

The rest of this section will be taken up with estimating the derivatives of ℓ_1 and ℓ_2. To estimate the first and second derivatives of ℓ_1 we need to introduce and investigate an auxiliary function. This is done in the following.

Lemma 8.67 *For any* $p \in L_0$, *set*

$$\psi(p) = \int_1^\infty (\varphi_r * \varphi_r)(p)\,\frac{dr}{r}.$$

Then the integral defining $\psi(p)$ *converges absolutely and the resulting function of* p *is smooth on all of* L_0 *with support contained in the interval* $[-2, 2]$. *Moreover,* $\|\psi\|_1 \leq 2\,\|\varphi\|_1\|\varphi\|_\infty$ *and* $\|\psi'\|_1 \leq \|\varphi'\|_1\|\varphi\|_\infty$.

Proof Since $|(\varphi_r * \varphi_r)(p)| \leq \|\varphi_r\|_1\|\varphi_r\|_\infty/\sqrt{2\pi} = \|\varphi\|_1\|\varphi\|_\infty/(\sqrt{2\pi}\,r)$, the absolute convergence is obvious.

Given $p, h \in L_0$, clearly we have

$$(\star)\ \frac{\psi(p+h)-\psi(p)}{h} = \int_1^\infty \int_{-\infty}^\infty \frac{\varphi_r(p+h-q)-\varphi_r(p-q)}{h}\varphi_r(q)\,\frac{dq}{\sqrt{2\pi}}\,\frac{dr}{r}$$

Note that $(\varphi_r)' = (\varphi')_r/r$. Thus by the Mean Value Theorem,

$$\left|\frac{\varphi_r(p+h-q)-\varphi_r(p-q)}{h}\varphi_r(q)\right| \leq \|(\varphi_r)'\|_\infty|\varphi_r(q)| = \frac{\|\varphi'\|_\infty}{r^2}|\varphi_r(q)|.$$

Also

$$\begin{aligned}\int_1^\infty \int_{-\infty}^\infty \frac{\|\varphi'\|_\infty}{r^2}|\varphi_r(q)|\,\frac{dq}{\sqrt{2\pi}}\,\frac{dr}{r} &\leq \frac{\|\varphi'\|_\infty}{\sqrt{2\pi}}\int_1^\infty \frac{\|\varphi_r\|_1}{r^3}\,dr \\ &= \frac{\|\varphi'\|_\infty\|\varphi\|_1}{\sqrt{2\pi}}\int_1^\infty \frac{dr}{r^3} \\ &= \frac{\|\varphi'\|_\infty\|\varphi\|_1}{2\sqrt{2\pi}} \quad < \quad \infty.\end{aligned}$$

Thus we may let $h \to 0$ in $(\star)$ and apply Lebesgue's Dominated Convergence Theorem [RUD, 1.34] to conclude that ψ is differentiable at p with the following formula for the derivative:

$$(\star\star)\ \psi'(p) = \int_1^\infty ((\varphi')_r * \varphi_r)(p)\,\frac{dr}{r^2}.$$

An obvious induction argument using the same sort of reasoning to get from $\psi^{(n)}$ to $\psi^{(n+1)}$ as we just used to get from $\psi = \psi^{(0)}$ to $\psi' = \psi^{(1)}$ now shows that ψ is infinitely differentiable with

$$\psi^{(n)}(p) = \int_1^\infty ((\varphi^{(n)})_r * \varphi_r)(p) \frac{dr}{r^{n+1}}$$

for every $p \in L_0$ and every nonnegative integer n.

We now turn to showing that $\mathrm{spt}(\psi) \subseteq [-2, 2]$ ... which is not at all obvious. So let $p \notin [-2, 2]$ and $\varepsilon > 0$. Since $|p| > 2$ and ψ is continuous at p, we may choose and fix a positive integer n such that $2 + 1/n < |p|$ and $\sup\{|\psi(p) - \psi(q)| : |p - q| < 1/n\} < \varepsilon$. Let f_n be any nonnegative Lipschitz function supported on $[-1/n, 1/n]$ with integral 1, e.g., $f_n(q) = \max\{n(1 - n|q|), 0\}$.

Consider the following chain of equalities, immediately after which a justification will be given for each of its lines:

$$\begin{aligned}
0 &= \sqrt{2\pi} \int_0^\infty (f_n * \varphi_r * \varphi_r)(p) \frac{dr}{r} \\
&= \sqrt{2\pi} \int_1^\infty (f_n * \varphi_r * \varphi_r)(p) \frac{dr}{r} \\
&= \int_1^\infty \int_{-\infty}^\infty f_n(p - q)\, (\varphi_r * \varphi_r)(q)\, dq \frac{dr}{r} \\
&= \int_{-\infty}^\infty f_n(p - q) \left\{ \int_1^\infty (\varphi_r * \varphi_r)(q) \frac{dr}{r} \right\} dq \\
&= \int_{-\infty}^\infty f_n(p - q) \psi(q)\, dq.
\end{aligned}$$

[The first line follows from Calderón's Formula (8.65) since $f_n(p) = 0$; the second from the fact that $p \notin [-2 - 1/n, 2 + 1/n] \supseteq \mathrm{spt}(f_n * \varphi_r * \varphi_r)$ for $0 < r < 1$; the third from the definition of a convolution; the fourth from Fubini's Theorem [RUD, 8.8] with the absolute convergence of the double integral a consequence of (a) of Lemma 8.64; and the last from the definition of ψ.]

Thus

$$\begin{aligned}
|\psi(p)| &= \left| \int_{-\infty}^\infty f_n(p - q) \psi(p)\, dq \right| \\
&= \left| \int_{-\infty}^\infty f_n(p - q) [\psi(p) - \psi(q)]\, dq \right| \\
&\le \sup\{|\psi(p) - \psi(q)| : |p - q| < 1/n\} \int_{-1/n}^{1/n} f_n(p - q)\, dq \\
&< \varepsilon.
\end{aligned}$$

Since $\varepsilon > 0$ was otherwise arbitrary, we conclude that $\psi(p) = 0$.

Having shown that $\mathrm{spt}(\psi) \subseteq [-2, 2]$, the first line of the proof shows that

$$\begin{aligned} \|\psi\|_1 = \int_{-2}^{2} |\psi(p)|\, dp &\leq \int_{-2}^{2} \int_{1}^{\infty} |(\varphi_r * \varphi_r)(p)| \frac{dr}{r}\, dp \\ &\leq \int_{-2}^{2} \int_{1}^{\infty} \frac{\|\varphi\|_1 \|\varphi\|_\infty}{\sqrt{2\pi}\, r} \frac{dr}{r}\, dp \leq 2\, \|\varphi\|_1 \|\varphi\|_\infty. \end{aligned}$$

Since $\mathrm{spt}(\psi') \subseteq [-2, 2]$ also and we have a representation $(\star\star)$ for ψ' analogous to our definition of ψ, a similar argument will establish the claimed estimate on $\|\psi'\|_1$. □

In this last proof we had to justify a differentiation under a double integral. In what follows we will need to differentiate into a convolution, i.e., differentiate under a single integral. Justifying this is done in the following.

Lemma 8.68 *Suppose that $f \in L^\infty(\mathbb{R})$ is Lipschitz on $\mathbb{R}$ with bound M and that $g \in L^1(\mathbb{R})$. Then $f * g$ has a bounded derivative everywhere on $\mathbb{R}$ with $(f * g)' = f' * g$ and $\|(f * g)'\|_\infty \leq M\|g\|_1$.*

Note that in this lemma we may take g to be any smooth function with compact support. Also, by the Mean Value Theorem, we may take f to be any smooth function with compact support (and in this situation $M = \|f'\|_\infty$).

Proof Recall from Lemma 4.15 that f' exists and is bounded by M $\mathcal{L}^1$-a.e. on $\mathbb{R}$. Thus we need only show that $(f * g)'$ exists and equals $f' * g$ everywhere on $\mathbb{R}$.

For any p and h,

$$\frac{(f * g)(p + h) - (f * g)(p)}{h} = \int_{-\infty}^{\infty} \frac{f(p + h - q) - f(p - q)}{h}\, g(q)\, \frac{dq}{\sqrt{2\pi}}.$$

For any p, q, and h,

$$\left| \frac{f(p + h - q) - f(p - q)}{h}\, g(q) \right| \leq M|g(q)| \in L^1(\mathbb{R}).$$

Lebesgue's Dominated Convergence Theorem [RUD, 1.34] now finishes off the proof. □

Recall that in Construction 8.12 we made the convention that $U_a = (-a, a)$ for $a > 0$.

Proposition 8.69 *The function ℓ_1 is twice differentiable on U_3 with*

$$\|\ell_1'\|_{U_3} \leq 142\, \|\varphi\|_1 \|\varphi\|_\infty\, \alpha_0 \text{ and } \|\ell_1''\|_{U_3} \leq 71\, \|\varphi'\|_1 \|\varphi\|_\infty\, \alpha_0.$$

Thus ℓ_2 is Lipschitz on U_3 with bound $71\{1 + 2\, \|\varphi\|_1 \|\varphi\|_\infty\}\, \alpha_0$.

Proof We may write $\ell_1 = \ell_a + \ell_b$ where

$$\ell_a(p) = \int_1^\infty (\ell * \varphi_r * \varphi_r)(p)\,\frac{dr}{r} \text{ and } \ell_b(p) = \int_0^1 (\{\mathcal{X}_{L_0\setminus U_4}(\ell * \varphi_r)\} * \varphi_r)(p)\,\frac{dr}{r}.$$

Since $\mathrm{spt}(\ell_b) \subseteq \{L_0 \setminus U_4\} + [-1, 1] = L_0 \setminus U_3$, it suffices to prove the assertions of the proposition concerning ℓ_1 for ℓ_a instead.

By (a) of Lemma 8.64 and Fubini's Theorem [RUD, 8.8], $\ell_a = \ell * \psi$ where ψ is as in Lemma 8.67. Since $\ell \in L^\infty(\mathbb{R})$ is Lipschitz on $\mathbb{R}$ with bound $71\alpha_0$ by Construction 8.66 and $\psi \in L^1(\mathbb{R})$ by Lemma 8.67, we may apply the last lemma to conclude that ℓ_a is differentiable on $\mathbb{R}$, that $\ell_a' = \ell' * \psi$ there, and, with another invocation of Lemma 8.67, that $\|\ell_a'\|_\infty \leq 71\alpha_0\|\psi\|_1 \leq 71\alpha_0 \cdot 2\,\|\varphi\|_1\|\varphi\|_\infty$.

Since convolution is commutative, $\ell_a' = \psi * \ell'$. Clearly $\psi \in L^\infty(\mathbb{R})$ is Lipschitz on $\mathbb{R}$ with bound $\|\psi'\|_\infty$ (by the Mean Value Theorem) and $\ell' \in L_c^\infty(\mathbb{R}) \subseteq L^1(\mathbb{R})$, so we may again apply the last lemma to conclude that ℓ_a' is differentiable on $\mathbb{R}$ and that $\ell_a'' = \psi' * \ell'$ there. Invoking Lemma 8.67 one last time, we see that $\|\ell_a''\|_\infty \leq \|\psi'\|_1\|\ell'\|_\infty \leq \|\varphi'\|_1\|\varphi\|_\infty \cdot 71\alpha_0$.

From the Mean Value Theorem and our just proven bound on ℓ_1', it follows that ℓ_1 is Lipschitz on U_3 with bound $142\,\|\varphi\|_1\|\varphi\|_\infty\,\alpha_0$. The assertion of the proposition concerning ℓ_2 now follows from Construction 8.66. □

Rather than the obvious L^∞ estimate on ℓ_2' over U_3 which would follow from the last proposition, we turn out to need a more subtle L^2 estimate on ℓ_2' over U_3. For this we require the following.

Lemma 8.70 *Suppose U is a bounded open subinterval of $\mathbb{R}$. Then $C_c^\infty(U)$ is dense in $L^2(U)$.*

Proof In a finite measure space uniform convergence implies L^2 convergence. Thus, by the coincidence of $\mathcal{H}^1$ and $\mathcal{L}^1$ for linear sets (from section 2.1), Proposition 5.22, and Proposition 5.23, it suffices to show that $C_c^\infty(U)$ is uniformly dense in $C_c(U)$. So suppose that $f \in C_c(U)$.

First, let $\widetilde{\varphi}$ denote the function gotten from Lemma 8.34 when $a = b = 0$ and $l = 1$. Set $g = \sqrt{2\pi}\,\widetilde{\varphi}/\|\widetilde{\varphi}\|_1$. Note that $g \in C_c^\infty(\mathbb{R})$. Indeed, $\mathrm{spt}(g) \subseteq [-1, 1]$. Also, $g \geq 0$ on $\mathbb{R}$ and $\int_{\mathbb{R}} g(y)\,dy/\sqrt{2\pi} = 1$. In consequence, $g_r \in C_c^\infty(\mathbb{R})$, $\mathrm{spt}(g_r) \subseteq [-r, r]$, $g_r \geq 0$ on $\mathbb{R}$, and $\int_{\mathbb{R}} g_r(y)\,dy/\sqrt{2\pi} = 1$ for any $r > 0$.

Second, note that $\mathrm{spt}(f * g_r) \subseteq U$ whenever $0 < r < \mathrm{dist}(\mathrm{spt}(f), \mathbb{R} \setminus U)$. Thus, by Lemma 8.68, $f * g_r = g_r * f \in C_c^\infty(U)$ for these same r.

Third, note that for any $x \in \mathbb{R}$ and $r > 0$,

$$\begin{aligned} |f(x) - (f * g_r)(x)| &= \left|\int_{-\infty}^\infty \{f(x) - f(x-y)\}g_r(y)\,\frac{dy}{\sqrt{2\pi}}\right| \\ &\leq \sup\{f(x) - f(x-y)| : |y| \leq r\}. \end{aligned}$$

Thus, taking the supremum over all $x \in \mathbb{R}$, we see that for all $r > 0$,

$$\|f - (f * g_r)\|_\infty \le \operatorname{osc}(f, \mathbb{R}; r).$$

Fourth and last, note that any continuous function compactly supported on $\mathbb{R}$ is uniformly continuous on $\mathbb{R}$. Thus $\operatorname{osc}(f, \mathbb{R}; r) \downarrow 0$ as $r \downarrow 0$ and so we are done. □

Proposition 8.71 $\displaystyle\int_{U_3} |\ell_2'(p)|^2\, dp \le C_7(\varphi) \int_0^1 \int_{U_4} \gamma^2(p; r)\, dp\, \frac{dr}{r}.$

Recall that the quantity $C_7(\varphi)$ was defined back in Section 8.6.

Proof From Proposition 8.69 and Lemma 4.15, we see that ℓ_2' exists $\mathcal{L}^1$-a.e. on U_3 and is bounded there. Let f be any element of $C_c^\infty(U_3)$. Then, by the Mean Value Theorem and Proposition 8.69 again, $f\ell_2$ is Lipschitz on U_3 with bound $\|f'\|_\infty \|\ell_2\|_\infty + 71\{1 + 2\|\varphi\|_1 \|\varphi\|_\infty\}\alpha_0 \|f\|_\infty$. Thus, by Lemma 4.15 again, $(f\ell_2)'$ exists $\mathcal{L}^1$-a.e. on U_3 and $f\ell_2$ can be recovered from it by integration there. Clearly $(f\ell_2)' = f\ell_2' + f'\ell_2$ $\mathcal{L}^1$-a.e. on U_3.

In what follows we will have two long chains of in/equalities. Immediately after each chain, a justification will be given for each of its lines.

$$
\begin{aligned}
(\star)\quad \left|\int_{U_3} f(p)\, \ell_2'(p)\, dp\right|^2 &= \left|\int_{U_3} f'(p)\, \ell_2(p)\, dp\right|^2 \\
&= \left|\int_{-\infty}^{\infty} f'(p) \left\{\int_0^1 \int_{-\infty}^{\infty} \varphi_r(p-q) \cdot \{\mathcal{X}_{U_4}(\ell * \varphi_r)\}(q)\, \frac{dq}{\sqrt{2\pi}}\, \frac{dr}{r}\right\} dp\right|^2 \\
&= \left|\int_0^1 \int_{-\infty}^{\infty} \left\{\int_{-\infty}^{\infty} \varphi_r(q-p)\, f'(p)\, \frac{dp}{\sqrt{2\pi}}\right\} \cdot \{\mathcal{X}_{U_4}(\ell * \varphi_r)\}(q)\, dq\, \frac{dr}{r}\right|^2 \\
&= \left|\int_0^1 \int_{-\infty}^{\infty} \{\varphi_r * f'\}(q) \cdot \{\mathcal{X}_{U_4}(\ell * \varphi_r)\}(q)\, dq\, \frac{dr}{r}\right|^2 \\
&\le \int_0^1 \int_{-\infty}^{\infty} [r(\varphi_r * f')(q)]^2\, dq\, \frac{dr}{r} \leftarrow \text{(I)!} \\
&\quad \times \int_0^1 \int_{U_4} \left[\frac{(\ell * \varphi_r)(q)}{r}\right]^2 dq\, \frac{dr}{r}. \leftarrow \text{(II)!}
\end{aligned}
$$

[The first line follows from an integration by parts that is justified by the opening paragraph of the proof; the second line from the fact that $f' = 0$ off U_3, the definition of ℓ_2, the commutativity of convolution, and the definition of convolution; the third line from Fubini's Theorem [RUD, 8.8] and the evenness of φ_r with the absolute convergence needed to justify Fubini's Theorem [RUD, 8.8] a consequence of (b) of Lemma 8.64 and the commutativity of convolution again; the fourth from the definition of convolution; and the last from the Cauchy–Schwarz Inequality [RUD, 3.5].]

$$
\begin{aligned}
(\dagger)\ (\mathrm{I}) &= \int_0^1 \int_{-\infty}^{\infty} \{r\,\widehat{\varphi}(\xi r)\}^2\,|i\,\xi\,\widehat{f}(\xi)|^2\,d\xi\,\frac{dr}{r} \\
&= \int_{-\infty}^{\infty} \left\{ \int_0^1 \{\xi r\,\widehat{\varphi}(\xi r)\}^2\,\frac{dr}{r} \right\} |\widehat{f}(\xi)|^2\,d\xi \\
&= \int_{-\infty}^{\infty} \left\{ \int_0^{|\xi|} \{s\,\widehat{\varphi}(s)\}^2\,\frac{ds}{s} \right\} |\widehat{f}(\xi)|^2\,d\xi \\
&\le \left\{ \int_0^{\infty} s\,\{\widehat{\varphi}(s)\}^2\,ds \right\} \int_{-\infty}^{\infty} |\widehat{f}(\xi)|^2\,d\xi \\
&= \frac{\|\varphi\|_\infty^2 + \|\varphi''\|_\infty^2}{2} \int_{U_3} |f(p)|^2\,dp.
\end{aligned}
$$

[The first line follows from Plancherel's Theorem (4.10), the fact that the Fourier transform of a convolution is the product of the transforms, the last fact noted in Stipulation 8.61, and Proposition 4.9; the second from Fubini's Theorem [RUD, 8.8]; the third from the change of variables $s = |\xi|r$ and the evenness of $\widehat{\varphi}$; the fourth is clear; and the last from Lemma 8.60, Plancherel's Theorem (4.10) again, and the fact that $f = 0$ off U_3.]

Recall that φ is even with mean value zero. From this it easily follows that $(l * \varphi_r)(q) = 0$ for any linear function l, any point $q \in U_4$, and any $r \in (0, 1)$. Hence

$$
\begin{aligned}
\left| \frac{(\ell * \varphi_r)(q)}{r} \right| &= \left| \frac{(\{\ell - l\} * \varphi_r)(q)}{r} \right| \\
&\le \int_{-\infty}^{\infty} \frac{|\{\ell - l\}(q - p)||\varphi_r(p)|}{r} \frac{dp}{\sqrt{2\pi}} \\
&\le \frac{\|\varphi\|_\infty}{\sqrt{2\pi}} \frac{1}{r} \int_{I(q;r)} \frac{|\ell(t) - l(t)|}{r}\,dt
\end{aligned}
$$

and so, taking the infimum over all linear l, we conclude that

$$
\left| \frac{(\ell * \varphi_r)(q)}{r} \right| \le \frac{\|\varphi\|_\infty}{\sqrt{2\pi}}\,\gamma(q; r)
$$

for all $q \in U_4$ and $r \in (0, 1)$. Thus

$$
(\ddagger)\ (\mathrm{II}) \le \frac{\|\varphi\|_\infty^2}{2\pi} \int_0^1 \int_{U_4} \gamma^2(q; r)\,dq\,\frac{dr}{r}.
$$

From $(\star)$, $(\dagger)$, and $(\ddagger)$, we obtain

$$(\star\star)\quad \left|\int_{U_3} f(p)\,\ell_2'(p)\,dp\right|^2 \le C_7(\varphi)\left\{\int_{U_3} |f(p)|^2\,dp\right\}\left\{\int_0^1\int_{U_4} \gamma^2(q;r)\,dq\,\frac{dr}{r}\right\}$$

for any $f \in C_c^\infty(U_3)$. By the last lemma, $(\star\star)$ extends to any $f \in L^2(U_3)$. So setting $f = \ell_2'$ in $(\star\star)$, performing a division, and changing two qs to ps, we are done. □

8.24 The Smallness of K_3

To show that K_3 has small measure it is necessary to introduce a special maximal function and some other auxiliary notions. Thus the following ...

Definition 8.72 (A Special Maximal Function) *Given a function f defined on a closed subinterval I of L_0, recall from Section 7.5 that the average value of f on I is given by*

$$\langle f\rangle_I = \frac{1}{|I|}\int_I f(p)\,dp.$$

We now define the oscillation of f on I to be

$$\mathrm{osc}_I(f) = \sup\{|f(p) - \langle f\rangle_I| : p \in I\}.$$

(The reader should not confuse this with the notion, introduced in Section 3.2, of the oscillation of a function on a set with a given gauge.)

Given a function f defined on the interval $I(p_0;1) \subseteq L_0$, define an uncentered maximal function of the integral average of the scaled pointwise oscillation of f at p_0 by setting

$$(\mathcal{N}f)(p_0) = \sup\left\{\frac{1}{|I|}\int_I \frac{|f(p) - \langle f\rangle_I|}{|I|}\,dp : p_0 \in I, \text{a closed interval with } |I| \le 1\right\}.$$

The mysterious parameter θ_0, about which we could say little when we chose it in Section 8.6, now makes its first substantial appearance. It and the maximal function just introduced are needed to define a subset H_{θ_0} of L_0 whose linear Lebesgue measure will control the μ-measure of K_3.

Recall that in Construction 8.12 we made the convention that $U_a = (-a, a)$ for $a > 0$.

Definition 8.73 (A Technical Set) $H_{\theta_0} = \{p \in U_2 : (\mathcal{N}\ell_2)(p) \le \theta_0^2\,\alpha_0\}$.

The smallness of K_3 will be easily established after the seven lemmas that follow have been verified!

Lemma 8.74 *For any $p_0 \in L_0$ and any $r \in (0, \theta_0]$ such that $I(p_0; r) \cap H_{\theta_0} \neq \emptyset$, let $L(p_0)$ be the line that is the graph of the linear function*

$$p \mapsto \ell_1'(p_0)(p - p_0) + \ell(p_0).$$

(Note that $I(p_0; r) \subseteq U_3$ since $I(p_0; r)$ intersects $H_{\theta_0} \subseteq U_2$ and $|I(p_0; r)| = 2r \leq 2\theta_0 \leq 1$ by our choice of θ_0. Thus L_0 is well defined, i.e., $\ell_1'(p_0)$ exists, by Proposition 8.69.)

Then

$$\sup\{\mathrm{dist}(p + i\,\ell(p), L(p_0)) : p \in I(p_0; r)\} \leq C_8(\varphi)\,\theta_0\,\alpha_0\,r.$$

Recall that the quantity $C_8(\varphi)$ was defined back in Section 8.6.

Proof For the sake of concision, denote $I(p_0; r)$ by I and $71\{1 + 2\,\|\varphi\|_1 \|\varphi\|_\infty\}$ by C. Then, given $p \in I$, it suffices to show that $|\ell(p) - \{\ell_1'(p_0)(p - p_0) + \ell(p_0)\}| \leq \{16\sqrt{C} + (71/2)\|\varphi'\|_1 \|\varphi\|_\infty\}\,\theta_0\,\alpha_0\,r$. To the end we need the following.

Claim. $\{\mathrm{osc}_I(\ell_2)\}^2 \leq 16C\,\alpha_0 \int_I |\ell_2(p) - \langle \ell_2 \rangle_I|\,dp \leq 64C\,\theta_0^2\,\alpha_0^2\,r^2$.

We need only focus on the first inequality of the Claim since the second inequality follows easily from our choice of θ_0 and the Definitions of the maximal function $\mathcal{N}$ and the set H_{θ_0} (8.72 and 8.73). Pick $q \in I$ such that $|\ell_2(q) - \langle \ell_2 \rangle_I| = \mathrm{osc}_I(\ell_2)$. Set $J = \{p \in I : |p - q| \leq \mathrm{osc}_I(\ell_2)/(2C\,\alpha_0)\}$. Since $I \subseteq U_3$, we may apply Proposition 8.69 to conclude that for any $p \in J$, $|\ell_2(p) - \ell_2(q)| \leq C\,\alpha_0\,|p - q| \leq \mathrm{osc}_I(\ell_2)/2$ and so $|\ell_2(p) - \langle \ell_2 \rangle_I| \geq |\ell_2(q) - \langle \ell_2 \rangle_I| - |\ell_2(q) - \ell_2(p)| \geq \mathrm{osc}_I(\ell_2) - \mathrm{osc}_I(\ell_2)/2 = \mathrm{osc}_I(\ell_2)/2$. Thus

$$\int_I |\ell_2(p) - \langle \ell_2 \rangle_I|\,dp \geq \frac{\mathrm{osc}_I(\ell_2)}{2}|J|.$$

Suppose, on the one hand, $\mathrm{osc}_I(\ell_2)/(2C\,\alpha_0) \leq |I|/2$. Then J contains an interval of length $\mathrm{osc}_I(\ell_2)/(2C\,\alpha_0)$, either $[q - \mathrm{osc}_I(\ell_2)/(2C\,\alpha_0), q]$ or $[q, q + \mathrm{osc}_I(\ell_2)/(2C\,\alpha_0)]$, and so

$$\int_I |\ell_2(p) - \langle \ell_2 \rangle_I|\,dp \geq \frac{\{\mathrm{osc}_I(\ell_2)\}^2}{4C\,\alpha_0}.$$

This rearranges to what we want with a factor of 4 to spare.

Suppose, on the other hand, $\mathrm{osc}_I(\ell_2)/(2C\,\alpha_0) > |I|/2$. Then J contains an interval of length $|I|/2$, either $[q - |I|/2, q]$ or $[q, q + |I|/2]$, and so

$$\int_I |\ell_2(p) - \langle \ell_2 \rangle_I|\,dp \geq \frac{\mathrm{osc}_I(\ell_2)}{4}|I|.$$

Appealing to the definition of average value and to Proposition 8.69 again, we have

$$\int_I |\ell_2(p) - \langle \ell_2 \rangle_I|\, dp \leq \int_I \frac{1}{|I|} \int_I |\ell_2(p) - \ell_2(\tilde{p})|\, d\tilde{p}\, dp$$

$$\leq \int_I \frac{1}{|I|} \int_I C\,\alpha_0\, |I|\, d\tilde{p}\, dp = C\,\alpha_0\, |I|^2.$$

From the last two displayed inequalities it follows that

$$\operatorname{osc}_I(\ell_2)^2 \leq \frac{16}{|I|^2} \left\{ \int_I |\ell_2(p) - \langle \ell_2 \rangle_I|\, dp \right\}^2 \leq 16C\,\alpha_0 \int_I |\ell_2(p) - \langle \ell_2 \rangle_I|\, dp$$

and so the Claim is proven.

From Construction 8.66, Taylor's Theorem, the Claim, and Proposition 8.69, we may now conclude that

$$|\ell(p) - \{\ell_1'(p_0)(p - p_0) + \ell(p_0)\}|$$

$$\leq |\ell_2(p) - \ell_2(p_0)| + |\ell_1(p) - \{\ell_1'(p_0)(p - p_0) + \ell_1(p_0)\}|$$

$$\leq 2\operatorname{osc}_I(\ell_2) + \frac{\|\ell_1''\|_{U_3}}{2} |p - p_0|^2$$

$$\leq 16\sqrt{C}\,\theta_0\,\alpha_0\, r + \frac{71\, \|\varphi'\|_1 \|\varphi\|_\infty\, \alpha_0}{2} r^2$$

for any $p \in I$. Since $r \leq \theta_0$, the lemma is proven. □

Lemma 8.75 *Suppose $z \in K_3$ and L is a line such that $\beta_1^L(z; h(z)) < 3\varepsilon_0$. Then $\angle(L, L_0) > \alpha_0/2$.*

Proof From the Definition of K_3, K_2, and K_1 (8.41), we get a point $w \in K$, a number $s \in [h(z)/4, 3h(z)/4]$, and a line L_* such that $z \in B(w; s/2)$, $\delta(w; s) \geq 2\delta_0$, $\beta_1^{L_*}(w; s) < \varepsilon_0$, and $\angle(L_*, L_0) \geq \alpha_0$. Clearly $|w - z| \leq s/2 \leq (k_0/2)s$ and $\beta_1^L(z; s) \leq \{h(z)/s\}^2 \beta_1^L(z; h(z)) \leq 4^2(3\varepsilon_0)$. We may thus apply (c) of Proposition 8.7 to conclude that $\angle(L_*, L) \leq C_1\,\varepsilon_0$. We are now done by Condition (3) for the selection of ε_0. □

Lemma 8.76 *Suppose $B(z; r) \in S_0$, $r \geq \theta_0$, and L is a line such that $\beta_1^L(z; r) < 3\varepsilon_0$. Then $\angle(L, L_0) \leq \alpha_0/6$.*

Proof Consider the case $r \leq 1$ first. There is a unique nonnegative integer N such that $e^{-(N+1)} < r \leq e^{-N}$. Since $r \geq \theta_0$, it follows that $N \leq \ln(1/\theta_0)$. By (c) of Corollary 8.18, $B(z; e^{-n}) \in S_0$ for $n = 1, 2, \ldots, N$. Thus, by the Definitions of S_{total} and S_0 (8.13 and 8.16), we have $\delta(z; e^{-n}) \geq \delta_0$ and we may choose lines L_n with $\beta_1^{L_n}(z; e^{-n}) < 3\varepsilon_0$ for these n. (Do not confuse these lines with the lines L_n of Construction 8.32!) Note too that $\delta(z;r) \geq \delta_0$. For our baseline L_0 we have $\beta_1^{L_0}(z;e^{-0}) < \varepsilon_0$ by Proposition 8.14. Thus $\beta_1^{L_{n-1}}(z;e^{-n}) \leq \{e^{-(n-1)}/e^{-n}\}^2 \beta^{L_{n-1}}(z;e^{-(n-1)}) < e^2(3\varepsilon_0)$ for each $n = 1, 2, \ldots, N$, and, similarly, $\beta_1^{L_N}(z;r) < e^2(3\varepsilon_0)$. We may thus apply (c) of Proposition 8.7 $N+1$ times to conclude that

$$\angle(L, L_0) \leq \angle(L, L_N) + \sum_{n=1}^{N} \angle(L_n, L_{n-1}) \leq \{1+N\}\, C_1\, \varepsilon_0 \leq \{1+\ln(1/\theta_0)\}\, C_1\, \varepsilon_0.$$

We are now done with the case $r \leq 1$ by condition (5) for the selection of ε_0.

So turn to the case $r > 1$. By Proposition 8.14, $\delta(z;1) > \delta_0$ and $\beta_1^{L_0}(z;1) < \varepsilon_0$. Note that $r < 3$ since $B(z;r) \in S_0$. Hence $\beta_1^L(z;1) \leq \{r/1\}^2 \beta_1^L(z;r) \leq 3^2(3\varepsilon_0)$. We may thus apply (c) of Proposition 8.7 to conclude that $\angle(L, L_0) \leq C_1\, \varepsilon_0$. We are now done with the case $r > 1$ by condition (3) for the selection of ε_0. □

Definition 8.77 (A Very Good Set) *Recalling the set G of Definition* 8.48, *we define a new set*

$$G^* = \left\{ z \in G : \mu(G \cap B(z;r)) \geq \left\{ 1 - \frac{c_2(M, \delta_0)}{2M} \right\} \mu(B(z;r)) \text{ for all } r \in (0, 1/2] \right\}.$$

Lemma 8.78 (The Smallness of $K \setminus G^*$) $\mu(K \setminus G^*) \leq 4 \times 10^{-5}$.

Proof Since $K \setminus G^* = \{K \setminus G\} \cup \{G \setminus G^*\}$, by the weaker conclusion of Proposition 8.49 it suffices to show that $\mu(G \setminus G^*) \leq 2 \times 10^{-5}$. Without loss of generality, $G \setminus G^* \neq \emptyset$. For each $z \in G \setminus G^*$, there exists a nontrivial closed ball B_z centered at z and of radius at most 1/2 such that $\mu(G \cap B_z) < \{1 - c_2/(2M)\}\mu(B_z)$. Applying (a) of Besicovitch's Covering Lemma (5.26) to the collection $\{B_z : z \in G \setminus G^*\}$, we extract a subcollection $\mathcal{B}$ which still covers $G \setminus G^*$ but, in addition, now has a controlled overlap of at most 125. Note that for any $B \in \mathcal{B}$ we have $\mu(G \cap B) < \{2M/c_2\}\mu((K \setminus G) \cap B)$. Thus

$$\mu(G \setminus G^*) \leq \sum_{B \in \mathcal{B}} \mu(G \cap B) < \frac{2M}{c_2} \sum_{B \in \mathcal{B}} \mu((K \setminus G) \cap B) \leq \frac{250M}{c_2} \mu(K \setminus G).$$

By the stronger conclusion of Proposition 8.49 we are done. □

Lemma 8.79 *If* $z \in K_3 \cap G^*$, *then* $h(z) \in (0, \theta_0)$ *and* $I(\pi(z); h(z)) \subseteq U_2 \setminus H_{\theta_0}$.

Proof By the Definition of K_3 and K_0 (8.41 and 8.23) and (b) of Proposition 8.21, $h(z) > 0$. Then by (b) of Corollary 8.18, $B(z, h(z)) \in S_0$. From the Definitions of S_{total} and S_0 (8.13 and 8.16), there is a line $L_{h(z)}$ such that $\beta_1^{L_{h(z)}}(z, h(z)) < 3\varepsilon_0$. Lemmas 8.75 and 8.76 now force $h(z) < \theta_0$.

Then by (b) of Corollary 8.18, $B(z, \theta_0) \in S_0$. From the Definitions of S_{total} and S_0 (8.13 and 8.16), there is a line L_{θ_0} such that $\beta_1^{L_{\theta_0}}(z, \theta_0) < 3\varepsilon_0$.

At this point we need the following.

Claim. Suppose $r \in [h(z), \theta_0]$ is such that $I(\pi(z), r) \cap H_{\theta_0} \neq \emptyset$ and L is a line such that $\beta_1^L(z; r)) < 3\varepsilon_0$. Then $\angle(L, L(\pi(z))) \leq \alpha_0/6$.

To see this first consider any $\zeta \in G \cap B(z; r)$. On the one hand, from the Definition of G (8.48) and (a) and (b) of Proposition 8.21 it follows that

$$|\zeta - \{\pi(\zeta) + i\,\ell(\pi(\zeta))\}| \leq \sqrt{\varepsilon_0}\, d(\zeta) \leq \sqrt{\varepsilon_0}\{d(z) + |\zeta - z|\} \leq \sqrt{\varepsilon_0}\{h(z) + r\} \leq 2\sqrt{\varepsilon_0}\, r.$$

On the other hand, from Lemma 8.74 it follows that $\text{dist}(\pi(\zeta) + i\,\ell(\pi(\zeta)), L(\pi(z))) \leq C_8\, \theta_0\, \alpha_0\, r$. Hence $\text{dist}(\zeta, L(\pi(z))) \leq \{2\sqrt{\varepsilon_0} + C_8\, \theta_0\, \alpha_0\} r \leq 2C_8\, \theta_0\, \alpha_0\, r$ by condition (6) for the selection of ε_0. Since $\zeta \in G \cap B(z; h(z))$ was otherwise arbitrary, we have shown

$$(\star)\quad \sup\{\text{dist}(\zeta, L(\pi(z))) : \zeta \in G \cap B(z; r)\} \leq 2C_8\, \theta_0\, \alpha_0\, r.$$

Next note that $\delta(z; r) \geq \delta_0$ due to (b) of Corollary 8.18 and the Definitions of S_{total} and S_0 (8.13 and 8.16). Thus we may get two balls B_1 and B_2 by applying Lemma 8.5 to the ball $B = B(z; r)$. (Do not confuse these two balls with the balls B_1 and B_2 of Construction 8.31!) Set

$$G_j = \{\zeta \in B(z; r) \cap B_j : \text{dist}(\zeta, L) < (6\varepsilon_0/c_2) r\}$$

for $j = 1, 2$. Then

$$\mu(B(z; r) \cap B_j \setminus G_j) \leq \int_{B(z;r)\cap B_j \setminus G_j} \frac{\text{dist}(\zeta, L)}{(6\varepsilon_0/c_2) r}\, d\mu(\zeta) \leq \frac{c_2 r}{6\varepsilon_0} \beta_1^L(z; r) < \frac{c_2}{2} r$$

and so $\mu(G_j) = \mu(B(z; r) \cap B_j) - \mu(B(z; r) \cap B_j \setminus G_j) > c_2 r - (c_2/2) r = (c_2/2) r$. It now follows from (3) of Reduction 8.3 that

$$\mu(G_j) > \{c_2/(2M)\}\mu(B(z; r)).$$

Moreover, since $z \in G^*$,

$$\mu(G \cap B(z; r)) \geq \{1 - c_2/(2M)\}\mu(B(z; r)).$$

Hence

$$\begin{aligned}\mu(G_j \cap G \cap B(z; r)) &= \mu(G_j) + \mu(G \cap B(z; r)) - \mu(G_j \cup \{G \cap B(z; r)\}) \\ &> \mu(B(z; r)) - \mu(G_j \cup \{G \cap B(z; r)\}) \;\geq\; 0.\end{aligned}$$

We may thus choose $z_j \in G_j \cap G \cap B(z; r)$ for $j = 1, 2$. Note that $|z_1 - z_2| \geq \text{dist}(B_1, B_2) > 10c_1 r$. Let L_* denote the line containing z_1 and z_2.

If L is parallel to L_*, then we trivially have $\angle(L, L_*) \leq \alpha_0/12$. So we may as well assume L and L_* have an intersection point w. Then $|w - z_j| > 5c_1 r$ for some j. Since this z_j is in G_j, by the way G_j was defined we then have

$$\begin{aligned}\angle(L, L_*) \leq \frac{\pi}{2} \sin \angle(L, L_*) &= \frac{\pi}{2} \frac{\text{dist}(z_j, L)}{|w - z_j|} \leq \frac{\pi}{2} \frac{(6\varepsilon_0/c_2)r}{5c_1 r} \leq \frac{2\varepsilon_0}{c_1 c_2} \\ &= \frac{2C_1\,\varepsilon_0}{10^5} \leq \frac{\alpha_0}{12}\end{aligned}$$

by the specification of C_1 in Proposition 8.7 and condition (3) for the selection of ε_0.

Similarly, if $L(\pi(z))$ is parallel to L_*, then we trivially have $\angle(L(\pi(z)), L_*) \leq \alpha_0/12$. So we may as well assume $L(\pi(z))$ and L_* have an intersection point w. Then $|w - z_j| > 5c_1 r$ for some j. Since this z_j is in $G \cap B(z; r)$, by ($\star$) we then have

$$\begin{aligned}\angle(L(\pi(z)), L_*) \leq \frac{\pi}{2} \sin \angle(L(\pi(z)), L_*) &= \frac{\pi}{2} \frac{\text{dist}(z_j, L(\pi(z)))}{|w - z_j|} \leq \frac{\pi}{2} \frac{2C_8\,\theta_0\,\alpha_0\,r}{5c_1 r} \\ &\leq \frac{C_8\,\theta_0\,\alpha_0}{c_1} \leq \frac{\alpha_0}{12}\end{aligned}$$

by our choice of θ_0.

With the last two paragraphs the Claim has been established.

Suppose that we had $I(\pi(z), h(z)) \cap H_{\theta_0} \neq \emptyset$. We could then apply the Claim twice, once to $r = h(z)$ with $L = L_{h(z)}$ and again to $r = \theta_0$ with $L = L_{\theta_0}$, to conclude that $\angle(L_{h(z)}, L(\pi(z))) \leq \alpha_0/6$ and $\angle(L_{\theta_0}, L(\pi(z))) \leq \alpha_0/6$. But by Lemma 8.76 with $r = \theta_0$ and $L = L_{\theta_0}$, $\angle(L_{\theta_0}, L_0) \leq \alpha_0/6$. Adding these three inequalities up, we would then have $\angle(L_{h(z)}, L_0) \leq \alpha_0/2$. This conclusion contradicts Lemma 8.75 with $L = L_{h(z)}$ and so $I(\pi(z), h(z)) \cap H_{\theta_0} = \emptyset$.

The point $\pi(z)$ is in $U_{1/2}$ by Construction 8.12. The number $h(z)$ has been shown to be at most θ_0 and so at most $1/2$ by our choice of θ_0. Thus it is clear that $I(\pi(z); h(z))$ is contained in U_2 and so we are done. □

Lemma 8.80 $\mu(K_3 \cap G^*) \leq 2{,}001\, M\, \mathcal{L}^1(U_2 \setminus H_{\theta_0})$.

Proof Without loss of generality, $\mu(K_3 \cap G^*) \neq \emptyset$. By the Definition of K_3 and K_0 (8.41 and 8.23) and (b) of Proposition 8.21, $h(z) > 0$ for each $z \in K_3 \cap G^*$. Applying (b) of Besicovitch's Covering Lemma (5.26) to the collection $\{B(z; 2h(z)) : z \in K_3 \cap G^*\}$, we extract countable subcollections $\mathcal{B}_1, \mathcal{B}_2, \ldots \mathcal{B}_{2001}$ that collectively cover $K_3 \cap G^*$ with each subcollection consisting of pairwise disjoint balls. For each n, set $\mathcal{I}_n = \{I(\pi(z); h(z)) : B(\pi(z); 2h(z)) \in \mathcal{B}_n\}$.

Claim. For each n, the collection $\mathcal{I}_n$ consists of pairwise disjoint intervals.

To see this suppose that $B(z; 2h(z))$ and $B(w; 2h(w))$ are distinct, and thus disjoint, elements of $\mathcal{B}_n$. Using this disjointness, the Definitions of G and G^* (8.48 and 8.77), Proposition 8.39, and (b) of Proposition 8.21, we see that

$$\begin{aligned}
2h(z) + 2h(w) &< |z - w| \\
&\leq |z - \{\pi(z) + i\,\ell(\pi(z))\}| + |\{\pi(z) + i\,\ell(\pi(z))\} \\
&\quad - \{\pi(w) + i\,\ell(\pi(w))\}| + |\{\pi(w) + i\,\ell(\pi(w))\} - w| \\
&\leq \sqrt{\varepsilon_0}\, d(z) + |\pi(z) - \pi(w)| + |\ell(\pi(z)) - \ell(\pi(w))| + \sqrt{\varepsilon_0}\, d(w) \\
&\leq \sqrt{\varepsilon_0}\, h(z) + \{1 + 71\,\alpha_0\}|\pi(z) - \pi(w)| + \sqrt{\varepsilon_0}\, h(w).
\end{aligned}$$

Upon rearrangement we obtain

$$|\pi(z) - \pi(w)| > \frac{2 - \sqrt{\varepsilon_0}}{1 + 71\,\alpha_0}\{h(z) + h(w)\} \geq h(z) + h(w)$$

with the last inequality holding by condition (1) for the selection of ε_0 and our choice of α_0. Thus $I(\pi(z); h(z))$ and $I(\pi(w); h(w))$ are disjoint and the Claim has been established.

From the covering property of our subcollections $\mathcal{B}_n$, (3) of Reduction 8.3, the obvious fact that $|B(\pi(z); 2h(z))|/2 = |I(\pi(z); h(z))|$, Lemma 8.79, and the Claim, we have

$$\begin{aligned}
\mu(K_3 \cap G^*) &\leq \sum_{n=1}^{2001} \sum_{B \in \mathcal{B}_n} \mu(B) \leq \sum_{n=1}^{2001} \sum_{B \in \mathcal{B}_n} M|B|/2 \\
&= M \sum_{n=1}^{2001} \sum_{I \in \mathcal{I}_n} |I| \leq 2{,}001 M\, \mathcal{L}^1(U_2 \setminus H_{\theta_0})
\end{aligned}$$

and so are done. □

Lemma 8.81 $\mathcal{L}^1(U_2 \setminus H_{\theta_0}) \leq \dfrac{4{,}000}{\theta_0^4\,\alpha_0^2} \displaystyle\int_{U_3} |\ell_2'(p_0)|^2\, dp_0.$

Proof From the Definition of H_{θ_0} (8.73) it clearly follows that

$$(\star)\ \mathcal{L}^1(U_2 \setminus H_{\theta_0}) \leq \frac{1}{\theta_0^4\,\alpha_0^2} \int_{U_2} (\mathcal{N}\ell_2)^2(p_0)\, dp_0.$$

For any $p_0 \in U_2$, let I be a closed interval containing p_0 with $|I| \leq 1$. Note that $I \subseteq I(p_0; |I|) \subseteq U_3$. From Proposition 8.69 and Lemma 4.15, we see that ℓ_2' exists $\mathcal{L}^1$-a.e. on U_3 and that ℓ_2 can be recovered from it by integration there. Using the definition of average value, it then follows that

$$\begin{aligned}\frac{1}{|I|}\int_I \frac{|\ell_2(p)-\langle \ell_2\rangle_I|}{|I|}\,dp &\le \frac{1}{|I|^3}\int_I\int_I |\ell_2(p)-\ell_2(\tilde p)|\,d\tilde p\,dp\\ &= \frac{1}{|I|^3}\int_I\int_I \left|\int_{\tilde p}^{p} \ell_2'(q)\,dq\right| d\tilde p\,dp\\ &\le \frac{1}{|I|}\int_{I(p_0;|I|)} |\ell_2'(q)|\,dq.\end{aligned}$$

Thus, defining

$$f(p)=\begin{cases}\ell_2'(p) & \text{when } p\in U_3\\ 0 & \text{when } p\in L_0\setminus U_3\end{cases}$$

and using $\mathcal{M}f$ as shorthand for $\mathcal{M}_{\mathcal{L}^1}(|f|\,d\mathcal{L}^1)$, the maximal function of $|f|\,d\mathcal{L}^1$ with respect to $\mathcal{L}^1$ as defined just before Proposition 5.27, we see that for every $p_0\in U_2$,

$$(\mathcal{N}\ell_2)(p_0)\le 2\sup_{0<r\le 1}\frac{1}{\mathcal{L}^1(I(p_0;r))}\int_{I(p_0;r)}|\ell_2'|\,d\mathcal{L}^1\le 2(\mathcal{M}f)(p_0).$$

Our $\mathcal{M}f$ is the usual Hardy–Littlewood maximal function as defined on page 138 in [RUD]. Hence, invoking the proof of [RUD, 8.18] and Proposition 5.27 (see the Remark to follow), we obtain

$$\begin{aligned}(\star\star)\quad \int_{U_2}(\mathcal{N}\ell_2)^2(p_0)\,dp_0 &\le 4\int_{L_0}(\mathcal{M}f)^2(p_0)\,dp_0\\ &\le 4(8\times 125)\int_{L_0}|f(p_0)|^2\,dp_0\\ &= 4{,}000\int_{U_3}|\ell_2'(p_0)|^2\,dp_0.\end{aligned}$$

Because of $(\star)$ and $(\star\star)$, we are done. □

Remark. The proof of [RUD, 8.18] shows that for all measurable functions f we have

$$\int_{\mathbb{R}}(\mathcal{M}f)^2\,d\mathcal{L}^1\le 8A\int_{\mathbb{R}}|f|^2\,d\mathcal{L}^1$$

provided A is a constant such that

$$\mathcal{L}^1(\{\mathcal{M}f>t\})\le\frac{A}{t}\int_{\mathbb{R}}|f|\,d\mathcal{L}^1$$

for all measurable functions f and all positive numbers t. Our invocation of Proposition 5.27 above allowed us to set $A = 125$. But this use of Proposition 5.27, whose proof relies on the difficult Besicovitch Covering Lemma (5.26), can be avoided by using [RUD, 7.4], whose proof relies on an almost trivial covering lemma [RUD,7.3]. If we do this, we may then set $A = 3$!

We have now assembled all the machinery needed to show that K_3 is small.

Proposition 8.82 (The Smallness of K_3) $\mu(K_3) \leq 5 \times 10^{-5}$.

Proof Since $K_3 \subseteq \{K \setminus G^*\} \cup \{K_3 \cap G^*\}$, by Proposition 8.78 it suffices to show that $\mu(K_3 \cap G^*) \leq 10^{-5}$. Stringing Lemma 8.80, Lemma 8.81, Proposition 8.71, and Proposition 8.59 together, we see that

$$\mu(K_3 \cap G^*) \leq 8{,}004{,}000\, M \,\frac{C_6\, C_7}{\theta_0^4\, \alpha_0^2}\, \varepsilon_0^2.$$

By condition (7) for the selection of ε_0, we are done. □

To recapitulate: In Construction 8.36 we finished the definition of a function ℓ which was later shown to be Lipschitz in Proposition 8.39. The beginning of the definition of ℓ in Construction 8.25 makes it clear that the graph of ℓ contains a subset K_0 of the support K of our measure. Propositions 8.42, 8.43, 8.50, and 8.82 make it clear that the total measure of $K \setminus K_0$ is at most 10^{-4}. Since our measure has total mass greater than 1, we have produced a Lipschitz graph covering at least 99.99% of the mass of our measure. Thus we have established Reduction 8.3 and so David and Léger's curvature theorem (8.1). With this, the raison d'être of the book, to prove Vitushkin's conjecture for removable sets, is completed!

Postscript: Tolsa's Theorem

The author hopes to have convinced the reader who has persisted to this point that Vitushkin's Conjecture is true. But Vitushkin's Conjecture arose as a proper part of Painlevé's Problem and is certainly not all of it. What is left of Painlevé's Problem is the task of determining the removability of those compact subsets K of $\mathbb{C}$ for which $\dim_{\mathcal{H}}(K) = 1$ but $\mathcal{H}^1(K) = \infty$. Can we extend our affirmative resolution of Vitushkin's Conjecture to any of these sets? The answer, taking as given all the difficult work that we have done up till now, is a cheap "yes." One only needs to combine Proposition 1.7 with Vitushkin's Conjecture Resolved (6.5) to obtain the following.

Theorem P.1 *Suppose K is a compact subset of $\mathbb{C}$ that can be written as a countable union of compact sets K_n such that $\mathcal{H}^1(K_n) < \infty$ for each n. Then K is nonremovable if and only if $\mathcal{H}^1(K \cap \Gamma) > 0$ for some rectifiable curve Γ.*

Let us use the phrase *Vitushkin's σ-Finite Conjecture* for the assertion that any compact subset of the complex plane which is σ-finite for linear Hausdorff measure is nonremovable if and only if it hits some rectifiable curve in a set of positive linear Hausdorff measure. An overly hasty reader might jump to the conclusion that we have just affirmatively resolved Vitushkin's σ-Finite Conjecture. Not quite! Recall that a set is σ-finite for a measure if it can be written as a countable union of measurable sets that are assigned finite mass by the measure. Using the Inner Regularity of $\mathcal{H}^1$ for Some Sets (5.21), we see that a compact subset K of $\mathbb{C}$ that is σ-finite for $\mathcal{H}^1$ is one that can be written as a countable union of compact sets K_n such that $\mathcal{H}^1(K_n) < \infty$ for each n ... *and a Borel set E such that $\mathcal{H}^1(E) = 0$.* So the possible presence of this $\mathcal{H}^1$-null set E results in an annoying gap here between what we can prove and what we would like to prove!

Changing gears a bit, we now wish to describe and discuss a conjecture attributed to Mark Melnikov in [MAT5] and [MAT6]. *Melnikov's Conjecture* states that a compact subset of the complex plane is nonremovable if and only if it supports a nontrivial positive Borel measure with finite Melnikov curvature and linear growth. Note that the backward implication of this conjecture is a simple consequence of Melnikov's Lower Capacity Estimate (4.3), so that the status of the forward implication is what is of interest here. Recall that we have had two natural conjectured answers to Painlevé's Problem before and that each was slain by a counterexample

J.J. Dudziak, *Vitushkin's Conjecture for Removable Sets*, Universitext,
DOI 10.1007/978-1-4419-6709-1, © Springer Science+Business Media, LLC 2010

(Sections 2.4 and 4.7). This conjecture will not share their fate; it will be shown to be true with the aid of a profound and difficult result about analytic capacity due to Xavier Tolsa!

Is this affirmative resolution of Melnikov's Conjecture a complete solution to Painlevé's Problem? It clearly represents great progress on Painlevé's Problem and certainly gives a criterion for removability that does not involve analytic function theory. However, a complete solution should be a purely "geometric" criterion. Is Melnikov's criterion purely "geometric"? The answer to this question probably differs from mathematician to mathematician. For what it is worth, the author's judgement here is negative. Although he is willing to declare the notion of a rectifiable curve and the notion of Hausdorff measure ... even *generalized* Hausdorff measure (see the last paragraph of Chapter 4) ... as purely "geometric," he is not willing to so classify the notion of an arbitrary measure with linear growth. Thus, in his view, one has left the land of the purely "geometric" when one considers criteria couched in terms of such measures and their Melnikov curvatures. However, even if one differs on this point with the author, one must admit that Melnikov's characterization of removable sets is not very easy to use ... Chapter 8 is surely testimony to that.

On the positive side, we shall see that Melnikov's characterization of removable sets leads to an affirmative resolution of Vitushkin's σ-Finite Conjecture – thus getting rid of that annoying gap mentioned a few paragraphs ago.

Strictly speaking, Tolsa's theorem is outside of the self-imposed scope of this book – the presentation of a complete self-contained proof of Vitushkin's Conjecture. Also, the inclusion of a proof of this theorem would be a very long and difficult addition to an already very long and difficult book. However, as indicated above, Tolsa's theorem definitely sheds light on that part of Painlevé's Problem not covered by Vitushkin's Conjecture, so it would be inappropriate to say nothing about it at all. As a compromise, the author has thus decided to append this postscript, the rest of which will proceed as follows: first, Tolsa's theorem will be stated but not proved; second, Tolsa's theorem will be shown to imply the reversibility of Melnikov's Lower Capacity Estimate (4.3); third, this reversibility will be shown to imply Melnikov's Conjecture; and fourth, Melnikov's Conjecture will be shown to imply Vitushkin's σ-Finite Conjecture. The reader wishing to see a proof of Tolsa's theorem should consult the original sources: [TOL3], with [TOL1] and [TOL2] being preparatory to it, or [TOL4].

We note that Tolsa's theorem has many more consequences than just the ones of interest to us here, the most notable being the establishment of the subadditivity of analytic capacity, a very old open problem from rational approximation theory (see [VIT2]).

To state Tolsa's theorem we first need to make some observations about Cauchy transforms and introduce a new capacity.

Let μ be a finite positive Borel measure on $\mathbb{C}$. By Fubini's Theorem [RUD, 8.8] and Lemma 6.8, for any $R > 0$ we have

$$\int_{B(0;R)} \int_{\mathbb{C}} \frac{1}{|\zeta - z|} \, d\mu(\zeta) \, d\mathcal{L}^2(z) = \int_{\mathbb{C}} \int_{B(0;R)} \frac{1}{|\zeta - z|} \, d\mathcal{L}^2(z) \, d\mu$$
$$\leq 2\pi R \, \mu(\mathbb{C}) < \infty.$$

We conclude that the integral defining $\hat{\mu}(z)$, the Cauchy transform of μ at z, converges absolutely for $\mathcal{L}^2$-a.e. $z \in \mathbb{C}$.

For any compact subset K of $\mathbb{C}$, define

$$\gamma_+(K) = \sup\{\mu(K) : \mu \in \mathcal{M}_+(K) \text{ is such that } |\hat{\mu}| \leq 1 \ \mathcal{L}^2\text{-a.e. on } \mathbb{C}\}$$

where $\mathcal{M}_+(K)$ denotes the set of finite positive Borel measures supported on K (note that all these measures are regular by [RUD, 2.18]). Since $\hat{\mu}$ is continuous on $\mathbb{C}^* \setminus \mathrm{spt}(\mu)$, $\|\hat{\mu}\|_{\mathbb{C}^* \setminus \mathrm{spt}(\mu)} \leq \|\hat{\mu}\|_\infty$ (the L^∞ norm here is with respect to $\mathcal{L}^2$, planar Lebesgue measure). Also, $\hat{\mu}'(\infty) = -\mu(K)$. Thus we always have $\gamma_+(K) \leq \gamma(K)$. Tolsa's theorem reverses this modulo a constant!

Theorem P.2 (Tolsa) *There exists an absolute constant $C > 0$ such that*

$$\gamma(K) \leq C \, \gamma_+(K)$$

for any compact subset K of $\mathbb{C}$.

We will now use Tolsa's theorem to reverse Melnikov's Lower Capacity Estimate (4.3). To do so however, two preparatory lemmas are needed.

Lemma P.3 *Suppose K is a compact subset of $\mathbb{C}$ and $\mu \in \mathcal{M}_+(K)$ is such that $|\hat{\mu}| \leq 1$ $\mathcal{L}^2$-a.e. on $\mathbb{C}$. Then μ has linear growth with bound 1.*

Proof Consider any closed ball $B(z_0; R)$ and any number $\varepsilon > 0$.

Claim. $\displaystyle\int_0^{2\pi} |\hat{\mu}(z_0 + re^{i\theta})| \, d\theta \leq 2\pi$ for $\mathcal{L}^1$-a.e. $r \in [R, R + \varepsilon)$.

To see this, given $\delta > 0$, set $E_\delta = \{r \in [R, R+\varepsilon) : \int_0^{2\pi} |\hat{\mu}(z_0 + re^{i\theta})| \, d\theta \geq 2\pi(1+\delta)\}$ and $\widetilde{E}_\delta = \{z_0 + re^{i\theta} : r \in E_\delta \text{ and } \theta \in [0, 2\pi)\}$. It suffices to show that $\mathcal{L}^1(E_\delta) = 0$. Invoking polar coordinates, our hypothesis on $\hat{\mu}$, and polar coordinates once again, we have

$$2\pi(1+\delta) \int_{E_\delta} r dr \leq \int_{E_\delta} \int_0^{2\pi} |\hat{\mu}(z_0 + re^{i\theta})| \, d\theta \, r dr \ = \ \iint_{\widetilde{E}_\delta} |\hat{\mu}| \, d\mathcal{L}^2$$
$$\leq \iint_{\widetilde{E}_\delta} d\mathcal{L}^2 \ = \ \int_{E_\delta} \int_0^{2\pi} d\theta \, r dr \ = \ 2\pi \int_{E_\delta} r dr.$$

Since $\int_{E_\delta} r dr$ is clearly finite and nonnegative, this inequality forces us to conclude that $\int_{E_\delta} r dr$ must be 0. Thus $\mathcal{L}^1(E_\delta) = 0$, establishing the Claim.

Clearly $\mu(\partial B(z_0; \rho)) > 0$ for at most countably many $\rho > 0$. This fact and the Claim allow us to choose $\rho \in [R, R+\varepsilon)$ such that

$$(\star)\ \mu(\partial B(z_0;\rho)) = 0 \text{ and } (\star\star) \int_0^{2\pi} |\hat{\mu}(z_0 + \rho e^{i\theta})|\, d\theta \leq 2\pi.$$

Let C denote the counterclockwise circular boundary of $B(z_0; \rho)$. Then for μ-a.e. $\zeta \in \mathbb{C}$,

$$\frac{1}{2\pi i}\int_C \frac{1}{z-\zeta}\, dz = \mathcal{X}_{B(z_0;\rho)}(\zeta)$$

by $(\star)$ and [RUD, 10.11]. From this, Fubini's Theorem [RUD, 8.8], and $(\star\star)$, we see that

$$\begin{aligned}
\mu(B(z_0; R)) \leq \mu(B(z_0;\rho)) &= \left| \int \left\{ \frac{1}{2\pi i}\int_C \frac{1}{z-\zeta}\, dz \right\} d\mu(\zeta) \right| \\
&= \left| \frac{-1}{2\pi i}\int_C \hat{\mu}(z)\, dz \right| \leq \frac{1}{2\pi}\int_C |\hat{\mu}(z)|\, |dz| \\
&= \frac{1}{2\pi}\int_0^{2\pi} |\hat{\mu}(z_0 + \rho e^{i\theta})|\, \rho\, d\theta \leq \rho < R + \varepsilon.
\end{aligned}$$

Since $\varepsilon > 0$ was otherwise arbitrary, we conclude that $\mu(B(z_0; R)) \leq R$ and so are done. □

The next lemma's proof essentially just recycles three arguments from Chapter 6.

Lemma P.4 *Suppose K is a compact subset of $\mathbb{C}$ and $\mu \in \mathcal{M}_+(K)$ is such that $|\hat{\mu}| \leq 1$ $\mathcal{L}^2$-a.e. on $\mathbb{C}$. Then $c^2(\mu) \leq 864\, \mu(K)$.*

Proof Because of the last lemma, if one takes the proofs of Propositions 6.9 and 6.11 and replaces all occurrences of $h\, d\mathcal{H}^1_K$ and M by $d\mu$ and 1, respectively, then, with a few minor changes in justifications at some points, one will have an argument showing that

$$\left| \int_{\mathbb{C}\setminus B(z;\varepsilon)} k_\Phi(\zeta, z)\, d\mu(\zeta) \right| \leq 12$$

for every $z \in \mathbb{C}$, $\varepsilon > 0$, and Lip(1)-function Φ. Setting $\Phi \equiv \lambda$ where $\lambda > 0$, we have

$$\left| \int_{\mathbb{C}\setminus B(z;\varepsilon)} k_\lambda(\zeta, z)\, d\mu(\zeta) \right| \leq 12$$

for every $z \in \mathbb{C}$, $\varepsilon > 0$, and $\lambda > 0$.

For each $z \in \mathbb{C}$ and $\lambda > 0$, the function $\zeta \in \mathbb{C} \mapsto k_\lambda(\zeta, z) \in \mathbb{C}$ is in $L^\infty(\mu)$ by (d) of Proposition 6.10. We may thus invoke Lebesgue's Dominated Convergence Theorem [RUD, 1.34] to let $\varepsilon \downarrow 0$ and conclude that

$$\left| \int_{\mathbb{C}} k_\lambda(\zeta, z)\, d\mu(\zeta) \right| \leq 12$$

for every $z \in \mathbb{C}$ and $\lambda > 0$. Clearly then

$$\|\mathcal{K}_\lambda\, \mathcal{X}_{\mathbb{C}}\|^2_{L^2(\mu)} = \int_{\mathbb{C}} \left| \int_{\mathbb{C}} k_\lambda(\zeta, z)\, d\mu(\zeta) \right|^2 d\mu(\zeta) \leq 12^2 \mu(\mathbb{C})$$

for every $\lambda > 0$.

This last inequality is just ($\star$) in the proof of Proposition 6.15 with $\Phi \equiv 0$, $U = \emptyset$, and $N = 12$. Repeating the reasoning of the proof of Proposition 6.15 from ($\star$) onward, we conclude that

$$c^2(\mu) = \iiint_{\mathbb{C}^3} c^2(\zeta, \eta, \xi)\, d\mu(\zeta)\, d\mu(\eta)\, d\mu(\xi) \leq 6(12^2)\mu(\mathbb{C}).$$

Since $6(12^2) = 864$ and μ is supported on K, we are done. □

Given two nonnegative quantities $Q_1(K)$ and $Q_2(K)$ defined for all compact subsets K of $\mathbb{C}$, we say that $Q_1(K)$ and $Q_2(K)$ are *comparable*, and write $Q_1(K) \asymp Q_2(K)$, if there exists an absolute constant $C \geq 1$ such that $C^{-1}\, Q_1(K) \leq Q_2(K) \leq C\, Q_1(K)$ for every compact subset K of $\mathbb{C}$.

Theorem P.5 (Reversibility of Melnikov's Lower Capacity Estimate)

$$\gamma(K) \asymp \sup \left\{ \frac{M\, \mu^2(K)}{M^2\, \mu(K) + c^2(\mu)} : \mu \in \mathcal{M}_+(K) \textit{ has linear growth with bound } M \right\}.$$

Proof Let K be a compact subset of $\mathbb{C}$.

By Tolsa's Theorem (P.2), $\gamma(K) \leq C\, \gamma_+(K)$ for some absolute constant C.

Elementary algebra shows that if $c^2(\mu) \leq 864\, \mu(K)$, then $\mu(K) \leq 865\, \mu^2(K)/\{\mu(K) + c^2(\mu)\}$. Thus the definition of $\gamma_+(K)$ and the last lemma show that

$$\gamma_+(K) \leq 865 \sup \left\{ \frac{\mu^2(K)}{\mu(K) + c^2(\mu)} : \mu \in \mathcal{M}_+(K) \text{ has linear growth with bound } 1 \right\}.$$

Obviously this last supremum is dominated by the supremum in the theorem's enunciation.

Finally, by Melnikov's Lower Capacity Estimate (4.3),

$$\sup \left\{ \frac{M\, \mu^2(K)}{M^2\, \mu(K) + c^2(\mu)} : \mu \in \mathcal{M}_+(K) \text{ has linear growth with bound } M \right\}$$
$$\leq 375{,}000\, \gamma(K).$$

Clearly we are done. □

The following result is more in the nature of a corollary to the last theorem, but we label it a theorem due to its importance.

Theorem P.6 (Melnikov's Conjecture Resolved) *A compact subset of the complex plane is nonremovable if and only if it supports a nontrivial positive Borel measure with finite Melnikov curvature and linear growth.*

Returning to the concern of the first few paragraphs of this postscript, the concluding result of this book is the following.

Theorem P.7 (Vitushkin's σ-Finite Conjecture Resolved) *Suppose K is a compact subset of $\mathbb{C}$ that is σ-finite for $\mathcal{H}^1$. Then K is nonremovable if and only if $\mathcal{H}^1(K \cap \Gamma) > 0$ for some rectifiable curve Γ.*

Proof As noted in the paragraph after Theorem P.1, $K = \bigcup_n K_n \cup E$ where each K_n is a compact set with $\mathcal{H}^1(K_n) < \infty$ and E is a Borel set with $\mathcal{H}^1(E) = 0$.

Suppose that K is nonremovable. Then by the hard half of Melnikov's Conjecture Resolved (P.6), K supports a nontrivial positive Borel measure μ with finite Melnikov curvature and linear growth. Denote the linear growth bound of μ by M and let $\varepsilon > 0$. From the definition of Hausdorff measure and the fact that any set can be contained in a ball whose radius is the diameter of the set, we obtain a countable collection of balls $\{B_n\}$ which cover E such that $\sum_n \mathrm{rad}(B_n) < \varepsilon/M$. Clearly then $\mu(E) \leq \sum_n \mu(B_n) \leq M \sum_n \mathrm{rad}(B_n) < \varepsilon$. Since $\varepsilon > 0$ is otherwise arbitrary, we must have $\mu(E) = 0$. But then $\mu(K_n) > 0$ for some n and so for this n, μ restricted to K_n is a nontrivial positive Borel measure supported on K_n with finite Melnikov curvature and linear growth. Invoking Melnikov's Lower Capacity Estimate (4.3), we conclude that K_n is nonremovable. Since $\mathcal{H}^1(K_n) < \infty$, Vitushkin's Conjecture Resolved (6.5) now implies that $\mathcal{H}^1(K \cap \Gamma) \geq \mathcal{H}^1(K_n \cap \Gamma) > 0$ for some rectifiable curve Γ. This establishes the forward implication.

The backward implication follows trivially from the Denjoy Conjecture Resolved (4.27) and does not need the σ-finiteness assumption at all. □

What is left of Painlevé's Problem now is the task of determining the removability of those compact subsets of the complex plane that are non-σ-finite for linear Hausdorff measure but have Hausdorff dimension one. Of course Melnikov's Conjecture Resolved (P.6) gives us an answer of sorts, but more "geometric", or at least more tractable, criteria are wanted. The paper [MAT4] provides a good example of what is meant here. The existence of a necessary and sufficient "geometric" criterion for all the remaining sets seems doubtful.

Bibliography

[AHL] L. Ahlfors, *Bounded analytic functions*, Duke Math. J., Vol. 14 (1947), 1–11. (Section 1.2)

[BES1] A. S. Besicovitch, *On the fundamental geometrical properties of linearly measurable plane sets of points*, Math. Ann., Vol. 98 (1927), 422–464. (Section 6.6)

[BES2] A. S. Besicovitch, *On the fundamental geometrical properties of linearly measurable plane sets of points II*, Math. Ann., Vol. 115 (1938), 296–329. (Section 6.6)

[BES3] A. S. Besicovitch, *On the fundamental geometrical properties of linearly measurable plane sets of points III*, Math. Ann., Vol. 116 (1939), 349–357. (Section 6.6)

[BES4] A. S. Besicovitch, *On approximation in measure to Borel sets by F_σ-sets*, J. London Math. Soc., Vol. 29 (1954), 382–383. (Section 5.4)

[CAL] A. P. Calderón, *Cauchy integrals on Lipschitz curves and related operators*, Proc. Nat. Acad. Sci. USA, Vol. 74 (1977), 1324–1327. (Section 3.1)

[CHRIST] M. Christ, *A $T(b)$ theorem with remarks on analytic capacity and the Cauchy integral*, Colloq. Math., Vol. 60/61 (1990), 601–628. (Section 6.5)

[DAV1] G. David, *Unrectifiable 1-sets have vanishing analytic capacity*, Rev. Mat. Iberoamericana, Vol. 14 (1998), 369–479. (Section 6.5)

[DAV2] G. David, *Analytic Capacity, Cauchy Kernel, Menger Curvature, and Rectifiability* in *Harmonic Analysis and Partial Differential Equations*, University of Chicago Press (1999), 183–197. (Preface)

[DAV3] G. David, *Analytic capacity, Calderón-Zygmund operators, and rectifiability*, Publ. Mat., Vol. 43 (1999), 3–25. (Preface)

[DM] G. David and P. Mattila, *Removable sets for Lipschitz harmonic functions in the plane*, Rev. Mat. Iberoamericana, Vol. 16 (2000), 137–215. (Section 6.5)

[DS1] G. David and S. Semmes, *Singular integrals and rectifiable sets in $\mathbb{R}^n$: Au delà des Graphes Lipschitziens*, SMF No. 193 in Astérisque, Vol. 193 (1991). (Sections 8.3 and 8.8)

[DS2] G. David and S. Semmes, *Analysis of and on uniformly rectifiable sets*, Mathematical Surveys and Monographs, Vol. 38, American Mathematical Society (1993). (Sections 8.3 and 8.8)

[DEN] A. Denjoy, *Sur les fonctions analytiques uniformes à singularités discontinues*, C. R. Acad. Sci. Paris, Vol. 149 (1909), 258–260. (Section 3.1)

[FALC] K. J. Falconer, *The Geometry of Fractal Sets*, Cambridge University Press (1985). (Preface and Sections 2.1, 4.7, 5.1, 5.2, 5.3, 6.5, and 6.6)

[FISH] S. Fisher, *On Schwarz's lemma and inner functions*, Trans. Amer. Math. Soc., Vol. 138 (1969), 229–240. (Section 1.2)

[FRAZ] M. W. Frazier, *An Introduction to Wavelets Through Linear Algebra*, Springer-Verlag (2001). (Section 8.22)

[FROST] O. Frostman, *Potential d'équilibre et capacité des ensembles avec quelques applications à la théorie des fonctions*, Meddel. Lunds. Univ. Mat. Sem., Vol. 3 (1935), 1–118. (Section 2.3)

[GAM1] T. W. Gamelin, *Uniform Algebras*, Prentice-Hall (1969). (Sections 1.2 and 3.3)

[GAM2] T. W. Gamelin, *Lectures on $H^\infty(D)$*, Notas de Mathematica Universidad Nacional de La Plata, Argentina, No. 21 (1972). (Section 3.1)

[GS] P. R. Garabedian and M. Schiffer, *On existence theorems of potential theory and conformal mapping*, Ann. of Math., Vol. 52 (1950), 164–187. (Section 3.1)

[GAR1] J. Garnett, *Positive length but zero analytic capacity*, Proc. Amer. Math. Soc., Vol. 21 (1970), 696–699. (Section 2.4)

[GAR2] J. Garnett, *Analytic capacity and measure*, Lecture Notes in Math., Vol. 297, Springer-Verlag (1972). (Preface and Sections 1.2, 2.4, and 3.1)

[JON1] P. W. Jones, *Square functions, Cauchy integrals, analytic capacity, and harmonic measure*, Lecture Notes in Math., Vol. 1384, Springer-Verlag (1989), 24–68. (Section 2.4)

[JON2] P. W. Jones, *Rectifiable sets and the traveling salesman problem*, Invent. Math., Vol. 102 (1990), 1–15. (Section 8.3)

[JM1] P. W. Jones and T. Murai, *Positive analytic capacity but zero Buffon needle probability*, Pacific J. Math., Vol. 133 (1988), 99–114. (Section 6.6)

[JM2] H. Joyce and P. Mörters, *A set with finite curvature and projections of zero length*, J. Math. Anal. and Appl., Vol. 247 (2000), 126–135. (Section 4.7)

[KK] R. Kannan and C. K. Krueger, *Advanced Analysis on the Real Line*, Springer-Verlag (1996). (Section 2.1)

[LÉG] J. C. Léger, *Menger curvature and rectifiability*, Ann. of Math., Vol. 149 (1999), 831–869. (Sections 6.5 and 8.1)

[MARSH] D. E. Marshall, *Removable sets for bounded analytic functions*, Lecture Notes in Math., Vol. 1574, Springer-Verlag (1994), 141–144. (Preface and Section 3.1)

[MARST] J. M. Marstrand, *Some fundamental geometrical properties of plane sets of fractional dimensions*, Proc. London Math. Soc. (3), Vol. 4 (1954), 257–302. (Section 6.6)

[MAT1] P. Mattila, *A class of sets with positive length and zero analytic capacity*, Ann. Acad. Sci. Fenn. Ser. A I Math., Vol. 10 (1985), 387–395. (Section 2.4)

[MAT2] P. Mattila, *Smooth maps, null-sets for integralgeometric measure and analytic capacity*, Ann. of Math., Vol. 123 (1986), 303–309. (Section 6.6)

[MAT3] P. Mattila, *Geometry of Sets and Measures in Euclidean Spaces*, Cambridge University Press (1995). (Preface and Sections 4.7 and 5.5)

[MAT4] P. Mattila, *On the analytic capacity and curvature of some Cantor sets with non-σ-finite length*, Publ. Mat., Vol. 40 (1996), 195–204. (Section 4.7 and Postscript)

[MAT5] P. Mattila, *Singular integrals, analytic capacity and rectifiability*, J. Fourier Anal. Appl., Vol. 3 (1997), 797–812. (Preface and Postscript)

[MAT6] P. Mattila, *Rectifiability, analytic capacity, and singular integrals*, Proc. of the ICM, Doc. Math., Extra Vol. II (1998), 657–664. (Preface and Postscript)

[MMV] P. Mattila, M. Melnikov, and J. Verdera, *The Cauchy integral, analytic capacity, and uniform rectifiability*, Ann. of Math., Vol. 144 (1996), 127–136. (Section 6.5)

[McM] T. J. McMinn, *Linear measures of some sets of the Cantor type*, Proc. Cambridge Philos. Soc., Vol. 53 (1957), 312–317. (Sections 2.1 and 2.4)

[MEL] M. Melnikov, *Analytic capacity: discrete approach and curvature of measure*, Sbornik Mathematics, Vol 186 (1995), 827–846. (Sections 3.1, 4.2, and 6.5)

[MV] M. Melnikov and J. Verdera, *A geometric proof of the L^2 boundedness of the Cauchy integral on Lipschitz graphs*, Internat. Math. Res. Notices, Vol. 7 (1995), 325–331. (Sections 3.1 and 4.2)

[MUR] T. Murai, *Construction of H^1 functions concerning the estimate of analytic capacity*, Bull. London Math. Soc., Vol. 19 (1986), 154–160. (Section 2.4)

[NT] F. Nazarov and S. Treil, *The hunt for a Bellman function*, Algebra i Analiz, Vol. 8 (1996), 32–162. (Section 7.6)

[NTV] F. Nazarov, S. Treil, and A. Volberg, *The Tb-theorem on non-homogeneous spaces that proves a conjecture of Vitushkin*, Preprint No. 519, Center de Recerca Matemàtica, Barcelona, 2002. (Sections 6.5 and 7.1)

[PAIN] P. Painlevé, *Sur les lignes singulières des fonctions analytiques*, Annales de la Faculté des Sciences de Toulouse (1888). (Preface and Section 2.2)

[PAJ1] H. Pajot, *Sous-ensembles de courbes Ahlfors-régulières et nombres de Jones*, Publ. Mat., Vol. 40 (1996), 497–526. (Section 8.3)

[PAJ2] H. Pajot, *Analytic capacity, rectifiability, menger curvature and the Cauchy integral*, Lecture Notes in Math., Vol. 1799, Springer-Verlag (2002). (Preface)

[PAJ3] H. Pajot, *Capacité analytique et le problème de Painlevé*, Séminaire Bourbaki, Astérisque, Vol. 299 (2005), 301–328. (Preface)

[POM] Ch. Pommerenke, *Über die analytische Kapazität*, Archiv der Math., Vol. 11 (1960), 270–277. (Section 1.2)

[ROG] C. A. Rogers, *Hausdorff Measures*, Cambridge University Press (1970). (Preface and Section 4.7)

[RUD] W. Rudin, *Real and Complex Analysis, 3rd Edition*, McGraw-Hill Book Company (1987). (Preface and Many Sections)

[TOL1] X. Tolsa, *L^2-boundedness of the Cauchy integral operator for continuous measures*, Duke Math. J., Vol. 98 (1999), 269–304. (Postscript)

[TOL2] X. Tolsa, *On the analytic capacity γ_+*, Indiana Math. J., Vol. 51 (2002), 317–343. (Postscript)

[TOL3] X. Tolsa, *Painlevé's problem and the semiadditivity of analytic capacity*, Acta Math., Vol. 190 (2003), 105–149. (Postscript)

[TOL4] X. Tolsa, *Painlevé's problem and analytic capacity*, Lecture notes of a minicourse given at El Escorial (2004), 1–28. (Preface and Postscript)

[TOL5] X. Tolsa, *Finite curvature of arc length measure implies rectifiability: a new proof*, Indiana Math. J., Vol. 54 (2005), 1075–1105. (Section 6.5)

[VER] J. Verdera, *Removability, capacity, and approximation*, Complex Potential Theory NATO ASI Series, Kluwer Academic Publishers (1994), 419–473. (Preface)

[VIT1] A. G. Vitushkin, *Example of a set of positive length but zero analytic capacity*, Dokl. Akad. Nauk. SSSR, Vol. 127 (1959), 246–249 (Russian). (Section 2.4)

[VIT2] A. G. Vitushkin, *Analytic capacity of sets and problems in approximation theory*, Russian Math. Surveys, Vol. 22 (1967), 139–200. (Sections 1.2, 6.6, and Postscript)

[YB] I. M. Yaglom and V. G. Boltyanski, *Convex Figures*, Holt, Rinehart and Winston (1961). (Section 1.2)

[ZALC] L. Zalcman, *Analytic capacity and rational approximation*, Lecture Notes in Math., Vol. 50, Springer-Verlag (1968). (Section 1.2)

Symbol Glossary and List

What follows is a symbol glossary for some basic notions that are *not* defined or explained in the text – the author believes most of the associated notation to be either fairly standard or rather obvious, but in many cases what goes without saying goes much better when actually said! After that we have a symbol list for some basic and technical notions that *are* defined or explained, sometimes rather casually, in the text. Notation which is used in only one section of the book is not listed here. Thus, for example, most of the notation from Section 2.4 is not in the symbol list.

Symbol Glossary

$\#E$ denotes the cardinality of a set E (with $\#E = \infty$ when E is infinite).

$\mathcal{X}_E$ denotes the characteristic function of a set E.

$\operatorname{Re} z$, $\operatorname{Im} z$, $|z|$, and $\bar{z}$ denote the real part, imaginary part, absolute value, and complex conjugate of a complex number z, respectively.

$[a, b]$, (a, b), $[a, b)$, and $(a, b]$ denote the closed, open, and two half-open intervals with endpoints a and b, respectively.

$B(z; r)$ denotes the closed ball in $\mathbb{C}$ with center z and radius r.

$\alpha E + \beta = \{\alpha x + \beta : x \in E\}$.

$E + F = \{x + y : x \in E \text{ and } y \in F\}$.

$\operatorname{dist}(x, E) = \inf\{\operatorname{dist}(x, y) : y \in E\}$.

$\operatorname{dist}(E, F) = \inf\{\operatorname{dist}(x, y) : x \in E \text{ and } y \in F\}$.

$\operatorname{int} E$, $\operatorname{cl} E$, and ∂E denote the interior, closure, and boundary of subset E of a topological space respectively.

$\|f\|_\infty$ denotes either the supremum norm of a function f with respect to an understood set or the essential supremum norm of f with respect to an understood measure (we rely upon context to determine which is meant).

Given a measure μ, we denote the total variation of μ by $|\mu|$. Given measures μ and ν, $\mu \ll \nu$ means that $|\mu|$ is absolutely continuous with respect to $|\nu|$ and $\mu \perp \nu$ means that $|\mu|$ and $|\nu|$ are mutually singular.

Symbol List

Notation	Section (Page)
$\mathbb{C}$	1.1 (1)
$H^\infty(V)$	1.1 (1)
$\mathbb{C}^*$	1.1 (1)
$\mathbb{R}$	1.1 (3)
$f(\infty)$	1.1 (4)
$f'(\infty)$	1.1 (5)
$\gamma(K)$	1.2 (9)
$\widehat{K}$	1.2 (10)
$\|E\|$	1.2 (13)
$\mathcal{L}^1(E)$	1.2 (14)
$\mathcal{L}^2(E)$	1.2 (15)
$\hat{\mu}(z)$	1.2 (16)
$\mathcal{H}^s_\delta(E)$	2.1 (19)
$\mathcal{H}^s(E)$	2.1 (19)
$\dim_{\mathcal{H}}(E)$	2.1 (20)
$\mathcal{H}^1_\infty(K)$	2.2 (24)
$\mathrm{rad}(B)$	2.3 (26)
$\mathcal{H}^s_\infty(K)$	2.3 (26)
$n(\Gamma; z)$	2.4 (31)
$l(Q)$	2.4 (32)
$B_j = B(c_j; r_j)$	3.1 (40)
$K = \bigcup_{j=1}^N B_j$	3.1 (40)
$U = \mathbb{C}^* \setminus K$	3.1 (40)
$\Gamma_j = \partial B_j$	3.1 (40)
$\Gamma = \bigcup_{j=1}^N \Gamma_j$	3.1 (40)
$\Delta = \mathrm{int}\, K$	3.1 (40)
z_α	3.1 (40)
E_α	3.1 (40)
$C(X)$	3.1 (40)
$A(U)$	3.1 (40)
$A_0(U)$	3.1 (40)
$\alpha(K)$	3.1 (40)
$H_0^\infty(U)$	3.1 (40)

$\gamma(K)$	3.1 (40)
$\alpha_2(K)$	3.1 (40)
$\mathbb{D}$	3.2 (42)
$\text{osc}(f, E; \varepsilon)$	3.2 (43)
$\text{spt}(f)$	3.2 (43)
$\text{spt}(\mu)$	3.2 (43)
$L^\infty(\Gamma)$	3.2 (45)
$L^1(\Gamma)$	3.2 (45)
f^*	3.2 (45)
$f_\alpha(z)$	3.2 (46)
$\mathbb{T}$	3.3 (47)
dm	3.3 (47)
$\mu \perp A(\mathbb{D})$	3.3 (48)
ψ^*	3.4 (50)
$\psi(z)$	3.5 (52)
γ	3.7 (53)
$d\zeta_\gamma$	3.7 (53)
R_+	3.7 (53)
R_-	3.7 (53)
R	3.7 (53)
z^*	3.7 (54)
$U(\alpha_0)$	3.8 (59)
$Z(g; E)$	3.9 (62)
$c(\zeta, \eta, \xi)$	4.1 (69)
$\sum_{\mathcal{P}erm} f(\zeta, \eta, \xi)$	4.1 (70)
$c(\mu)$	4.2 (71)
$L^p(E)$	4.3 (78)
$\hat{f}(\zeta)$ for $f \in L^1(\mathbb{R})$	4.3 (79)
$\hat{f}$ for $f \in L^2(\mathbb{R})$	4.3 (80)
$l(\gamma)$	4.5 (86)
$l_\gamma(E)$	4.5 (90)
$\mathcal{P}(X)$	5.1 (105)
$\mathcal{B}(X)$	5.1 (105)
$\mathcal{M}_\mu$	5.1 (106)
μ_A	5.4 (123)
$C_c(X)$	5.4 (123)

$K = \mathrm{spt}(\mu)$	8.2 (224)
$\delta(B)$	8.2 (225)
$\delta(z; r)$	8.2 (225)
$\mathrm{cen}(B)$	8.2 (225)
$c_1(M, \delta)$	8.2 (225)
$c_2(M, \delta)$	8.2 (225)
$c_3(M, \delta)$	8.2 (227)
$c_4(M, \delta)$	8.2 (227)
$\beta_p(B)$	8.3 (228)
$\beta_p(z; r)$	8.2 (228)
$\beta_p^L(B)$	8.3 (228)
$\beta_p^L(z; r)$	8.3 (228)
$c_5(M, \delta)$	8.3 (229)
$C_1(M, \delta)$	8.3 (229)
$c(B)$	8.4 (232)
$c(z; r)$	8.4 (232)
$\Delta(B)$	8.4 (232)
$\Delta(z; r)$	8.4 (232)
$C_2(M, \delta, k)$	8.4 (232)
$\tilde{\delta}(B)$	8.5 (237)
$\tilde{\delta}(z; r)$	8.5 (237)
$C_3(M, \delta, k, k^*)$	8.5 (237)
$C_4(M, \delta)$	8.6 (239)
$C_5(M, \delta)$	8.6 (239)
$C_6(M, \delta)$	8.6 (239)
$C_7(\varphi)$	8.6 (239)
$C_8(\varphi)$	8.6 (239)
z_0	8.7 (241)
L_0	8.7 (241)
U_a	8.7 (241)
π	8.7 (241)
$L_0^\perp$	8.7 (241)
$\pi^\perp$	8.7 (241)
S_{total}	8.8 (241)
$h(z)$	8.8 (242)
S_0	8.8 (242)

$d(z)$	8.9 (245)
K_0	8.9 (246)
ℓ on $\pi(K_0)$	8.9 (247)
$D(p)$	8.10 (248)
I_n	8.10 (249)
p_n	8.11 (250)
z_n	8.11 (250)
r_n	8.11 (250)
B_n	8.11 (250)
L_n	8.11 (251)
ℓ_n	8.11 (251)
$C^\infty(U)$	8.12 (253)
ℓ on U_5	8.12 (256)
$\Gamma = \{p + i\ell(p) : p \in U_5\}$	8.12 (256)
K_1	8.14 (262)
K_2	8.14 (262)
K_3	8.14 (262)
H	8.16 (264)
$I(p; r)$	8.17 (267)
G	8.17 (269)
$\gamma(p; r)$	8.19 (273)
$\gamma^l(p; r)$	8.19 (273)
$\tilde{\gamma}(p; r)$	8.19 (273)
$\tilde{\gamma}^L(p; r)$	8.19 (273)
$\mathcal{N}(p; r)$	8.20 (275)
$C_c^\infty(U)$	8.22 (287)
$L_c^\infty(U)$	8.22 (287)
$L_c^1(U)$	8.22 (287)
φ_r	8.22 (290)
$(f * g)(x)$	8.22 (290)
ℓ on L_0	8.23 (295)
ℓ_1	8.23 (295)
ℓ_2	8.23 (295)

Index

LaVergne, TN USA
24 September 2010

198384LV00002B/3/P

9 781441 967084